Recent Developments in

Gravitation

Cargèse 1978

NATO ADVANCED STUDY INSTITUTES SERIES

A series of edited volumes comprising multifaceted studies of contemporary scientific issues by some of the best scientific minds in the world, assembled in cooperation with NATO Scientific Affairs Division.

Series B: Physics

RECENT VOLUMES IN THIS SERIES

Volume 39 – Hadron Structure and Lepton–Hadron Interactions – *Cargèse* 1977
edited by Maurice Lévy, Jean-Louis Basdevant, David Speiser, Jacques Weyers, Raymond Gastmans, and Jean Zinn-Justin

Volume 40 – Kinetics of Ion–Molecule Reactions
edited by Pierre Ausloos

Volume 41 – Fiber and Integrated Optics
edited by D. B. Ostrowsky

Volume 42 – Electrons in Disordered Metals and at Metallic Surfaces
edited by P. Phariseau, B. L. Györffy, and L. Scheire

Volume 43 – Recent Advances in Group Theory and Their Application to Spectroscopy
edited by John C. Donini

Volume 44 – Recent Developments in Gravitation – *Cargèse* 1978
edited by Maurice Lévy and S. Deser

Volume 45 – Common Problems in Low- and Medium-Energy Nuclear Physics
edited by B. Castel, B. Goulard, and F. C. Khanna

Volume 46 – Nondestructive Evaluation of Semiconductor Materials and Devices
edited by Jay N. Zemel

Volume 47 – Site Characterization and Aggregation of Implanted Atoms in Materials
edited by A. Perez and R. Coussement

Volume 48 – Electron and Magnetization Densities in Molecules and Crystals
edited by P. Becker

Volume 49 – New Phenomena in Lepton-Hadron Physics
edited by Dietrich E. C. Fries and Julius Wess

This series is published by an international board of publishers in conjunction with NATO Scientific Affairs Division

A Life Sciences B Physics	Plenum Publishing Corporation London and New York
C Mathematical and Physical Sciences	D. Reidel Publishing Company Dordrecht and Boston
D Behavioral and Social Sciences	Sijthoff International Publishing Company Leiden
E Applied Sciences	Noordhoff International Publishing Leiden

Recent Developments in
Gravitation
Cargèse 1978

Edited by
Maurice Lévy
Laboratory of Theoretical Physics and High Energies
Université Pierre et Marie Curie
Paris, France

and

S. Deser
Brandeis University
Waltham, Massachusetts

PLENUM PRESS • NEW YORK AND LONDON
Published in cooperation with NATO Scientific Affairs Division

Library of Congress Cataloging in Publication Data

Nato Advanced Study Institute on Recent Developments in Gravitation, Cargèse, Corsica, 1978.
Recent developments in gravitation.

(NATO advanced study institutes series: Series B, Physics; v. 44)
"Proceedings of the 1978 Cargèse Summer Institute on Recent Developments in Gravitation, held in Cargèse, Corsica, July 10–29, 1978, and sponsored in part by NATO."
Includes index.
1. Gravitation–Congresses. I. Lévy, Maurice 1922- II. Deser, Stanley. III. North Atlantic Treaty Organization. IV. Title. V. Series.
QC178.N34 1978 531'.14 79-9174

ISBN-13:978-1-4613-2957-2 e-ISBN-13:978-1-4613-2955-8
DOI: 10.1007/978-1-4613-2955-8

Proceedings of the 1978 Cargèse Summer Institute on
Recent Developments in Gravitation, held in Cargèse, Corsica,
July 10–29, 1978, and sponsored in part by NATO.

Softcover reprint of the hardcover 1st editon 1979
A Division of Plenum Publishing Corporation
227 West 17th Street, New York, N.Y. 10011

Preface

The theory of General Relativity, after its invention by Albert Einstein, remained for many years a monument of mathematical speculation, striking in its ambition and its formal beauty, but quite separated from the main stream of modern Physics, which had centered, after the early twenties, on quantum mechanics and its applications.

In the last ten or fifteen years, however, the situation has changed radically. First, a great deal of significant experimental data became available. Then important contributions were made to the incorporation of general relativity into the framework of quantum theory. Finally, in the last three years, exciting developments took place which have placed general relativity, and all the concepts behind it, at the center of our understanding of particle physics and quantum field theory. Firstly, this is due to the fact that general relativity is really the "original non-abelian gauge theory," and that our description of quantum field interactions makes extensive use of the concept of gauge invariance. Secondly, the ideas of supersymmetry have enabled theoreticians to combine gravity with other elementary particle interactions, and to construct what is perhaps the first approach to a more finite quantum theory of gravitation, which is known as supergravity.

Since many physicists are now involved in one way or another with gravitation, and since there had not been any international summer institute on the theory of gravitation for approximately ten years, it seemed to the organizers of the 1978 Cargèse Summer Institute particularly appropriate to review the whole field, starting with the classical theory of gravitation, then covering the quantum formulation, and arriving finally at the very recent developments of supergravity. A review was also included of the present experimental status of the theory of gravitation.

Thus, the 1978 Summer Institute brought together the "classical" specialists of general relativity who were interested in learning more about recent developments and the elementary particle physicists who wanted to familiarize themselves with the mathematical and experimental framework of gravitation. It was, in all respects, very successful, showing the validity of the original idea.

Thanks to the active cooperation of all the lecturers we were able to bring quickly together the texts of all the courses given at Cargèse. This enables us to present here a relatively complete review of many aspects of the theory of gravitation. This book should be extremely useful therefore to all the physicists who want to learn about recent developments, or to the specialists in other branches of physics who want to incorporate general relativity into their work.

This Summer Institute was made possible by a grant from the Scientific Committee of NATO, within its Advanced Study Institutes Programme. It is a pleasure to thank its members for their encouragement and understanding. We would also like to thank three organizations in France which helped us financially: the Délégation Générale à la Recherche Scientifique et Technique (DGRST), the Centre National de la Recherche Scientifique (CNRS) and the Institute National d'Astronomie et de Géophysique (INAG). Our thanks are also due to the National Science Foundation in the United States, to the University of Nice and to the "Université Pierre et Marie Curie" in Paris, who all helped in one way or another.

Finally, we would like to thank all those who worked very hard to organize the Institute, particularly Dr. Jean-Louis Basdevant and Miss Marie-France Hanseler, and of course, all the lecturers, without whose efforts and enthusiasm nothing would have been possible.

Maurice Lévy

Stanley Deser

CONTENTS

PART I: CLASSICAL RELATIVITY

PART II: QUANTUM GRAVITY

Part I
Classical Relativity

STATUS OF EXPERIMENTAL GRAVITATION

Bruno Bertotti

Istituto di Fisica Teorica
Universita di Pavia
Via Bassi, 6 - 27100 Pavia, Italy

I. INTRODUCTION

The gist of what I am going to say is the following. 1. In recent years, a great wealth of experiments and observations related to gravity have been made and striking agreement with the theory of general relativity has been found. 2. They have been planned and interpreted within theoretical schemes which have either a limited degree of generality (like PPN) or are incomplete, qualitative, or inconsistent. 3. Some basic questions concerning the very foundations of gravitation have not yet been asked with sufficient strength and originality. Therefore it is possible (this is however a personal view) that some basic changes will be needed if experiments under very different conditions or with a higher accuracy are performed.

I hope that this radical point of view will at least make my talks interesting; in fact a mere review of the current experiments of gravity, while it is really fascinating for an experimental physicist because of the extraordinary ingenuity and sophistication of the techniques used, tends to be boring for a theoretician: it appears to him as a sequence of great efforts whose net result is to show that the Einstein is right to, say, 0.5% rather than 2%. Therefore it is important to stress for a theoretical audience the problems of interpretation and the "paradigms" (in the sense of T. S. Kuhn (1962)) one relies upon more or less unconsciously.

2. THE PPN SCHEME

A great part of the experimental program in gravitation during the last ten years has been carried out under the framework and the paradigm of the "Parametrized Post Newtonian" (PPN) framework (see, e.g., Will 1974a). In this scheme special relativity is taken for granted: locally space time is a four dimensional manifold and one seeks all possible information about its metric of Lorentzian signature.

The microscopic laws of special relativity (and quantum theory) are assumed to determine the local structure of macroscopic gravitating bodies; in particular, the mass density ρ, the pressure p, the mean velocity $\underline{v}$ and the stress tensor π. The usual slow motion approximation (a being the size of the body)

$$v^2 \sim \frac{m}{a} \sim \frac{p}{\rho} \sim \frac{\pi}{\rho} \ll 1 \tag{1}$$

is adopted. One then constructs, with ρ, p, $\underline{v}$ and π (but not their derivatives!) as sources, dimensionless integrals having scalar, vector or tensor character; the scalar is needed up to $O(v^4)$, the vector up to $O(v^3)$, the tensor up to $O(v^2)$. They yield, respectively, the components g_{00}, $g_{0\alpha}$ and $g_{\alpha\beta}$ of the metric.[1] The kernels of these integrals are chosen to be rational functions of the distance $|\underline{r}-\underline{r}_i|$ between the point $\underline{r}$ where the field is wanted and the point $\underline{r}_i$ where the source is, and its space derivatives; they must have the correct dimension and must not contain any additional dimensional constant. They are, of course, the natural generalization of Newton's potential $|\underline{r}-\underline{r}_i|^{-1}$. After the appropriate condition of Lorentz invariance has been imposed, it turns out that the most general metric satisfying these requirements, to the post-newtonian approximation, depends only on seven dimensionless parameters.

$$\beta,\ \gamma,\ \zeta_1,\ \zeta_2,\ \zeta_3,\ \zeta_4,\ \zeta_w\ .$$

The equations of motion for a test body are derived from the geodesic principle; finite (in size and in mass) bodies are governed by the requirement that their energy-momentum be conserved:

$$T^{ij}{}_{/j} = 0 \tag{2}$$

The acceleration of a slow test body differs from its newtonian form by terms of relative order - v^2.

---

(1) Greek indices refer to coordinates x^α which are cartesian to lowest approximation and range from 1 to 3. Latin indices run from 0 to 3. The velocity of light is unity.

In the theory of general relativity it is always possible to write down the law of conservation of energy and momentum in an integral form by adding the pseudotensor t^{ij} to the energy momentum tensor T^{ij}, so that from (2) it follows

$$\partial_j \left(T^{ij} + t^{ij} \right) = 0; \tag{3}$$

in our case t^{ij} can be found only if $\zeta_1 = \zeta_2 = \zeta_3 = \zeta_4 = \zeta_w = 0$. Thus this class of "conservative" theories depends only upon two parameters (γ and β); however, for bodies with stationary and spherically symmetric structure, the ζ parameters have no effect upon the motion in any case.

In this scheme the dynamics of gravitating bodies is the same in all inertial frames; but there is no a priori reason to exclude a privileged state of motion determined by distant matter or by the background radiation. If one does so, three new parameters α_1, α_2 and α_3 are added to the list, bringing to 10 their total number; moreover, the metric contains the velocity $\underline{w}$ with respect to the cosmological standard of rest, to be determined directly or by a best fit.

Essentially every known metrical theory of gravity falls in this class and corresponds to a well defined point in the ten-dimensional parameter space. General relativity has

$$\beta = \gamma = 1 \ , \ \zeta_1 = \zeta_2 = \zeta_3 = \zeta_4 = \zeta_w = \alpha_1 = \alpha_2 = \alpha_3 = 0. \tag{4}$$

The great advantage of this scheme consists in the fact that most experiments are done with slowly moving bodies (in particular, planets and artificial satellites); they can be described precisely and lead to a definite result for each point in parameter space. Therefore a least square fit or in the more complex cases, a filtering process, will then produce an estimate of a function of the parameters. For example, the recent lunar laser experiment (Williams et al., 1976) has given for the parameter

$$\eta = 4\beta - \gamma - 3 - \alpha_1 + \frac{2}{3}\alpha_2 - \frac{2}{3}\zeta_1 - \frac{1}{3}\zeta_2 - \frac{13}{3}\zeta_w \tag{5}$$

the value $\eta = 0.0 \pm 0.03$. η vanishes in Einstein's theory.

The basic parameters are, of course, those that describe a spherically symmetric metric expanded in powers of $m/r = GM/c^2r$:

$$ds^2 = \left(1 - 2\frac{m}{r} + 2\beta\frac{m^2}{r^2} + \ldots\right)dt^2 - \left(1 + 2\gamma\frac{m}{r} + \ldots\right)\left(dx^2 + dy^2 + dz^2\right). \tag{6}$$

Note how simple and straightforward our research strategy now becomes. In view of the fact that observations agree with the general relativistic predictions to 1-2%, and paying our debt to beauty, we take $\beta = \gamma = 1$ as reference and consider the plane

$$\left(-\log\left|\frac{\gamma+1}{2}-1\right|, -\log\left|\frac{2-\beta+2\gamma}{3}-1\right|\right),$$

which I called the plane of theories (Bertotti 1977). A given theory - that is to say, a pair (γ,β) - is represented in this plane by a point (or by a line if it has a free parameter); the point at infinity corresponds to Schwarzschild's metric. I have preferred the pair ((γ+1)/2, (2-β+2γ)/3) to (γ,β) because light deflection and the advance of the periastron are proportional, respectively, to (γ+1/2 and (2-β+2γ)/3. Near the origin we have those metrics which differ appreciably from Schwarzschild's; the purpose of our present programme is to test the validity of general relativity to higher and higher and higher accuracy or, to express it in Popper's (1959) language, to falsify alternative theories which differ from general relativity by less and less and are represented in our plane by points further and further from the origin. To do this, we note that an experiment will produce, by the method of confidence intervals a region in the plane where the real value is expected to fall with a given confidence level; for example, if the variance and the correlation coefficient are, respectively, σ_γ, σ_β and ρ and (β^*,γ^*) is the average, the "1 σ" confidence level is given by the interior of the ellipse

$$\left(\frac{\beta-\beta^*}{\sigma_\beta}\right)^2 - 2\rho\,\frac{\beta-\beta^*}{\sigma_\beta}\,\frac{\gamma-\gamma^*}{\sigma_\gamma} + \left(\frac{\gamma-\gamma^*}{\sigma_\gamma}\right)^2 = 1. \qquad (7)$$

For $\sigma_\gamma = 10^{-2}$, $\sigma_\beta = 2\cdot10^{-2}$, $\beta^* = \gamma^* = 1$ and $\rho = 0$, this curve is indicated in Fig. 1 (boundary of "viable" theories). All theories lying to the left of or below it can conventionally be regarded as falsified.

If the domain of falsified points includes infinity, general relativity has to be abandoned; theorists would soon embark on extensive developments and applications of the theories which are still viable, in order to understand the new physical effects and to formulate a new, acceptable theory of gravity. A new reference point (β_1, γ_1)$\neq$(1,1) will be adopted and the game of aiming at a theory and trying to shoot it down will continue with a different "plane of theories."

The history of experimental gravity within the PPN scheme is described by the progression with time of the boundary of viable theories. For simplicity, let us consider the progression of the intersection of its asymptotes. During the last century only the

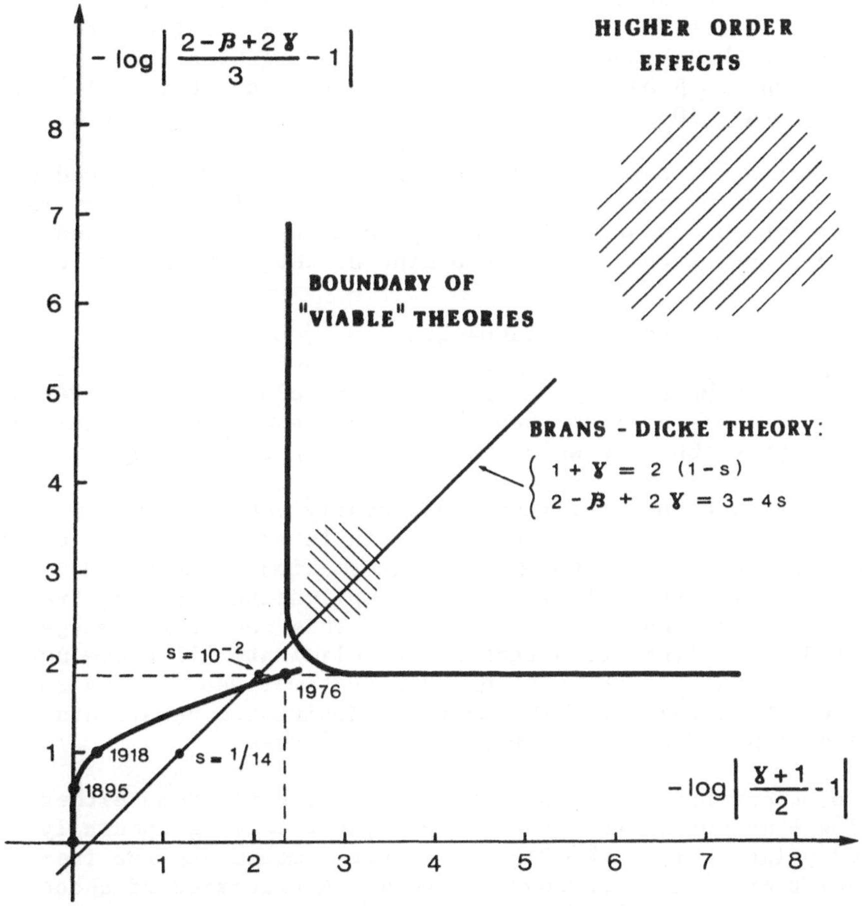

Fig.1 - The "plane of theories", with general relativity as the reference point at infinity. The boundary of viable theories has asymptotes

$$\left(-\log\frac{\sigma_\gamma}{2}\,,\,-\log\frac{\sigma_\beta+2\sigma_\gamma}{3}\right).$$

perihelion advance was measured, but later the curve was displaced to the right by the eclipse expeditions. Today radioastronomical observations give a better accuracy for the deflection than for β. The theory of Brans and Dicke (1961) has already been falsified for the values $s \sim 1/14$ of the coupling parameter usually proposed (Dicke and Peebles 1965). Effects corresponding to higher order terms in the metric (6) require an accuracy in β and γ of order m/r and lie in the shaded region in the upper right corner. To get corrections to β and γ proportional to the cosmic velocity an accuracy of order 10^{-3} will be needed (shaded region in the centre).

The planning of our expensive experiments is greatly aided by these considerations: one can discuss an exploration strategy in the space of theories and give a money value to a given region. Gravitational research has become a kind of geographical exploration.

3. TECHNIQUES

In the PPN scheme every known experiment can be reinterpreted and used to falsify a given theory; but the main results come from five important techniques which is interesting to review.

Radar observations of planets, especially Venus and Mercury, can obtain an accuracy up to 30 m in a single observation; the basic problem then is how to extract information about the motion of the center of gravity from all the complications arising from the surface topography near the equator. This requires a large amount of data analysis to determine and eliminate all these unwanted parameters. As these techniques are improved and the data accumulate, the final precision in the relativistic parameters, in particular β and γ, will improve in the future.

Radio observation from spacecraft are currently done either by ranging - measuring the round trip light time of a specially modulated pulse - or by the Doppler effect - measuring the frequency shift of a transponded radio beam. A precision of about 1 m is currently possible for an interplanetary orbit; for the Doppler effect, an effective frequency stability of about 10^{-12}, corresponding to .03 cm sec^{-1}, is usually achieved. However, the implementation of hydrogen masers at the receiving stations will increase this figure by at least two orders of magnitude in the near future. To assess the importance of these accuracies, note that the nonsecular displacement due to relativistic effects on a solar orbit is of order $m_\odot = GM_\odot/c^2 \sim 10^5$ cm; and that the perturbation in velocity at a distance r is of order

$$\sqrt{\frac{m_\odot}{r}}\,\frac{m_\odot}{r}c \sim 3\cdot 10^{-2}\left(\frac{1\,AU}{r}\right)^{3/2} \text{cm sec}^{-1}.$$

This shows vividly that rather small errors in β and γ are within our experimental possibilities. However, spacecraft are subject to nongravitational accelerations - in particular, the unmodelled part of the solar radiation pressure - which make it rather difficult to observe their relativistic orbital perturbations. Nongravitational forces can be eliminated with special systems (drag-free spacecraft, accelerometer on board or the twin probe method) or with a planetary lander. These techniques have so far been tested only for a near earth orbit and have reached an accuracy in the compensation of nongravitational forces of about 10^{-8} cm sec^{-2}. Note that the acceleration we want to measure is

$$\frac{GM_\odot}{r^2}\frac{m_\odot}{r} \sim \frac{c^2}{r}\left(\frac{m_\odot}{r}\right)^2 \sim 7\cdot 10^{-9}\left(\frac{1\,AU}{r}\right)^3,$$

so that this is quite sufficient near the Sun.

Important experiments have been made with ordinary satellites by measuring the relativistic correction to the light transit time near conjunction; this correction is appreciably greater than the expected value $m_\odot/c = 5\mu$ sec and leads to a very good measurement of γ, the leading parameter in the photon trajectory.

This research program will reach a climax if the Solar Probe will be launched. This is a mission - studied both by NASA and ESA - aimed at sending a special, drag-free spacecraft very near the Sun. It has been shown that it is possible to achieve almost a parabolic orbit, with a perihelion of $4R_\odot = 0.02$ AU. Such a mission is the only way to make a satisfactory measurement of the quadrupole moment of the Sun (which produces a force $\sim r^{-4}$), besides giving a very good estimate of the relativistic parameters. This will at last clear up a fundamental query about the structure and the history of the Sun, does its interior rotate faster than its surface? A positive answer would agree with the extraction of angular momentum from the Sun accomplished by the solar wind. Since the solar quadrupole moment contributes to the perihelion advance, its knowledge is an essential part of any precise measurement of β and γ.

The third technique, radio interferometry of a celestial radio source, is well developed for astronomical reasons; when two sources are near the Sun their interferometric patterns are affected by the gravitational deviation of the rays. For example, in the latest measurements by Sramek and Fomalont (1977) a baseline of 35 km was used, corresponding to an angular accuracy of about 10^{-6} rad at 8.1 GHz. The relativistic deflection for a ray grazing the Sun is 1.75" $=8.5\cdot 10^{-6}$ rad$=4M_\odot/R_\odot$; of course statistics offers much scope for improvement.

The astronauts of Apollo 14 have placed a corner reflector on the Moon, with which it is possible to measure, by laser ranging, the Earth-Moon distance with an accuracy of less than 10 cm. This experiment has tested an important feature of relativistic theories of gravity, their effect on the three body problem;the other tests, in fact are really concerned only with the gravitational field of the Sun. The analysis of these measurements is complex, because they are the outcome of a host of different effects, including the continental drift; as K. Nordvedt (1977) has shown, this sets a rather stringent upper limit to the parameter η (eq. (5)).

Finally, one must mention the outstanding discovery of the binary pulsar PSR 1913+16, which shows a periodic variation of the period, indicating that the pulsar is a planet revolving around another (neutron?) star. The pulse and its changes are determined so accurately that one is not only able to estimate the keplerian elements (period 27900 sec, eccentricity 0.615, semimajor axis about 10^{10} cm, etc.), but the periastron advance was measured to be $4^{o}.22\pm 0^{o}.04\ y^{-1}$ (Taylor et al., 1976). If this advance is entirely due to relativistic effects, in any given theory of gravity one can deduce from it the total mass of the system, which is not known directly from the Newtonian orbit: a unique example in which general relativity is actually used to determine with great precision ($\sim$1%) an astrophysical quantity. Several other relativistic effects are important for this pulsar and have been summarized very well in Will (1978); since the smallness parameter m/r is about ten times larger than for Mercury, this is an intrinsically better system to test the relativistic theories of gravity.

4. THE PRINCIPLE OF EQUIVALENCE

The experimental and theoretical research carried out under the PPN scheme is a beautiful example of "normal science" in the Kuhn (1962) sense. Under this paradigm science proceeds by orderly accumulation of new results and by resolution of well defined problems according to precise prescriptions and models accepted by the scientific community and codified in textbooks. This paradigm is based upon the metric picture of space-time and relies upon familiar and intuitive tools like proper time chronometry and riemannian indefinite manifolds. Only the actual form of the metric and the field equations are subject to doubt and falsification.

A typical feature of normal science is its power of discovering its own anomalies and limitations by carrying out to the extreme the process of accumulation and refinement (see Kuhn (1962) especially Ch. VI); but within the PPN paradigm there may be still a long way to go before this. Its prescriptions are so general that it is not easy to find ways to falsify it. A crisis will occur by intrinsic development when the accuracy of the experiments reaches the post-post-newtonian level ($\Delta\gamma\sim\Delta\beta\sim 10^{-7}$); at this point the number of free parameters will grow so much as to make the method rather difficult to handle, if not altogether useless. Our present experimental possibilities are very far from this level, but unless a new, interesting theory is proposed which lies in the allowed region of parameter space, there seems to be little point in wandering with our research programs through a featureless desert. It seems that the PPN scheme has lost its great driving force and new experiments and frameworks have to be devised.

Broadly speaking, we should question the principle of equivalence and the metric structure of spacetime; as R.H.Dicke has pointed out very forcefully already in 1962, this principle, which in general relativity is transfigured into an abstract and simple mathematical property, is in reality a very complex statement and depends upon the separate verification of different physical assumptions concerning the effect of gravity upon electromagnetic, strong and weak interactions. In Einstein's theory these laws are universal, in the sense that any experiment conducted in a sufficiently small region of spacetime gives the same result, irrespective of the place and the time, and shows no effect of gravitation. The main question is, under which circumstances and at which order of magnitude can we expect a violation of the universality of local physics? Recalling Mach (1893), it is the motion of distant galaxies which determine the apparent forces and make us giddy in a merry-go-round; can one really think that local laws reveal absolutely no trace of what happens in the Cosmos?

Of course in the PPN scheme we can have violations of the principle of equivalence. This happens when gravitation is described by other fields besides the metric tensor g_{ij}; then in an inertial frame we can have locally

$$g_{ij} = \mathrm{diag}\,(1,-1,-1,-1) \text{ and } \partial_r g_{ij} = 0 ,$$

but each event is still distinguished by the value of other scalars. The current research on the principle of equivalence, however, is based upon more general and more empirical assumptions. Models of local physical laws are considered in which there are some dimensionless free parameters α_i, such as the fundamental microscopic constants, the dielectric constant and the magnetic susceptibility of vacuum, etc. Possible temporal variations of such parameters linked with the expansion of the Universe are measured by the dimensionless coefficients

$$\gamma_i = H^{-1} \frac{d \ln \alpha_i}{dt} , \tag{8}$$

where $H = (1/8 \cdot 10^{10} y)^{-1}$ is Hubble's constant (Sandage and Tamman 1975). The spacial variability produced by a mass M at a distance r is supposedly determined by the gravitational potential

$$U = \frac{GM}{c^2 r} = \frac{m}{r} . \tag{9}$$

Since usually $U \ll 1$ we can describe the dependence of α_i on U by means of the coefficients of the series expansion

$$\alpha_i = \alpha_{i0} (1 + \Gamma_i U + \Gamma_i' U^2 + \ldots) ; \tag{10}$$

note that in this view near a black hole or for cosmological phenomena, when $U \sim 1$, local physics should show drastic changes.

There is now a large amount of experimental evidence concerning γ_i and Γ_i; in no case has a discrepancy with the general relativistic values been found and in many cases upper limits appreciably less than unity have been established. We shall review in the next section a few points of this evidence; but we must face the fact that we have no theoretical or experimental answers to our query about the limits of validity of the principle of equivalence.

In gravitation there are three fundamental potential anomalies. First, the absolute and privileged character of proper time, which seems to betray a vicious circle (a proper clock is the only means to measure nongravitational forces, upon which, on the other hand, its very functioning is based) (see, for a similar criticism to Newton's absolute time, Mach (1960) p. 273). Secondly, the riddle of inertia: distant masses and their particular (e.g., isotropic) disposition should play an essential role in the local laws of physics; in my opinion we do not have yet a satisfactory dynamical theory valid also when only a few masses are present, without any unphysical boundary condition. Thirdly, the puzzle of the large dimensionless numbers and the cosmological coincidences. One wonders whether in these riddles there are the seeds of a crisis; note that a complete theory in which the principle of equivalence is violated drastically and "ab initio" is very difficult to imagine. It is possible that ordinary dynamics would show up in such a theory only when a stable and very massive cosmological background is assumed. A prerelativistic example of this cosmic approximation has been developed in detail by Barbour (1974), Barbour and Bertotti (1977) and Bertotti and Easthope (1978). The laws of (gravitation) physics are described at two different levels. At the fundamental level ("protophysics") the equations of motion are expressed in terms of relative, invariant quantities referring to two bodies (like the distance in a threedimensional euclidean space); moreover, there is no privileged way to measure time (in other words, time is just any ordering parameter). These laws of physics bear no resemblance with the usual ones. However, Newtonian gravity is recovered when most of the masses have a cosmological distribution with a mass M and a radius R; more precisely, the corrections are of order m/M or r/R, where m is a local mass and r a local distance. This mathematical limit $m/M \to 0$ is a distinctive feature of this scheme.

5. COSMOLOGICAL VIOLATIONS OF THE PRINCIPLE OF EQUIVALENCE

Our information about possible cosmological changes of the "constants" of physics comes (Fig. 2): a) from precision experiments in the laboratory or in the solar system; b) from the remote past of the earth or the stars in our galaxy; c) from the radiation coming from distant galaxies.

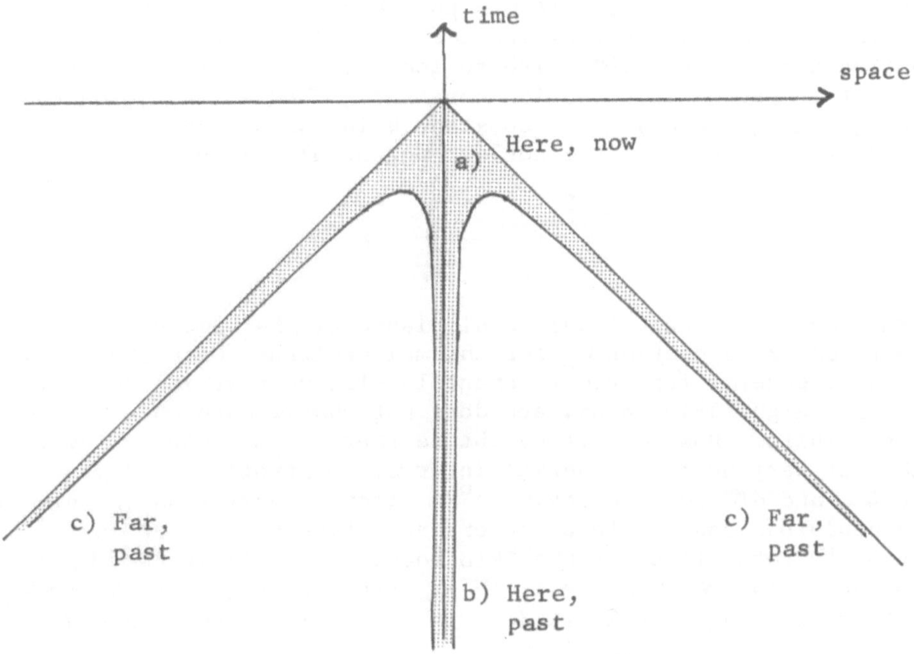

Fig. 2 - The three domains of the Universe directly accessible to our observations.

Direct comparison between different frequency standards (method a)), each of which depends upon a different combination of fundamental constants, will be made possible in the next few years by improvements in the hydrogen maser and the use of superconductive resonant cavity (Fig. 3). The main change in planetary motion inducted by a change in G is a secular increase of the semimajor axis of the orbit. From the study of the observations of the inner planets it was possible to set an upper limit to γ_G of order unity.

Much more effective are the measurements based upon b). The study of the evolution of the Sun, the stars and the Earth if G was larger in the past is an important and complex branch of astrophysics and geophysics on which I cannot dwell here: but, again, $|\gamma_G| \lesssim 1$ (see later). Nuclear decays of charged particles are very sensitive to the fine structure constant α through the Coulomb screening factor. If α was different in the past, the isotope composition of several elements would be different. For instance, from Rhenium it can be concluded $|\gamma_\alpha| < 10^{-3}$.

A significant improvement in these estimates was made possible by the discovery of the natural reactor at Oklo in Gabon, which attained criticality about $1.8 \cdot 10^9$ years ago (I.A.E.A. 1975). The rocks there contain rare earths which are fission products; by measuring their abundances it was possible to estimate separately both the neutron flux during the active phase and the neutron absorption cross sections σ_n with an accuracy, say, of 50%. Suppose now that the strong interaction constant g_s had at that time the value $g_s + \Delta g_s$, every other constant being unchanged. This changes the depth E_o of the nuclear potential well by

$$\frac{\Delta E_o}{E_o} = \frac{\Delta g_s}{g_s} , \qquad (11)$$

so that every resonant level is displaced by the same account. The absorption cross section σ_n for thermal neutrons, with energy 0.025 eV, is in general far from resonant levels, so that its value is not very large (5-10 barns) and does not change much when the resonances shift. However, if by chance there is a resonant level E_r within or very near the thermal interval, a strong enhancement of both σ_n and $d\sigma_n/dE_r$ ensues. ^{149}Sm offers a very good example of this behaviour and, in fact, enters as a fission product in the rare earth composition of the Oklo rocks (Fig. 4): it has $\sigma_n = 6 \cdot 10^4$ barn and $d\sigma_n/\sigma_n dE_2 = 17\ eV^{-1}$. From the requirement $|\Delta\sigma_n| \lesssim \sigma_n$ we get an upper limit ΔE_r in the shift of the resonance

$$\Delta E_r = \frac{\sigma_n}{d\sigma_n/dE_r} \sim 0.05\,eV ;$$

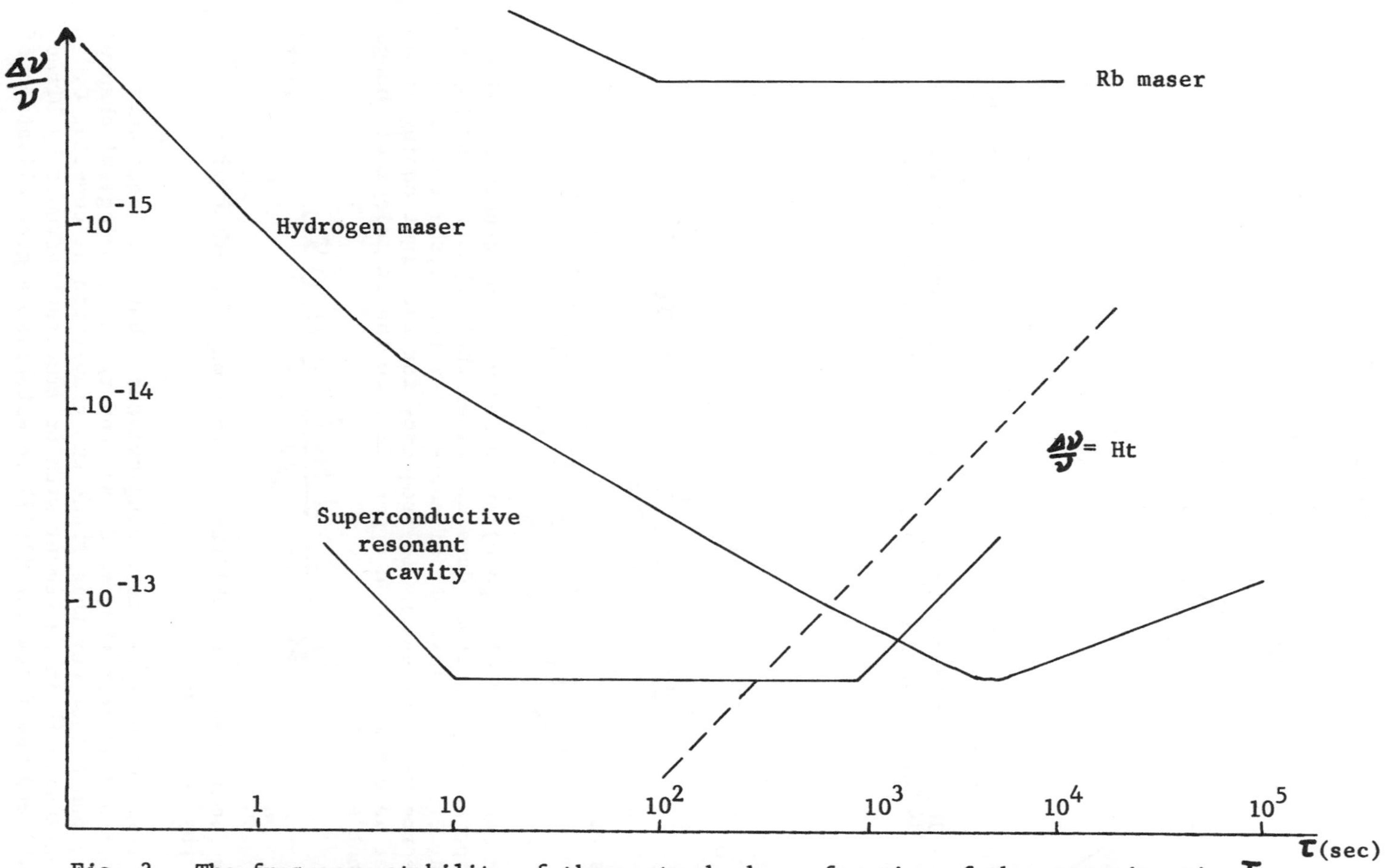

Fig. 3 - The frequency stability of three standards as function of the averaging time τ. The dotted line indicates the expected level for cosmological discrepancies between frequency standards.

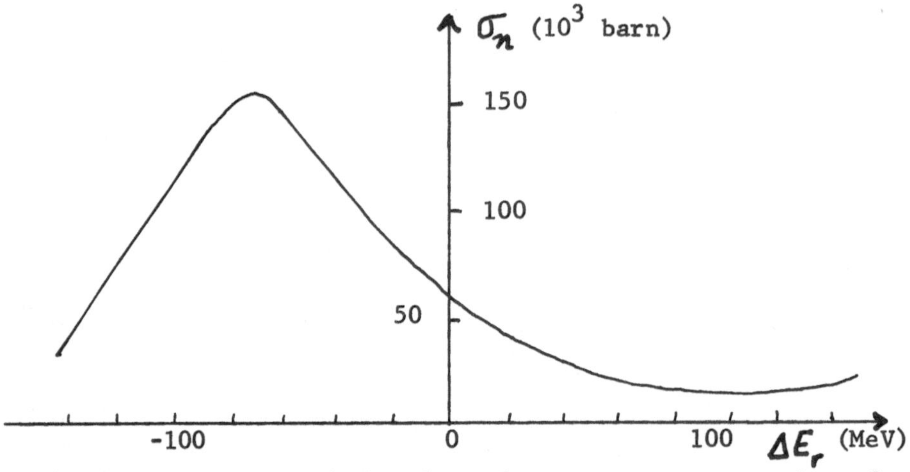

Fig. 4 - The Variation of the thermal neutron cross section of ^{149}Sm produced by a shift ΔE_r of all resonances.

since E_o is, say, 50 MeV, from (11) we get

$$\frac{\Delta g_s}{g_s} \lesssim \frac{1}{20 \cdot 50 \cdot 10^6} = 10^{-9}$$

that is to say,

$$|\gamma_{g_s}| = \frac{\Delta g_s}{g_s} \frac{2 \cdot 10^{10}}{1.8 \cdot 10^9} \lesssim 10^{-8}\,!$$

The measurements of type c) are based upon the hypothesis that the propagation of photons does not alter their spectral and physical characteristics (except, of course, for the redshift). If we measure the ratio of energy to frequency for the light coming from a quasar we can establish an upper limit to the cosmological change of Planck's constant:

$$\gamma_h = \frac{1}{H} \frac{d \ln h}{dt} \lesssim 0.08\,. \qquad (12)$$

Similar limits can be established in this way for other microscopic physical quantities.

In conclusion, there is strong evidence that the "constants" of atomic and nuclear physics do not undergo a cosmological change; however, one cannot say this about the gravitational constant G. This conclusion is in agreement with an important attempt by Dirac (1973a,b) and Canuto et al. (1977) to establish a generalization of

GR in which all clocks are equally privileged. One looks for field quantities and field equations which are invariant under a conformal transformation

$$ds_E \rightarrow ds_A = \beta^{-1}(x^i)ds_E \tag{13}$$

and reduce to the GR values in the "Einstein's gauge" $\beta = 1$. Every measurement carried out with any atomic time standard corresponds to another gauge. The theory predicts that all microscopic measurements agree with each other, but in general disagree with measurements based on gravitational clocks (like the moon); however, it is incomplete since it does not prescribe a rule for the determination of the atomic gauge. In a cosmological approximation the choice

$$\beta \sim 1/\mathcal{H}t \tag{14}$$

can explain the mysterious cosmic coincidences

$$\frac{e^2}{Gm_e m_p} \sim 2\cdot 10^{38} \sim \sqrt{\frac{M}{m_p}} \sim t_0 \frac{m_e c^3}{e^2}, \tag{15}$$

where M is the mass of the Universe and t_0 its age. In this theory M is not a constant and there is creation of matter. If new particles appear uniformly throughout space ("additive creation"), the masses of celestial bodies do not change appreciably; but one can also consider the hypothesis of "multiplicative creation," proportional to the matter density at each point. In such case every macroscopic mass is proportional to t^2. From the past thermal history of the Earth one can exclude the first case, while multiplicative creation is compatible with all the observations.

6. SPATIAL VIOLATIONS OF THE PRINCIPLE OF EQUIVALENCE

Very strong limits upon the constants Γ_i follow from the Eötvös experiment, which tests the independence of gravitational acceleration from the composition of the body. To see this, note that the total mass energy m of a body is the sum of its rest mass m_0 and all the contributions m_r of the different forms of binding energy; if a dependence from the gravitational potential (10) is assumed,

$$m = m_0 + \Sigma_r m_{0r}(1 + \Gamma_r U + O(U^2)) \tag{16}$$

depends on position. This produces additional terms in the weight-to-mass ratio

$$g = \frac{P}{m} = \frac{1}{m}\frac{dmU}{dz} = \frac{dU}{dz}\left(1 + \Sigma_r \frac{m_{r0}}{m}\Gamma_r U\right). \tag{17}$$

Two bodies with different ratios m_r/m will experience different accelerations. For example, for the electrostatic binding energy one can conclude from the experimental results $\Gamma_{el} < 0.03$.

In this model it is to be expected that the gravitational redshift will also show discrepancies from Einstein's prediction

$$\frac{\Delta\nu}{\nu} = \Delta U . \tag{18}$$

For an excitation which uses the r-th interaction one can show (Nordtvedt 1975a), on the basis of conservation of energy, that the frequency at infinity of a photon emitted where the gravitational potential is U is displaced by

$$\frac{\Delta\nu}{\nu} = U(1 + \Gamma_r U). \tag{19}$$

A new, second order term appears in the redshift; notice that for a solar probe at $4R_\odot$, $(M_\odot/4R_\odot)^2 = 3\cdot10^{-13}$, well within the H-maser accuracy.

The heuristic assumption (10) has been incorporated by Lightman and Lee (1973) into a consistent, but approximate theory, which consists in a generalization of electromagnetism based on four dimensionless scalars: the dielectric constant ϵ, the magnetic susceptibility of vacuum μ and two other, K and T, which appear in a generalization of the special relativistic expression of the mass-energy of a body. These four scalars, equal to unity far away from the masses, are functions of the gravitational potential U (assumed to be time independent):

$$\epsilon(U) = 1 + U\epsilon'(0) + \frac{1}{2}U^2\epsilon''(0) + \dots, \tag{20}$$

etc. Note that when this expansion is made, as one must do to carry out any calculation, one confines himself to weak violations of the equivalence principle and shrinks from considering the basic question, what is electromagnetism in a strong gravitational field? This Lagrangian scheme includes as a particular case, all metric theories with a line element of the form

$$ds^2 = T^2(U)dt^2 - K(U)(dx^2 + dy^2 + dz^2); \tag{21}$$

in order for this to happen, it is necessary and sufficient that

$$\epsilon = \mu = \sqrt{K/T} . \tag{22}$$

When this relation is not fulfilled, the equations of motion of a small body cannot be expressed as geodesic equations; the theory is not of the metric type and violates the principle of equivalence. The Eötvös experiment, as repeated by Roll et al. (1964) and Braginsky (1974) sets rather stringent limits to the violation of (22).

All frequency standards based upon electromagnetic interactions can be described precisely in this formalism. For example, for an H-maser the gravitational shift for a photon coming from a place of gravitational potential $\mathcal{V}$ reads

$$\frac{\Delta\nu}{\nu} = \frac{1}{\sqrt{T}} \frac{T\epsilon^2}{K} \frac{\epsilon}{\mu} - 1 \tag{23}$$

and take the conventional value $T^{-1/2}$ in the metric case (22)(Will 1974b). For a small $\mathcal{V}$ one can write, in general,

$$\frac{\Delta\nu}{\nu} = (1 + \delta\mathcal{V}) + O(\mathcal{V}^2) . \tag{24}$$

A first order redshift experiment is, therefore, a test of the metric character of electromagnetism.

On June 18th, 1976 a rocket from Virginia flew a hydrogen maser for about two hours at an altitude of about 10,000 km (Vessot and Levine 1977); the analysis of the observed frequency shift has confirmed the metric prediction with an accuracy $\delta \sim 2\cdot 10^{-4}$. This extraordinary result can no doubt be improved when an H-maser will be flown on a spacecraft.

7. ISOTROPY

The construction of a hypothetical theory of gravitation which violates "ab initio" the principle of equivalence and describes a basic influence of distant matter upon local physics must satisfy a very stringent experimental constraint, <u>the extraordinary degree of isotropy of the inertial mass</u>. A simple expression of a mass tensor generated gravitationally by a mass M_o at a distance r_o is

$$m_o \frac{GM_o}{c^2 r_o} \underset{\sim}{r_o}\, \underset{\sim}{r_o} = m_o \underline{\underline{\Delta}} \;;$$

such a tensor might appear in the Newtonian expression of the dynamical law

$$\underset{\sim}{F} = m_o (\underline{\underline{I}} + \underline{\underline{\Delta}}) \cdot \underline{a} \; . \tag{25}$$

For the Galaxy $GM_o/c^2r_o^2$ is about $5\cdot10^{-7}$ and this would produce unacceptable disturbances in the metric of the planets and in the dynamics of the binary pulsar PSR 1913-16 (Nordtvedt 1975b).

For electromagnetic and nuclear forces the anisotropy of inertial mass is subject to much more severe constraints. A quantum mechanical system corresponding to a classical system with kinetic energy

$$T = \frac{1}{2} m_{\alpha\beta} v^{\alpha} v^{\beta} = \frac{1}{2} m^{-1\alpha\beta} p_{\alpha} p_{\beta} \qquad (26)$$

gives rise to a Schrödinger equation in which the Laplacian is replaced by $m^{-1\alpha\beta}\partial_{\alpha}\partial_{\beta}$; this makes the energy levels dependent upon the orientation of the system with respect to the external mass which determines the anisotropy. The rotating earth will modulate any spectroscopic transition with the period of a sidereal day. The most precise experiment of this kind was made by Drever (1961) who, using a nuclear level of Li^7, established the limit

$$\Delta m/m < 5\cdot 10^{-23} \;!! \qquad (27)$$

(see also Dicke, 1964). It seems, therefore, that the rule for calculating the kinetic energy of a material point must be the same as that for calculating the distance between two points. It should also be mentioned that Drever's experiment refers to the <u>spatial</u> isotropy of the mass and says nothing about its scalar character with respect to transformations which involve time.

It seems, therefore, that although the Universe around us has a very anisotropic appearance, a local dynamical system does not see this anisotropy to an extraordinary degree of accuracy. At this point, as a conjectural conclusion, a question is naturally raised, is the Universe as a whole <u>really anisotropic</u>? In other words, perhaps our direct vision is myopic, while a dynamical system may be able to probe the Universe much deeper. The extraordinary degree of isotropy of the background radiation (Smoot et al. 1977) is certainly a puzzle in conventional cosmology, which allows a bewildering variety of solutions. Perhaps the isotropy of the Universe is a fundamental law of nature.

REFERENCES

Barbour, J.(1974). Nature 249, 328.

Barbour, J. and Bertotti, B. (1977). Nuovo Cimento 38B , 1

Bertotti, B. (1977). in "Experimental Gravitation" (B.Bertotti, editor), p.17. Rome, Accademia dei Lincei.

Bertotti, B. and Easthope, P. (1978). Int.J.of Theor.Phys.

Braginsky, V.B. (1974). Rendiconti S.I.F., Course LVI, p.252.

Brans, C. and Dicke, R.H. (1961). Phys.Rev. 124, 925.

Canuto, V., Adams, P.J., Hsieh, S.H. and Tsiang, E. (1977). Phys. Rev. D16, 1643.

Dicke, R.H. (1962). Rendiconti S.I.F., Course XX, p.1.

Dicke,R.H. (1964). "The Theoretical Significance of Experimental Relativity". New York, Gordon and Breach.

Dicke,R.H. and Peebles,P.J. (1965). Space Science Reviews 4, 419.

Dirac, P.A.M. (1973a). Proc. R.Soc.Lond. A333, 403.

Dirac, P.A.M. (1973b). Proc.R.Soc.Lond. A338, 439.

Drever, R.W.P. (1961). Phil.Mag. 6, 683.

I.A.E.A.(1975). "The Oklo phenomenon". SM-204/38.

Kuhn, T.S.(1962). "The Structure of Scientific Revolutions". Chicago, University of Chicago Press.

Lightman, A.P. and Lee, D.L. (1973). Phys.Rev. 8, 364.

Mach, E. (1893). "Die Mechanik in ihrer Entwicklung historish-kritisch dargestellt". Leipzig. English edition of 1960 "The Science of Mechanics: a Critical and Historical Account of its Development". La Salle, Open Court.

Nordtvedt, K.Jr. (1975a). Phys.Rev.D11, 245.

Nordtvedt, K.Jr. (1975b). Ap.J. 202, 248.

Nordtvedt, K.Jr.. (1977) in "Experimental Gravitation" (B.Bertotti, editor), p.183.

Popper, K.R. (1959). "The Logic of Scientific Discovery". London, Hutchinson.

Roll, P.G., Krotkov, R.and Dicke, R.H. (1964). Annals of Phys. 26, 442.

Sandage, A.and Tammann, G.A. (1975). Ap.J. 97, 265.

Smoot, G.F., Gorenstein, M.V. and Muller, R.A. (1977). Phys.Rev. Lett. 39, 898.

Sramek, R. and Fomalont, E. (1977) in "Experimental Gravitation" (B.Bertotti, editor), p.136. Rome, Accademia dei Lincei.

Taylor, J.H. and others (1976). Ap.J. 206, L53.

Vessot, R.F.C. and Levine, M.W. (1977). Proc. Intern. Symposium "Experimental Gravitation". Rome, Accademia dei Lincei, p.371.

Will, C.M. (1974a) Rendiconti S.I.F., Course LVI, p.1. New York, Academic Press.

Will, C.M. (1974b), Phys.Rev. D10, 2330.

Will, C.M. (1979). In "Einstein Centenary Volume" (W.Israel and S. Hawking, editors). Cambridge, Cambridge University Press.

Williams, J.G. and others (1976) Phys.Rev.Lett. 11, 551.

AN INTRODUCTION TO GENERAL RELATIVITY AND ITS RECENT ACHIEVEMENTS

Yvonne Choquet-Bruhat

Université P. et M. Curie, I.M.T.A.

4 Place Jussieu, 75007 Paris

INTRODUCTION

I shall speak in these lectures of classical general relativity, trying to start from its foundations to arrive at some of its most recent achievements.

General relativity was devised by Einstein as a theory for the classical gravitational field, but in doing so he also gave to this field the fundamental role of determining what is a time-like or a spatial separation of events, and therefore the causal structure of space time.

The guiding lines which led Einstein to general relativity are well known. They were on the one hand the theory of special relativity, on the other hand the equivalence principle.

In special relativity space and time are no more independent entities, as they were in the mechanics of Galilée-Newton. They are linked through the space-time Minkowski metric

$$ds^2 = -(dx^o)^2 + \sum_{i=1}^{3} (dx^i)^2$$

which is invariant by the Lorentz group.

The equivalence principle, equality of the gravitational and inertial mass, then led Einstein to the hypothesis that the motion of a particle falling freely in a gravitational field should follow a geodesic of some curved space time; this space time should look like the Minkowski space time in a small neighborhood of any of its points. Einstein furthermore limited the possible physical models for this curved space time by the "Einstein equations" which link the curvature to the energy sources. This is "General Relativity". We shall give briefly the mathematical apparatus necessary to set precisely the theory.

I EINSTEINIAN SPACE TIME MANIFOLDS

1 Space Time Manifold

In special relativity the Minkowski metric is put on $\mathbb{R}^4$. This set will be generalized in G.R. to a 4-dimensional differentiable manifold V which is equivalent to $\mathbb{R}^4$ only Locally T, namely

Definition An n-dimensional differentiable (C^∞) manifold V is a set which can be covered by a family of subsets U_i such that each U_i is in bijective correspondence with an open set Ω_i of $\mathbb{R}^n$:

$$U_i \to \Omega_i \equiv \varphi_i(U_i) \subset \mathbb{R}^n \quad , \text{ by } \quad x \mapsto \varphi_i(x) = (x^1,\ldots,x^n)$$

with the property that, for all i, j :

$$\varphi_i \circ \varphi_j^{-1} \; : \varphi_j(U_i \cap U_j) \to \varphi_i(U_i \cap U_j)$$

is a C^∞ diffeomorphism between open sets of $\mathbb{R}^n$.

The family of pairs (U_i, φ_i) are said to be an atlas of V, each pair being a chart and $\varphi_i(x) = (x^1,\ldots,x^n)$ local coordinates of x. Another pair (U,φ) is compatible with the atlas if, when added to the previous family, we still have an atlas, which is then said to be equivalent to the first one. Equivalent atlases define the same differentiable structure.

At each point x of the differentiable manifold V there is a tangent space T_xV, n-dimensional vector space which may be defined as the space of equivalence classes of tangent curves drawn on V through x.

From this tangent space one defines, as for any vector space, the space of tensors at x. A p-covariant tensor u at x is a p-linear form on T_xV :

$$u_x \;:\; T_xV \times \;\ldots\; \times T_xV \to \mathbb{R}$$

$$\text{by } (v_{(1)}, \;\ldots\;, v_{(p)}) \mapsto u_x(v_{(1)}, \;\ldots\;, v_{(p)})$$

A p-contravariant tensor is a p-linear form on the cotangent space $T_x^* V$ (dual of T_xV).

A frame in the tangent space is defined by the choice of n

independent vectors at x. A "natural" frame associated with a system of coordinates is defined by the curves where only one of the coordinates varies. Frames at a point are in bijective correspondance with the linear group $G\ell(n)$ (the correspondence is non canonical, depending on the choice of one frame).

A, for instance p-covariant, tensor field on V is an assignment $x \mapsto u_x$, p covariant tensor at x, for each $x \in V$.

2 Metric

A metric : is a covariant 2-tensor field on V which is symmetric and non degenerate, i.e. at each point $x \in V$, g_x is a non degenerate quadratic form in the tangent space :

$$(v_x, w_x) \mapsto g_x(v_x, w_x) \in \mathbb{R} \quad , \quad v_x, w_x \in T_xV$$

A metric gives a canonical correspondence - and thus an identification - between contravariant and covariant vectors (and therefore tensors) : a vector v has contravariant component (v^i) and covariant components (v_i) with (1) $v^i = g^{ij} v_j$.

In General Relativity the tangent space at each point of the space time manifold is the Minkowski space time, the quadratic form has signature (- + + +). $g_x(v_x, v_x)$ can be written :

$$g_x(v_x, v_x) = - (v^o)^2 + \sum_{i=1}^{3}(v^i)^2$$

where (v^o, v^i) are the components of v_x with respect to a frame of a special family, called orthonormal frames - these frames are related to each other by a Lorentz transformation. It is usual to write the metric in the natural frame, in a chart :

$$ds^2 = g_{\alpha\beta} \, dx^\alpha \, dx^\beta$$

The metric defines on V a field of cones $\Gamma_x \subset T_xV$, $g_x(v,v) = 0$, which determine, as in special relativity, the time-like $(g_x(v,v) < 0)$, null $(g_x(v,v) = 0)$ and spatial directions. If the cones Γ_x can be split globally and continuously into the 1/2 cones Γ_x^+ (future) and Γ_x^- (past) the manifold is said to be time orientable. We shall always suppose it time oriented. The future of a space-time event x is now the set of points of V which can be joined to x by a future oriented time-like or null curve $\widehat{xy}$. A general space time (V,g) can present paradoxal behavior, like closed time like paths where the future meets the past. These pathologies are removed by global hypothesis on the metric, like global hyperbolicity which excludes in particular the occurrence of closed, or almost closed, time like curves.

----

(1) sum over repeated indices, Einstein's convention.

The metric is used to define the length of a curve γ joining 2 points a and b (proper time[1] along that curve if it is time like) :

$$(2\text{-}1)\qquad \ell(\gamma) = \int_a^b ds = \int_{t_o}^{t_1} \{ g_x (\frac{dx}{dt}, \frac{dx}{dt})\}^{1/2} dt$$

$$\gamma : t \mapsto x(t) \quad , \quad a = x(t_o) \quad , \quad b = x(t_1) .$$

The geodesics are local extrema of that length[2]; they are the trajectories of the free particles.
The geodesics satisfy second order differential equations on V, the Euler equations of the lagrangian (2-1). These equations can be written as a first order system on the tangent space to V, $TV = \bigcup_{x\in V} T_x V$:

$$\frac{dx}{dt} = v \quad , \quad \frac{dv}{dt} = - \Gamma(v,v)$$

with, by a classical computation, in local coordinates

$$\Gamma(v,v) = (\Gamma^k_{ij} v^i v^j)$$

$$\Gamma^k_{ij} = \frac{1}{2} g^{k\ell} (\partial_i g_{jk} + \partial_j g_{ki} - \partial_k g_{ji})$$

Note [3] that $(v, -\Gamma(v,v))$ is a vectoe field on TV (called the spray of the metric g, or geodesic spray) but that $\Gamma(v,v)$ is not a vector field on V, $\Gamma = (\Gamma^k_{ij})$ is not a tensor field on V.

A geodesic γ is also called an "autoparallel" curve, its tangent vector is said to be "parallel transported" along γ .

3 Riemannian Connexion

If we want to compare tangent vectors to V at different points we must have a definition of parallel transport along an arbitrary curve. We say, that v is parallel transported along the curve γ ,

----

(1) Proportional to the time measured by a standard clock following this trajectory.
(2) In fact local maxima for time like geodesics.
(3) If $x^i \mapsto x^{j'}(x^i)$ is a change of coordinates in V, the corresponding change in TV is :

$$(x^i, v^i) \mapsto (x^{j'}(x^i) \, , \, v^{j'} = \frac{\partial x^{j'}}{\partial x^i} (x^\ell) v^i)$$

thus, if (v,V) is a tangent vector to TV :

$$V^{j'} = v^i \frac{\partial v^{j'}}{\partial x^i} + V^i \frac{\partial v^{j'}}{\partial v^i}$$

which gives the following law of transformation of Γ :

$$\Gamma^{k'}_{i'j'} = \frac{\partial x^{k'}}{\partial x^k} \frac{\partial x^i}{\partial x^{i'}} \frac{\partial x^j}{\partial x^{j'}} \Gamma^k_{ij} + \frac{\partial^2 x^k}{\partial x^{i'} \partial x^{j'}} \frac{\partial x^{k'}}{\partial x^k}$$

$t \mapsto x(t)$, with tangent vector u, if it is satisfies the differential relations on TV :

$$\frac{dv}{dt} = - \Gamma(u,v) \quad , \quad \frac{dx}{dt} = u$$

The covariant derivative of a vector field v on V in the direction u is the vector field obtained by comparing the value v_t of this vector at a given point $x(t) \in \gamma$ with the parallel translated $\tau_t^{t+h} v_{t+h}$ from a nearly point x(t+h) along the curve γ with tangent vector u, namely :

$$\nabla_u v = \lim_{h = 0} \frac{1}{h} \left[\tau_t^{t+h} v_{t+h} - v_t\right]$$

which gives immediately, in local coordinates :

$$(\nabla_u v)^k = \frac{dv^k}{dt} + \Gamma^k_{ij} u^i v^j = u^i \left(\frac{\partial v^k}{\partial x^i} + \Gamma^k_{ij} v^j \right)$$

The covariant derivatives of contravariant tensor fields are obtained by the Leibniz rule applied to the tensor product :

$$\nabla_u (T \otimes S) = \nabla_u T \otimes S + T \otimes \nabla_u S$$

and analogously for covariant tensors, after using the law :

$$\nabla_u < v,w > = < \nabla_u v, w > + < v, \nabla_u w >$$

if w is a 1-form (covariant vector).
$\nabla_u T$ being a tensor field of the same type as T, depending linearly on u, we can write

$$\nabla_u T = < u, \nabla T > \qquad \text{(duality)}$$

where ∇T is a tensor, 1-covariant in addition to the type of T. An easy computation shows that if, for instance

$$T = (T_{i_1 \ldots i_p}{}^{j_1 \ldots j_q})$$

$$\nabla T = (\nabla_\ell T_{i_1 \ldots i_p}{}^{j_1 \ldots j_q})$$

$$= (\partial_\ell T_{i_1 \ldots i_p}{}^{j_1 \ldots j_q} + \Gamma^h_{\ell i_1} T_{h i_2 \ldots i_p}{}^{j_1 \ldots j_q} + \ldots$$

$$+ \Gamma^h_{\ell i_p} T_{i_1 \ldots h}{}^{j_1 \ldots j_q} - \Gamma^h_{\ell j_1} T_{i_1 \ldots i_p}{}^{h j_2 \ldots j_q} - \Gamma^h_{\ell j_q} T_{i_1 \ldots i_p}{}^{j_1 \ldots h}$$

In particular we have the fundamental property :

$$\nabla g = 0$$

This property can be taken as the characterization, of metric connexions among objects called connexions[1] used on differentiable manifolds to define parallel transport. The Γ we have used is a special metric connexion, called the riemannian connexion, which has moreover a vanishing torsion $(\Gamma^k_{ij} - \Gamma^k_{ji} = 0)$.

4 Curvature - Bianchi identities.

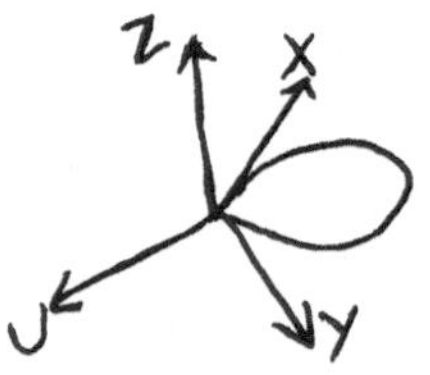

If a vector Z is parallel transported along a closed curve the result is in general different from Z : this difference is characteristic of the curvature of the connexion. An infinitesimal loop being given by 2 vectors X and Y, the curvature tensor corresponds to the following map [2] :

$$(Z,X,Y) \mapsto U \quad \text{by} \quad U = \nabla_X \nabla_Y Z - \nabla_Y \nabla_X Z - \nabla_{[X,Y]} Z$$

where $[X,Y] = \mathcal{L}_X Y = - [Y,X]$ is the Lie bracket of the vectors X and Y, vector given in local coordinates by :

$$[X,Y] = \left(X^i \frac{\partial Y^j}{\partial x^i} - Y^i \frac{\partial X^j}{\partial x^i} \right)$$

Thus the curvature tensor is a 3-covariant, 1 contravariant tensor; for connexions with vanishing torsion and in particular for the riemannian connexion $(\Gamma^k_{ij} = \Gamma^k_{ji})$ one has, in local coordinates :

$$\nabla_k \nabla_\ell v^i - \nabla_\ell \nabla_k v^i \equiv R^i{}_{jk\ell}\, v^j$$

and therefore the Riemann curvature tensor is equal to :

$$(4\text{-}1) \qquad R^i{}_{jk\ell} = \partial_k \Gamma^i_{\ell j} - \partial_\ell \Gamma^i_{kj} + \Gamma^i_{km} \Gamma^m_{\ell j} - \Gamma^i_{\ell m} \Gamma^m_{kj}$$

It can be proved that a riemannian manifold is locally flat (i.e. admits an atlas such that in every chart

$ds^2 = \sum_{i=0}^{n} \pm (dx^i)^2$) if and only if its curvature tensor vanishes.

Ricci tensor and scalar curvature

They are obtained by contraction of the curvature tensor as follows :

----

(1) For general connexions cf B. Carter, this volume or |8|

(2) The term $\nabla_{[X,Y]} Z$ is added to insure the linearity.

(3) Note that $\nabla_{\ell_\ell} Z = (\nabla_\ell \; Z^i)$ but that, $\nabla_{\ell_k} \nabla_{\ell_\ell} \; Z = (\nabla_k \nabla_\ell \; Z^i + \Gamma^m_{k\ell} \nabla_m Z^i)$

Ricc(g), the Ricci tensor, is given in local coordinates by :

$$R_{j\ell} \equiv R^i{}_{ji\ell} = \partial_i \Gamma^i_{j\ell} - \partial_\ell \Gamma^\ell_{j\ell} + \Gamma^i_{j\ell} \Gamma^k_{ik} - \Gamma^i_{jk} \Gamma^k_{\ell i}$$

it is a symmetric 2-tensor.
R(g), the scalar curvature is :

$$R \equiv g^{j\ell} R_{j\ell} .$$

Bianchi identities

The curvature tensor of a riemannian connexion satisfies the identities
a) Algebraic identities :

$$(4\text{-}2\ a) \qquad R^i{}_{jk\ell} = - R^i{}_{j\ell k}$$

$$(4\text{-}2\ b) \qquad \oint_{j,k\,\ell} R^i{}_{jk\ell} = 0$$

$$(4\text{-}2\ c) \qquad R_{ij,k\ell} = R_{k\ell,ij}$$

b) Bianchi identities

$$(4\text{-}3) \qquad \oint_{h,k,\ell} \nabla_h R^i{}_{j,k\ell} = 0$$

From these one deduces the following identities, which play a key role in Einstein's theory of gravitation :

$$(4\text{-}4) \qquad \nabla_h (R^{hi} - \frac{1}{2} g^{hi} R) \equiv 0$$

5 The Einstein Equations.

Einstein, inspired in a sense by Mach's ideas thought that the curvature of the hyperbolic metric of space time should be linked to the energy sources. The corresponding equations must be approximated, in the case of weak gravitational fields and small velocities, by the Poisson equation satisfied by the Newtonian potential, which is a partial differential equation of second order. This fact and the identity (4-4) led Einstein to the following equations to be satisfied by the "gravitational potential", the metric tensor g :

$$\mathrm{Ricc}(g) - \frac{1}{2} g\, R(g) = \Lambda g + kT$$

that is, in coordinates

$$(5\text{-}1) \qquad R_{\alpha\beta} - \frac{1}{2} g_{\alpha\beta} R = \Lambda g_{\alpha\beta} + kT_{\alpha\beta}$$

Λ is the "cosmological constant", usually taken to be zero while $T_{\alpha\beta}$ is a phenomelogical symmetric 2-tensor describing the (non gravitational) energy sources. It is called the "stress-energy tensor" being the 4-dimensional relativistic synthesis of the classical stress tensor and of the energy density. As a consequence of the identity (4-4) it satisfies the "conservation laws" :

$$(5\text{-}2) \qquad \nabla_\alpha T^{\alpha\beta} = 0$$

In the case of electromagnetic energy sources due to an electromagnetic field $F = (F_{\alpha\beta})$, with $dF = 0$, ($\oint_{\alpha,\beta,\gamma} \nabla_\alpha F_{\beta\gamma} = 0$) T is the Maxwell stress energy tensor.

$$\tau_{\alpha\beta} = \frac{1}{4} g_{\alpha\beta} F^{\lambda\mu} F_{\lambda\mu} - F_\alpha{}^\lambda F_{\beta\lambda}$$

and the equation (5-2) gives the Maxwell equation

$$\nabla_\alpha F^{\alpha\beta} = 0$$

In the case of a perfect isentropic fluid

$$T_{\alpha\beta} = (\rho+p)\, u_\alpha u_\beta + p\, g_{\alpha\beta}$$

(ρ and p density and pressure, u_α four velocity, $u^\alpha u_\alpha = -1$).
Equation (5-2) gives the "equations of motion" and "conservation of energy" :

$$\nabla_\alpha((\rho+p)\, u^\alpha) - u^\alpha \partial_\alpha p = 0$$

$$(\rho+p)\, u^\alpha \nabla_\alpha u_\beta + (g^{\alpha\beta} + u^\alpha u^\beta)\partial_\beta p = 0$$

6 Integration of Einstein's Equations.

An Einsteinian space time is a 4 dimensional riemannian manifold (V,g), with g a hyperbolic metric satisfying the Einstein equations (5-1), with energy sources satisfying (5-2). Many exact solutions have been found fulfilling these requirements. In particular the Schwarzchild solution

$$(6\text{-}1) \quad ds^2 = \left(1 - \frac{2m}{r}\right) dt^2 - \left(1 - \frac{2m}{r}\right)^{-1} dr^2 - r^2(d\theta^2 + \sin^2\theta\, d\varphi^2)$$

satisfies the equation

$$R_{\alpha\beta} = 0$$

and is interpreted, for $r > 2m$, as the static gravitational field of a spherical mass. It can be completed into a regular space time $(\mathbb{R}^4, g)$, with g given by (6-1) for $r > r_o > 2m$ and g solution of equations (5-1) with non vanishing $T_{\alpha\beta}$ ("interior" Schwarzchild solution) when $r < r_o$; it gives then, for instance, a model for the solar system. The occurrence of strange phenomena in the metric (6-1) for $r = 2m$, and the fact that no time like or null curve entering the region $r < 2m$ can ever leave it again is reflected in the name "black holes" given to a model represented by the metric (6-1) considered to be still valid in a region where $r < 2m$, and interpreted as the gravitational field of a "collapsed" object.

Cosmological models are space times (V,g) with many symmetries. The most popular are the Robertson-Walker universes

$$ds^2 = dt^2 - R^2(t)\, d\sigma^2$$

with $d\sigma^2$ the metric of a 3-dimensional manifold with constant curvature (3-sphere, flat 3-space or 3-pseudo-sphere, with negative curvature). Such a model is an expanding universe if R(t) is increasing. Under some physically reasonable assumptions on the positivity of the energy of the sources these models have a singularity ($R(t_o)=0$ for some t_o) : in an expanding universe it is interpreted as the initial "big bang". The fact of possessing a singularity seems a general property of realistic space times (Penrose-Hawking singularity theorems cf |6|.

In the next chapter we shall study the general problem of integration of the Einstein equations. We shall restrict ourselves, for simplicity, to the exterior case ($T_{\alpha\beta} = 0$) where the main features of the properties of the solutions already appear.

II CAUCHY PROBLEM

Hamiltonian Formalism

Introduction

A classical (non quantum) theory is considered to be well posed if it is deterministic : given initial data at some "instant of time" they must determine the entire evolution of the system. In relativity an "instant of time" is a 3-dimensional, space like submanifold. To analyse what the initial data are for the gravitational field we first give an exposition of the "3+1 formalism", which rewrites Einstein's equation in space time as an evolution system for the (intrinsic and extrinsic) geometry of a 3-dimensional manifold, geometry which is always astrained to satisfy some constraints.

1 Slicings. Lapse and Shift. Intrinsic and Extrinsic Geometry.

The 3+1 formalism supposes that the space-time manifold V admits a slicing by a family of 3-dimensional space like submanifolds, that is V is diffeomorphic to a product $S \times I$ with :

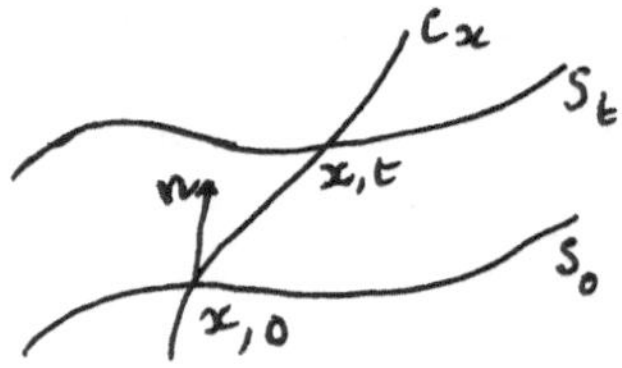

$f = V \to S \times I$, diffeomorphism

$S_t = f^{-1}(S \times \{t\})$ space like submanifolds.

The lines $C_x \equiv f^{-1}(\{x\} \times I)$ are in general supposed time like, but this hypothesis is not necessary to write the equations which follow, which are just identities in riemannian geometry.

We denote by α_t , lapse function, the (orthogonal) projection on n (unit normal to S_t) of the tangent vector to C_x at (x,t) and by β_t, shift vector, its (parallel to n) projection on the tangent space to S_t.
We denote by γ_t the positive definite riemannian metric induced on S_t by the space time metric g.
In local coordinates with $x^o = t$ and (x^i), i = 1,2,3 coordinates of x, the above choice corresponds to the decomposition of the metric g :

$$g_{\alpha\beta}\, dx^\alpha\, dx^\beta = -\,(\alpha^2 - \beta^i\, \beta_i)\, dt^2 + 2\, \beta_i\, dx^i\, dt + \gamma_{ij}\, dx^i\, dx^j$$

with

$$\alpha = (-\, g^{oo})^{-1/2} \ , \quad \beta_i = g_{oi},\ \gamma_{ij} = g_{ij},\ \beta^i = \gamma^{ij}\, \beta_j \ .$$

Besides the metric γ_t we introduce the "extrinsic curvature" or "second fundamental form" K_t of S_t, which describes the infinitesimal change of the metric γ_t with t. This symmetric 2-tensor, which depends only on the family of submanifolds S_t (and not on the choice of the time like lines) is equal to $-\frac{1}{2}\frac{\partial \gamma}{\partial s}$ when the time like lines are the geodesics of g normal to the S's. Its expression for an arbitrary choice of lapse and shift is

$$K_t = -\frac{1}{2}\, \Pi \mathcal{L}_n g \ , \quad (K_{ij} = -\frac{1}{2}(\nabla_i\, n_j + \nabla_j\, n_i) = \Gamma^o_{ij}\, n_o)$$

when Π is the orthogonal projection on S_t. The definition of K_t may be written as follows :

$$(1) \quad \frac{\partial \gamma}{\partial t} = -\, 2\alpha\, K_t + \mathcal{L}_\beta \gamma$$

with $\mathcal{L}_\beta \gamma$ the Lie derivative[(1)] of γ with respect to β .

----

(1) $(\mathcal{L}_\beta\, \gamma)_{ij} = D_i\, \beta_j + D_j\, \beta_i$, D covariant derivative in the metric γ.

2 Decomposition of Ricc(g)

A space time vector v is characterized at each space time point (x,t) by its projections $v^{\perp}$ (orthogonal on n and v' (parallel to n) on the tangent space to S_t. An analogous decomposition is valid for an arbitrary tensor. Straight forward computations (done for instance using Cartan calculus, cf |2| or in local coordinates |4|) gives the following equations

$$(2) \qquad \mathrm{Ricc}(g) \equiv \mathrm{Ricc}(\gamma) + K \,\mathrm{tr}\, K + 2\, K \times K - \frac{1}{\alpha}\mathrm{Hess}\,\alpha$$

$$- \frac{1}{\alpha}\partial_o K + \mathcal{L}_\beta K \qquad \text{(contravariant tensors on S)}$$

where $(K \times K)^{ij} \equiv K_\ell{}^i \quad K^{\ell j}$

While the following combinations are found to be dependant(1) only on γ and K :

$$(3) \quad 2S^{\perp\perp} \equiv R(\gamma) - K.K + (\mathrm{tr}\, K)^2 \equiv -\mathcal{H} \quad \text{(scalar)}$$

$$(4) \quad S_i^{\perp} = - D.K + D\,\mathrm{tr}\, K = - \mathcal{J}/2 \quad \text{(covariant vector)}$$

where D is the covariant derivative and the dot a contraction in the metric γ .

On the other hand the Bianchi identities, $\nabla_\lambda S^{\lambda\mu} \equiv 0$, give the following homogeneous first order system for $\mathcal{J}$ and $\mathcal{H}$ (with $(\mathrm{div}\, S)^{\perp} = \nabla_\lambda S^{\lambda o}$, $(\mathrm{div}\, S)' = (\nabla_\lambda S^{\lambda i})$)

$$(5) \quad 2\alpha^2(\mathrm{div}\, S)^{\perp} \equiv \partial_o \mathcal{H} + \mathrm{div}(\alpha^2 \mathcal{J}) - \mathrm{tr}\, K\, \alpha^2 \mathcal{H} - L_\beta \mathcal{H}$$

$$(6) \quad 2\alpha(\mathrm{div}\, S)' = \partial_o \mathcal{J} - L_\beta \mathcal{J} + d\alpha, \mathcal{H}$$

The algebraic decomposition of Ricc(g), into the set of equations (2), (3), (4) indicates that the relevant Cauchy data for the gravitational field are the properly riemannian metric γ, on a 3-manifold Σ , and a second order symmetric tensor K, with γ and K satisfying the equations (3) and (4) (constraints).

The Cauchy problem with initial data set (Σ, γ ,K) is the problem of finding a space time (V,g) into which these data can be imbedded in the following sense :

1) there exists a submanifold S of V diffeomorphic to Σ , by a diffeomorphism Λ

2) The image by Λ of the metric induced by g on S, and of the extrinsic curvature of S in (V,g) are respectively γ and K.

----

(1) $S = \mathrm{Ricc}(g) - \frac{1}{2} g\, R(g)$, cf I, $S^{\perp\perp} = S^{\lambda\mu} n_\lambda n_\mu = \alpha^2 S^{oo}$, $S_i^{\perp} = (S_\lambda^\alpha \gamma^\lambda{}_\mu n_\alpha) = - S_i^o \alpha)$

To solve this problem we take for V the product $\Sigma \times R$, and put $\Sigma \times \{0\} = S$. Formally the dynamical system (1), (2) determines (uniquely) γ_t, K_t from their value $\gamma_o = \gamma$ and $K_o = K$ for $t = 0$, once the lapse α_t and the shift β_t are given, arbitrarily (but for $\alpha_t \neq 0$) on $\Sigma \times R$. For each t the metric g is deduced from γ_t and α_t, β_t - equation (1) insures then that K_t is indeed the extrinsic curvature of $S_t = \Sigma \times \{t\}$ in (V,g). On the other hand the identities (5), (6) show, at least formally[1] that γ_t, K_t satisfy the constraints for each t if they do so for t = 0.

The freedom in the choice of lapse and shift (i.e the apparent non uniqueness of the solution) corresponds to the fact that two isometric space times are geometrically (and physically) the same. We shall show in the next chapter that the considered Cauchy problem is, in a rigorous sense, well posed : it has one, and only one, solution (up to isometries) in satisfactory functional spaces.

3 Hamiltonian Interpretation.

The decomposition given above of the Einstein equations led to the writing of these equations as a dynamical system in an appropriate phase space, the space of pairs (γ, K) satisfying the constraints. This space can be shown to be, in the neighborhood of almost every point, an infinite dimensional manifold $\mathcal{C}$ (cf |3|,|7|).

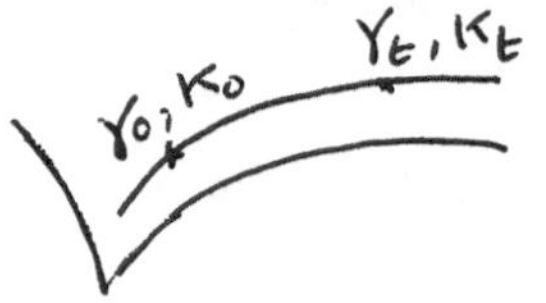

A trajectory of the dynamical system is a curve on $\mathcal{C}$.
These results can also be obtained by an hamiltonian formalism, deduced from the lagrangian of the Einstein equations[2]

$$(3\text{-}1) \qquad \mathcal{L}(g) = \int_V R(g)\, d\mu\, (g)$$

If $V = \Sigma \times I$

$$\mathcal{L}(g) = \int_I \int_{\Sigma_t} \alpha\, R(g)\, d\mu(\gamma)\, dt$$

Arnowitt, Deser and Misner (cf |4|) have shown that

$$\alpha R(g) \simeq (\gamma \operatorname{tr} K - K) . \frac{\partial \gamma}{\partial t} - (\alpha \mathcal{H} + \beta \mathcal{J})$$

where the sign $\simeq$ means modulo a space time divergence, which does not contribute to the Euler equations when boundary values are kept

------

(1) They also give a rigorous proof, in the real analytic case, for t small enough (cf |1|)

(2) $d\mu(g)$, $d\mu(\gamma)$ volume elements of g, γ .

fixed. We can therefore consider as lagrangian of our variational problem

$$\int_I L\, dt \quad , \quad L = \int_{\Sigma_t} \{ (\gamma \operatorname{tr} K - K) . \frac{\partial \gamma}{\partial t} - (\alpha \mathcal{H} + \beta \mathcal{J}) \} \, d\mu(\gamma)$$

This lagrangian L depends on α, β, γ as tensor fields on S_t, and on the derivatives $\frac{\partial \gamma}{\partial t}$. It is degenerate in the sense that $\frac{\partial \alpha}{\partial t}, \frac{\partial \beta}{\partial t}$ do not appear. However one can introduce conjugate variables to γ by setting

$$P = \frac{\partial L}{\partial(\partial \gamma / \partial t)} \qquad \text{(functional derivative)},$$

we see that

$$P \equiv \gamma \operatorname{tr} K - K$$

And L may be written in the standard way

$$\text{(3-2)} \qquad L \equiv (P , \frac{\partial \gamma}{\partial t})_{L^2(\Sigma)} - H$$

where (,) denotes the $L^2(\Sigma)$ scalar product and H is the hamiltonian :

$$\text{(3-3)} \qquad H = \int_{\Sigma_t} (\alpha \mathcal{H} + \beta \mathcal{J}) \, d\mu(\gamma)$$

The Euler equations of (3-1) relative to γ take then the form

$$\text{(3-4 a)} \qquad \frac{dP}{dt} = - \frac{\delta H}{\delta \gamma}$$

while the derivative $\frac{\delta H}{\delta P}$ is such that :

$$\text{(3-4 b)} \qquad \frac{d\gamma}{dt} = \frac{\delta H}{\delta P} .$$

Equations (3-4 a) and (3-4 b) are equivalent to the dynamical equations (2) and (1) - they appear in hamiltonian form.

The Euler equations corresponding to α and β are equivalent to the constraints, $\mathcal{H} = 0$ and $\mathcal{J} = 0$.

Note : the derivatives $\frac{\delta H}{\delta \gamma}$ and $\frac{\delta H}{\delta P}$ are by definition such that the differential of H, linear operator acting on the "variations" $h = \delta\gamma$, $p = \delta P$, is given by :

$$dH . (h , p) = (\frac{\delta H}{\delta \gamma}, h)_{L^2(\Sigma)} + (\frac{\delta H}{\delta P}, p)_{L^2(\Sigma)}$$

but, when $\mathcal{H} = \mathcal{J} = 0$,

$$dH . (h,p) = \int_{\Sigma} \{\alpha \, d\mathcal{H}(h,p) + \beta \, d\mathcal{J}(h,p)\} \, d\mu(\gamma)$$

where $d\mathcal{H}$, $d\mathcal{J}$, linear differential operators acting on the pair (h,p), are the "linearized constraints" operators. Denoting by $d\mathcal{H}^*$, $d\mathcal{J}^*$ the L^2 duals of the operators, acting respectively on scalars and vectors we have :

$$dH.(h,p) = \int_\Sigma (h,p).\ (d\mathcal{H}^*(\alpha) + d\mathcal{J}^*(\beta)\ d\mu(\gamma)$$

If we denote by Φ the pair $(\mathcal{H}, \mathcal{J})$ we can write the (empty space) Einstein equations in compact form :

$$\Phi = 0 \quad \text{(constraints)}$$

and

$$\left(\frac{d\gamma}{dt}, \frac{dP}{dt}\right) = J\ d\ \Phi^*\ (\alpha\ , \beta\ )$$

where J is the symplectic matrix

$$J = \begin{pmatrix} 0 & 1 \\ -1 & 0 \end{pmatrix}$$

The fact that the constraints are a closed system under the Poisson brackets (another aspect of the Bianchi identities, cf |5|) insures at least formally the conservation of the condition $\Phi = 0$ on the trajectories of the dynamical system, if it is satisfied initially.

III CAUCHY PROBLEM, EXISTENCE AND UNIQUENESS THEOREMS.

The analysis given in chapter II gives a good geometrical insight into the structure of Einstein equations, but to prove existence and uniqueness theorems in functional spaces more relevant to physics than the (real) analytical ones one must use another kind of argument, namely the choice of a special gauge called the harmonic gauge. With this choice one can also show the propagation of the gravitational field with a finite speed, in vacuo the speed of light. We will in this chapter give some ideas of the proofs.

1 Harmonic Gauge

The harmonic gauge was first defined with respect to the euclidean metric, and used as a coordinate condition to prove the existence and uniqueness theorems (|2|). It was defined with respect to an arbitrary metric, and used as a gauge condition (for diffeomorphisms) by Hawking and Ellis |6| - it permits a more elegant presentation of the proofs.

<u>Definition</u> A metric g is said to be harmonic with respect to a given metric e (e can be of arbitrary signature but we take it positive for later use) if the following vector F vanishes :

$$F(g,e) \equiv g.(\Gamma(g) - \Gamma(e)) = 0, \quad (F^\rho \equiv g^{\lambda\mu}(\Gamma^\rho_{\lambda\mu} - \bar{\Gamma}^\rho_{\lambda\mu}) = 0)$$

The usefulness of this "harmonic gauge" condition results from the next lemma :

Lemma The Ricci tensor of a metric g can be written :

$$\mathrm{Ricc}(g) \equiv \mathrm{Ricc}(g,e) + L(g,e)$$

with, D denoting the covariant derivative in the metric e :

$$L(g,e) \equiv \frac{1}{2} g_{(.}D_{.)}F\ (g,e)\ , \quad (L^{\alpha\beta} \equiv \frac{1}{2} g^{\lambda\beta} D_\lambda F^\alpha + \frac{1}{2} g^{\lambda\alpha} D_\lambda F^\beta)$$

$$\mathrm{Ricc}(g,e) = \frac{1}{2} g.D^2 g + P(g)(Dg,Dg) + \frac{1}{2} g_{(.}g_{.)}\ \mathrm{Riem}\ (e)$$

$$(R^{\alpha\beta}_{(e)} \equiv \frac{1}{2} g^{\lambda\mu} D^2_{\lambda\mu}\ g^{\alpha\beta} + P^{\alpha\beta,\lambda\mu}_{\nu\rho,\sigma\tau}\ D_\lambda g^{\nu\rho} D_\mu\ g^{\sigma\tau} + \frac{1}{2} g^{\alpha\beta} g^{\lambda\mu}\ \bar{R}^{\nu}_{\beta,\lambda\alpha}$$

with $P^{\alpha\beta,\lambda\mu}_{\nu\rho,\sigma\tau}$ a rational function of g, multiplied by $(\det g)^m$.

2 Existence Theorem

We deduce from the lemma that, in an harmonic gauge the Einstein equations $\mathrm{Ricc}(g) = 0$ take the form of a well posed quasi linear second order system, hyperbolic for a hyperbolic metric g, since each component $R^{\alpha\beta}_{(e)}$ admits as principal part the same second order operator, $\frac{1}{2} g^{\lambda\mu} D^2_{\lambda\mu}$ acting on the corresponding component $g^{\alpha\beta}$. This hyperbolic character, and the corresponding relevant properties of the Bianchi identities, enable one to show that the preceding analysis of the Cauchy problem is indeed true, namely

Theorem : Let (Σ,γ,K) be an initial data set for Einstein's equations satisfying the constraints with, $\gamma \in C^0_b(\Sigma)$ a uniformly properly riemannian metric(1) , $\bar{D}\gamma \in H_s(\Sigma)$, $K \in H_s(\Sigma)$. There exists a space-time (V,g) solution of the corresponding Cauchy problem if $s \geqslant 2$; its maximal globally hyperbolic extension is unique up to isometries if $s \geqslant 3$.
The metric g is defined on a product $V = \Sigma \times I$; it is continuous and has a restriction to each $S_t = \Sigma \times \{t\}$ whose (generalized) derivatives(2), $D^\alpha g$, of order $|\alpha| \leqslant s$, are, uniformly and measurably in t, square integrable on S_t. We say that $g \in E_s$, and we set

----

(1) The covariant derivative $\bar{D}$ is taken with respect to a "background" metric $\bar{e}$ given on Σ , H_s is the usual Sobolev space of tensor fields with generalized derivatives of order $\leqslant$ s square integrable on Σ . Remember that $H_s(\Sigma) \subset C^0_b(\Sigma)$ and is an algebra if $s \geqslant 2$ (if dimension of $\Sigma = 3$, $\bar{e}$ has a non zero injectivity radius and its curvature satisfies some boundedness conditions).
(2) $\Sigma \times I$ is endowed with the background metric $\bar{e} \times 1$, D is the corresponding covariant derivative. $|u|$ denotes the e-norm at a point of the tensor u.

$$||g||_{E_s} = \operatorname{Sup}(||g||_{C_b^o(V)} , ||Dg||_{\tilde{E}_{s-1}})$$

with

$$||Dg||_{\tilde{E}_{s-1}} = \operatorname*{Sup}_{t \in I} \{ \int_{S_t} \sum_{|\alpha| \leqslant s-1} |D^{\alpha+1} g|^2 \, d\mu(\bar{e}) \}^{1/2}$$

The proof goes in two steps :
(i) existence for the system in the harmonic gauge
(ii) verification of the harmonicity condition.

i) Let (Σ,γ,K) be an initial data set for the Einstein equations (cf § II) satisfying the constraints. We deduce from it Cauchy data for the hyperbolic system on $\Sigma \times R$

$$\operatorname{Ricc}(g,e) = 0$$

that is

$$g|_\Sigma = \varphi \qquad , \qquad \partial_o \, g|_\Sigma = \psi$$

by choosing arbitrarily the lapse α and shift β on Σ (for instance $\alpha = 1$ and $\beta = 0$), using the definitions of γ and K, and also the e-harmonicity condition on Σ , $F(g,e)_{|\Sigma} = 0$. The existence of a hyperbolic metric g, solution of $\hat{\operatorname{Ricc}}(g,e) = 0$, with the Cauchy data (φ, ψ) on a manifold Σ , can be obtained in the reasonably large functional space E_s on $\Sigma \times I$ (with I some interval of $\mathbb{R}$) through the energy estimates. We give the idea of these estimates for a simple example : the wave equation for a scalar field Φ in a curved space time, which reads :

$$(2\text{-}1) \qquad \Box_g \Phi = \nabla_\alpha \nabla^\alpha \Phi = \rho \, .$$

The stress energy tensor of this scalar field Φ is :

$$T_{\mu\nu} \equiv \partial_\mu \Phi \, \partial_\nu \Phi - \frac{1}{2} g_{\mu\nu} \, \partial_\rho \Phi \, \partial^\rho \Phi$$

the energy momentum vector, relative to an observer whose trajectory is tangent to the time like vector X is :

$$P_\nu \equiv T_{\mu\nu} \, X^\mu$$

while the energy density on a space like section with unit normal n is :

$$\mathcal{E} \equiv P_\nu \, n^\nu \equiv T(X,n) = \hat{\gamma}(\text{ grad } \Phi, \text{ grad } \Phi)$$

where $\hat{\gamma}$ turns out to be the positive definite metric :

$$\hat{\gamma}^{\mu\nu} \equiv X^{(\mu} n^{\nu)} - \frac{1}{2} g(X,n) \, g^{\mu\nu} \, .$$

The energy on the slice S_t (space like section) is :

$$E(t) \equiv \int_{S_t} \mathcal{E}(t)\, d\mu(\gamma)$$

where γ is the metric induced on S_t by g.
We deduce from the Stokes formula that if grad Φ is uniformly square integrable[(1)] on S_t , then

$$\int_{\Sigma\times(o,t)} \nabla_\nu P^\nu\, d\mu(g) = \int_{\Sigma\times(o,t)} (-\mathcal{L}_X\Phi\, \Box\Phi + \frac{1}{2}\, T.\mathcal{L}_X g)\, d\mu(g)$$

$$= E(t) - E(o) \quad ,$$

$$(\mathcal{L}_X\, \Phi = X^\mu\, \partial_\mu\, \Phi, \quad T.\mathcal{L}_X\, g = T^{\mu\nu}\, (\nabla_\mu\, X_\nu + \nabla_\nu\, X_\mu)\ )$$

and therefore the following fundamental energy estimate :

$$\text{(2-2)} \qquad E(t) \leqslant E(o) + C_1 \int_o^t \sqrt{E(\tau)}\ ||\rho(\tau)||_{L_2(S_\tau)} d\tau$$

$$+ C_2 \int_o^t E(\tau) \sup_{S_\tau} |L_X g|\, d\tau$$

where C_1 and C_2 are constants depending on g, X and n. From the integral inequality (2-2) one deduces that E(t) is uniformly bounded on a finite interval I of $\mathbb{R}$, if it is so of $||\rho(t)||_{L_2(S_t)}$ and $\sup_{S_t} |L_X g|$.

Bounds for the integral on S_t of the square of the $\hat{\gamma}$ norm of the higher derivatives of Φ can also be obtained, under relevant hypothesis on the derivatives of ρ and g, by derivating the equation (2-1). One then finds an a priori bound in E_s of the solutions Φ of the equation (2-1) with Cauchy data on $S_o = \Sigma \times \{o\}$

$$\Phi_{|S_o} = \varphi, \quad \partial_t\Phi_{|S_o} = \psi \quad , \ \varphi \in C_b^o\, , \ \bar{D}\varphi \in H_s(\Sigma), \ \psi \in H_{s-1}(\Sigma)$$

This a priori bound, together with some standard methods of functional analysis, leads to the proof of the existence theorem.

In the case of the e-harmonic Einstein equations (or more generally of a quasi linear second order system); one applies the previous existence result (slightly generalized) to the linear system obtained by replacing in the first member all terms $D^\alpha g, |\alpha| = 0$ or $|\alpha| = 1$, by $D^\alpha G$, with G a given hyperbolic metric, $G \in E_s$, $s \geqslant 3$: this hypothesis on G insures the verification of the hypothesis made in the linear case to obtain the a priori bounds on Φ and its derivatives of order $s \leqslant 3$. These a priori bounds also enable one

----

(1) Some weak assumptions about the regularity of g and S_t have also to be made.

to prove the existence of a fixed point of the map $G \mapsto g$, if the interval I is small enough. This fixed point is a solution, in E_s, of the given Cauchy problem for Ricc(g,e) = 0.

ii) We show that the solution of Ricc(g,e) = 0 with Cauchy data satisfying the constraints and $F(g,e)|_{S_o} = 0$ is indeed a metric in harmonic gauge, solution of Ricc(g) = 0 by noting that the Bianchi identities are, for such a solution, a linear hyperbolic system for F(g,e) with vanishing Cauchy data on S_o.

3 Uniqueness Theorem.

The uniqueness in the e-harmonic gauge comes as a corollary of the proof of the existence. The general local uniqueness is proved by showing, for every metric $g \in E_s$ $(s \geqslant 4)$, the existence of a diffeomorphism f which transforms this metric into a metric in e-harmonic gauge. The global uniqueness, in the class of maximal extensions of globally hyperbolic space times is more delicate to prove; it appeals to the geometric properties of such space times and of the boundaries of their globally hyperbolic subsets.

An important property of these globally hyperbolic solutions, which can be proved with a refinement of the energy estimates is the "domain of dependance". The gravitational field g, at a point (x,t) of the space time V, depends only on the Cauchy data which are in the past of (x,t), past determined as we have said before, by the field of cones Γ_x^-. As a consequence we have the finite speed of propagation of the gravitational field, the speed of light.

Bibliography (books : see other references therein)

|1| A. Lichnerowicz "Les théories relativistes de la gravitation et de l'électromagnétisme" Masson 1955.

|2| Y. Choquet-Bruhat "The Cauchy problem" in "Gravitation, an introduction to current research" L. Witten ed., J. Wiley 1962.

|3| Y. Choquet-Bruhat and J. York "The Cauchy problem" Einstein Centenary Volume, GRG Society and A. Held, ed. Plenum, 1979.

|4| R. Arnowitt, S. Deser and C. Misner "The dynamics of General Relativity" In "Gravitation, and introduction to current research" L. Witten ed. J. Wiley 1962.

|5| B. DeWitt "Dynamical theory of fields" Gordon and Breach 1967.

|6| S. Hawking and G. Ellis "The global structure of space time" Cambridge University press 1973.

|7| A. Fisher and J. Marsden "The initial value problem and the dynamical formulation of General Relativity" in Einstein Centenary volume, Hawking and Israël Edr Cambridge University press 1979.

|8| Y. Choquet-Bruhat, C. DeWitt-Morette and M. Dillard-Bleick "Analysis, Manifolds and Physics" North-Holland 1977.

UNDERLYING MATHEMATICAL STRUCTURES OF CLASSICAL GRAVITATION THEORY

Brandon Carter

Groupe d'Astrophysique Relativiste

Observatoire de Paris, 92190 MEUDON (France)

Introduction

If I had been asked five years or so ago to prepare a course on recent developments in classical gravitation theory, I would not have hesitated in choosing the classical theory of black holes as a central topic of discussion. However the most important developments in gravitation theory during the last three or four years have not been in the classical domain at all but rather in the problems of quantisation of—or in— curved space-time. Associated with these developments (partly as a cause and partly as a consequence) has been a strong trend towards integration of gravitation theory with the main stream of theoretical physics, from which it had hitherto been rather isolated. The mahayana of this unification has been the generalised theory of gauge fields, which has made spectacular progress on the testing ground of weak interactions. Under these circumstances it seemed that instead of concentrating on the comparatively minor and specialised developments that have in fact recently taken place in the classical domain, it would be more appropriate for the present school to provide an introductory review of the general principles of classical gauge theory as a solid foundation underlying the more genuinely recent — and therefore quantum — developments to be described in subsequent courses.

The present course therefore consists basically of a very general treatment of connection and curvature in a Lie group fibre bundle followed by a number of illustrative and, I trust, useful applications. In accordance with the prevailing spirit of unity I have tried to present a conceptually and notationally coherent scheme that is equally suitable for such diverse applications as flat space Yang-Mills theory or pure Riemannian geometry, which I hope will be useful to physicists who are familiar with the one but less acquainted with the other. In order to do this I have avoided the use of the most extremely condensed notation schemes that have been developed for specialised purposes but that tend to be ambiguous outside their originally intended context. The course is entirely self contained in the sense that, except for the sections (1 and 6) in which the relevant essentials of differential manifold and Lie group theory are recalled , none of the main results is dependent on the use of theorems quoted without proof from external sources.

1. Exterior calculus and Lie differentiation on a differentiable manifold.

Nearly all branches of theoretical physics except those involving relativistic gravitational effects are based on the use of a 4-dimensional spacetime endowed *a priori* with a Minkowski structure, as expressible in terms of a *flat* Lorentzian pseudo-metric, which means in particular that space-time is affine in the sense that relative to any arbitrarily chosen origin there is a natural linear (ie vector space) structure. Provided such an affine structure is taken for granted then general tensor fields can be (partially) differentiated to give other tensor fields, and they can also be integrated over arbitrary surfaces to give tensorial results. However it has been generally recognised since the time of Einstein that in order to incorporate relativistic gravitational effects the underlying structures of physics should be treated more democratically, the Lorentzian pseudo-metric being merely one among other dynamically variable fields, so that in order to reconstruct space-time one should start simply with a 4-dimensional *differentiable manifold*, with (at the outset) neither metric, affine, or even projective structure. We recall that a manifold is defined as being *locally* isomorphic to a vector space $\mathcal{J}$ in the sense that any point $x \in \mathcal{M}$ belongs to a neighbourhood $\mathcal{N}_\alpha$ that is related to a neighbourhood in $\mathcal{J}$ by an isomorphism $\mathcal{S}_\alpha$ say which can be used to define a set of local co-ordinates x^μ on $\mathcal{N}_\alpha$ by specifying x^μ to be the components of the image $\mathcal{S}_\alpha x$ in $\mathcal{J}$ with respect to some fixed basis of $\mathcal{J}$. The *differentiability* of the manifold is expressed by the requirement that wherever two such neighbourhoods $\mathcal{N}_\alpha$ and $\mathcal{N}_\beta$ have a non empty overlap $\mathcal{N}_\alpha \cap \mathcal{N}_\beta$ the corresponding maps $\mathcal{S}_\alpha \circ \mathcal{S}_\beta^{-1}$ from neighbourhoods of $\mathcal{J}$ onto neighbourhoods of $\mathcal{J}$ should be correspondingly differentiable, but *not* necessarily linear or even affine. Under these circumstances general tensors at different - even nearby - points of $\mathcal{M}$ can not be compared in a gauge invariant manner, ie in a way that does not depend on the particular choice of the mapping $\mathcal{S}_\alpha$ or equivalently of the local co-ordinates x^μ . It follows that tensorial differentiation and integration is only possible in certain special cases of which the most important is the class of *differential forms*, i.e fully antisymmetric covariant tensor fields, to which one can apply the exterior analysis scheme developed by Cartan.

To describe the essential formulae of the Cartan scheme that will be needed in our subsequent work we start by recalling the standard definition of the exterior product of a p-form $\underline{\omega}$ with components $\omega_{\mu_1 \cdots \mu_p}$ and a q-form $\underline{\Omega}$ with components $\Omega_{\mu_1 \cdots \mu_q}$ as being the (p + q)-form $\underline{\omega} \wedge \underline{\Omega}$ with components given by

$$(\underline{\omega} \wedge \underline{\Omega})_{\mu_1 \cdots \mu_p \mu_{p+1} \cdots \mu_{p+q}} = \frac{(p+q)!}{p!\, q!}\, \omega_{[\mu_1 \cdots \mu_p} \Omega_{\mu_{p+1} \cdots \mu_{p+q}]} \qquad (1\cdot1)$$

where we introduce the use of the square brackets around indices to denote antisymmetric averaging over permutations (i.e weighted averaging with weight ± 1 according as the permutation is even or odd); we shall similarly use round brackets to denote ordinary (symmetric) averaging over permutations. Introducing the standard abbreviation ∂_μ to denote the (co-ordinate gauge dependent) partial differentiation operator $\partial / \partial x^\mu$ we can analogously define the exterior derivative of $\underline{\omega}$ to be the (p + 1) form $\partial \wedge \underline{\omega}$ with components given by

$$(\partial \wedge \underline{\omega})_{\mu_1 \mu_2 \cdots \mu_{p+1}} = (p+1)\, \partial_{[\mu_1} \omega_{\mu_2 \cdots \mu_{p+1}]} \;, \qquad (1\cdot2)$$

the co-ordinate gauge independence of the result, i.e the fact that it really does specify a genuine covariant (p + 1)st order tensor, being a non-trivial consequence of the antisymmetrisation. It has become a widely used convention in recent years to use the abbreviation d in place of the more self-explanatory compound symbol $\partial \wedge$ for exterior differentiation, but in the present work we shall always retain the wedge symbol $\wedge$ explicitly whenever antisymmetrisation is implied, partly because we shall also be considering non-antisymmetric operations later on, and partly because we shall need the symbol d for its more traditional use to denote tangent space elements.

It is evident from the commutation property of partial differential operators that an exact form, i.e one that is itself an exterior derivative, has the cohomology property of being necessarily closed in the sense that its own exterior derivative necessarily vanishes, i.e

$$\partial \wedge \partial \wedge \underline{\omega} = 0 \quad . \qquad (1\cdot3)$$

The well known but less trivial converse property that a closed form is exact, i.e.

$$\partial \wedge \underline{\Omega} = 0 \;\Rightarrow\; \exists\, \underline{\omega} : \quad \underline{\Omega} = \partial \wedge \underline{\omega} \qquad (1\cdot4)$$

is in general valid only in a small neighbourhood but not globally, which is the reason why it is interesting to study differential cohomology classes which are subsets of closed differential forms such that the difference between any pair of members of a given class is exact. [These classes evidently have a natural Abelian group structure under addition; indeed by taking into account scalar multiplication they can be considered as forming a vector space.] Cohomology theory is dual to ordinary homology theory which is based on the analogous property of the boundary ∂S of a submanifold S which is also necessarily closed in the sense that $\partial\partial S = 0$. The duality is realised through the possibility of integrating a p-form over any orientable p-dimensional submanifold which (unlike more general kinds of tensor integration) can be achieved without reference to any affine or other kind of superstructure on the manifold. This integration can be conceived intuitively (with at least sufficient rigour to fix the normalisation) in terms of the limiting process

$$\int_S \underline{\omega} \,|\, \vec{dS} = \underset{dS \to 0}{\mathcal{L}t} \sum_{dS} \underline{\omega} \,|\, \vec{dS} \tag{1.5}$$

where

$$\vec{dS} = \vec{dx}_{(1)} \wedge \cdots \wedge \vec{dx}_{(p)} \tag{1.6}$$

is a p-vector (i.e antisymmetric contravariant tensor) constructed from a (suitably oriented) basis of tangentvectors (with co-ordinate components $dx_{(1)}^{\mu}, \ldots, dx_{(p)}^{\mu}$) spanning an element dS in a cellular approximation of S , and where a vertical bar is used to denote inner multiplication of a multivector and a form as defined by

$$\underline{\omega} \,|\, \vec{dS} = \frac{1}{p!} \omega_{\mu_1 \cdots \mu_p} dS^{\mu_1 \cdots \mu_p} . \tag{1.7}$$

The Cartan integration defined in this way has the Stokes property that for any p-form $\underline{\omega}$ and any (p + 1) - surface Σ ,

$$\int_{\partial\Sigma} \underline{\omega} \,|\, \vec{dS} = \int_{\Sigma} \partial \wedge \underline{\omega} \,|\, \vec{d\Sigma} \tag{1.8}$$

where $S = \partial\Sigma$ is the (appropriately oriented) boundary p-surface of Σ . (This theorem can be derived directly from the less primitive but more intuitively obvious Green theorem(1.25) to be described later on in this section It follows that when restricted to "closed" or more strictly compact submanifolds which are characterised by having zero boundary, the Cartan integral determines a (linear) function on the cohomology classes, since any two integrands from the same class give the same integral. Dually, a fixed closed p-form gives the same result when integrated over different orientable p-surfaces in the same homology class.

Whenever one is given a linear (e.g tensorial) structure X over a manifold $\mathcal{M}$ that is rectractable (in the sense that a suitably well behaved automorphism mapping $f : \mathcal{M} \to \mathcal{M}$ induces a corresponding retraction mapping $f : X(fx) \to fX(x)$ of X onto a new structure of the same kind), and in particular in the case of differential form, then there exists a corresponding Lie derivative $\vec{\xi} \pounds X$ of X associated with each (suitably well behaved) vector field $\vec{\xi}$ on $\mathcal{M}$. The Lie derivative arises from a consideration of the family of automorphism mappings $f(t)$ defined to consist of transporting each point $x \in \mathcal{M}$ by a parameter distance t allong the parametrised integral curves of $\vec{\xi}$ i.e. along the solution curves of the differential equations expressible in local co-ordinates by

$$\frac{dx^\mu}{dt} = \xi^\mu \ . \tag{1.9}$$

In terms of these automorphisms the Lie derivative can be defined by

$$\vec{\xi} \pounds X = \frac{d}{dt}\Big(f(t)X\Big)\Big|_{t=0} \ . \tag{1.10}$$

The classic example is the Lie bracket of two vector fields, $\vec{\xi}$ and $\vec{\eta}$ say, which is defined by

$$\vec{\xi} \pounds \vec{\eta} = [\vec{\xi}, \vec{\eta}] = -\vec{\eta} \pounds \vec{\xi} \ , \tag{1.11}$$

and given in terms of co-ordinates by

$$[\vec{\xi}, \vec{\eta}]^\mu = \xi^\nu \partial_\nu \eta^\mu - \eta^\nu \partial_\nu \xi^\mu \ . \tag{1.12}$$

In the case of a differential p-form $\underline{\omega}$ the normalisation convention (1.1) for the exterior derivative allows one to express the Lie derivative by the compact Cartan formula

$$\vec{\xi} \pounds \underline{\omega} = \vec{\xi}\cdot(\partial \wedge \underline{\omega}) + \partial \wedge (\vec{\xi}\cdot\underline{\omega}) \ . \tag{1.13}$$

(With different normalisation conventions such as the one used by Hawking and Ellis 1973, p-dependent normalisation factors would need to be included here). We have here introduced the use of a dot, $\cdot$, to denote the inner multiplication procedure consisting of contraction of a single pair of adjacent indices (as contrasted with the $|$ multiplication introduced in (1.7), which involves contraction of all possible adjacent indices) so that for example $\vec{\xi}\cdot\underline{\omega}$ is the (p-1) form with components given by

$$(\vec{\xi}\cdot\underline{\omega})_{\mu_1\cdots\mu_{p-1}} = \xi^\nu \omega_{\nu\mu_1\cdots\mu_{p-1}} \quad . \tag{1.14}$$

The Cartan formula can be used directly to evaluate the rate of change of an integral over a submanifold Σ as it is parametrically displaced along the integral curves of a vector field $\vec{\xi}$. It is evident from the fact the Cartan integration procedure involves no affine or other superstructure on $\mathcal{M}$ that we will simply have

$$\frac{d}{dt}\int_\Sigma \underline{\omega}\,|\,d\vec{\Sigma} = \int_\Sigma \vec{\xi}\,\pounds\,\underline{\omega}\,|\,d\vec{\Sigma} \tag{1.15}$$

so that by (1.13) and (1.8) we obtain the result

$$\frac{d}{dt}\int_\Sigma \underline{\omega}\,|\,d\vec{\Sigma} = \int_\Sigma \vec{\xi}\cdot(\partial\wedge\underline{\omega})\,|\,d\vec{\Sigma} + \int_{\partial\Sigma} \vec{\xi}\cdot\underline{\omega}\,|\,d\vec{S} \tag{1.16}$$

If $\underline{\omega}$ is closed then the integral will be conserved (in the sense that the right hand side of (1.16) will vanish) whenever Σ is compact or more generally whenever $\vec{\xi}$ is such as to leave the boundary invariant under the displacement.

Before going on to describe the more elaborate kinds of analysis that can be carried out on a manifold endowed with the heavier kinds of superstructure (such as connections or Riemannian metrics) that will be treated in later sections, we shall briefly touch on what can be done with the most primitive kind of structure over and above mere differentiability, which is that of a volume measure $\underline{\varepsilon}$, which can be represented in an n-dimensional manifold as an n-form. When it exists such a measure can be used to define an adjoint (n-q) form $*\vec{\beta}$ associated with any q-vector (i.e antisymmetric contravariant tensor) $\vec{\beta}$ by the formula

$$*\vec{\beta} = \vec{\beta}\,|\,\underline{\varepsilon} \tag{1.17}$$

which in terms of components is

$$*\beta_{\mu_1\cdots\mu_{n-q}} = \frac{1}{q!}\,\varepsilon_{\nu_1\cdots\nu_q\mu_1\cdots\mu_{n-q}}\,\beta^{\nu_1\cdots\nu_q} \tag{1.18}$$

We shall use an upstairs asterisk to denote the inverse operation, sending p-forms onto (n-p)-vectors, as defined by

$$*(^*\underline{\omega}) = \underline{\omega} \tag{1.19}$$

The adjoint operator defined in this way can be used to express the contraction of a q-vector $\vec{\beta}$ with a p-form $\underline{\omega}$ in terms of purely exterior multiplication by the formula

$$\vec{\beta} | \underline{\omega} = {}_*({}^*\underline{\omega} \wedge \vec{\beta}) \tag{1.20}$$

if $p \geq q$, and by the formula

$$\vec{\beta} | \underline{\omega} = {}^*(\underline{\omega} \wedge {}_*\vec{\beta}) \tag{1.21}$$

if $p \leq q$. This last formula suggests the appropriateness of defining the inner derivative or divergence of a p-vector to be

$$\mathrm{div}\, \vec{\beta} = {}^*(\partial \wedge {}_*\vec{\beta}) \tag{1.22}$$

whenever a fundamental measure $\underline{\varepsilon}$ is present. If in addition the more elaborate structures needed to define a covariant differentiation operator ∇ with coordinate components ∇_μ are present consistently with the measure, in the sense that $\nabla \underline{\varepsilon} = 0$, then it can be seen from (1.21) that the components of the divergence $\mathrm{div}\, \vec{\beta}$ will simply be $\nabla_\nu \beta^{\mu_1 \cdots \mu_{q-1} \nu}$ so that it will be consistent to use the alternative definition

$$\mathrm{div}\, \vec{\beta} = (-1)^{q-1}\, \nabla \cdot \vec{\beta} \tag{1.23}$$

When only a measure is present this last formula can be considered merely as a definition of the operator $\nabla\cdot$ acting on multivectors. (Many mathematical textbooks denote the divergence operator simply by δ, but in physical contexts this latter symbol is usually reserved for the Kronecker matrix and the Dirac distribution).

The preceeding definitions can be used to translate the earlier formulae involving exterior derivations into terms of the divergences with which physicists are usually more accustomed to work, always subject to the proviso that a volume measure $\underline{\varepsilon}$ has been given. To start with we can construct the integral of an (n-p) - vector, $\vec{\beta}$, over a p - surface S as

$$\int_S \vec{\beta} |_* d\vec{S} = \int_S {}_*(d\vec{S} \wedge \vec{\beta}) = (-1)^{n(p-1)} \int_S {}_*\vec{\beta} \,/\, d\vec{S} \tag{1.24}$$

and it can then be seen with the aid of (1.20) and (1.22) that the Stokes formula (1.8) can be replaced by the equivalent Green's formula

$$\int_{\partial\Sigma} \vec{\beta} \,|\, {*d\vec{S}} = (-1)^n \int_{\Sigma} \operatorname{div} \vec{\beta} \,|\, {*d\vec{\Sigma}} \tag{1.25}$$

Since (1.13) implies

$$\vec{\xi} \,\pounds\, \underline{\varepsilon} = \underline{\varepsilon} \operatorname{div} \vec{\xi} \tag{1.26}$$

it can be seen that the analogue of (1.15) is the less simple formula

$$\frac{d}{dt} \int_{\Sigma} \vec{\beta} \,|\, {*d\Sigma} = \int_{\Sigma} (\vec{\xi} \,\pounds\, \vec{\beta} + \vec{\beta} \operatorname{div} \vec{\xi}) \,|\, {*d\Sigma} \tag{1.27}$$

Now with further use of (1.20) and (1.22) it can be seen that the analogue of the Cartan formula (1.13) is the less well known expression

$$\vec{\xi} \,\pounds\, \vec{\beta} = \operatorname{div} (\vec{\beta} \wedge \vec{\xi}) + (\operatorname{div} \vec{\beta}) \wedge \vec{\xi} - \vec{\beta} \operatorname{div} \vec{\xi} \tag{1.28}$$

for the Lie derivative of any q - vector $\vec{\beta}$.
Combining these and using the Green theorem (1.25) we obtain for the analogue of (1.16) the variation law

$$\frac{d}{dt} \int_{\Sigma} \vec{\beta} \,|\, {*d\vec{\Sigma}} = \int_{\Sigma} (\operatorname{div} \vec{\beta}) \wedge \vec{\xi} \,|\, {*d\vec{\Sigma}} + (-1)^n \int_{\partial\Sigma} \vec{\beta} \wedge \vec{\xi} \,|\, {*d\vec{S}} \tag{1.29}$$

which shows that if the variation leaves the boundary invariant then the integral will be conserved for any multivector $\vec{\beta}$ whose divergence vanishes.

2. Principle, Fundamental and other Associated Group Representations.

A representation of an abstract group $\mathcal{G}$ may be defined as a realisation of the group in terms of some set of automorphism mappings of a representation space $\mathcal{V}$ onto itself, subject to the obvious requirement that the group multiplication is defined by the natural composition of mappings. In other words a representation is given formally by a mapping

$$rep : \quad S \mapsto rep\, S \quad , \quad S \in \mathcal{G} \tag{2.1}$$

with a corresponding mapping

$$\begin{aligned} rep : \quad & \mathcal{G} \times \mathcal{V} \longrightarrow \mathcal{V} \\ & (S, v) \mapsto rep\, S\, v \quad , \quad v \in \mathcal{V} \end{aligned} \tag{2.2}$$

where the mapping

$$rep\, S \qquad \mathcal{V} \rightarrow \mathcal{V} \tag{2.3}$$

is subject to the requirement that the composition of two such mappings satisfies

$$rep\, S_1 \circ rep\, S_2 = rep\,(S_1 S_2) \tag{2.4}$$

We shall be largely but not exclusively concerned with linear representations i.e representations for which $\mathcal{V}$ has a vector space structure in terms of which each automorphism map $rep\, S$ is required to be linear.

An important example of a non linear representation is the principle or progressive representation of $\mathcal{G}$ over itself, as specified by ordinary left multiplication, i.e by

$$pr : \quad S \mapsto pr\, S \quad , \quad S \in \mathcal{G} \tag{2.5}$$

where $pr\, S$ is the mapping defined by

$$pr\, S : \quad T \mapsto ST \quad , \quad T \in \mathcal{G} \tag{2.6}$$

This representation must be distinguished from the <u>retrograde</u> representation defined in terms of right multiplication by

$$\mathrm{ret} : S \mapsto \mathrm{ret}\, S \qquad S \in \mathcal{G} \qquad (2{\cdot}5)$$

where $\mathrm{ret}\, S$ is the mapping defined by

$$\mathrm{ret}\, S : T \mapsto T S^{-1} \qquad T \in \mathcal{G} \qquad (2{\cdot}6)$$

A third important representation of $\mathcal{G}$ over itself is the <u>group adjoint</u> representation

$$\mathrm{ad} : S \mapsto \mathrm{ad}\, S \qquad S \in \mathcal{G} \qquad (2{\cdot}7)$$

where $\mathrm{ad}\, S$ is the mapping defined by

$$\mathrm{ad}\, S : T \mapsto S T S^{-1} \qquad T \in \mathcal{G} \qquad (2{\cdot}8)$$

i.e by

$$\begin{aligned} \mathrm{ad}\, S &= \mathrm{pr}\, S \circ \mathrm{ret}\, S \\ &= \mathrm{ret}\, S \circ \mathrm{pr}\, S \end{aligned} \qquad (2{\cdot}9)$$

It is to be noted that (unlike pr and ret) ad is <u>not</u> necessarily faithful, i.e different elements S may give rise to the same mapping $\mathrm{ad}\, S$.

Apart from these three examples we shall mainly be concerned with <u>linear</u> representations, which will usually be built up from some <u>fundamental</u> representation, rep^1 say, in which $\mathcal{G}$ acts on same particularly simple or convenient vector space $\mathcal{U}$, most frequently characterised by the condition that it should have the minimum dimension compatible with faithfulness of the representation. When dealing with a fundamental representation we shall usually drop the prefixe rep^1, thereby implicitly identifying the elements of $\mathcal{G}$ with their fundamental representatives. Thus if $\mathcal{U}$ is the fundamental representation space we write

$$\mathrm{rep}^1 S\, u = S\, u \qquad u \in \mathcal{U} \qquad (2{\cdot}10)$$

i.e in terms of a basis in which u has components u^m we can think of S as being defined in terms of its matrix components by

$$\mathrm{rep}^1 S : \quad u^m \mapsto S^m{}_n u^n \tag{2.11}$$

Starting from a fundamental vector space $\mathcal{U}$ we can go on to construct associated vector spaces $\mathcal{U}_*, \mathcal{U}\otimes\mathcal{U}, \ldots,$ by taking duals, tensor products, etc., and over each such associated space we have an associated representation (i.e a representation of the same group $\mathcal{G}$) defined in the obvious natural manner. Thus for example on the dual space $\mathcal{U}_*$ of covectors π with components π_m we have a representation rep_1 defined by

$$\mathrm{rep}_1 S : \quad \pi_m \mapsto \pi_n S^{-1n}{}_m \tag{2.12}$$

while on the product space $\mathcal{U}\otimes\mathcal{U}$ of contravariant tensors $\vec{q}$ with components q we have a representation rep^2 defined by

$$\mathrm{rep}^2 S : \quad q^{mn} \mapsto S^m{}_k S^n{}_l q^{kl} \tag{2.13}$$

If we are dealing with a complex representation then we also have an associated representation $\mathrm{rep}^{\dot{1}}$ on the complex conjugate space $\overline{\mathcal{U}}$ whose effect on an element v with components $v^{\dot{m}}$ is

$$\mathrm{rep}^{\dot{1}} S : \quad v^{\dot{m}} \mapsto \bar{S}^{\dot{m}}{}_{\dot{n}} v^{\dot{n}} , \tag{2.14}$$

the dots on the indices serving to indicate that their transformation law is not of the ordinary kind.

In cases where the context allows, no ambiguity, we shall not bother to write in the distinguishing suffices 1, $\dot{1}$, 2 etc but will use the simplified notation $\mathrm{rep}\, S$ to indicate whichever kind of representation mapping is obviously appropriate.

3. The concept of a Fibre Bundle.

Having recalled the concept of a manifold in section 1 and of a group representation in section 2 we are now ready to describe the way in which these two concepts may be combined in the more sophisticated concept of a fibre bundle, of which the classic example is the set of all tangent vectors to a given base manifold $\mathcal{M}$.

Quite generally a set $\mathcal{B}$ is said to have the structure of a fibre bundle with base manifold $\mathcal{M}$ and representation rep over a representation space $\mathcal{V}$ whenever there is a fundamental bundle projection

$$\pi : \mathcal{B} \rightarrow \mathcal{M} \tag{3.1}$$

which is such that $\mathcal{M}$ can be covered by a family of neighbourhoods $\mathcal{N}_\alpha$ (which might be the same as those used to specify its own manifold structure) with the following properties. Firstly the bundle has a local product structure, in the sense that for each of the neighbourhoods $\mathcal{N}_\alpha$ there exists an isomorphism

$$\pi^{-1}\mathcal{N}_\alpha \longleftrightarrow \mathcal{N}_\alpha \times \mathcal{V} \tag{3.2}$$

(where $\pi^{-1}\mathcal{N}_\alpha$ denotes the set of points $b \in \mathcal{B}$ such that $\pi b \in \mathcal{N}_\alpha$) expressible by

$$p \longmapsto (\pi b, J_\alpha p) \quad , \quad p \in \pi^{-1}\mathcal{N}_\alpha \tag{3.3}$$

where

$$J_\alpha : \pi^{-1}\mathcal{N}_\alpha \longrightarrow \mathcal{V} \tag{3.4}$$

is the gauge mapping associated with the neighbourhood $\mathcal{N}_\alpha$. This entails that the restriction

$$J_\alpha(x) : \pi^{-1}x \longrightarrow \mathcal{V} \quad , \quad x \in \mathcal{N}_\alpha \tag{3.5}$$

of the gauge mapping to the fibre $\pi^{-1}x$ over a point $x \in \mathcal{M}$ (ie to the set of points that project onto x) is an isomorphism between the fibre and the representation space.(It also entails that if $\mathcal{V}$ has a manifold structure then $\mathcal{B}$ will also have the structure of a manifold whose dimension is the sum of those of $\mathcal{M}$ and $\mathcal{V}$). The second (and final) condition that must be satisfied is that the local product patches must be related by well behaved gauge transformation the sense that for each overlap region $\mathcal{N}_\alpha \cap \mathcal{N}_\beta$ there is a mapping

$$\begin{aligned} S_{\alpha\beta} : \mathcal{N}_\alpha \cap \mathcal{N}_\beta &\rightarrow \mathcal{G} \\ x &\longmapsto S_{\alpha\beta}(x) \end{aligned} \quad \left\{ \begin{aligned} &x \in \mathcal{N}_\alpha \cap \mathcal{N}_\beta \\ &S_{\alpha\beta}(x) \in \mathcal{G} \end{aligned} \right. \tag{3.6}$$

(onto the group $\mathcal{G}$ that is represented) that determines the relationship between the gauge mappings $\mathbf{J}_\alpha$ and $\mathbf{J}_\beta$ according to

$$\mathrm{J}_\alpha(x) = \mathit{rep}\, \mathrm{S}_{\alpha\beta}(x) \circ \mathrm{J}_\beta(x) \tag{3.7}$$

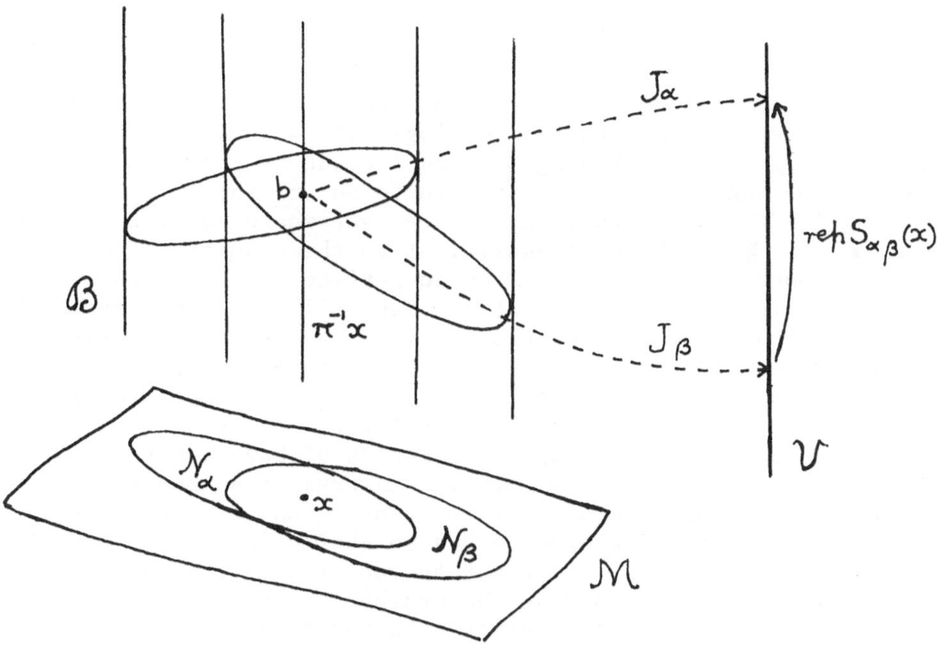

Just as two "atlasses" on an ordinary manifold $\mathcal{M}$ (i e coverings by neighbourhoods $\mathcal{N}_\alpha$ equipped with charts $\mathcal{I}_\alpha : \mathcal{N}_\alpha \to \mathcal{J}$ into the (tangent) vector space $\mathcal{J}$) are treated as equivalent if they are subsets of a larger atlas subject to the same axioms, so also in the case of fibre bundles two coverings by neighbourhoods $\mathcal{N}_\alpha$ with corresponding gauge maps J_α are to be treated as equivalent if they are subsets of a larger such covering.

The classic example of a fibre bundle is the set of all tangent vectors to a given base manifold $\mathcal{M}$. In this example each chart $\mathcal{I}_\alpha$ determines a corresponding set of local co-ordinates $x_\alpha{}^\mu$ on $\mathcal{N}_\alpha$ and hence a set of co-ordinate components $\xi_\alpha{}^\mu$ for any tangent vector $\vec{\xi}$. Thus in terms of any fixed basis on $\mathcal{J}$ (which in this example plays also the role of $\mathcal{V}$) the map $\mathcal{I}_\alpha$ determines a corresponding map J_α sending $\vec{\xi}$ into $\mathcal{J}$. When we change from old co-ordinates $x_\alpha{}^\mu$ to new co-ordinates

$x_\beta^{\ \mu}$, the corresponding guage change mapping $\mathrm{rep}\, S_{\alpha\beta}(x)$ is obviously determined in terms of the basis on $\mathcal{J}$ by the $GL(n)$ matrix with components $(\mathrm{rep}\, S_{\alpha\beta})^\mu{}_\nu$ given by the familiar formula $\partial x_\alpha^{\ \mu}/\partial x_\beta^{\ \nu}$.

The topologically interesting part of the bundle structure is contained in the overlap maps $S_{\alpha\beta}$. In fact it can be seen that for any covering $\mathcal{N}_\alpha$ of $\mathcal{M}$, an arbitrary specification of the $S_{\alpha\beta}$, subject to the obvious consistency condition

$$S_{\alpha\gamma} = S_{\alpha\beta} \circ S_{\beta\gamma} \tag{3.8}$$

wherever $\mathcal{N}_\alpha \cap \mathcal{N}_\beta \cap \mathcal{N}_\gamma$ is non empty, provides one sufficient knowledge to reconstruct a topologically unique corresponding bundle, by appropriate patching of direct products $\mathcal{N}_\alpha \times \mathcal{V}$. In particular the bundle structure is said to be trivial if there is a covering for which all the $S_{\alpha\beta}$ are everywhere equal to the unit element I of $\mathcal{G}$, which is equivalent to the condition that a single neighbourhood $\mathcal{N} = \mathcal{M}$ would suffice, i.e that the bundle has a direct product structure. Yang Mills theories based on $SU(2)$ or other groups were originally thought of in terms of trivial direct product structures, but more recent work, particularly on the theory of instantons, has drawn attention to the interest of considering non-trivial bundles structures (A very comprehensive introduction to the topological side of bundle structure theory has been given by Isham 1978 in the present summer school proceedings) . In the case of ordinary tensor bundles such as the tangent bundle described above one has no choice in the matter: if the manifold structure is itself non trivial (as is the case even in such an elementary example as the ordinary 2-sphere) then the natural tensor bundles will in general be non trivial also.

It frequently occurs that one is interested in simultaneous study not just of one single bundle over $\mathcal{M}$ but of a whole family of bundles with associated representations (in the sense of section 2) of the same group $\mathcal{G}$. The bundles themselves are said to be associated in the technical sense if and only if they can all be described in terms of the same covering $\mathcal{N}_\alpha$ of $\mathcal{M}$ with the same overlap maps $S_{\alpha\beta}$ from $\mathcal{N}_\alpha \cap \mathcal{N}_\beta$ to $\mathcal{G}$. The classic example of such a family of associated bundles is the set of $GL(n)$ bundles consisting of ordinary tensors of different kinds on $\mathcal{M}$.

4. The Retrograde action on the Associated Principle Bundle.

No matter what kind of $\mathcal{G}$ bundle $\mathcal{B}$ we may be given to start with, it is always of interest to consider its associated principle bundle $\mathcal{P}$, i.e. the bundle with the same overlap maps $S_{\alpha\beta}$ that is constructed using the principle representation pr of the bundle group $\mathcal{G}$ on itself (as defined in section 2).

Now in <u>any</u> bundle each gauge map J_α determines a corresponding action of $\mathcal{G}$, on the local patch $\pi^{-1}\mathcal{N}_\alpha$ given by

$$\left.\begin{array}{l}\mathcal{G}\times\pi^{-1}\mathcal{N}_\alpha \longrightarrow \pi^{-1}\mathcal{N}_\alpha \\ (S,p)\longmapsto J_\alpha^{-1}\circ \mathrm{rep}\, S\circ J_\alpha\end{array}\right\}\quad \begin{array}{l} S\in\mathcal{G}\\ p\in\pi^{-1}\mathcal{N}_\alpha\end{array} \tag{4.1}$$

However in a different gauge β instead of the map

$$J_\alpha^{-1}(x)\circ \mathrm{rep}\, S\circ J_\alpha(x)\ :\qquad \pi^{-1}x\longrightarrow\pi^{-1}x \tag{4.2}$$

we should have

$$J_\beta^{-1}(x)\circ \mathrm{rep}\, S\circ J_\beta(x) = J_\alpha^{-1}(x)\circ \mathrm{rep}\left(S_{\alpha\beta}(x)\circ S\circ S_{\alpha\beta}^{-1}(x)\right)\circ J_\alpha(x) \tag{4.3}$$

which will not be the same unless it happens that $S_{\alpha\beta}(x)$ commutes with S. Thus except in special cases, such as when the group is <u>abelian</u>, the <u>local</u> action of $\mathcal{G}$ on $\pi^{-1}\mathcal{N}_\alpha$ will not extend to any well defined global action of $\mathcal{G}$ on $\mathcal{B}$.

Nevertheless, in the particular case of a <u>principle</u> bundle, despite the fact there is in general no global action of $\mathcal{G}$ on $\mathcal{P}$ by <u>left</u> multiplication (in terms of which the principle representation is defined) it is <u>always</u> possible not withstanding to define a <u>globally</u> well behaved <u>retrograde</u> action of $\mathcal{G}$ on $\mathcal{P}$ in terms of right multiplication, by specifying that any element S in $\mathcal{G}$ induces the isomorphism

$$J_\alpha^{-1}(x)\circ \mathrm{ret}\, S\circ J_\alpha\ :\qquad \pi^{-1}x\longrightarrow\pi^{-1}x \tag{4.4}$$

on each of the fibres of $\mathcal{P}$. In this case (unlike that of the more obvious isomorphism defined in terms of $\mathrm{pr}\, S$ the result is <u>independent of the gauge</u>, α, in terms of which it specified, since for any $p\in\pi^{-1}x$ we have

$$\begin{aligned} J_\alpha^{-1}(x)\circ \mathrm{ret}\, S\circ J_\alpha(x)p &= J_\alpha^{-1}(x)\left(\left(J_\alpha(x)p\right)S^{-1}\right)\\ &= J_\beta^{-1}(x)\circ \mathrm{ret}\, S\circ J_\beta(x)\end{aligned} \tag{4.5}$$

by the group associativity law. Thus the local retrograde action, as defined on individual fibres by (4.4) , can be extended to give a canonical global retrograde action

$$\mathcal{R}et : \mathcal{G} \times \mathcal{P} \longrightarrow \mathcal{P} \tag{4.6}$$

on the principle bundle $\mathcal{P}$. The existence of this representation of $\mathcal{G}$ over $\mathcal{P}$ will play an important role in the discussion that follows.

In the case of ordinary tensors on $\mathcal{M}$, the associated principle $GL(n)$ bundle may be identified with the bundle of tangent frames or the bundle of cotangent frames, i.e. the bundles whose elements are the sets of n independent tangent vectors, or n independent 1-forms, at each point of $\mathcal{M}$. The retrograde action of a $GL(n)$ matrix has the effect of replacing the vectors of a given frame by new vectors whose components relative to that frame are specified by the matrix. This is to be contrasted with an ordinary (progressive) $GL(n)$ transformation, whose effect cannot be specified at all without reference to a particular choice of gauge (which fixes a choice of a preferred frame that is associated with the unit element) but whose effect on an individual member vector of a frame depends only on that vector itself, and not on the other ($n-1$) vectors with which it is associated.

5. Fields and Connection.

Nearly all physical theories are expressed in terms of fields, of ordinary tensors, or of other more elaborate structures, over a — usually four dimensional — base manifold $\mathcal{M}$, the term field being used very broadly by physicists to denote a mapping of $\mathcal{M}$ into ... virtually anything. (There is a regrettable custom among English -but not French- speaking pure mathematicians of using the same term field to denote the entirely different concept of a corps such as the real and complex number systems, $\mathbb{R}$ and $\mathbb{C}$). The most naive way to define e.g. a vector (or tensor) field is as a mapping of $\mathcal{M}$ directly into the relevant simple (or tensor product) vector space, but although one can- and physicists usually do- get away with this in non gravitational theories as expressed in terms of a flat, affinely structured space-time base manifold, it is inadequate to describe even such a simple concept as an ordinary tangent vector field on a 2 - sphere.

The simplest concept of a field that is applicable in the context of a general differentiable manifold is that of a section of a fibre bundle over the manifold. Formally, a section Φ of a fibre bundle $\mathcal{B}$ over $\mathcal{M}$ is a continuous (and, when appropriate differentiable) mapping

$$\Phi : \mathcal{B} \longrightarrow \mathcal{M} \tag{5.1}$$

such that

$$\pi \circ \Phi x = x \qquad x \in \mathcal{M} \tag{5·2}$$

i.e. it is a mapping that sends each point x of $\mathcal{M}$ onto some element of the fibre $\pi^{-1}x$ over that point.

It should be pointed out immediately that for many bundles no globally defined section can exist at all. For example when the tangent bundle of a manifold is restricted to the sub-bundle of non-zero tangent vectors, then a section can exist only when the Euler number of the base manifold is zero, which is the case, e.g for the 2-plane or the 2-torus but not for the 2-sphere.

In the case of a principle-bundle, $\mathcal{P}$, a section can exist if and only if the bundle is trivial, i.e. if and only if there exists a global gauge map $J : \mathcal{P} \rightarrow \mathcal{G}$. This can be seen by identifying the inverse image $J^{-1} I$ of the unity element $I \in \mathcal{G}$ with the section- if it exists- and then obtaining $J^{-1} S$ for an arbitrary element $S \in \mathcal{G}$ by applying the global retrograde action of S to the unity section.

Although the concept of a bundle section may be taken as a convenient precise definition of what is meant by an ordinary field in physics, it is necessary to introduce the slightly more elaborate concept of a connection to define the entities commonly known as gauge fields whose importance was first explicitly drawn to the attention of physicists by Yang and Mills. (Actually their use as primary field quantities had long previously been recognised as implicit in electromagnetic theory, while they were also already familiar as secondary, derived, fields in gravitation theory).

The necessity of introducing connections arises from the fact that nearly all physical theories involve differentiation of field quantities, and as we have already remarked in section 1, this can be done for entities such as tensor fields only in the presence of a flat affine structure, or some alternative structure that can be used for the same purpose of comparison of field values at (infinitesimally) neighbouring points, except in specialised cases such as the antisymmetrised differentiation of forms that has already beer described. The most rudimentary structure that can be used for comparison of field values, as one moves allong a curve λ in the base $\mathcal{M}$ of a bundle $\mathcal{B}$, is a general connection, which may be defined as a rule that associates with any parametrised curve

$$\lambda : t \mapsto x(t) \in \mathcal{M} \tag{5·3}$$

(where $t \in \mathbb{R}$ is the parameter) a corresponding family of maps

$$\Xi(t) : \pi^{-1}x(0) \rightarrow \pi^{-1}x(t) \tag{5·4}$$

of the form

$$\Xi(t) = J_\alpha^{-1}(x(t)) \circ \text{rep}\, T_\alpha(t) \cdot J_\alpha(x(0)) \tag{5.5}$$

for some corresponding gauge dependent curve

$$T_\alpha \;:\; t \mapsto T_\alpha(t) \in \mathcal{G} \tag{5.6}$$

in the group $\mathcal{G}$ of the bundle. If the gauge α is replaced by an alternative gauge β say, the group elements must evidently transform according to the rule

$$T_\beta(t) = S_{\alpha\beta}(x(t)) \circ T_\alpha(t) \cdot S_{\alpha\beta}^{-1}(x(0)) \tag{5.7}$$

in order that the map (5.3) should remain invariant.

Any parametrised curve of the form

$$t \mapsto \Xi(t)\, b \in \mathcal{B} \;, \quad b \in \pi^{-1}x(0) \tag{5.8}$$

as defined in terms of any fixed point b in the fibre over the initial point of λ is said to be a <u>horizontal lift</u> (with respect to the connection Ξ) of the curve λ .

The connection Ξ itself is said to be (at least locally) <u>integrable</u> if all horizontal curves through any given point $b \in \mathcal{B}$ lie in an (at least locally) well defined <u>section</u> through b , i.e. in the image of a fibre preserving map from $\mathcal{M}$ into $\mathcal{B}$. (An integrable connection is often described alternatively as being <u>flat</u>, the classic example being the integrable connection that is induced by the affine structure of ordinary flat-Euclidean or Minkowskian-space). An ordinary field is said to be <u>covariantly constant</u> if the section that it defines is <u>horizontal</u> in the sense that it contains all horizontal curves through any point within it; in practise it is rare that horizontal sections should exist at all for non-integrable connections.

It can be seen from the way it has been defined that the specification of a connection on a given bundle automatically induces on any associated bundle a corresponding <u>associated connection</u> characterised by the requirement that on each associated gauge patch α the group elements $T_\alpha(t)$ should be the <u>same</u> as in the original bundle for each fixed base curve λ . For any initially given connection the associated connection on the associated <u>principle</u> bundle $\mathcal{P}$ has the important property that the corresponding mappings <u>commute</u> with the canonical global retrograde action of the group on $\mathcal{P}$, as can be seen from the fact that for any element $p \in \pi^{-1}x(0) \subset \mathcal{P}$ and any $S \in \mathcal{G}$ one has

$$\mathcal{R}et\, S \circ \Xi(t)\, p = J_\alpha^{-1}(x(t)) \circ ret\, S \circ J_\alpha \circ \Xi(t)\, p$$

$$= J_\alpha^{-1}(x(t))\left(T_\alpha(t)(J_\alpha p)\, S^{-1}\right)$$

$$= \Xi(t) \circ \mathcal{R}et\, S\, p \qquad (5\cdot 9)$$

The general formulae listed so far in sections 2, 3, 4 and 5 have made no reference to differentiability either in the base manifold $\mathcal{M}$ or in the bundle manifold $\mathcal{B}$. However before we can use a connection to construct covariant derivatives we must first suppose the existence of a differentiable structure not only on $\mathcal{M}$ (as was already done in section 1) but also on $\mathcal{B}$. The latter requirement presupposes the existence of a differentiable structure on the bundle group, i. e. $\mathcal{G}$ can no longer be arbitrary but must be a Lie group. Before continuing with our description of the theory of connections in section 7, we therefore digress to recapitulate the relevant properties of Lie groups in section 6.

6. Lie groups and Algebras.

From this stage onwards we shall restrict our attention to cases for which the group $\mathcal{G}$ under consideration is a Lie group in the sense of having a (finite dimensional) differential manifold structure (as defined in section 1) over some vector space $\mathcal{A}$, which may be identified with the tangent space at the identity element $\mathbb{I} \in \mathcal{G}$, whose elements are the infinitesimal generators of one-parameter subgroups.

Now it is evident that (unlike the progressive and retrograde representations pr and ret) the group adjoint representation ad of $\mathcal{G}$ over itself (as defined by (2.8) or (2.9)) leaves the identity element of $\mathcal{G}$ invariant, i e

$$(ad\, S)\, \mathbb{I} = \mathbb{I} \qquad \forall S \in \mathcal{G} \qquad (6\cdot 1)$$

and hence (by retraction, as described in section 1) induces an isomorphism

$$@S : \mathcal{A} \to \mathcal{A} \qquad (6\cdot 2)$$

of the tangent space at $\mathbb{I}$. The corresponding action

$$@ : \mathcal{G} \times \mathcal{A} \to \mathcal{A} \qquad (6\cdot 3)$$

is known as the adjoint algebra representation of $\mathcal{G}$. It is to be noted that unlike the group adjoint representation ad from which it is derived, the adjoint algebra representation $@$ is a linear representation of $\mathcal{G}$.

Since they are (simply) transitive over $\mathcal{G}$, the effect of the progressive and retrograde actions pr and ret on $\mathcal{A}$ is to induce (instead of a new representation over $\mathcal{A}$ as was done by ad) families of respectively left and right invariant vector fields over $\mathcal{G}$. For each element $a \in \mathcal{A}$ we shall use the notation $\vec{a}_R$ to denote the corresponding left invariant vector field induced by transporting a under the action of pr and conversely we shall use the notation $\vec{a}_P$ to denote the corresponding right invariant vector field induced by ret . This apparent perversity of notation is motivated by the fact that each left invariant vector field $\vec{a}_R$ is an infinitesimal generator of the right action ret of the one parameter subgroup generated by a , while similarly each right invariant vector field $\vec{a}_P$ is an infinitesimal generator of the left action pr of the subgroup. This can be seen as a consequence of the obvious commutation law

$$pr\, S \circ ret\, T = ret\, T \circ pr\, S \qquad (6.4)$$

(for any $S, T \in \mathcal{G}$) which clearly implies that the left action of S is invariant under the right action of T and vice versa. The law (6.4) also directly implies the even more obvious fact that the property of right invariance is invariant under left action and vice versa. Thus the effect of a progressive (left) action on the (right invariant) vector fields generating the progressive action is to transform them onto themselves according to the adjoint representation, ie we have

$$pr\, S\, \vec{a}_P = ad\, S\, \vec{a}_P = \overrightarrow{(@S\, a)}_P \qquad (6.5)$$

and similarly under interchange of pr, P and ret, R.

Let us now consider a parametrised curve consisting of elements $S(t) \in \mathcal{G}$ ($t \in \mathbb{R}$, $S(0) = \mathbb{I}$) forming the one-parameter subgroup generated by an element $a \in \mathcal{A}$. In the manner described in section 1, any tensor field X on $\mathcal{G}$ will be mapped into another tensor field $pr\, S(t)\, X$ by (the retraction of) the progressive action of each element $S(t)$ so that by (1.10) we shall have

$$\vec{a}_P \pounds X = \frac{d}{dt}\left(pr\, S(t)\, X\right)\Big|_{t=0} \qquad (6.6)$$

and similarly with *ret*, R in place of *hr*, P. Applying this to the right invariant vector fields and using the definition (1.11) we see that we have

$$[\vec{a}_P , \vec{b}_R] = 0 \tag{6.7}$$

for any $a, b \in \mathcal{a}$. Since the Lie bracket of a pair of right invariant vector fields is itself right invariant one sees that there is a natural algebra structure on $\mathcal{a}$ with the product $[a,b]$ of two elements $a, b \in \mathcal{a}$ given by

$$[\vec{a}_P, \vec{b}_P] = \overrightarrow{[a,b]}_P \tag{6.8}$$

The same algebra could equally well have been defined in terms of left invariant vector fields, i.e. one has

$$[\vec{a}_R, \vec{b}_R] = \overrightarrow{[a,b]}_R \tag{6.9}$$

The equivalence of these two expressions follows from the fact that by combining (6.8) with (6.5) and (6.6) we can eliminate all mention of right as opposed to left, and express $[a, b]$ directly in terms of the one parameter subgroup $S(t) \in \mathcal{G}$ generated by a in the form

$$[a,b] = \frac{d}{dt}\left(@ S(t)\, b\right)\Big|_{t=0} \tag{6.10}$$

The antisymmetry of the Lie bracket (in (6.8) or (6.9)) immediately implies the corresponding <u>algebraic</u> antisymmetry property

$$[a, b] = -[b, a] \tag{6.11}$$

while the additional requirement characterising a <u>Lie</u> algebra, namely the Jacobi (non) associativity condition

$$[[a, b], c] + [[b, c], a] + [[c, a], b] = 0 \tag{6.12}$$

may be derived by applying the general identity

$$[\vec{a}_P \pounds , \vec{b}_P \pounds] = [\vec{a}_P , \vec{b}_P] \pounds \tag{6.13}$$

(which would relate the commutator of two Lie differentiations to Lie differentiation with respect to the Lie bracket for quite arbitrary vector fields in place of $\vec{a}_p$ and $\vec{b}_p$).

It can be seen that any linear representation rep of $\mathcal{G}$ over a vector space $\mathcal{V}$ induces a corresponding action

$$\mathit{rep} : \quad \mathcal{A} \times \mathcal{V} \rightarrow \mathcal{V}$$

$$(a, v) \mapsto (\mathit{rep}\, a)\, v \tag{6.14}$$

of the Lie algebra, for which the linear map

$$\mathit{rep}\, a : \quad \mathcal{V} \longrightarrow \mathcal{V} \tag{6.15}$$

is defined in terms of the image $(\mathit{rep}\, S)\, v$ — thought of as a field over $\mathcal{G}$ as $S \in \mathcal{G}$ varies for fixed v — by

$$(\mathit{rep}\, a)\, v = \vec{a}_p \pounds (\mathit{rep}\, S)\, v \Big|_{S=\mathbb{I}} \tag{6.16}$$

Furthermore it can be seen that such an action will satisfy the composition law

$$(\mathit{rep}\, a \circ \mathit{rep}\, b)\, v = (\vec{a}_p \pounds)(\vec{b}_p \pounds)(\mathit{rep}\, S)\, v \Big|_{S=\mathbb{I}} \tag{6.17}$$

Hence by (6.13) and (6.8) we see that in terms of the commutator product

$$[\mathit{rep}\, a, \mathit{rep}\, b] = \mathit{rep}\, a \circ \mathit{rep}\, b - \mathit{rep}\, b \circ \mathit{rep}\, a \tag{6.18}$$

the action defined by (6.16) may be interpreted as a representation of the Lie algebra, since it satisfies

$$[\mathit{rep}\, a, \mathit{rep}\, b] = \mathit{rep}\, [a, b] \tag{6.20}$$

In the particular case when we take $\mathcal{V}$ to be $\mathcal{A}$ and rep to be @ we obtain the adjoint representation of the Lie algebra, which may be seen (with the aid of (6.5) and (6.6)) to be given simply by

$$(@\, a)\, b = [a, b] \tag{6.21}$$

7. Differential Connections.

Whenever a fibre bundle $\mathcal{B}$ is constructed in terms of a Lie group $\mathcal{G}$ with a differentiable base $\mathcal{M}$ it makes sense to postulate that $\mathcal{B}$ itself should be differentiable by restricting the gauge mappings J_α and the transformation mappings $S_{\alpha\beta}$ to be differentiable.

An important property of a differentiable principle bundle $\mathcal{P}$ is the existence for each Lie algebra element a of a corresponding globally well defined vector field $\vec{a}_R$ over $\mathcal{P}$ that generates the global retrograde action (as defined in section 4) of the one-parameter subgroup generated by a . (There is however no analogue over $\mathcal{P}$ of the progressive action generators $\vec{a}_P$ that exist on $\mathcal{G}$). Each vector field $\vec{a}_R$ on $\mathcal{P}$ will evidently be vertical in the sense of being everywhere tangent to the fibres, and it will be left invariant under the (gauge dependent) progressive actions of $\mathcal{G}$ on local neighbourhoods. Under the global retrograde action (4.6) these vector fields will transform according to the law

$$\mathcal{R}et\, S\, \vec{a}_R = \overrightarrow{(@S\, a)}_R \qquad (7.1)$$

which (unlike its left analogue (6.5)) is unambiguously generalisea-ble from the Lie group $\mathcal{G}$ to the principle bundle $\mathcal{P}$.

A connection Ξ on a differentiable fibre bundle $\mathcal{B}$ is said to be a differential, or linear, connection if it satisfies the obvious requirement that the set of horizontal curves through each point p of the (differentiably) associated principle bundle, and hence also for all the other (differentiably) associated bundles, is tangent to a well defined linear subspace $\mathcal{H}(p)$ (isomorphic to the vector space $\mathcal{T}$ over which $\mathcal{M}$ is constructed) that is transverse to the fibres in the bundle tangent space at p , so that the bundle projection π induces a non-degenerate mapping of the linear subspace $\mathcal{H}(p)$ onto the tangent space $\mathcal{T}(x)$ at $x = \pi p$ in $\mathcal{M}$. (A(locally) integrable connection thus arises as the special case for which the horizontal subspaces are everywhere tangent to a congruence of (local) sections.) It is to be mentioned that many authors use the qualification linear in a more restricted sense with reference only to the natural $GL(n)$ (tangent frame) tensor bundles over $\mathcal{M}$, which is why we have preferred to use the qualification differential for connections in a more general context.

The existence of the Killing vector fields $\vec{a}_R$ (subject to (7.1)) on the principle bundle $\mathcal{P}$ makes it possible to think of any differential connection Ξ as specifying and spe-cified by a Lie algebra valued differential form, $\hat{\chi}$ say , over $\mathcal{P}$. (we shall use a hat , ^ , to distinguish forms on the bundle from forms on the base $\mathcal{M}$) This form is specifiable in terms of its contraction with an arbitrary tangent vector $\vec{X}$ to the bundle at an arbitrary point $p \in \mathcal{P}$ according to the following prescription based on the decomposition

$$\vec{X} = \vec{X}_{\parallel} + \vec{X}_{\perp} \tag{7.2}$$

where $\vec{X}_{\parallel}$ is vertical, i.e tangent to the fibre through p, while $\vec{X}_{\perp}$ is horizontal (with respect to Ξ) in the sense that it lies within the horizontal subspace $\mathcal{H}(p)$. The effect of contracting $\hat{\underline{\chi}}$ with the horizontal component is defined simply by

$$\hat{\underline{\chi}} \cdot \vec{X}_{\perp} = 0 \tag{7.3}$$

while its effect on the vertical component is given by

$$\hat{\underline{\chi}} \cdot \vec{X}_{\parallel} = a \tag{7.4}$$

where $a \in \mathcal{A}$ is the Lie algebra element for which

$$\vec{a}_R(p) = \vec{X}_{\parallel} \tag{7.5}$$

If the vector $\vec{X}_{\parallel}(p)$ is pulled back by a retrograde map $\mathcal{R}et\, S$, ($S \in \mathcal{G}$) to give a new vector $\vec{X}(p) = \mathcal{R}et\, S\, \vec{X}_{\parallel}(p)$ ($p = \mathcal{R}et\, S\, q$) then by the transformation law (7.1) we have

$$\hat{\underline{\chi}} \cdot (\mathcal{R}et\, S\, \vec{X}_{\parallel}) = @S\, (\hat{\underline{\chi}} \cdot \vec{X}_{\parallel}) \tag{7.6}$$

and since the commutation law (5.8) implies that horizontality is conserved by the retrograde action it can be seen that the corresponding variation law is given even more simply by

$$\hat{\underline{\chi}} \cdot (\mathcal{R}et\, S\, \vec{X}_{\perp}) = 0 \tag{7.7}$$

Combining (7.6) and (7.7) we see that when an arbitrary vector is transported analogously by $\mathcal{R}et\, S$, then its contraction with the connection form $\hat{\underline{\chi}}$ varies according to

$$\hat{\underline{\chi}} \cdot (\mathcal{R}et\, S\, \vec{X}) = @S\, (\hat{\underline{\chi}} \cdot \vec{X}) \tag{7.8}$$

which may be expressed more concisely without explicit reference to the vector $\vec{X}$ by

$$\mathcal{R}et\, S\, \hat{\underline{\chi}} = @S\, \hat{\underline{\chi}} \tag{7.9}$$

This is the characteristic defining property that must be posessed by a Lie algebra valued 1-form $\hat{\underline{\chi}}$ in order to specify a connection.

Before proceeding we shall introduce some algebraic machinery for use with differential forms such as $\hat{\underline{\chi}}$ whose values lie in a general algebra rather than in an ordinary corps of numbers, $\mathbb{R}$ or $\mathbb{C}$. For any two algebra valued one-forms, $\hat{\underline{\chi}}$ and $\hat{\underline{\omega}}$

say, the exterior product $\hat{\underline{\chi}} \wedge \hat{\underline{\omega}}$ is defined in terms of its contraction with an arbitrary pairs of vectors $\vec{X}$, $\vec{Y}$ by

$$\vec{X} \cdot (\hat{\underline{\chi}} \wedge \hat{\underline{\omega}}) \cdot \vec{Y} = (\hat{\underline{\chi}} \cdot \vec{X})(\hat{\underline{\omega}} \cdot \vec{Y}) - (\underline{\chi} \cdot \vec{Y})(\hat{\underline{\omega}} \cdot \vec{X}) \qquad (7.10)$$

whereas by contrast their commutator, $[\hat{\underline{\chi}}, \hat{\underline{\omega}}]$ might be defined by

$$\vec{X} \cdot [\hat{\underline{\chi}}, \hat{\underline{\omega}}] \cdot \vec{Y} = [\hat{\underline{\chi}} \cdot \vec{X}, \hat{\underline{\omega}} \cdot \vec{Y}] \qquad (7.11)$$

By combining the two kinds of permutation involved, one can go on to construct the exterior commutator $[\hat{\underline{\chi}} \wedge \hat{\underline{\omega}}]$ which is defined by

$$[\hat{\underline{\chi}} \wedge \hat{\underline{\omega}}] = [\hat{\underline{\chi}}, \hat{\underline{\omega}}] - [\hat{\underline{\omega}}, \hat{\underline{\chi}}] \qquad (7.12)$$

or equivalently

$$[\hat{\underline{\chi}} \wedge \hat{\underline{\omega}}] = \hat{\underline{\chi}} \wedge \hat{\underline{\omega}} + \hat{\underline{\omega}} \wedge \hat{\underline{\chi}} \qquad (7.11)$$

If the algebra valued 1-form $\hat{\underline{\omega}}$ were replaced by an algebra valued 2-form, $\hat{\underline{\Omega}}$ say, then instead of (7.13) we would have

$$[\hat{\underline{\chi}} \wedge \hat{\underline{\Omega}}] = \hat{\underline{\chi}} \wedge \hat{\underline{\Omega}} - \hat{\underline{\Omega}} \wedge \hat{\underline{\chi}} \qquad (7.12)$$

(Quite generally the formula (7.11) will be valid whenever the product of the orders of the forms involved is odd, whereas (7.12) will be valid whenever the product of the orders is even). In particular, for any algebra valued 1-form at all, and in particular for a connection 1-form we shall have, identically,

$$[\hat{\underline{\chi}}, \hat{\underline{\chi}}] = \hat{\underline{\chi}} \wedge \hat{\underline{\chi}} = \tfrac{1}{2}[\hat{\underline{\chi}} \wedge \hat{\underline{\chi}}] \qquad (7.13)$$

which in general is not zero. By the Jacobi identity, however, the triple exterior product of a single algebra valued 1-form will always vanish, i.e

$$[[\hat{\underline{\chi}} \wedge \hat{\underline{\chi}}] \wedge \hat{\underline{\chi}}] = 0 \qquad (7.14)$$

We conclude this section by remarking that the defining relation (7.9) for a connection form on a principe bundle can be converted into differential form in terms of the generating fields $\vec{a}_R$ of the global retrograde action in the form

$$\vec{a}_R \pounds \hat{\chi} = - @a\, \hat{\underline{\chi}} = -[a, \hat{\underline{\chi}}] \qquad (7.15)$$

(the negative sign arising from the retrograde character of $\mathcal{R}et$).

8. The Curvature of a Differential Connection.

By the defining property (7.4) it can be seen that when a curvature form $\hat{\underline{\chi}}$ is contracted with the representative at any point of the vector field $\vec{a}_R$ generating the global retrograde action on a principle bundle of the Lie subgroup generated by an algebra element a , the result is given simply by

$$\vec{a}_R \cdot \hat{\underline{\chi}} = a \tag{8.1}$$

It follows from (7.12) that when $\vec{a}_R$ is contracted with the algebra valued 2-form $[\hat{\underline{\chi}} \wedge \hat{\underline{\chi}}]$ the result is the algebra valued 1-form given by

$$\vec{a}_R \cdot [\hat{\underline{\chi}} \wedge \hat{\underline{\chi}}] = 2[a, \hat{\underline{\chi}}] \tag{8.2}$$

Now when applied to the vector field $\vec{a}_R$ and the 1-form $\hat{\underline{\chi}}$, the Cartan identity (1.13) takes the form

$$\vec{a}_R \pounds \hat{\underline{\chi}} = \vec{a}_R \cdot (\partial \wedge \hat{\underline{\chi}}) + \partial(\vec{a}_R \cdot \hat{\underline{\chi}}) \tag{8.3}$$

and it follows from (8.1) that the algebra valued scalar field $\vec{a}_R \cdot \hat{\underline{\chi}}$ is everywhere equal to the fixed element $a \in \mathcal{A}$ so that we have

$$\partial(\vec{a}_R \cdot \hat{\underline{\chi}}) = 0 \tag{8.4}$$

Hence by combining (8.2) and (8.3) we see that the characteristic differential equation(7.15) that must be satisfied by a connection form $\hat{\underline{\chi}}$ on a principle bundle is expressible equivalently in the form

$$\vec{a}_R \cdot (\partial \wedge \hat{\underline{\chi}}) = -\frac{1}{2} \vec{a}_R \cdot [\hat{\underline{\chi}} \wedge \hat{\underline{\chi}}] \tag{8.5}$$

This result, which is known as the Cartan-Maurer theorem, can be expressed more concisely by introducing the Lie algebra valued curvature 2-form, $\hat{\underline{\Omega}}$, on the principle bundle $\mathcal{P}$, according to the specification

$$\hat{\underline{\Omega}} = \partial \wedge \hat{\underline{\chi}} + \frac{1}{2} [\hat{\underline{\chi}} \wedge \hat{\underline{\chi}}] \tag{8.6}$$

In terms of this $\mathfrak{a}$-valued 2-form the Cartan-Maurer theorem may be stated simply as

$$\underline{\Omega} \cdot \vec{X}_{\parallel} = 0 \tag{8.7}$$

where $\vec{X}_{\parallel}$ is an arbitrary vector tangent to the fibres (since at any point $p \in \mathcal{P}$ such a vector can always be taken to have the form $\vec{a}_R(p)$ for some appropriate choice of a).

By using the differential identity

$$\partial \wedge [\hat{\underline{\chi}} \wedge \hat{\underline{\chi}}] = -2[\hat{\underline{\chi}} \wedge (\partial \wedge \hat{\underline{\chi}})] \tag{8.8}$$

in conjunction with the algebraic identity (7.14) (which itself resulted from the Jacobi identity) and the Poincaré closure identity (1.3), one can deduce that for <u>any</u> 1-form $\hat{\underline{\chi}}$, a 2-form of the form (8.6) must satisfy the generalised <u>Bianchi identity</u>

$$\partial \wedge \hat{\underline{\Omega}} + [\hat{\underline{\chi}} \wedge \hat{\underline{\Omega}}] = 0 \tag{8.9}$$

Since $\hat{\underline{\chi}}$ is not arbitrary but subject to the Lie variation formula (7.15) we can go on to deduce an analogous Lie variation formula for the curvature using the relevant form

$$\vec{a}_R \pounds \hat{\underline{\Omega}} = \vec{a}_R \cdot (\partial \wedge \hat{\underline{\Omega}}) + \partial \wedge (\vec{a}_R \cdot \hat{\underline{\Omega}}) \tag{8.10}$$

of the Cartan identity (1.13). We have already seen that the variation condition (7.15) on $\hat{\underline{\chi}}$ is equivalent to the Cartan Maurer property of the curvature which is expressible, analogously to (8.1) as

$$\vec{a}_R \cdot \hat{\underline{\Omega}} = 0 \tag{8.11}$$

Hence in (8.10) (as in (8.3)) the right hand side reduces to the first term so that substitution from the Bianchi identity (8.9) and further use of (8.1) and (8.11) leads finally to

$$\vec{a}_R \pounds \hat{\underline{\Omega}} = -[a, \hat{\underline{\Omega}}] = -@a\, \hat{\underline{\Omega}} \tag{8.12}$$

which is the analogue of (7.15).
The corresponding <u>finite</u> transformation law

$$\mathcal{R}\!\mathit{et}\, S\, \hat{\underline{\Omega}} = @S\, \hat{\underline{\Omega}} \tag{8.13}$$

which is of course a direct consequence of the analogous finite transformation law (7.9), expresses the condition that if $\hat{\underline{\Omega}}$ is contracted with a right invariant bivector field the resulting algebra valued scalar field over $\mathcal{P}$ transforms adjointly under the retrograde action on the fibres.

9. Curvature and Connection Forms on the Base Space.

The transformation property (8.13) (which it shares with the connection) together with the Cartan Maurer property (8.11) (which is stricter than the analogue, (8.1), for the connection) implies that the curvature form $\hat{\Omega}$ (unlike the connection form $\hat{\chi}$) determines, and can be specified in terms of an ordinary field, Ω say, over the base space. Explicitly this base curvature Ω is a section of an appropriate fibre bundle of adjointly transforming $\mathcal{a}$-valued 2-forms on $\mathcal{M}$, its construction being describable as follows. Far any particular gauge

$$J_\alpha : \pi^{-1}(\mathcal{N}_\alpha) \to \mathcal{G} \tag{9.1}$$

on the principle bundle $\mathcal{P}$ there is a corresponding canonical gauge section

$$J_\alpha^{-1}(\mathbb{1}) : \mathcal{N}_\alpha \to \mathcal{P} \tag{9.2}$$

(where $\mathbb{1}$ denotes the unit element of $\mathcal{G}$). The restriction of the 2-form $\hat{\Omega}$ on $\mathcal{P}$ to this section obviously determines (via the projection π) a corresponding 2-form Ω_α over the neighbourhood $\mathcal{N}_\alpha$ in the base space, and by the Cartan-Maurer property (8.7) the value that is obtained for this 2-form at a given point $x \in \mathcal{N}_\alpha$ depends only on the restriction $J_\alpha(x)$ of the gauge map (9.1) to the particular fibre $\pi^{-1}(x)$ over x and not on the way the gauge map affects nearly fibres. To see how $\Omega_\alpha(x)$ changes when we make a gauge transformation.

$$J_\alpha(x) \mapsto J_\beta(x) = \text{pr}\, S_{\beta\alpha}(x) \circ J_\alpha(x) \tag{9.3}$$

we use the fact that

$$\begin{aligned} J_\beta^{-1}(x)\,\mathbb{1} &= J_\alpha^{-1}(x)\, S_{\beta\alpha}^{-1}(x) \\ &= J_\alpha^{-1}(x)\,(\text{ret}\, S_{\beta\alpha}(x)\,\mathbb{1}) \end{aligned} \tag{9.4}$$

Thus the intersection with the fibre $\pi^{-1}(x)$ of the canonical gauge section is transported in a gauge change by the <u>retrograde</u> action of the relevant gauge transformation element $S_{\beta\alpha}(x)$. Hence, by considering the contraction of $\underline{\hat{\Omega}}$ with any pair of right invariant vector fields, we see from (8.13) that $\Omega_\alpha(x)$ transforms under the gauge change according to the <u>adjoint</u> representation, i.e we have

$$\underline{\Omega}_\beta = @ S_{\beta\alpha} \underline{\Omega}_\alpha \tag{9.5}$$

This means we can think of the contraction of $\underline{\Omega}_\alpha$ with any fixed bivector field on $\mathcal{M}$ as the α-gauge image of a well behaved section of the <u>associated adjoint bundle</u> i.e the associated bundle defined in terms of the adjoint representation $@$ of the group $\mathcal{G}$ of the original bundle over its own algebra $\mathcal{a}$. Thus $\underline{\Omega}_\alpha$ itself can be interpreted as the α-gauge image of a well behaved section, $\underline{\Omega}$, of a corresponding bundle of adjointly transforming $\mathcal{a}$-valued 2-forms.

If one tries, in an analogous manner, to express the connection 1-form $\underline{\hat{\chi}}$ in terms of a corresponding 1-form on the base $\mathcal{M}$, complications arise, due to the fact that instead of the Cartan-Maurer property (8.7), one has the less convenient vertical contraction property (8.1). Thus although $\underline{\hat{\chi}}$ does induce a well defined 1-form field, which we shall denote by $\underline{\omega}_\alpha$, via the canonical gauge section $J_\alpha^{-1}(\mathbb{1})$, this 1-form can <u>not</u> be interpreted as the α-gauge image of an ordinary field because it depends on the <u>angle</u> of the section, i.e on the way J_α varies over nearby fibres, not just on the restriction $J_\alpha(x)$ to a single fibre through a particular point under consideration. We can- somewhat artificially - improve the situation by making the decomposition

$$\underline{\hat{\chi}} = \underline{\hat{\chi}}_\alpha + \underline{\hat{\omega}}_\alpha \tag{9.6}$$

where $\underline{\hat{\chi}}_\alpha$ is the connection form that arises naturally from the gauge map J_α by setting $T_\alpha = \mathbb{1}$ in (5.5) (which is equivalent to requiring that the canonical gauge section $J_\alpha^{-1}(\mathbb{1})$ should appear horizontal with respect to $\hat{\chi}_\alpha$). Thus the algebra-valued 1-form $\hat{\omega}_\alpha$ defined on $\mathcal{P}$ by (9.6) is an example of a <u>difference between two connections</u>. Now since all connection 1-forms must satisfy (8.1) it follows that <u>any</u> $\underline{\omega}_\alpha$ that is the difference between two connection 1-forms must obey the analogue of the Cartan Maurer condition (8.11), i.e it must satisfy

$$\underline{\hat{\omega}}_\alpha \cdot \vec{X}_{\parallel} = 0 \tag{9.7}$$

for any vector $\vec{X}_{\parallel}$ tangent to the fibres, while similarly by (7.9) it must also obey

$$\mathcal{R}et\, S\ \underline{\hat{\omega}}_\alpha = @S\ \underline{\hat{\omega}}_\alpha \tag{9.8}$$

It therefore follows, by reasoning precisely analogous to that used in constructing $\underline{\Omega}$ from $\underline{\hat{\Omega}}$, that any difference of two connection 1-forms on $\mathcal{P}$ will induce a corresponding base field which will be a well behaved section of a bundle of adjointly transforming 1-forms. In particular, such an adjointly transforming 1-form field, $\underline{\omega}_\alpha$ say, over $\mathcal{N}_\alpha \subset \mathcal{M}$ will be induced by the field $\underline{\hat{\omega}}_\alpha$ over $\mathcal{P}$ that is defined by (9.6). Unlike the base space curvature 2-form $\underline{\Omega}$ however, the base space connection 1-form $\underline{\omega}_\alpha$ not only is limited to the neighbourhood $\mathcal{N}_\alpha$ (which nevertheless, for a topologically trivial bundle, might include the whole of $\mathcal{M}$) but also is gauge dependent in the more fundamental sense that it depends on the choice of the gauge map $\mathbf{J}_\alpha$

Despite this somewhat unsatisfactory feature, the gauge connection 1-form $\underline{\omega}_\alpha$ has the convenient property that it can be used directly to determine the corresponding gauge image, $\underline{\Omega}_\alpha$, of the base space curvature 2-form, without any need for reference to the bundle forms $\underline{\hat{\omega}}_\alpha$ and $\underline{\hat{\chi}}_\alpha$ (of which the latter is what has recently been characterised by Thierry-Mieg (1978) as the ghost form). This can easily be seen by introducing local coordinates $\{Y^i\}$ on $\mathcal{G}$ and $\{x^\mu\}$ on the patch $\mathcal{N}_\alpha \subset \mathcal{M}$, and letting $\{Y^i, x^\mu\}$ be a corresponding patch induced on $\mathcal{P}$ by the gauge map J_α so that the decomposed parts of the connection 1-form $\underline{\hat{\chi}}$ on $\mathcal{P}$ can be expressed in the form

$$\underline{\hat{\chi}}_\alpha = \hat{\chi}_{\alpha i}\, \partial Y^i \tag{9.9}$$

$$\underline{\hat{\omega}}_\alpha = \hat{\omega}_{\alpha\mu}\, \partial x^\mu \tag{9.10}$$

where $\hat{\chi}_{\alpha i}$ and $\hat{\omega}_{\alpha\mu}$ are Lie algebra valued component functions. In terms of these functions we obtain for the bundle curvature 2-form the component expression

$$\begin{aligned}\underline{\hat{\Omega}} = {} & \left(\frac{\partial \hat{\chi}_{\alpha j}}{\partial Y^i} + [\hat{\chi}_{\alpha i}, \hat{\chi}_{\alpha j}]\right) \partial Y^i \wedge \partial Y^j \\ & + \left(\frac{\partial \hat{\omega}_{\alpha\mu}}{\partial Y^i} + 2[\hat{\chi}_{\alpha i}, \hat{\omega}_{\alpha\mu}]\right) \partial Y^i \wedge \partial x^\mu \\ & + \left(\frac{\partial \hat{\omega}_{\alpha\nu}}{\partial x^\mu} + [\hat{\omega}_{\alpha\mu}, \hat{\omega}_{\alpha\nu}]\right) \partial x^\mu \wedge \partial x^\nu \end{aligned} \tag{9.11}$$

using the fact that the functions $\hat{\chi}_{\alpha i}$ are independent of the base co-ordinates x^μ. The Cartan Maurer property (8.7) now implies that all terms except the last must vanish. The vanishing of the first term is in fact equivalent to the condition

$$\partial \wedge \underline{\hat{\chi}}_\alpha + \frac{1}{2}[\underline{\hat{\chi}}_\alpha \wedge \underline{\hat{\chi}}_\alpha] = 0 \tag{9.12}$$

which expresses the integrability property of the (ghost) connection induced by the choice of the gauge α. From the last term in (9.11) one obtains the expression for the curvature 2-form induced on the manifold in the gauge α as

$$\underline{\Omega}_\alpha = \partial \wedge \underline{\omega}_\alpha + \frac{1}{2}[\underline{\omega}_\alpha \wedge \underline{\omega}_\alpha] \tag{9.13}$$

In view of the fact that it is valid for any gauge α, it is convenient and natural to express this formula in a compact shorthand notation as

$$\underline{\Omega} = \partial \wedge \underline{\omega} + \frac{1}{2}[\underline{\omega} \wedge \underline{\omega}] \tag{9.14}$$

in which explicit reference to the gauge α has been dropped. It is by this last formula that the curvature is commonly introduced by physicists (who are often thinking in terms of a single fixed gauge, so that α is not so much dropped as forgotten) since it allows them to ignore bundle structure as such. Again using the identity (7.14) in the form

$$[[\underline{\omega}_\alpha \wedge \underline{\omega}_\alpha] \wedge \underline{\omega}_\alpha] = 0 \tag{9.16}$$

we can go on as before to obtain the base space Bianchi identity,

$$\partial \wedge \underline{\Omega}_\alpha + [\underline{\omega}_\alpha \wedge \underline{\Omega}_\alpha] = 0 \tag{9.17}$$

This also may be written in shorthand notation as

$$\partial \wedge \underline{\Omega} + [\underline{\omega} \wedge \underline{\Omega}] = 0 \tag{9.18}$$

in which again, as in (9.14), it is to be remembered that $\underline{\omega}$ (unlike $\underline{\Omega}$) does not denote a well defined field in the ordinary sense of the word.

10. Covariant Differentiation.

The principal use of connections in physical theories is for the construction of gauge covariant derivatives of fields that are sections of vector bundles, i.e bundles constructed in terms of a linear representation, rep say, of $\mathcal{G}$ over a vector space $\mathcal{V}$ say. Let

$$\Phi : \mathcal{M} \rightarrow \mathcal{B} \tag{10.1}$$

be a section of such a bundle $\mathcal{B}$, and let

$$\Phi_\alpha = J_\alpha \circ \Phi : \mathcal{N}_\alpha \rightarrow \mathcal{V} \tag{10.2}$$

be its α-gauge image (which is what physicists working in a fixed gauge usually treat as being the same as the field itself). Since the image $\Phi_\alpha(x)$ of $x \in \mathcal{N}_\alpha \subset \mathcal{M}$ lies in the fixed linear space $\mathcal{V}$, one can proceed directly by partial differentiation to construct first its gradient $\partial \Phi_\alpha$, which will be a $\mathcal{V}$-valued 1-form over $\mathcal{N}_\alpha$, and hence also the corresponding differential

$$d\Phi_\alpha = dx^\mu \partial_\mu \Phi_\alpha = (\vec{\xi} \cdot \partial \Phi_\alpha)\, dt \tag{10.3}$$

associated with the infinitesimal displacement field

$$d\vec{x} = \vec{\xi}\, dt \tag{10.4}$$

generated by an arbitrary tangent vector field $\vec{\xi}$ on the base manifold $\mathcal{M}$ (We remind the mathematical reader that our d has nothing to do with exterior differentiation, which will always be indicated in longhand notation by $\partial\wedge$). The problem is that although it can be constructed unambiguously in terms of any given $\vec{\xi}$, the differential field $d\Phi_\alpha$ is not the α-gauge image of any well behaved (gauge invariantly specified) section $d\Phi$ of $\mathcal{B}$ as one might have hoped. When a differential connection is present however one can define the α-gauge image of a genuine (gauge) invariantly defined) section by taking the difference between the value of $\vec{\xi} \cdot \partial \Phi_\alpha$ that is actually obtained and the value that would be obtained if the section Φ were horizontal with respect to the connection.

A horizontal field $\Phi_\perp$ say through the point $b \in \mathcal{B}$ in question is required to satisfy

$$\Xi(t)\, b = \Phi_\perp(x(t)) \tag{10.5}$$

(in the notation of section 5) for any parametrised curve $x(t)$ with initial point $x(0) = \pi b$, which is equivalent by (5.5) to

$$rep\, T_\alpha(t) \circ J_\alpha(x) \circ \Phi(\pi b) = J_\alpha(x(t)) \circ \Phi_\perp(x(t)) \tag{10.6}$$

If we now let a be the Lie algebra vector that is tangent to the curve $T_\alpha(t)$ at the origin $\mathbb{1}$ in $\mathcal{G}$, then we can write the differential of (10.6) in the form

$$\mathrm{rep}\,(a\,dt)\circ J_\alpha(x)\circ\Phi(x) = \frac{d}{dt}\Big(J_\alpha(x(t))\circ\Phi_\perp(x(t))\Big)\Big|_{t=0} dt$$
$$= \left(\vec{\xi}\cdot\partial\Phi_{\perp\alpha}\right)dt \tag{10.7}$$

where $\vec{\xi}$ is the generating tangent vector to the parametrised curve $x(t)$ at $x(0)$. Now the algebra element a tangent to the Lie group curve $T_\alpha(t)$ will itself be completely determined in terms of the connection by the choice of the base space tangent vector $\vec{\xi}$. We may evaluate a in terms of the connection 1-form $\hat{\underline{X}}$ on the associated principle bundle $\mathcal{P}$, by considering the lift, $\vec{X}$ say, of $\vec{\xi}$ onto the associated canonical gauge section $J_\alpha^{-1}(\mathbb{1})$ in $\mathcal{P}$. (We use the same symbols π, J_α etc for the bundle projection, associated gauge maps, etc on all members of an associated family of bundles, relying on the context to indicate which particular member is under consideration in any instance). In other words $\vec{X}_\alpha$ is defined by the conditions that it be horizontal with respect to the gauge (or, in Thierry-Mieg's (1978) terminology, ghost) connection, $\hat{\underline{X}}_\alpha$, i.e

$$\vec{X}\cdot\hat{\underline{X}}_\alpha = 0 \tag{10.8}$$

and that its projection by π onto $\mathcal{M}$ should be $\vec{\xi}$. Now it is geometrically evident (see the accompanying diagram)

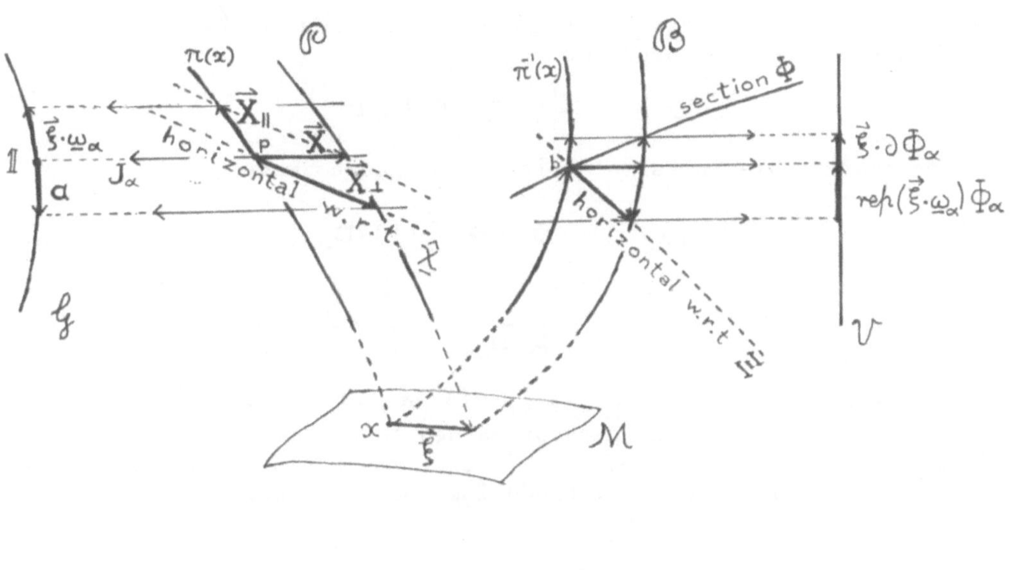

that the tangent vector a generating the parametrised curve $T_\alpha(t)$ in $\mathcal{G}$ will be determined from ξ via the decomposition (7.2) of its lift into the gauge section by

$$\vec{a}_R = -\vec{X}_{\parallel} \qquad (10\cdot 9)$$

which is equivalent, by (8.1) to

$$a = -\vec{X}\cdot\hat{\underline{\chi}} \ . \qquad (10\cdot 10)$$

Hence by (9.6) and (10.8) we finally obtain the formula

$$a = -\vec{X}\cdot\hat{\underline{\omega}}_\alpha = -\xi\cdot\underline{\omega}_\alpha \qquad (10\cdot 11)$$

for the value of a appearing in the equation (10.7) which therefore reduces to

$$rep(-\vec{\xi}\cdot\underline{\omega}_\alpha)\,\Phi_\alpha = \xi\cdot\partial\,\Phi_{\perp\alpha} \qquad (10\cdot 12)$$

thus providing for the required difference the expression

$$d\,\Phi_\alpha - d\,\Phi_{\perp\alpha} = \left(\vec{\xi}\cdot D\,\Phi_\alpha\right)dt \qquad (10\cdot 13)$$

where

$$\vec{\xi}\cdot D\Phi_\alpha = \vec{\xi}\cdot\partial\,\Phi_\alpha + rep(\vec{\xi}\cdot\underline{\omega}_\alpha)\,\Phi_\alpha \qquad (10\cdot 14)$$

The quantity $\vec{\xi}\cdot D\Phi_\alpha$ constructed in this way can evidently be thought of as the α-<u>gauge image</u>

$$\vec{\xi}\cdot D_\alpha\Phi_\alpha = (\vec{\xi}\cdot D\,\Phi)_\alpha = J_\alpha(\xi\cdot D\,\Phi) \qquad (10\cdot 15)$$

of a well behaved <u>section</u>, $\xi\cdot D\Phi$, of the bundle $\mathcal{B}$, the corresponding differential operator

$$\vec{\xi}\cdot D : \quad \Phi \longmapsto \vec{\xi}\cdot D\,\Phi \qquad (10\cdot 16)$$

being expressible in terms of its α-gauge image

$$\vec{\xi}\cdot D_\alpha : \quad \Phi_\alpha \longmapsto \vec{\xi}\cdot D\,\Phi_\alpha \qquad (10\cdot 17)$$

by

$$\xi\cdot D_\alpha = \xi\cdot\partial_\alpha + rep(\vec{\xi}\cdot\omega_\alpha) \qquad (10\cdot 18)$$

Since the linearity property $(\vec{\xi}+\vec{\eta})\cdot\underline{\omega} = \vec{\xi}\cdot\underline{\omega} + \vec{\eta}\cdot\underline{\omega}$ allows us to define the operator valued form $rep\,\underline{\omega}$ by setting $rep(\vec{\xi}\cdot\underline{\omega}) = \vec{\xi}\cdot rep\,\underline{\omega}$, we may avoid reference to any particular field $\vec{\xi}$ by introducing the α-image

$$D\,\Phi_\alpha = \partial\,\Phi_\alpha + (\mathit{rep}\,\underline{\omega}_\alpha)\,\Phi_\alpha \qquad (10{\cdot}19)$$

of the covariant derivative $D\Phi$, which is evidently a well defined section in the appropriate bundle of rep transforming $\mathcal{V}$-valued 1-forms over $\mathcal{M}$, the image of the corresponding differentiation operator D being

$$D_\alpha = \partial + \mathit{rep}\,\underline{\omega}_\alpha \qquad (10{\cdot}20)$$

It may be remarked at this point that since any gauge transformation

$$\Phi_\alpha \longmapsto \Phi_\beta = (\mathit{rep}\,S_{\beta\alpha})\,\Phi_\alpha \qquad (10{\cdot}21)$$

has an identical effect on $\vec{\xi}\cdot D\,\Phi_\alpha$, the base connection representative must evidently transform according to the slightly more complicated law

$$\mathit{rep}\,\underline{\omega}_\alpha \longmapsto \mathit{rep}\,\underline{\omega}_\beta = \mathit{rep}\,S_{\beta\alpha}\circ\left(\mathit{rep}\,\underline{\omega}_\alpha\circ \mathit{rep}\,S_{\alpha\beta} + \partial\,\mathit{rep}\,S_{\alpha\beta}\right) \qquad (10{\cdot}22)$$

This can be expressed concisely in terms of the covariant differentiation operator D_α by

$$\mathit{rep}\,\underline{\omega}_\alpha - \mathit{rep}\,\underline{\omega}_\beta = \mathit{rep}\,S_{\beta\alpha}\circ D_\alpha\,\mathit{rep}\,S_{\alpha\beta} \qquad (10{\cdot}23)$$

It is to be remembered here that for example if Φ transforms according to a fundamental contravariant representation, so that the symbol rep is interpretable more precisely (see sections 2, 12) as meaning rep^1 in all the preceeding equations of this section except the all purpose definitions (10.12) and (10.20), then $\mathit{rep}\,S_{\alpha\beta}$ itself will transform under the corresponding mixed representation, so that the rep implicit in D_α in (10.23) is to be interpreted as meaning rep^1_1 (whose effect is given in equation (12.8)).

The concept of curvature turns up spontaneously in the context of covariant differentiation when one considers the commutator of successive covariant differential operations. For any two vector fields $\vec{\xi}$ and $\vec{\eta}$ over $\mathcal{M}$, the operator product of $\vec{\xi}\cdot D$ and $\vec{\eta}\cdot D$ acting on Φ is given in terms of any chosen gauge α by

$$J_\alpha\circ(\vec{\eta}\cdot D)\circ(\vec{\xi}\cdot D) = \left(\vec{\eta}\cdot\partial + \mathit{rep}\,(\vec{\eta}\cdot\underline{\omega}_\alpha)\right)\circ\left(\vec{\xi}\cdot\partial + \mathit{rep}\,(\vec{\xi}\cdot\underline{\omega}_\alpha)\right)\circ J_\alpha \qquad (10{\cdot}24)$$

so that the commutator satisfies

$$J_\alpha\circ[\vec{\eta}\cdot D, \vec{\xi}\cdot D]\circ J_\alpha^{-1} = [\vec{\eta}\cdot\partial, \vec{\xi}\cdot\partial] + [\mathit{rep}(\vec{\eta}\cdot\underline{\omega}_\alpha), \mathit{rep}(\vec{\xi}\cdot\underline{\omega}_\alpha)]$$

$$+\,\vec{\eta}\cdot\partial\,\mathit{rep}\,(\vec{\xi}\cdot\underline{\omega}_\alpha) - \vec{\xi}\cdot\partial\,\mathit{rep}\,(\vec{\eta}\cdot\underline{\omega}_\alpha) \qquad (10{\cdot}2$$

Now (as a special case of the general Lie commutator formula (6.13)) the first term on the right is expressible in terms of the Lie brackets of the vector fields (as defined by (1.11)) in the form

$$[\vec{\eta}\cdot\partial, \vec{\xi}\cdot\partial] = [\vec{\eta}, \vec{\xi}]\cdot\partial \tag{10.26}$$

By the definition (7.12) af the algebra valued exterior commutator, the second term can be written as

$$[\mathrm{rep}(\vec{\eta}\cdot\underline{\omega}_\alpha), \mathrm{rep}(\vec{\xi}\cdot\underline{\omega}_\alpha)] = \tfrac{1}{2}\,\mathrm{rep}\,(\vec{\eta}\cdot[\underline{\omega}_\alpha\wedge\underline{\omega}_\alpha]\cdot\vec{\xi}) \tag{10.27}$$

Finally for the last term one has

$$\vec{\eta}\cdot\partial\,\mathrm{rep}\,(\vec{\xi}\cdot\underline{\omega}_\alpha) - \vec{\xi}\cdot\partial\,\mathrm{rep}(\vec{\eta}\cdot\underline{\omega}_\alpha) = \mathrm{rep}\,(\underline{\omega}_\alpha\cdot[\vec{\eta},\vec{\xi}] + \vec{\eta}\cdot(\partial\wedge\underline{\omega}_\alpha)\cdot\vec{\xi}) \tag{10.28}$$

Pulling these together, and using (19.13), we can express the result in terms of the representative acting on $\mathcal{V}$ of the contraction with $\vec{\xi}$ and $\vec{\eta}$ of the gauge image $\underline{\Omega}_\alpha$ of the base space curvature 2-form $\underline{\Omega}$: we obtain

$$[\vec{\eta}\cdot D, \vec{\xi} D]\,\Phi_\alpha = [\vec{\eta}, \vec{\xi}]\cdot D\Phi_\alpha + \mathrm{rep}(\vec{\eta}\cdot\underline{\Omega}_\alpha\cdot\vec{\xi})\,\Phi_\alpha \tag{10.29}$$

In cases where (in contrast with equations (10.21), (10.22), (10.23)) one is not concerned with the distinction between one particular gauge and another, we can use the shorthand notation introduced in (9.14), whereby the gauge label α is dropped, to summarise the foregoing results in the easily memorable formulae

$$D = \partial + \mathrm{rep}\,\underline{\omega} \tag{10.30}$$

and

$$[\vec{\eta}\cdot D, \vec{\xi}\cdot D] = [\vec{\eta}, \vec{\xi}]\cdot D + \mathrm{rep}(\vec{\eta}\cdot\underline{\Omega}\cdot\vec{\xi}) \tag{10.31}$$

In applying this last formula one can use the fact that it follows from (19.13) that the representative over $\mathcal{V}$ of the curvature gauge image $\underline{\Omega}_\alpha$ can be evaluated directly in terms of the representative of the gauge connection $\underline{\omega}_\alpha$ using the (shorthand) expression

$$\mathrm{rep}\,\underline{\Omega} = \partial\wedge\mathrm{rep}\,\underline{\omega} + \tfrac{1}{2}\,[\mathrm{rep}\,\underline{\omega}\wedge\mathrm{rep}\,\underline{\omega}] \tag{10.32}$$

(which a physicist working with a single fixed representation, rather than dealing collectively with a whole family of associated representations as we are doing here would abbreviate even further by dropping the rep symbols). By (9.17) the analogous direct expression of the Bianchi identity is

$$\partial\wedge \mathrm{rep}\,\Omega + [\mathrm{rep}\,\omega\wedge \mathrm{rep}\,\Omega] = 0 \qquad (10.26)$$

Since it can easily be seen that the condition for a connection to be integrable (in the sense described in section 5) is that the commutator of covariant differentials with respect to commuting vector fields should be zero, the formula (10.24) shows directly that the associated connection on a vector bundle will be integrable if and only if the relevant representative of Ω is zero, and hence, provided we are concerned with a faithful representation if and only if Ω itself vanishes.

11. Torsion and sub-covariant exterior differentiation.

Most simple field theories are implicitly or explicitly expressed in terms of an associated family of bundles that is constructed in terms of a Lie group $\mathcal{G}$ of the form

$$\mathcal{G} = GL(n)\times\tilde{\mathcal{G}} \qquad (11.1)$$

(where $\tilde{\mathcal{G}}$ could be a subgroup of $GL(n)$ -such as the Lorentz group, as discussed in section 18- but might also be an entirely independent Yang Mills group) in such a way that it includes the ordinary tangent space tensor bundles as special cases in which the action of the subgroup $\tilde{\mathcal{G}}$ is trivial. (Until recently gravitation theoristy tended usually to concentrate on the $GL(n)$ part, ignoring $\tilde{\mathcal{G}}$ except in so far as it is a Lorentzian subgroup or covering space thereof, while other phycisists usually concentrated only on the $\tilde{\mathcal{G}}$ part, ignoring the possibility of non-trivial tangent bundle structure). For any group of the form(11.1) the algebra splits up correspondingly as a direct sum of two mutually commuting subalgebras in the form

$$\mathfrak{a} = \mathfrak{gl}(n)\oplus\tilde{\mathfrak{a}}$$
$$\alpha = \ell + \tilde{\alpha}\ , \qquad [\ell,\tilde{\alpha}] = 0 \qquad (11.2)$$

where $\mathfrak{gl}(n)$ is the Lie algebra of $GL(n)$ and $\tilde{\mathfrak{a}}$ is the Lie algebra of $\tilde{\mathcal{G}}$.

The decomposition (11.2) can be applied in particular to the connection one form $\hat{\underline{\chi}}$ which splits up into a "linear", i.e $\mathfrak{gl}(n)$ part $\hat{\underline{\lambda}}$ say, and an $\tilde{\mathfrak{a}}$ part $\hat{\tilde{\underline{\chi}}}$ in the form

$$\hat{\underline{\chi}} = \hat{\underline{\lambda}} + \hat{\tilde{\underline{\chi}}}\ , \qquad [\hat{\underline{\lambda}},\hat{\tilde{\underline{\chi}}}] = 0 \qquad (11.3)$$

which implies a corresponding decomposition

$$\underset{\sim}{\omega}_\alpha = \underset{\sim}{\Gamma}_\alpha + \underset{\sim}{\tilde{\omega}}_\alpha \quad , \quad [\underset{\sim}{\Gamma}_\alpha , \underset{\sim}{\tilde{\omega}}_\alpha] = 0 \tag{11.4}$$

of the base space gauge connection 1-forms.
This in turn leads to a corresponding decomposition of the base space curvature 2-form into a $gl(n)$ part $\underset{=}{R}$ and an $\tilde{a}$ part $\underset{=}{\tilde{\Omega}}$ satisfying

$$\underset{=}{\Omega} = \underset{=}{R} + \underset{=}{\tilde{\Omega}} \quad , \quad [\underset{=}{R} , \underset{=}{\tilde{\Omega}}] = 0 \tag{11.5}$$

where (using the shorthand notation in which the gauge label is dropped)

$$\underset{=}{R} = \partial \wedge \underset{\sim}{\Gamma} + \underset{\sim}{\Gamma} \wedge \underset{\sim}{\Gamma} \tag{11.6}$$

this being the ordinary geometric (Riemann) part of the curvature, and where

$$\underset{=}{\tilde{\Omega}} = \partial \wedge \underset{\sim}{\tilde{\omega}} + \underset{\sim}{\tilde{\omega}} \wedge \underset{\sim}{\tilde{\omega}} \tag{11.7}$$

this being the $\tilde{\mathcal{G}}$ (e.g Yang Mills) part of the curvature.

It is a fairly standard practise to use the notation ∇ to denote tangentially covariant differentiation as given by

$$\nabla = \partial + \mathrm{rep}\, \underset{\sim}{\Gamma} \tag{11.8}$$

with suppression of an implicit gauge label α . This operator coincides with D when acting on a section of an ordinary tensor bundle, for which its result is also a well behaved bundle section. In general however it will give a result that is gauge dependent.

Taking the action of $GL(n)$ on ordinary contravariant tangent vectors to be the fundamental representation, for which the prefix rep may be dropped, we may express the effect of ∇ on an ordinary tangent vector field $\vec{\xi}$ in terms of a local coordinate system x^μ by

$$D_\nu \xi^\mu = \nabla_\nu \xi^\mu = \partial_\nu \xi^\mu + \Gamma_\nu{}^\mu{}_\rho \xi^\rho \tag{11.9}$$

Hence it can be seen from (1.12) that the Lie bracket of two vector fields $[\vec{\xi},\vec{\eta}]$ can be expressed in terms of covariant derivatives by

$$[\vec{\eta},\vec{\xi}] = \vec{\eta}\cdot D\cdot\vec{\xi} - \vec{\xi}\cdot D\cdot\vec{\eta} - \vec{\eta}\cdot\underset{=}{\vec{\Theta}}\cdot\vec{\xi} \tag{11.10}$$

where $\underset{=}{\vec{\Theta}}$ is a tangent-vector valued 2-form called the torsion with mixed contravariant and antisymmetric covariant coordinates indices, whose components are given by

$$\Theta_{\nu\mu}{}^{\rho} = 2\Gamma_{[\nu}{}^{\rho}{}_{\mu]} \tag{11.11}$$

Since we evidently have

$$[\vec{\eta}\cdot D, \vec{\xi}\cdot D] = \eta^{\mu}\xi^{\nu}(D_{\mu}\circ D_{\nu} - D_{\nu}\circ D_{\mu}) + (\eta^{\mu}D_{\mu}\xi^{\nu} - \xi^{\mu}D_{\mu}\eta^{\nu})D_{\nu} \tag{11.12}$$

it follows that in terms of the definition

$$(D\wedge D)_{\mu\nu} = D_{\mu}\circ D_{\nu} - D_{\nu}\circ D_{\mu} \tag{11.13}$$

the commutator formula (10.30) may be expressed without reference to the particular fields $\vec{\eta}$, $\vec{\xi}$ in the form

$$D\wedge D = rep\,\underline{\Omega} - \underline{\vec{\Theta}}\cdot D \tag{11.14}$$

The torsion may also be used to express the Jacobi identity(10.26) in terms of covariant derivatives. In terms of partial derivatives (10.26) is equivalent to the coordinate component formula

$$\partial_{[\mu}\Omega_{\nu\rho]} + [\omega_{[\mu}, \Omega_{\nu\rho]}] = 0 \tag{11.15}$$

which can be rewritten as

$$D_{[\mu}\Omega_{\nu\rho]} + \Theta_{[\mu\nu}{}^{\sigma}\Omega_{\rho]\sigma} = 0 \tag{11.16}$$

Whenever one is working with generalised a let us say $\mathcal{V}$ -valued, differential p-forms (such as $\underline{\Omega}$) on which $\mathcal{G}$ and $\mathcal{A}$ are represeated by the naturally appropriate action, rep say, over the relevant space $(\Lambda^{p}\mathcal{J})\otimes\mathcal{V}$, it is convenient to introduce a modified representation, $\widetilde{rep}$ say, that is defined to have the same effect as rep on the factor space $\mathcal{V}$ but to reduce to the identity on the (p-times repeated) factor space $\mathcal{J}$ (which we recall denotes the base manifold tangent space). Thus in the particular case where $\mathcal{V}$ is a $\mathcal{G}$ tensor space (i.e when the $GL(n)$ action on $\mathcal{V}$ is the trivial identity action) then for any $a\in\mathcal{A}$ we shall have $\widetilde{rep}\, a = rep\,\tilde{a}$. This modified representation can be used to define a very useful operation that we shall refer to as exterior subcovariant differentiation, given by

$$\tilde{D}\wedge = (\partial + \widetilde{rep}\,\underline{\omega})\wedge \tag{11.17}$$

that acts on generalized differential forms.
Despite the fact that it is defined in terms of an "incorrect" representation, this differentiation procedure is genuinely

covariant in the sense that when acting on a $\mathcal{V}$ -valued p-form $\underline{\Psi}$ say that is a well behaved fibre bundle section, the result $\tilde{D}\wedge\underline{\Psi}$ will also be a well behaved section of a corresponding (higher order bundle). This essential property can be seen directly from the relationship between exterior subcovariant differentiation and ordinary exterior covariant differentiation which may be expressed for any $\mathcal{V}$ -valued p-form $\underline{\Psi}$ in terms of the $\mathcal{T}$ valued torsion 2-form $\underline{\vec{\Theta}}$ by the general formula

$$\tilde{D}\wedge\underline{\Psi} = D\wedge\underline{\Psi} + \wedge(\underline{\vec{\Theta}}\cdot\underline{\Psi}) \tag{11.18}$$

where $\wedge(\underline{\vec{\Theta}}\cdot\underline{\Psi})$ denotes the (p + 1)-form whose components are given by

$$\wedge(\Theta\cdot\Psi)_{\mu_1\mu_2\mu_3\cdots\mu_{p+1}} = \frac{p(p+1)}{2!}\,\Theta_{[\mu_1\mu_2|}{}^{\nu}\,\Psi_{\nu|\mu_3\cdots\mu_{p+1}]} \tag{11.19}$$

Since both terms on the right hand side of (11.18) are genuinely covariant, it follows that exterior subcovariant differentiation is itself genuinely covariant. In view of this satisfactory property and the fact that it is comparatively easy to evaluate, many recent papers work entirely in terms of exterior sub-covariant differentiation, often neglecting to distinguish it from ordinary exterior covariant differentiation, with which however it may strictly be identified only if (as can be seen from (11.18) the torsion is known to be zero.

It can be seen directly from the defining formula (11.17)that the product of two successive exterior subcovariant differentiations is given simply by

$$\tilde{D}\wedge\tilde{D}\wedge = \mathrm{rep}\,\underline{\Omega}\,\wedge \tag{11.20}$$

which is a generalisation of the Poincare property (1.3) of ordinary exterior differentiation . We can also directly obtain the Bianchi identity for both the total curvature and the decomposed parts in the form

$$\tilde{D}\wedge\underline{\Omega} = 0 \tag{11.21}$$

and

$$\tilde{D}\wedge\underline{\mathcal{R}} = 0\ , \qquad \tilde{D}\wedge\underline{\tilde{\Omega}} = 0 \tag{11.22}$$

Provided we continue to treat the tangent space representation of $GL(n)$ as fundamental as in (11.9)then it is consistent with the definition (11.6) to use the notation $\mathcal{R}$ to denote the Riemann tensor as conventionally defined to be the tangent space representative of the curvature, i.e we set

$$(\mathrm{rep}\ \Omega)\,\vec{\xi} = \underline{\mathcal{R}}\,\vec{\xi} \tag{11.23}$$

for any tangent vector $\vec{\xi}$. Then for any three vectors $\vec{\xi}_{(1)}, \vec{\xi}_{(2)}, \vec{\xi}_{(3)}$ we can use (10.30) and (11.10) to obtain

$$\left(\vec{\xi}_{[(1)}\cdot\underline{\mathcal{R}}\cdot\vec{\xi}_{(2)}\right)\vec{\xi}_{(3)} = \left(\vec{\xi}_{[(1)}\cdot D\right)\left(\vec{\xi}_{(2)}\cdot\underline{\vec{\Theta}}\cdot\vec{\xi}_{(3)]}\right) + \vec{\xi}_{[(1)}\cdot\underline{\vec{\Theta}}\cdot\left[\vec{\xi}_{(2)},\vec{\xi}_{(3)]}\right] \tag{11.24}$$

with the aid of the identity (6.13) in the form

$$\left[\left[\vec{\xi}_{[(1)}, \vec{\xi}_{(2)}\right], \vec{\xi}_{(3)]}\right] = 0 \tag{11.25}$$

Working this out using (11.10) again we obtain

$$\left(\vec{\xi}_{[(1)}\cdot\underline{\mathcal{R}}\cdot\vec{\xi}_{(2)}\right)\vec{\xi}_{(3)]} = -\vec{\xi}_{[(1)}\cdot\left(\vec{\xi}_{(2)}\cdot D\ \underline{\vec{\Theta}}\right)\cdot\vec{\xi}_{(3)]} - \vec{\xi}_{[(1)}\cdot\underline{\vec{\Theta}}\cdot\left(\vec{\xi}_{(2)}\cdot\underline{\vec{\Theta}}\cdot\vec{\xi}_{(3)]}\right) \tag{11.26}$$

which is equivalent in terms of base space co-ordinate components to

$$\begin{aligned}\mathcal{R}_{[\mu\nu}{}^{\lambda}{}_{\rho]} &= D_{[\mu}\Theta_{\nu\rho]}{}^{\lambda} + \Theta_{[\mu\nu|}{}^{\tau}\,\Theta_{\tau|\rho]}{}^{\lambda} \\ &= \tilde{D}_{[\mu}\Theta_{\nu\rho]}{}^{\lambda}\end{aligned} \tag{11.27}$$

12. Adjoint, Fundamental, and Associated Lie Algebra Representations.

Up to this stage we have been working in very abstract terms for the sake of maximum generality and formal simplicity. In order to make contact with the more concrete formulations in terms of explicit components with which physicists have traditionally preferred to work, let us now make a rather more detailed examination of the most important kinds of Lie algebra representation.

By definition, each Lie algebra vector a is the generator of a one parameter subgroup of elements

$$S(t) = \mathrm{exp}\,(at) \quad \in \mathcal{G} \tag{12.1}$$

where t is the parameter, which is specified so as to satisfy

$$S(t_1)\,S(t_2) = S(t_1+t_2) \tag{12.2}$$

this additivity property justifying the notation exp by analogy with the corresponding property of the ordinary exponential function. It follows (c.f. equations (6.6) and (6.16)) that for any linear representation of $\mathcal{G}$ acting on a vector space $\mathcal{V}$ the corresponding representation of the algebra a is specified by

$$\mathrm{rep}\,a\ v = \frac{d}{dt}\left(\mathrm{rep}\,S(t)\ v\right)\Big|_{t=0} \tag{12.3}$$

Hence if v has components v^a in some given base, the corresponding components of $rep\, a$ will be given in terms of those of $rep\, S(t)$ by

$$(rep\, a)^a{}_b = \frac{d}{dt}(rep\, S(t))^a{}_b\Big|_{t=0} \tag{12.4}$$

It is well known that just as in the case of ordinary numbers such a relation can be inverted in terms of the familiar (and always convergent) expansion

$$(rep\, S(t))^a{}_b = \delta^a{}_b + (rep\, a)^a{}_b + \frac{t^2}{2!}(rep\, a)^a{}_c(rep\, a)^c{}_b + \cdots \tag{12.5}$$

As for the group itself so also for the algebra we shall drop the prefix rep in the case of a fundamental representation. Thus if u^m are the components of a vector u in the fundamental representation space $\mathcal{U}$ we have

$$rep^1 : \quad u^m \mapsto a^m{}_n u^n \tag{12.6}$$

By comparison with (2.13) we see that if π_m are components of a convector π in the dual space $\mathcal{U}_*$ we have the corresponding associated reprentation

$$rep_1 : \quad \pi_m \mapsto -\pi_n a^n{}_m \tag{12.7}$$

We can proceed in the obvious way to build up representations over tensor product spaces. Thus for example if $T^m{}_n$ are components of a mixed tensor T in the tensor product space $\mathcal{U} \times \mathcal{U}_*$ we have

$$rep^1_1 : \quad T^m{}_n \mapsto a^m{}_\ell T^\ell{}_n - T^m{}_\ell a^\ell{}_n \tag{12.8}$$

An important example of a (non fundamental) linear representation that exists for any Lie group is the adjoint algebra representation which we have chosen to denote by the symbol @ . Let us introduce a set of basis vectors $\lambda_K \in \mathcal{A}$ for the Lie algebra, so that an arbitrary vector $a \in \mathcal{A}$ may be expressed in the form

$$a = a^K \lambda_K \tag{12.9}$$

where the a^K are scalar components. Let us introduce the use of a dagger prefix as an abbreviation to denote the adjoint map representative of a Lie algebra vector, i.e let us write

$$@ : \quad a \mapsto @a = {}^\dagger a \tag{12.10}$$

so that the adjoint action of a Lie algebra vector a on another Lie algebra vector b is expressed as

$$^{\dagger}a\, b = [a, b] = [a, b]^{K} \lambda_{K} \tag{12·11}$$

It follows that in terms of components we have

$$@a : b^{K} \longrightarrow {}^{\dagger}a^{K}{}_{L}\, b^{L} \tag{12·12}$$

the component matrix ${}^{\dagger}a^{K}{}_{L}$ of the adjoint action being given explicitly by

$$^{\dagger}a^{K}{}_{L} = @_{ML}{}^{K} a^{M} \tag{12·13}$$

where $@_{ML}{}^{K}$ are the <u>structure constants</u> associated with the basis λ_{K} as defined by

$$[\lambda_{K}, \lambda_{L}] = @_{KL}{}^{M} \lambda_{M} \tag{12·14}$$

Evidently the structure constants may also be specified directly in terms of the component matrices of the adjoint action of the basis vectors by

$$@_{KL}{}^{M} = ({}^{\dagger}\lambda_{K}\, \lambda_{L})^{M} = {}^{\dagger}\lambda_{K}{}^{M}{}_{L} \tag{12·15}$$

In addition to the obvious antisymmetry property

$$@_{KL}{}^{M} = @_{[KL]}{}^{M} \tag{12·16}$$

it follows from the Jacobi identity that the structure constants satisfy

$$@_{[JK}{}^{M} @_{L]M}{}^{N} = 0 \tag{12·17}$$

The structure constants may be used to define a covariant tensor

$$\gamma_{KL} = \tfrac{1}{2} @_{KN}{}^{M} @_{ML}{}^{N} \tag{12·18}$$

on the algebra, which obviously satisfies the symmetry condition

$$\gamma_{KL} = \gamma_{(KL)} \tag{12·19}$$

If the component matrix γ_{KL} is non-singular -in which case the group is said to be semi-simple - this tensor γ may be used as a pseudo-metric for raising and lowering Lie algebra indices, and even if it is singular it can still be used for index lowering. Thus for example we can always define a fully covariant structure tensor given by

$$@_{JKL} \equiv \gamma_{LN} @_{JK}{}^{N}$$

$$= \tfrac{1}{2} \left(@_{JQ}{}^{N} @_{KN}{}^{M} @_{LM}{}^{Q} - @_{KQ}{}^{N} @_{JN}{}^{M} @_{LMQ} \right) \qquad (12.20)$$

It is evident from the form of this last expression, which is obtained by substituting (12.18) and using (12.17)) that this tensor is unaffected by cyclic interchange of indices and hence is totally antisymmetric, i.e

$$@_{JKL} = @_{[JKL]} \qquad (12.21)$$

It follows that the covariant tensors obtained by lowering the indices of the adjoint algebra matrices are also antisymmetric, i.e

$$^{+}a_{KL} = {}^{+}a_{[KL]} \qquad (12.22)$$

It is to be noted that the Lie co-algebra vector obtained simply by lowering the index of an ordinary Lie algebra vector a may be expressed directly in terms of the corresponding adjoint action operator by the formula

$$a_K = -\tfrac{1}{2} @_{KL}{}^{M}\, {}^{+}a_M{}^{L} \qquad (12.23)$$

as may be seen directly from (12.18). <u>If</u> the group under consideration were semi-simple so that we had an inverse pseudo metric with components γ^{KL} that could be used for index raising then the formula (12.23) could be used directly to solve (12.13) for a^K as in terms of $a^K{}_L$. However in non semi-simple cases the mapping $a \mapsto {}^{+}a$ can not be inverted.

The structure constants are so described in recognition of the fact that they are <u>invariant</u> under adjoint algebra transformations as a consequence of the Bianchi identities. Explicitly one has (c.f (12.6)(12.7)(12.8))

$$(\mathrm{rep}\, a)\, @_{KL}{}^{M} \equiv {}^{+}a^M{}_N @_{KL}{}^{N} - {}^{+}a^N{}_K @_{NL}{}^{M} - {}^{+}a^N{}_L @_{KN}{}^{M}$$

$$= 3\, a^J @_{[KJ}{}^{N} @_{L]N}{}^{M} = 0 \qquad (12.24)$$

using (12.13) and (12.12). It follows that compound tensors built purely from the structure constants must also be invariant. This applies in particular to the pseudo-metric, which therefore has the invariance property

$$(\mathrm{rep}\, a)\, \gamma_{KL} \equiv -{}^{+}a^N{}_K \gamma_{NL} - a^N{}_L \gamma_{KN} = 0 \qquad (12.25)$$

as could be seen directly by combining (12.19) and (12.22). This last result can be expressed more briefly in index free notation as,

$$(\mathrm{rep}\, a)\, \gamma = 0 \qquad (12.26)$$

while in view of (12.15) the preceeding invariance condition (12.24) can be expressed as

$$(\mathrm{rep}\ a)\ {}^{\dagger}\lambda = 0 \tag{12.27}$$

treating ${}^{\dagger}\lambda$ as an algebra automorphism valued algebra covector. An analogous result holds for λ itself when considered as an algebra covector whose values are automorphisms of the fundamental representation space, with components $\lambda_K{}^m{}_n$ so that the fundamental representative of a general Lie algebra vector a has matrix components given by

$$a^m{}_n = a^K \lambda_K{}^m{}_n \tag{12.28}$$

To demonstrate this invariance we start by evaluating the fundamental matrix components of the image under the adjoint action of a on another Lie algebra vector b which gives

$$\begin{aligned}(\dagger a)\, b^m{}_n = (\mathrm{rep}\ a)\, b^m{}_n &= [a, b]^m{}_n \\ &= a^m{}_l b^l{}_n - a^l{}_n b^m{}_l \\ &= (a^m{}_l \lambda_K{}^l{}_n - a^l{}_n \lambda_K{}^m{}_l)\, b^K \end{aligned} \tag{12.29}$$

However we also have

$$(\dagger a)\, b^m{}_n = {}^{\dagger}a^L{}_K\, b^K \lambda_L{}^m{}_n \tag{12.30}$$

By comparing these expressions and noting that they are valid for arbitrary b^K we deduce that the algebra basis components $\lambda_K{}^m{}_n$ must satisfy

$$(\mathrm{rep}_{1}^{1}\, a)\, \lambda_K{}^m{}_n + ({}^{\dagger}a)\, \lambda_K{}^m{}_n = 0 \tag{12.31}$$

which written out in full gives

$$a^m{}_l \lambda_K{}^l{}_n - a^l{}_n \lambda_K{}^m{}_l - {}^{\dagger}a^L{}_K \lambda_L{}^m{}_n = 0 \tag{12.32}$$

On the other hand, in index free notation this reduces to the compact form

$$(\mathrm{rep}\ a)\, \lambda = 0 \tag{12.33}$$

which expresses the required invariance property of λ as a mixed fundamental tensor valued Lie algebra covector.

13. Metrics and Sympletics.

In a very wide range of circumstances one has to deal with a fundamental representation space $\mathcal{U}$ that is endowed with a group invariant <u>parenthesis</u> i.e. a bilinear product mapping the direct product $\mathcal{U} \times \mathcal{U}$ onto either the real numbers $\mathbb{R}$ or, in the case of complex vector spaces, onto the complex numbers $\mathbb{C}$. The product of any two vectors v and u in $\mathcal{U}$ can be expressed as

$$(v, u) = v \cdot q \cdot u = v^m q_{mn} u^n \tag{13.1}$$

where q with components q_{mn} is a covariant tensor in $\mathcal{U}_* \otimes \mathcal{U}_*$ which we shall refer to as the <u>parenthetic</u>. Whenever a preferred parenthetic is given it is customary to use it to define an index lowering operation

$$q : \mathcal{U} \longrightarrow \mathcal{U}_* \tag{13.2}$$

given by

$$v^m \longmapsto v_m = v^n q_{nm} \tag{13.3}$$

or in index free notation

$$v \longmapsto \underline{v} = v \cdot q \tag{13.4}$$

so that the parenthesis is expressible by

$$(v, u) = \underline{v} \cdot u \tag{13.5}$$

(Our use of the algebra tensor γ in the previous section was an example of this procedure). Provided that the parenthetic is non-singular we can define the <u>transposed</u> <u>inverse</u> tensor $(q^{-1})^T$, or in more compact notation q^{-T}, with components q^{mn}, by the condition

$$q^{\ell n} q_{\ell m} = q_m{}^n = \delta^n_m \tag{13.6}$$

(where δ^n_m is the Kronecker delta) which can be used to define a corresponding index raising operation

$$q^{-1} : \mathcal{U}_* \longrightarrow \mathcal{U} \tag{13.7}$$

whereby a covector $\pi \in \mathcal{U}_*$ with components π_m transforms according to

$$\begin{aligned} \pi &\longmapsto \vec{\pi} = q^{-T} \cdot \pi \\ \pi_m &\longmapsto \pi^m = q^{mn} \pi_n \end{aligned} \tag{13.8}$$

These rules are such as to ensure that we have

$$\underline{(\vec{v})} = v \qquad \underline{(\vec{\pi})} = \pi \tag{13.9}$$

and

$$\underline{(q^{-T})} = q \qquad \vec{q} = q^{-T} \tag{13.10}$$

as required for notational consistency. The condition that the parenthesis be group invariant is expressible by

$$S^k{}_m S^\ell{}_n q_{k\ell} = q_{mn} \tag{13.11}$$

where $S^m{}_n$ is the matrix representative on $\mathcal{V}$ of any group element $S \in \mathcal{G}$, i.e. by

$$S_{\ell m} S^{\ell n} = \delta^n_m \tag{13.12}$$

In terms of the algebra the analogous invariance condition is

$$a^\ell{}_m q_{\ell n} + a^\ell{}_n q_{m\ell} = 0 \tag{13.13}$$

In nearly all cases of interest the parenthetic is required to satisfy the symmetry or antisymmetry condition

$$q_{mn} = \pm q_{nm} \tag{13.14}$$

which is equivalent to

$$q^m{}_n = \pm \delta^m_n \tag{13.15}$$

In the former (symmetric) case, for which q is describable as a (pseudo-)metric the maximal group leaving it invariant is one of the orthogonal groups $O(n-p, p)$ (or $O(n)\mathbb{C}$ if we are dealing with a complex vector space) where n is the dimension of $\mathcal{U}$ and $n-2p$ is the signature of the metric (which is an invariant in the real case). In the antisymmetric case q may be described as a symplectic, the maximal group leaving it invariant being the symplectic group $Sp(n)$ (or $Sp(n)\mathbb{C}$ in the complex case). The symplectic possibility occurs only if n is even since an antisymmetric matrix necessarily has even rank and hence can be non-singular only in an even number of dimension. In the symplectic case there is a natural measure (i.e alternating) tensor ε, given by the exterior product $\varepsilon = q \wedge q \wedge \cdots \wedge q$, whose obvious invariance automatically entails the condition that the symplectic transformations matrices $S^m{}_n$ must have unit determinant. However in the pseudo-metric case a natural measure tensor ε can be determined only to within an arbitrary choice of sign so that the orthogonal transformation matrices may have determinant equal to plus or minus unity, which means that the unit determinant matrices form the special subgroup $SO(n-p,p)$ (or $SO(n)\mathbb{C}$ in the complex case).

It is to be remarked that if the downstairs adjoint operation is defined by (1.17) then the upstairs (inverse) adjoint operation on a p-form $\underset{\sim}{\omega}$, as defined by (1.19) will be expressible explicitly directly in terms of the raised (i.e contravariant) counter part $\vec{\varepsilon}$ of the alternating n-form ε by

$$*\omega = \epsilon\{q\}\, \vec{\varepsilon} \,|\, \underset{\sim}{\omega} \tag{13.16}$$

with

$$\epsilon\{q\} = \frac{|q|}{\|q\|} \tag{13.17}$$

where $|q|$ denotes the determinant of the component matrix q_{mn} and $\|q\|$ denotes the modulus thereof, which implies that $\epsilon\{q\}$ is an invariant with value always equal to plus or minus unity, being always positive in the sympletic and positive definite metric cases, but negative for a four dimensional Lorentzian pseudometric. This relation can be derived directly from the useful general formula

$$\varepsilon^{k_1\dots k_p \ell_1\dots\ell_{n-p}}\varepsilon_{m_1\dots m_p \ell_1\dots\ell_{n-p}} = \epsilon\{q\}\, p!\,(n-p)!\; \delta^{k_1}_{[m_1}\cdots\delta^{k_p}_{m_p]} \tag{13.18}$$

which holds for any partial contraction of the raised and lowered versions of the alternating tensor.

The main application that we have in mind in the present section is to the ordinary special Lorentz group $SO(3,1)$ and two other groups sharing the same algebra namely the complex three dimensional rotation group $SO(3)\mathbb{C}$, and the complex two dimensional symplectic group $Sp(2)\mathbb{C}$ which is more widely known as the special linear group $SL(2)\mathbb{C}$ consisting of complex 2-dimensional matrices with unit determinant. In two dimensions the speciality condition (i.e the preservation of an antisymmetric measure tensor) is precisely equivalent to the symplecticity condition (i.e preservation of an antisymmetric parenthetic) so that we have

$$SL(2)\mathbb{C} \equiv Sp(2)\mathbb{C} \tag{13.19}$$

(and similarly for the real case with which however we shall not be concerned). The three groups named above are related by

$$Sp(2)\mathbb{C}/Z_2 \equiv SO(3)\mathbb{C} \equiv SO(3,1)/Z_2 \tag{13.20}$$

where the quotient operations are performed by treating each representation matrix as equivalent to its negative. The special Lorentz group $SO(3,1)$ consists of two disconnected components, the identity (i.e proper orthochronous) component being identical to $SO(3)\mathbb{C}$, so that we may express the group as a whole as the direct product

$$SO(3,1) \equiv SO(3)\mathbb{C} \times Z_2 \tag{13.21}$$

However since $Sp(2)\mathbb{C}$ — like $SO(3)\mathbb{C}$ — is connected (indeed — unlike $SO(3)\mathbb{C}$ — it is actually simply connected) the quotient operation on the left of (13.20) cannot be analogously multiplied out.

It is evident quite generally that whenever the parenthetic satisfies the symmetry or antisymmetry condition (13.14), the invariance condition (13.13) on the algebra reduces respectively to the antisymmetry or symmetry condition

$$a_{m\ell} \pm a_{\ell m} = 0 \tag{13.22}$$

i.e the generators of the appropriate invariance group consist of all linear transformations whose fully covariant components satisfy (13.22). This means that a basis for the Lie algebra can be considered as being generated by any set of independent linear operators λ_K, $K = 1, \cdots, \frac{1}{2}n(n\mp 1)$ whose matrix components satisfy

$$\lambda_K{}^{\ell}{}_m = \lambda_{Km}{}^{\ell} \tag{13.23}$$

and in terms of which we can construct scalar quantities $@_{KL}{}^M$ according to the specification

$$[\lambda_K, \lambda_L]^{\ell}{}_m = @_{KL}{}^M \lambda_M{}^{\ell}{}_m \tag{13.24}$$

The elements λ_K and the corresponding quantities $@_{KL}{}^M$ can strictly speaking be identified with a basis of the Lie algebra and with the corresponding structure constants only in the case of real orthogonal or symplectic groups, but they may nevertheless be treated entirely analogously to the true basis vectors (of which there would actually be twice as many) and the true structure constants even in the complex case. Thus all the equations of the preceeding section 12 will still be valid when appropriately reinterpreted in the context of complex vector spaces, and in particular the quantities $@_{KL}{}^M$ will still satisfy the Jacobi identity (12.17) and can still be used to define a group invariant pseudo-metric tensor γ_{KL} on the Lie algebra (which by construction is symmetric in both the orthogonal and symplectic cases) in accordance with (12.18). Moreover, provided that $n \geq 3$ in the orthogonal case, or that $n \geq 2$ in the symplectic case, the non-degeneracy of the parenthetic q_{mn} entails the non-degeneracy of γ_{KL} which can therefore be used for raising as well as lowering of Lie algebra indices.

Much of our following work will be based on the useful lemma, which may be verified by working in any conveniently chosen canonical base (I known of no more elegant method) to the effect that the algebra metric as defined by (12.18) may be obtained directly in terms of the traces of the products of the basis matrices according to the formula

$$\gamma_{KL} = -\tfrac{1}{2}(n \mp 2)\, tr(\lambda_K \cdot \lambda_L) \tag{13.25}$$

where, as throughout this section, the upper sign (negative in this example) refers to the orthogonal case and the lower sign to the symplectic case, and where the dot denotes contraction of adjacent indices which here implies ordinary matrix multiplication. Using γ_{KL} as a pseudo-metric for raising and lowering indices we may rewrite (13.25) as

$$\tfrac{1}{2}(\pm n - 2)\, \lambda_K{}^{\ell m} \lambda^L{}_{\ell m} = \delta^L_M \tag{13.26}$$

which shows that the λ^L form a basis of the coalgebra that, to within a dimension dependent normalisation factor, is dual to the original basis λ_K . It follows that we may invert the relationship to obtain

$$(\pm n - 2)\, \lambda_K{}^{k\ell} \lambda^K{}_{mn} = \delta^k_m \delta^\ell_n \mp \delta^k_n \delta^\ell_m \tag{13.27}$$

whose contraction has the form

$$(n \mp 2)\, \lambda_K \cdot \lambda^K = (1 \mp n)\, \mathbb{1} \tag{13.28}$$

The relation (13.26) may be used for example to solve the equation (12.9) for the vectorial components a^K of a in terms of the matrix components $a^m{}_n$ the result being

$$a^K = -\tfrac{1}{2}(n \mp 2)\, \lambda^K \tag{13.29}$$

By writing (12.31) in the form

$$\dagger a^L{}_K \lambda_{L\ell m} = a_{\ell m} \lambda_K{}^m{}_n \mp a_{nm} \lambda_K{}^m{}_\ell \tag{13.30}$$

we can obtain an analogous solution for the adjoint components in terms of the fundamental components in the form

$$\dagger a^L{}_K = (n \mp 2)\, \lambda_K{}^\ell{}_m \lambda^{L\,m}{}_k\, a^k{}_\ell \tag{13.31}$$

by again using (13.26). Alternatively with the aid of (13.27) we can use (13.30) to obtain the fundamental components in terms of the adjoint components in the form

$$a^{k}{}_{\ell} = -\lambda_{K}{}^{k}{}_{m}\lambda^{L\,m}{}_{\ell}\,{}^{+}a^{K}{}_{L} \qquad (13\cdot 32)$$

Since (13.25) implies

$$(n \mp 2)\ \mathrm{tr}\left(\lambda_{K}\cdot\lambda_{L}\cdot\lambda^{M}\right) = -@_{KL}{}^{M} \qquad (13\cdot 33)$$

combining (13.29) and (13.32) gives

$$a^{K} = -\tfrac{1}{2}\,@^{KLM}\,{}^{+}a_{LM} \qquad (13\cdot 34)$$

(This last relation is of course obtainable directly from (12.23) whenever γ is non-singular, and we can see that (13.32) also is valid in the general semi-simple case, not just for orthogonal and simplectic groups).

4. Pseudo-Riemannian Structures: the Decomposition of the Curvature tensor.

The most ancient application of the machinery developped in the preceeding sections is of course to the case of a (pseudo-) Riemannian structure (which in a less traditional but more systematic nomenclature system would be described as an almost (pseudo-) metric structure (where "almost" is to be understood as an abbreviation for "not necessarily integrable") on an n-dimensional manifold $\mathcal{M}$. Such a structure may be defined most directly in terms of the natural tangent or cotangent bundles and their associated principle frame bundle over the base manifold $\mathcal{M}$ as consiting of a (pseudo-)metric tensor field g , with components $g_{\mu\nu}$ in local co-ordinates, (i.e a non-degenerate section of the associated bundle of symmetric covariant tensors) together with a connection such that the covariant derivative of g is zero, i.e

$$Dg \equiv \nabla g = 0 \qquad (14\cdot 1)$$

or in terms of local coordinates

$$D_{\rho}\,g_{\mu\nu} \equiv \nabla_{\rho}\,g_{\mu\nu} = 0 \qquad (14\cdot 2)$$

where we introduce the standard convention of using a semi-colon suffix, $;\rho$, as an abbreviation to denote the effect of the operator ∇_{ρ} . We recall the definition (11.23) of the Riemann tensor $\mathcal{R}$ as the tangent space representative of the curvature. This means that the 2-form valued components obtained from the action of the curvature on a tangent vector field $\vec{\xi}$, with coordinate component ξ^{ρ} , is given by

$$(\text{rep}\,\underline{\Omega})\,\xi^\rho = (\underline{\mathcal{R}}\,\vec{\xi})^\rho = \underline{\mathcal{R}}^\rho{}_\sigma\,\xi^\sigma \tag{14.3}$$

By (14.1) and (11.14) we shall have

$$(\nabla\wedge\nabla + \vec{\underline{\Theta}}\cdot\nabla)\,g \equiv (\text{rep}\,\underline{\Omega})\,g = 0 \tag{14.4}$$

i.e in terms of components

$$(\text{rep}\,\underline{\Omega})\,g_{\rho\sigma} \equiv -\underline{\mathcal{R}}^\tau{}_\rho g_{\tau\sigma} - \underline{\mathcal{R}}^\tau{}_\sigma g_{\rho\tau} \equiv -2\,\underline{\mathcal{R}}_{(\rho\sigma)} = 0 \tag{14.5}$$

This means (c.f. equation (13.13)) that $\underline{\mathcal{R}}$ is a generator of the orthogonal tangent space transformation group leaving g invariant. Thus, applying the formalism of the preceeding section, we may write

$$\underline{\mathcal{R}} = \underline{\mathcal{R}}^K \lambda_K \tag{14.6}$$

where the components $\underline{\mathcal{R}}^K$ are two forms, whose co-ordinate components $\mathcal{R}_{\mu\nu}{}^K$ therefore satisfy

$$\mathcal{R}_{(\mu\nu)}{}^K = 0 \tag{14.7}$$

and where the λ_K are a basis of generators of the orthogonal transformation group leaving g invariant, whose co-ordinate components $\lambda_K{}^\rho{}_\sigma$ are such that when lowered using the pseudo-metric they also satisfy the antisymmetry condition

$$\lambda_{K(\rho\sigma)} = 0 \tag{14.8}$$

Now since the component 2-forms $\underline{\mathcal{R}}^K$ obey the same antisymmetry condition it follows that they too can be expanded in terms of the λ_K in the form

$$\mathcal{R}_{\mu\nu}{}^L = \mathcal{R}^{KL}\lambda_{L\mu\nu} \tag{14.9}$$

Thus in terms of the complete set of local co-ordinate indices the Riemann tensor components are given by

$$\mathcal{R}_{\mu\nu}{}^\rho{}_\sigma = \mathcal{R}_{\mu\nu}{}^L\lambda_L{}^\rho{}_\sigma = \mathcal{R}^{KL}\lambda_{K\mu\nu}\lambda_L{}^{\rho\sigma} \tag{14.10}$$

Conversely with the aid of (13.26) we may obtain the algebra components in terms of the co-ordinate components using the relation

$$\mathcal{R}_{KL} = \tfrac{1}{4}(n-2)^2\,\mathcal{R}_{\mu\nu}{}^\rho{}_\sigma\,\lambda_K{}^{\nu\mu}\lambda_L{}^\sigma{}_\rho \tag{14.11}$$

Now it follows from (11.27) that when the pseudo-Riemannian symmetry condition (14.5) is satified, which is equivalent to

$$\mathcal{R}_{\mu\nu\rho\sigma} = \mathcal{R}_{\mu\nu[\rho\sigma]} \tag{14.12}$$

then we shall have

$$\mathcal{R}_{\mu\nu\rho\sigma} - \mathcal{R}_{\rho\sigma\mu\nu} = (\tilde{D}\wedge\vec{\underline{\Theta}})_{\mu\nu[\sigma\rho]} - (\tilde{D}\wedge\vec{\underline{\Theta}})_{\sigma\rho[\mu\nu]} \qquad (14\cdot13)$$

which implies the relation

$$\mathcal{R}_{[KL]} + (\tilde{D}\wedge\vec{\underline{\Theta}})_{[KL]} = 0 \qquad (14\cdot14)$$

on the algebra components, where

$$(\tilde{D}\wedge\vec{\underline{\Theta}})_{KL} = \frac{1}{4}(n-2)^2(\tilde{D}\wedge\vec{\underline{\Theta}})_{\mu\nu\sigma}{}^{\rho}\,\lambda_K{}^{\nu\mu}\,\lambda_L{}^{\sigma}{}_{\rho} \qquad (14\cdot15)$$

In the torsion free case, i.e when

$$\vec{\underline{\Theta}} = 0 \qquad (14\cdot16)$$

we shall simply have

$$\mathcal{R}_{\mu\nu\rho\sigma} = \mathcal{R}_{\rho\sigma\mu\nu} \qquad (14\cdot17)$$

which is equivalent to the symmetry condition

$$\mathcal{R}_{[KL]} = 0 \quad . \qquad (14\cdot18)$$

Under these circumstances the Ricci tensor as defined by

$$R_{\mu\nu} = \mathcal{R}_{\rho\mu}{}^{\rho}{}_{\nu} \qquad (14\cdot19)$$

will satisfy the corresponding symmetry condition

$$R_{[\mu\nu]} = 0 \quad . \qquad (14\cdot20)$$

Let us introduce the notation

$$\breve{R}_{\mu\nu} = R_{\mu\nu} - \frac{1}{n}\Lambda g_{\mu\nu} \qquad (14\cdot21)$$

for the <u>trace-free</u> part of the Ricci tensor,where

$$\Lambda = R_{\mu}{}^{\mu} = \mathcal{R}_{\rho\mu}{}^{\rho\mu} \qquad (14\cdot22)$$

is the Ricci scalar, and let us make the definitions

$$E_{\mu\nu}{}^{\rho\sigma} = \frac{4}{n-2}\, g^{[\rho}{}_{[\mu}\breve{R}^{\sigma]}{}_{\nu]} \qquad (14\cdot23)$$

$$\gamma_{\mu\nu}{}^{\rho\sigma} = \frac{2}{n-2}\, g^{[\rho}{}_{[\mu}\, g^{\sigma]}{}_{\nu]} \qquad (14\cdot24)$$

The latter notation is motivated by the fact that by (14.24)

$$\gamma_{\mu\nu\rho\sigma} = \gamma_{KL} \lambda^K{}_{\mu\nu} \lambda^L{}_{\rho\sigma} \tag{14.25}$$

In an analogous manner we may write

$$E_{\mu\nu\rho\sigma} = E_{KL} \lambda^K{}_{\mu\nu} \lambda^L{}_{\rho\sigma} \tag{14.26}$$

where by (13.28)

$$E_{KL} = -(n-2)\, \breve{R}^{\mu}{}_{\nu} \lambda_K{}^{\nu}{}_{\rho} \lambda_L{}^{\rho}{}_{\mu} \tag{14.27}$$

The tensors so introduced obviously share with the torsion free Riemann tensor the property of vanishing under triple antisymmetrisation, i.e

$$E_{\mu[\nu\rho\sigma]} = 0 = \gamma_{\mu[\nu\rho\sigma]} \tag{14.28}$$

The tensor E is so contrived that when one makes the decomposition

$$\mathcal{R}_{\mu\nu\rho\sigma} = C_{\mu\nu\rho\sigma} + E_{\mu\nu\rho\sigma} + \frac{n-2}{n(n-1)} \gamma_{\mu\nu\rho\sigma} \tag{14.29}$$

the Weyl tensor,

$$C_{\mu\nu\rho\sigma} = C_{KL} \lambda^K{}_{\mu\nu} \lambda^L{}_{\rho\sigma} \tag{14.30}$$

that is left over as a residue not only inherits from the Riemann tensor the property analogous to (14.28), i.e

$$C_{\mu[\nu\rho\sigma]} = 0 \tag{14.31}$$

but in addition has the property of being entirely trace free, i.e.

$$C_{\rho\mu}{}^{\rho}{}_{\nu} = 0 \tag{14.32}$$

(from which the vanishing of its other traces is evident by the manifest symmetry properties). In terms of the algebra components the above decomposition takes the form

$$\mathcal{R}_{KL} = C_{KL} + E_{KL} + \frac{n-2}{n(n-1)} \Lambda \gamma_{KL} \tag{14.33}$$

wherin the components are all symmetric with traces given by

$$C_K{}^K = 0 = E_K{}^K , \qquad \mathcal{R}_K{}^K = \frac{n-2}{2} \Lambda \tag{14.34}$$

15. The 4-dimensional Case.

Let us now specialise the considerations of the previous section to the physically most important four dimensional case. Thus it is to be taken that

$$n = 4 \tag{15.1}$$

throughout the present section.
In this case the alternating measure tensor determined (to within an ambiguity of sign corresponding to a choice of orientation) by the (pseudo-)metric is of 4th order and hence can be expressed in the form

$$\varepsilon_{\mu\nu\rho\sigma} = \varepsilon_{KL}\,\lambda^{K}{}_{\mu\nu}\,\lambda^{L}{}_{\rho\sigma} \tag{15.2}$$

with

$$\varepsilon_{KL} = \varepsilon_{\mu\nu\rho\sigma}\,\lambda_{K}{}^{\mu\nu}\,\lambda_{L}{}^{\rho\sigma} \tag{15.3}$$

which obviously entails the symmetry (not antisymmetry) property

$$\varepsilon_{KL} = \varepsilon_{(KL)} \tag{15.4}$$

It is to be recalled that for any two form $\underline{a}$ with components $a_{\rho\sigma}$ the adjoint is defined by

$$*a^{\mu\nu} = \frac{1}{2}\,\varepsilon^{\mu\nu\rho\sigma}\,a_{\rho\sigma} \tag{15.5}$$

Since a two form $\underline{a}$ can always be expressed in terms of algebra components a^{K} in the form

$$a_{\rho\sigma} = a^{K}\,\lambda_{K\rho\sigma} \tag{15.6}$$

while similarly

$$*a^{\mu\nu} = *a^{K}\,\lambda_{K}{}^{\mu\nu} \tag{15.7}$$

we see we that we can use the algebra component form ε_{KL} of the alternating tensor to express the adjoint operation compactly in the form

$$*a^{K} = \frac{1}{2}\,\varepsilon^{KL}\,a_{L} \tag{15.8}$$

It can be seen that the identity (13.18) with p = 2 entails a corresponding identity

$$\varepsilon^{KL}\,\varepsilon_{LM} = 4\,\epsilon\{g\}\,\delta^{K}_{M} \tag{15.9}$$

The identity (13.18) with p = 4 can be used to show that the double (left and right) dual of the Riemann tensor satisfies

$$\epsilon\{g\}\,{}^{*}\mathcal{R}^{*}{}_{KL} = \mathcal{R}_{KL} - 2\,E_{KL} \tag{15.10}$$

It is convenient when working in four dimensions to introduce a (possibly complex) tensor

$$\eta^K{}_L = \epsilon\{g\}\, \varepsilon^K{}_L \tag{15.11}$$

so that $\eta^K{}_L$ is real and coincides with $\varepsilon^K{}_L$ when the signature is ++++ or ++-- but it is purely imaginary in the Lorentzian case +++-) and to define an involutive adjoint mapping

$$a \mapsto {\star a} \ , \qquad {\star a}^K = \tfrac{1}{2}\eta^K{}_L a^L \tag{15.12}$$

that maps orthogonal group generators (and hence also the corresponding covariant 2-forms or contravariant bivectors) onto —possibly complex—orthogonal group generators (or 2-forms or bivectors as the case may be) and that satisfies the convenient involutive property that

$$\star\star a \equiv a \tag{15.13}$$

In terms of this modified adjoint we can express (15.10) in the equivalent form

$$E_{KL} = \tfrac{1}{2}\left(\mathcal{R}_{KL} - {\star\mathcal{R}\star}_{KL}\right) \tag{15.14}$$

which implies by that the Weyl tensor is given by

$$C_{KL} = \tfrac{1}{2}\left(\mathcal{R}_{KL} + {\star\mathcal{R}\star}_{KL}\right) - \tfrac{1}{6}\Lambda\gamma_{KL} \tag{15.15}$$

with (in this 4-dimensional case)

$$\Lambda = \mathcal{R}_K{}^K \tag{15.16}$$

The representation of the Riemann and Weyl tensors as second order algebra tensors rather than in the more traditional way as fourth order tangent space tensors was exploited in the four dimensional case by Petrov when he originally introduced his now well known algebraic classification scheme. However it is possible in the four dimensional case (though not in general) to use an even more efficient representation scheme (pioneered most notably by Debever and his co-workers) that has acquired a particular interest in recent years because of its relevance to the theory of instantons. This scheme is based on the fact that with any orthogonal group generator a we can associate a pair of (possibly complex) generators

$$\left.\begin{aligned} {}^{+}a &= \tfrac{1}{2}\left(a + \star a\right) \\ {}^{-}a &= \tfrac{1}{2}\left(a - \star a\right) \end{aligned}\right\} \tag{15.17}$$

which belong respectively to the subspaces of self-adjoint and anti-self adjoint generators in the sense that

$$\left.\begin{aligned} * {}^{+}a &= {}^{+}a \\ * {}^{-}a &= -{}^{-}a \end{aligned}\right\} \qquad (15\cdot18)$$

these subspaces and the corresponding + and - projections being obviously invariant under the action of the relevant (special) orthogonal group. This invariance property is manifested in terms of the algebra by the fact that since for any two elements we have

$$[a, *b] = *[a, b] \qquad (15\cdot19)$$

it follows that the adjoint action of any element a on a self adjoint element ${}^{+}b$ is given by

$$[a, {}^{+}b] = {}^{+}[a, b] \qquad (15\cdot20)$$

and similarly

$$[a, {}^{-}b] = {}^{-}[a, b] \qquad (15\cdot21)$$

which implies the existence of (closed) commuting self adjoint and self adjoint algebras satisfying

$$\left.\begin{aligned} [{}^{+}a, {}^{+}b] &= {}^{+}[a, b] \\ [{}^{-}a, {}^{-}b] &= {}^{-}[a, b] \\ [{}^{+}a, {}^{-}b] &= 0 \end{aligned}\right\} \qquad (15\cdot22)$$

This reflects the fact that on the global level the complex 4-dimensional rotation group is equivalent, modulo an identification of each transformation matrix with its negative, to a direct product of 3-dimensional complex rotation groups, i.e we have

$$SO(4)\mathbb{C}/Z_2 \equiv SO(3)\mathbb{C} \times SO(3)\mathbb{C} \qquad (15\cdot23)$$

In the case of the real rotation groups that preserve metrics with signature ++++ or ++-- the self adjoint and anti self adjoint generators form real 3-dimensional subspaces of the full 6-dimensional algebra so that the direct product relation is preserved in the form

$$SO(4)/Z_2 \equiv SO(3) \times SO(3) \qquad (15\cdot24)$$

$$SO(2,2)/Z_2 \equiv SO(2,1) \times SO(2,1) \qquad (15\cdot25)$$

In the physically most interesting case for which the metric has the Lorentz signature +++-, the self adjoint generators are not elements of a real 3-dimensional subspace but of a complex 3-dimensional subspace with a natural isomorphism (performed by taking the real part) onto the whole of the real 6-dimensional algebra, so that we have the simple relation

$$SO(3,1)/Z_2 \equiv SO(3)\,\mathbb{C} \qquad (15.26)$$

to which reference was made in section 13.

This special feature of 4-dimensional rotation groups makes it possible to work to a large extent with representations of only 3-dimensions, which is not merely better than working with the full adjoint representation dimension, namely 6, but even better than working with the fundamental dimension 4 (indeed it is the possibility of getting down from 6 to 3 that provides part of the motivation for introducing the adjoint representation in the first place To exploit this possibility one introduces a basis $^{+}Z_{\alpha}$, $\alpha = 1,2,3,$ for the vector space of self-adjoint generators. Such a basis may be constructed directly from the first three members of an ordinary six dimensional algebra basis λ_K simply by setting

$$^{+}Z_{\alpha} = \delta^{K}_{\alpha}\ {}^{+}\lambda_K \qquad (15.27)$$

provided that the elements λ_1, λ_2, λ_3 are taken linearly independent not only of each other (as any basis elements must be) but also of all their adjoints. Such a subset of elements λ_1, λ_2 λ_3 also determines a corresponding basis

$$^{-}Z_{\dot\alpha} = \delta^{K}_{\dot\alpha}\ {}^{-}\lambda_K \qquad (15.28)$$

for the vector space of anti-self adjoint generators. The dot over the index has been introduced to indicate that the relevant components transform differently under change of basis from those with undotted indices. In the Lorentzian $SO(3,1)$ case there is a canonical correspondance $Z_\alpha \mapsto {}^{-}Z_{\dot\alpha} = (\overline{^{+}Z_\alpha})$ so that a change of basis $Z_\alpha \longmapsto {}^{+}S_\alpha{}^\beta Z_\beta$ entails a corresponding change of basis ${}^{-}Z_{\dot\alpha} \mapsto {}^{-}S_{\dot\alpha}{}^{\dot\beta}\,{}^{-}Z_{\dot\beta}$ with ${}^{-}S_{\dot\alpha}{}^{\dot\beta} = (\overline{^{+}S_\alpha{}^\beta})$.
In the $SO(4)$ and $SO(2,2)$ cases the relationship between $^{+}Z_\alpha$ and $^{-}Z_{\dot\alpha}$ is <u>not</u> canonical (since in these cases the relevant λ_1, λ_2, λ_3 cannot be recovered uniquely from a knowledge of the $^{+}Z_\alpha$ alone) and hence the ${}^{-}S_{\dot\alpha}{}^{\dot\beta}$ can be chosen independently of the ${}^{+}S_\alpha{}^\beta$ It follows immediately from the rules (15.22) that the new basis vectors satisfy the commutation laws

$$\left.\begin{aligned} [{}^{+}Z_\alpha, {}^{+}Z_\beta] &= {}^{+}@_{\alpha\beta}{}^{\gamma}\,{}^{+}Z_\gamma \\ [{}^{-}Z_{\dot\alpha}, {}^{-}Z_{\dot\beta}] &= {}^{-}@_{\dot\alpha\dot\beta}{}^{\dot\gamma}\,{}^{-}Z_{\dot\gamma} \\ [{}^{+}Z_\alpha, {}^{-}Z_{\dot\beta}] &= 0 \end{aligned}\right\} \qquad (15.29)$$

where, for any basis λ_K satisfying (15.27) and (15.28), the new structure constants are identical with the corresponding subset of the ordinary structure constants, i.e.

$$^{+}@_{\alpha\beta}{}^{\gamma} = \delta^{K}_{\alpha}\,\delta^{L}_{\beta}\,\delta^{\gamma}_{M}\,@_{KL}{}^{M} \tag{15.30}$$

and similarly for the dotted case.

The self-adjoint and anti-self adjoint subsets are in fact closed not only under the commutator product operation but even under direct matrix multiplication, since it can in fact be checked (by working in any convenient—e.g orthonormal—basis) that the matrix product of $^{+}z_{\alpha}$ and $^{-}z_{\beta}$ is given simply by

$$^{+}z_{\alpha}\cdot{}^{+}z_{\beta} = \tfrac{1}{2}\,{}^{+}@_{\alpha\beta}{}^{\gamma}\,{}^{+}z_{\gamma} - \tfrac{1}{4}\,{}^{+}\gamma_{\alpha\beta}\,\mathbb{1} \tag{15.31}$$

(and similarly for $^{-}z_{\dot\alpha}$ and $^{-}z_{\dot\beta}$) where $\mathbb{1}$ is the 4-dimensional unit tensor with components δ^{μ}_{ν}, and where $^{+}\gamma_{\alpha\beta}$ is defined (analogously to the ordinary algebra metric γ_{KL}) by

$$^{+}\gamma_{\alpha\beta} = \tfrac{1}{2}\,@_{\alpha\gamma}{}^{\delta}\,@_{\delta\beta}{}^{\gamma} \tag{15.32}$$

which implies that for any basis λ_K satisfying (15.27) we shall have (note the 1/2 factor)

$$^{+}\gamma_{\alpha\beta} = \tfrac{1}{2}\,\delta^{K}_{\alpha}\,\delta^{L}_{\beta}\,\gamma_{KL} \tag{15.33}$$

and similarly for $^{-}\gamma_{\dot\alpha\dot\beta}$. Using $^{+}\gamma_{\alpha\beta}$ for raising and lowering undotted Greek indices we obtain from (15.31) the formulae

$$^{+}@_{\alpha\beta}{}^{\gamma}\,{}^{+}z^{\alpha}\cdot{}^{+}z^{\beta} = {}^{+}z^{\gamma} \tag{15.34}$$

and

$$^{+}z^{\alpha}\cdot{}^{+}z_{\alpha} = -\tfrac{3}{4}\,\mathbb{1} \tag{15.35}$$

while taking the trace of (15.31) gives

$$^{+}z_{\alpha}{}^{\mu\nu}\,{}^{+}z^{\beta}{}_{\mu\nu} = \delta^{\beta}_{\alpha} \tag{15.36}$$

Precisely similar relations hold for the dotted (anti-self adjoint) basis vectors, and for the last one we also have a corresponding mixed contraction formula which is simply

$$^{+}z_{\alpha}{}^{\mu\nu}\,{}^{-}z^{\beta}{}_{\mu\nu} = 0 \tag{15.37}$$

The relation (15.36) is analogous to (13.26) and can be inverted to give

$$ {}^{+}Z_{\alpha}{}^{\mu\nu}\,{}^{+}Z^{\alpha}{}_{\rho\sigma} = \tfrac{1}{2}\left(\delta^{\mu}_{[\rho}\,\delta^{\nu}_{\sigma]} + \tfrac{1}{2}\eta^{\mu\nu}{}_{\rho\sigma}\right) \tag{15.38} $$

The corresponding formula for dotted indices differs by a sign in the last term, thus having the form

$$ {}^{-}Z_{\dot\alpha}{}^{\mu\nu}\,{}^{-}Z^{\dot\alpha}{}_{\rho\sigma} = \tfrac{1}{2}\left(\delta^{\mu}{}_{[\rho}\,\delta^{\nu}{}_{\sigma]} - \eta^{\mu\nu}{}_{\rho\sigma}\right) \tag{15.39} $$

so that by addition one obtains

$$ {}^{+}Z_{\alpha}{}^{\mu\nu}\,{}^{+}Z^{\alpha}{}_{\mu\nu} + {}^{-}Z_{\alpha}{}^{\mu\nu}\,{}^{-}Z^{\alpha}{}_{\mu\nu} = \delta^{\mu}{}_{[\rho}\,\delta^{\nu}{}_{\sigma]} \tag{15.40} $$

This shows that (as we could also have seen from (15.36) and (15.37)) if we take ${}^{+}\vec{Z}_{\alpha}$ and ${}^{-}\vec{Z}_{\alpha}$ conjointly as a basis for the set of bivectors, then the 2-forms ${}^{+}\underline{Z}^{\alpha}$ and ${}^{-}\underline{Z}^{\alpha}$ conjointly form the corresponding dual basis. This means that if we expand an arbitrary generator a in the form

$$ a = a^{\alpha}\,{}^{+}Z_{\alpha} + a^{\dot\alpha}\,{}^{-}Z_{\dot\alpha} \tag{15.41} $$

then the components will be given by

$$ a^{\alpha} \doteq {}^{+}Z^{\alpha}{}_{\mu\nu}\,a^{\mu\nu}\,, \quad a^{\dot\alpha} = {}^{-}Z^{\dot\alpha}{}_{\mu\nu}\,a^{\mu\nu} \tag{15.42} $$

As an application we can make the expansion

$$ \gamma_{KL}\,\lambda^{K}\lambda^{L} = \gamma_{\alpha\beta}\,{}^{+}Z^{\alpha}\,{}^{+}Z^{\beta} + \gamma_{\dot\alpha\dot\beta}\,{}^{-}Z^{\dot\alpha}\,{}^{-}Z^{\dot\beta} \tag{15.43} $$

where the absence of mixed terms (of the form $\gamma_{\dot\alpha\beta}\,{}^{-}Z^{\dot\alpha}\,{}^{+}Z^{\beta}$, $\gamma_{\alpha\dot\beta}\,{}^{+}Z^{\alpha}\,{}^{-}Z^{\dot\beta}$), expresses the obvious double self duality property

$$ \gamma = {}^{*}\gamma^{*} \tag{15.44} $$

The non identically vanishing components are given by

$$ \gamma_{\alpha\beta} = \gamma_{\mu\nu\rho\sigma}\,{}^{+}Z_{\alpha}{}^{\mu\nu}\,{}^{+}Z_{\beta}{}^{\rho\sigma} \tag{15.45} $$

and similarly for $\gamma_{\dot\alpha\dot\beta}$. Since by (14.24) the coordinates components of γ are simply given by

$$ \gamma_{\mu\nu}{}^{\rho\sigma} = \delta^{\rho}{}_{[\mu}\,\delta^{\sigma}{}_{\nu]} \tag{15.46} $$

in the 4-dimensional case, we obtainfrom (15.36) the solution

$$\gamma_{\alpha\beta} = {}^{+}\gamma_{\alpha\beta} \quad , \quad \gamma_{\dot{\alpha}\dot{\beta}} = {}^{-}\gamma_{\alpha\beta} \tag{15.47}$$

Since by (15.13) the Weyl tensor obviously has the same double self duality property as γ , i.e

$$C = *C* \tag{15.48}$$

we can make the similar expansion

$$C_{KL}\lambda^{K}\lambda^{L} = C_{\alpha\beta}\,{}^{+}Z^{\alpha}\,{}^{+}Z^{\beta} + C_{\dot{\alpha}\dot{\beta}}\,{}^{-}Z^{\dot{\alpha}}\,{}^{-}Z^{\dot{\beta}} \tag{15.49}$$

The symmetry between K and L equivalent to the conditions

$$C_{\alpha\beta} = C_{\beta\alpha} \quad , \quad C_{\dot{\alpha}\dot{\beta}} = C_{\dot{\beta}\dot{\alpha}} \tag{15.50}$$

and in the Lorenztian (+++-) case the additional requirement that the Weyl tensor be real is equivalent to

$$\bar{C}_{\alpha\beta} = C_{\dot{\alpha}\dot{\beta}} \tag{15.51}$$

Since by (15.14) the remaining term in the decomposition of the Riemann tensor satisfies the antidouble self duality condition

$$E = -*E* \tag{15.52}$$

it will have an expansion of the form

$$E_{KL}\lambda^{K}\lambda^{L} = E_{\alpha\dot{\beta}}\,{}^{+}Z^{\alpha}\,{}^{-}Z^{\dot{\beta}} + E_{\dot{\alpha}\beta}\,{}^{-}Z^{\dot{\alpha}}\,{}^{+}Z^{\beta} \tag{15.53}$$

(from which the terms $E_{\alpha\beta}\,{}^{+}Z^{\alpha}\,{}^{+}Z^{\beta}$ and $E_{\dot{\alpha}\dot{\beta}}\;Z^{\dot{\alpha}}\,{}^{-}Z^{\dot{\beta}}$ are absent) the symmetry property being equivalent to

$$E_{\alpha\dot{\beta}} = E_{\dot{\beta}\alpha} \tag{15.54}$$

while in the Lorentzian case the additional reality condition is

$$E_{\alpha\dot{\beta}} = \overline{(E_{\alpha\dot{\beta}})} \tag{15.55}$$

which by (15.54) is the same as the hermiticity condition $\overline{(E_{\alpha\dot{\beta}})} = E_{\beta\dot{\alpha}}$
Reassembling the pieces, we see that the Riemann tensor has the form

$$\underline{R} = {}^{+}\underline{R} + {}^{-}\underline{R} \tag{15.56}$$

with

$${}^{+}\underline{R} = \underline{R}_{\alpha}\,{}^{+}Z^{\alpha} \quad , \quad {}^{-}\underline{R} = \underline{R}_{\alpha}\,{}^{-}Z^{\alpha} \tag{15.57}$$

where

$$\underline{\mathcal{R}}_\alpha = (C_{\alpha\beta} + \tfrac{1}{6}\Lambda\,\gamma_{\alpha\beta})\,{}^{+}\underline{Z}^{\alpha} + E_{\alpha\dot\beta}\,{}^{-}\underline{Z}^{\dot\beta} \qquad (15{\cdot}58)$$

and similarly for $\mathcal{R}_{\dot\alpha}$ (which in the Lorentzian case will be the same as $\overline{(\mathcal{R}_\alpha)}$). Thus instead of the traditional 4th order 4-dimensional tensor representation, the Riemann tensor has been expressed in the Lorentzian case in terms of a scalar, a trace free complex symmetric 3×3 matrix, and a hermitian 3×3 matrix, while in the $SO(4)$ and $SO(2,2)$ cases it is expressed in terms of a scalar, two trace free real symmetric 3×3 matrices and one arbitrary real 3×3 matrix, which add up in all cases to 10 independent Weyl tensor components and 10 independent Ricci tensor components.

16. General Theory of Invariant, Curvature Integrals.

The examples treated in the previous section 15 can be used to provide illustrative applications, to be described in section 17, of the general theory of invariant curvature integrals which plays an important role in the study of the global topological properties of differentiable fibre bundles.

This theory is based on the class of covariant Lie algebra tensors c say that are invariant under the group adjoint action, $@$. If the tensor has components $c_{K_1 \dots K_r}$ with respect to a basis λ_K of the algebra (as introduced in section 12) then its invariance is defined by

$$c_{K_1\cdots K_r} = c_{L_1\cdots L_r}\, S^{-1\,L_1}{}_{K_1} \cdots S^{-1\,L_r}{}_{K_r} \qquad (16{\cdot}1)$$

where the transformation matrices $S^{-1\,L}{}_K$ are defined by

$$@S\;\lambda_K = \lambda_L\, S^{-1\,L}{}_K \qquad (16{\cdot}2)$$

Suppose that $\underline{\Psi}_{(1)}, \cdots, \underline{\Psi}_{(r)}$ are algebra valued differential forms (such as the curvature $\underline{\Omega}$) that map exterior products of base space tangent vectors into the adjoint algebra bundle. In any given gauge α the image $\underline{\Psi}_{(i)\alpha}$ of such a p-form may be written as

$$\underline{\Psi}_{(i)\alpha} = \underline{\Psi}^{K}_{(i)\alpha}\,\lambda_K \qquad (16{\cdot}3)$$

where the components $\underline{\Psi}^{K}_{(i)\alpha}$ are ordinary differential p-forms on $\mathcal{N}_\alpha$. Under a change of gauge determined by an element $S_{\beta\alpha} \in \mathcal{G}$ one will have

$$\underline{\Psi}_{(i)\alpha} \mapsto \underline{\Psi}_{(i)\beta} = S_{\beta\alpha}{}^{K}{}_{L}\,\underline{\Psi}^{L}_{(i)\alpha} \qquad (16{\cdot}4)$$

Since $S_{\beta\alpha}{}^{K}{}_{L} = S_{\alpha\beta}^{-1}{}^{K}{}_{L}$ with $S_{\alpha\beta} \in \mathcal{G}$, it can hence be seen from the invariance condition (16.1) that the ordinary scalar valued form $c(\underline{\Psi}_{(1)} \wedge \cdots \wedge \underline{\Psi}_{(r)})$ specified by

$$c(\underline{\Psi}_{(1)} \wedge \cdots \wedge \underline{\Psi}_{(r)}) = c_{K_1 \ldots K_r} \underline{\Psi}_{(1)}{}^{K_1} \wedge \cdots \wedge \underline{\Psi}_{(r)}{}^{K_r} \tag{16.5}$$

is a well behaved field, independent of the gauge in terms of which the components $\underline{\Psi}_{(i)}^{K}$ are implicitly specified (which justifies the use here of abbreviated gauge label free notation).

Now by decomposing the gauge connection 1-form $\underline{\omega}_\alpha$ in the form

$$\underline{\omega}_\alpha = \underline{\omega}_\alpha{}^{K} \lambda_K \tag{16.6}$$

where $\underline{\omega}_\alpha{}^{K}$ are ordinary (but gauge dependent) 1-forms, we may write the subcovariant exterior derivative (as introduced in section 11) of each of the $\underline{\Psi}_{(i)}$ fields in the form

$$D \wedge \underline{\Psi}_{(i)} = \partial \wedge \underline{\Psi}_{(i)} + \underline{\omega}^{K} \wedge (@\lambda_K)\, \underline{\Psi}_{(i)} \tag{16.7}$$

where again the implicit gauge label α has been suppressed. Moreover the invariance property (16.1) under the group adjoint action $@$ implies a corresponding invariance under the adjoint algebra action $@$, which is expressible by

$$0 = \sum_{(i)} c(\underline{\Psi}_{(1)} \wedge \cdots \wedge (@a)\underline{\Psi}_{(i)} \wedge \cdots \wedge \underline{\Psi}_{(r)}) \tag{16.8}$$

for $a \in \mathcal{a}$. This enables us to obtain a useful lemma, relating ordinary exterior differentiation of the scalar valued form $c(\underline{\Psi}_{(1)} \wedge \cdots \wedge \underline{\Psi}_{(r)})$ to the subcovariant exterior differentiation of the fields $\underline{\Psi}_{(i)}$ by the formula

$$\partial \wedge c(\underline{\Psi}_{(1)} \wedge \cdots \wedge \underline{\Psi}_{(r)}) = c(\tilde{D} \wedge \underline{\Psi}_{(1)} \wedge \cdots \wedge \underline{\Psi}_{(r)}) \tag{16.9}$$

The theory of integral topological invariants is based on the application of this lemma to any $2r$-form $\underline{W}(c, \underline{\Omega})$ obtained by substituting the curvature form $\underline{\Omega}$ in place of each $\underline{\Psi}_{(i)}$, $i = 1, \ldots, r$, i.e.

$$\begin{aligned} \underline{W}(c, \Omega) &= c(\underline{\Omega} \wedge \cdots \wedge \underline{\Omega}) \\ &= c_{K_1 \ldots K_r} \underline{\Omega}^{K_1} \wedge \cdots \wedge \underline{\Omega}^{K_r} \end{aligned} \tag{16.10}$$

where we may without loss of generality assume that c is restricted to satisfy the symmetry condition

$$c_{K_1 \ldots K_r} = c_{(K_1 \ldots K_r)} \tag{16.11}$$

The essential point is that subject to suitable boundary conditions (such as compactness or assymptotic flatness) the corresponding global integral

$$\langle \underline{W}, \Sigma \rangle = \int_\Sigma \underline{W}(c, \underline{\Omega}) | d\vec{\Sigma} \qquad (16.12)$$

over a $2r$ -surface Σ in $\mathcal{M}$ depends only on the topology in so much as it is unaffected by arbitrary continuous variation of the connection $\hat{\underline{\chi}}$ from which $\underline{\Omega}$ is constructed, as well as being unaffected by continuous (boundary condition preserving) deformations of Σ .

To demonstrate these properties we start by remarking that it is obvious from the lemma (16.9) using the form (11.21) of the Bianchi identity that $W(c, \underline{\Omega})$ is necessarily closed, i.e.

$$\partial \wedge \underline{W}(c, \underline{\Omega}) = 0 \qquad (16.13)$$

This property (which is of course trivial if $2r$ equals the dimension n of $\mathcal{M}$) is already sufficient to establish that the integral is unaffected by continuous deformations of Σ since if the deformation is generated by a vector field $\vec{\xi} = d\vec{x}/dt$ on $\mathcal{M}$ then in accordance with (1.16) we shall have

$$\frac{d}{dt} \langle \underline{W}, \Sigma \rangle = \int_{S=\partial\Sigma} \vec{\xi} \cdot \underline{W}(c, \underline{\Omega}) | d\vec{S} \qquad (16.14)$$

This expression obviously vanishes if either (a) Σ is compact so that the boundary S is empty; (b) $\underline{\Omega}$ is constrained to vanish in the limit on the boundary; or (c) the deformation $\vec{\xi}$ is constrained to leave the boundary S invariant.

More remarkable than this is the property that if we make a change in $\underline{\Omega}$ by altering the connection $\hat{\underline{\chi}}$ then the resulting change in $\underline{W}(c, \underline{\Omega})$ is not merely closed but exact. Since by (7.4) any (finite or infinitesimal) difference $\Delta\hat{\underline{\chi}}$ between two connection forms $\hat{\underline{\chi}}_{(1)}$ and $\hat{\underline{\chi}}_{(2)}$ on the associated principle bundle satisfies

$$\vec{X}_{\parallel} \cdot \Delta \hat{\underline{\chi}} = 0 \qquad (16.15)$$

for any vector parallel to the fibres, we see that the corresponding difference

$$\Delta \underline{\omega}_\alpha = \underline{\omega}_{\alpha(2)} - \underline{\omega}_{\alpha(1)} \qquad (16.16)$$

in the algebra valued 1-form $\underline{\omega}_\alpha$ induced on $\mathcal{M}$ in any given gauge α will in fact be the image of a well behaved field $\Delta\underline{\omega}$ which is a section of the adjointly transforming Lie algebra valued 1-forms (by the same line of reasoning that was used in section 9 to establish that Ω_α is the image of an analogously well behaved field Ω). The infinitesimal change $d\Omega$ in Ω resulting from an infinitesimal change $d\underline{\omega}$ in the gauge connection 1-form can be seen from (9.14) to be given simply in terms of subcovariant exterior differentiation by

$$d\Omega = \tilde{D}\wedge(d\underline{\omega}) \tag{16.17}$$

It now follows from the lemma (16.9) and the ever useful Bianchi identity(11.21) that the corresponding variantion $d\underline{W}(c,\underline{\Omega})$ satisfies the exactness condition

$$d\underline{W}(c,\underline{\Omega}) = \partial\wedge\underline{W}(rc, d\underline{\omega}) \tag{16.18}$$

where we make the definition

$$\underline{W}(c, d\underline{\omega}) = c_{K_1K_2\cdots K_r}(d\underline{\omega})^{K_1}\wedge\underline{\Omega}^{K_2}\wedge\cdots\wedge\underline{\Omega}^{K_r} \tag{16.19}$$

Hence using Stokes theorem (1.8) we obtain

$$d\int \underline{W}(c,\underline{\Omega})|d\vec{\Sigma} = \int_{S=\partial\Sigma} W(rc, d\underline{\omega})|dS \tag{16.20}$$

By integrating over infinitesimal variations (taking $\Delta\underline{\omega} = \int d\underline{\omega}$) we thus obtain the required invariance condition

$$\Delta\langle \underline{W}, \Sigma\rangle = 0$$

provided either: Σ is compact; $\underline{\Omega}$ is constrained to vanish in the limit on $\partial\Sigma$; or $\Delta\omega$ is constrained to vanish on $\partial\Sigma$.

The discovery of this correspondence between invariant symmetric Lie, co-algebra tensors, and integrals whose invariance implies that their value depend only on the topology, is attributeable to Weyl.

When $r = 1$ or $r = 2$ (so that $\underline{W}$ is respectively a 2-form or a 4-form) which are the only cases of interest when the base represents ordinary space time, one can go on to express the unvaried form $\underline{W}(c,\underline{\Omega})$ as an explicit exterior derivative within any given local gauge patch. (I do not know whether mathematicians have solved the problem of doing this generally for arbitrary values of r) A convenient way to do this is to start by expressing the curvature by the formula

$$\underline{\Omega} = \tfrac{1}{2}(\partial + \tilde{D})\wedge\underline{\omega} \tag{16.22}$$

(which is equivalent to (9.14)) and then going on to express the exterior product tensor $\underline{\Omega}\wedge\underline{\Omega}$ similarly in terms of sums of ordinary and subcovariant exterior derivatives, which can be achieved with the aid of the Bianchi identities in the formula

$$\underline{\Omega}\wedge\underline{\Omega} = \tfrac{1}{3}\tilde{D}\wedge\omega\wedge\Omega + \tfrac{1}{6}(\tilde{D}+\partial)\wedge\omega\wedge(\underline{\Omega}+\partial\wedge\underline{\omega}) \tag{16.23}$$

Using the lemma 16.9 by which subcovariant (and hence also as a trivial, special case, ordinary) exterior differentiation inside a contraction with c may be taken outside as ordinary exterior differentiation, we obtain for the case $r=1$,

$$c_K\underline{\Omega}^K = \partial\wedge c_K\underline{\omega}^K \tag{16.24}$$

and for the case $r=2$,

$$c_{KL}\underline{\Omega}^K\wedge\underline{\Omega}^L = \partial\wedge c_{KL}\underline{\omega}^K\wedge(\tfrac{2}{3}\underline{\Omega}^L+\tfrac{1}{6}\partial\wedge\underline{\omega}^L) \tag{16.25}$$

The possibility of doing this was drawn to my attention by Madore. I should be interested to know of a solution for higher r .

18. Examples of Invariant Curvature Integrals.

An obvious way to obtain a Lie coalgebra invariant in the case $r=2$ is to identify c_{KL} with the group metric components γ_{KL}. More such invariants can be obtained from corresponding Casimir operators, an rth order Casimir operator being defined in terms of an rth order contravariant Lie algebra tensor in terms of any given representation rep as $c^{K_1}\dots c^{K_r}\, rep\,\lambda_{K_1}\circ\cdots\circ rep\,\lambda_{K_r}$ The tensor $c^{K_1\dots K_r}$ may be taken to be symmetric without loss of generality since non-symmetric parts can be taken into account — using the commutation relations — in terms of lower order Casimir operators. In semi-simple cases, when γ is not singular there will be a 1-1 correspondence between Casimir operators and Weyl forms $\underline{W}(c,\underline{\Omega})$ obtained simply by lowering and raising algebra indices.

By working with an algebra base λ_K in a fundamental representation (as described in section 12) we see that a generally available sequence of suitable invariant Lie algebra invariants is given by

$$C_{K_1\dots K_r} = \delta^{\ell_1}_{[m_1}\cdots\delta^{\ell_r}_{m_r]}\,\lambda_{K_1}{}^{m_1}{}_{\ell_1}\cdots\lambda_{K_r}{}^{m_r}{}_{\ell_r} \tag{17.1}$$

so that one obtains a corresponding set of Weyl-forms

$$W(c,\underline{\Omega}) = \delta^{\ell_1}{}_{[m_1}\cdots\delta^{\ell_r}{}_{m_r]}\,\underline{\Omega}^{m_1}{}_{\ell_1}\wedge\cdots\wedge\underline{\Omega}^{m_r}{}_{\ell_r} \tag{17.2}$$

whose integrals (when suitably normalised) give rise to the well known Chern-Pontriagin numbers (see e.g Kobayashi and Nomizu, 1969).

An even more obvious general possibility is to use the sum-of-traces invariant

$$c_{K_1\dots K_r} = \delta^{\ell_1}{}_{m_1}\cdots\delta^{\ell_r}{}_{m_r}\,\lambda_{K_1}{}^{m_1}{}_{\ell_1}\cdots\lambda_{K_r}{}^{m_r}{}_{\ell_r} \tag{17.3}$$

but due to the redundancy inherent in fundamental matrix (compared to adjoint vectorial) representations these obvious simple possibilities are not always independent, and in particular the product of traces invariant (17.3) will vanish for the fundamental representations of all the orthogonal groups that were discussed in section 13. For these orthogonal groups in particular there is however another obvious possibility when $n = 2r$, namely to take

$$c_{K_1 \dots K_r} = \varepsilon^{l_1}{}_{m_1} \cdots {}^{l_r}{}_{m_r} \lambda_{K_1}{}^{m_1}{}_{l_1} \cdots \lambda_{K_r}{}^{m_r}{}_{l_r} \tag{17.4}$$

which gives rise to the Chern-Gauss-Bonnet invariants.

A particularly important special case is of course the case of Lorentz-signature pseudo-Riemannian geometry in 4-dimensions. Since the Lorentz group has just two independent 2nd order Casimir operators, there are just two independent symmetric Lie co-algebra tensors c_{KL} which may conveniently be taken to be γ_{KL} and the algebra version ε_{KL} of the alternating tensor, as defined by (15.3), and hence the corresponding Weyl 4-forms may be taken to be

$$\underline{W}_1 = \tfrac{1}{2}\, \underline{\mathcal{R}}_K \wedge \underline{\mathcal{R}}^K = -\tfrac{1}{2}\, \underline{\mathcal{R}}^l{}_m \wedge \underline{\mathcal{R}}^m{}_l = W_1 \underline{\varepsilon} \tag{17.5}$$

and

$$\underline{W}_2 = \tfrac{1}{2}\, \underline{\mathcal{R}}_K \wedge {}^*\underline{\mathcal{R}}^K = -\tfrac{1}{2}\, \underline{\mathcal{R}}^l{}_m \wedge {}^*\underline{\mathcal{R}}^m{}_l = W_2 \underline{\varepsilon} \tag{17.6}$$

where $\underline{\varepsilon}$ is the standard metric volume measure form (i.e the ordinary alternating tensor) and the scalars W_1 and W_2 are given by

$$W_1 = \tfrac{1}{2} \epsilon\{g\}\, \varepsilon^{MN} \mathcal{R}_{NK} \mathcal{R}_M{}^K = \epsilon\{g\}\, \mathcal{R}_{\mu\nu\rho\sigma} {}^*\mathcal{R}^{\mu\nu\rho\sigma} \tag{17.7}$$

and

$$W_2 = \tfrac{1}{4} \epsilon\{g\}\, \varepsilon^{MN} \varepsilon^{KL} \mathcal{R}_{MK} \mathcal{R}_{NL} = \epsilon\{g\}\, {}^*\mathcal{R}_{\mu\nu\rho\sigma} {}^*\mathcal{R}^{\mu\nu\rho\sigma}$$

$$= \mathcal{R}_{\mu\nu\rho\sigma} \mathcal{R}^{\mu\nu\rho\sigma} - 4\, R_{\mu\nu} R^{\mu\nu} + \Lambda^2 \tag{17.8}$$

(the last line being obtained with the aid of 13.18)
The corresponding integrals are respectively proportional to the relevant Pontriagin and generalised Gauss-Bonnet invariants .

The invariance of the integrals of Weyl forms (such as (16.29) and (16.30)) is equivalent to the condition that the Eulerian derivatives with respect to $\underline{\omega}$ of the corresponding scalars (W_1 and W_2 in this case) must vanish. Thus the existence of these forms reduces (in this case by two) the number of possible independent (in this case quadratic) Lagrangians that can be constructed from the curvature. This not only limits the number of ways one might try to construct classical theories, but also simplifies the renormalisation of quantised theories by limiting the number of possible counter terms in the manner described in accompanying courses of these proceedings.

The easiest invariant integrals to work out in full detail are of course those over two (rather than four dimensional surfaces) corresponding to the case $r = 1$. If c_K are elements of a first order Lie coalgebra element c the invariance condition (16.1) is equivalent to the requirement that the fundamental matrix representative $c^m{}_n$ say of c should commute with the representative $S^m{}_n$ of any $S \in \mathcal{G}$ and hence also with the representative $a^m{}_n$ of any $a \in \mathcal{Q}$, i.e

$$c^k{}_l\, a^l{}_m = a^k{}_l\, c^l{}_m \tag{17.8}$$

For any such matrix $c^m{}_n$ we have the useful lemma that for

$$S^l = (\exp a)^l = \delta^l{}_m + a^l{}_m + \frac{1}{2!} a^l{}_k a^k{}_m + \dots \tag{17.9}$$

then

$$\begin{aligned} & c^k{}_l\, S^{-1l}{}_m\, a^m{}_k \\ &= c^k{}_l \left(\delta^l{}_m - a^l{}_m + \frac{1}{2!} a^l{}_k a^k{}_m\right)\left(D a^m{}_k + \frac{1}{2!} D a^m{}_j a^j{}_k + \dots\right) \\ &= c^k{}_l\, D a^l{}_k \end{aligned} \tag{17.10}$$

are more concisely

$$\operatorname{tr} c \cdot S^{-1} \cdot DS = \operatorname{tr} c \cdot Da \tag{17.11}$$

whenever D is any (not necessarily covariant) differentiation operator obeying the Leibnitz rule $D\, x \cdot y = x \cdot (Dy) + (Dx) \cdot y$ This result follows from the well known properties of these (necessarily convergent) series for ordinary numbers and from the fact that the commutation property (17.8) allows the orders to be permuted at will, so that the ordinary algebraic structure survives, without any need to suppose that a commutes with Da . This lemma is very useful for the discussion of gauge transformations, since the general Law (10.23) implies that for any gauge transformation matrix $S_{\alpha\beta} \in \mathcal{G}$ that can be written in the form

$$S_{\alpha\beta} = \exp a_{\alpha\beta}$$

for some $a_{\alpha\beta} \in \mathcal{G}$ (which is possible not only — as always — near the origin $\mathbb{1}$, but even globally for the connected component of any compact Lie group, and also of the Lorentz group) the corresponding change in the gauge connection 1-form may be expressed as

$$\begin{aligned} \operatorname{tr} c \cdot (\omega_\beta - \omega_\alpha) &= \operatorname{tr} c \cdot D_\alpha a_{\alpha\beta} \\ &= \partial \operatorname{tr} c \cdot a_{\alpha\beta} \end{aligned} \tag{17.13}$$

the last step being an application of the previous lemma 16.9. treating $a_{\alpha\beta}$ as an algebra valued zero-form so that the distinction between $\tilde{D}$ and D does not arise.

Now by (16.24) one sees that the invariant integral of any Weyl 2-form $\underline{W} = c^{m}{}_{\ell}\,\underline{\Omega}^{\ell}{}_{m}$ over a compact two-surface S can be expressed, in terms of a suitable covering of S by a mesh whose cells are contained in individual gauge patches, as

$$\langle \underline{W}, S \rangle = \int_{\Sigma} tr\, c \cdot \underline{\Omega} \,|\, d\vec{S} = \sum_{\alpha} \int_{\partial(\alpha)} tr\, c \cdot \underline{\omega}_{\alpha} \,|\, d\vec{s} \qquad (17.14)$$

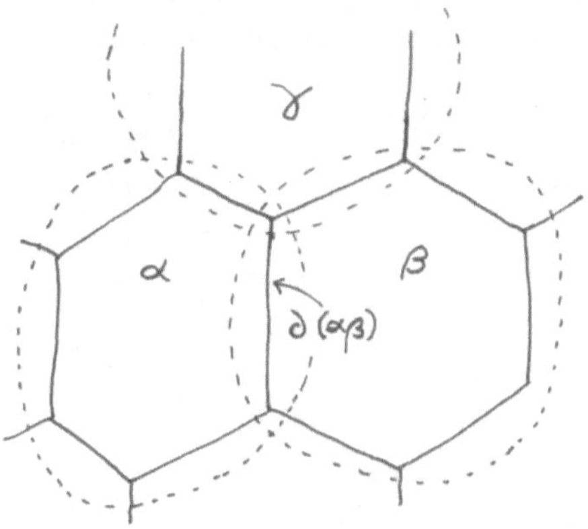

where $\partial(\alpha)$ is the boundary of the cell (which can of course be chosen to be unique) contained in the gauge patch α, and the summation is over all such cells. If we use $\partial(\alpha\beta)$ to denote the common boundary (with orientation as determined by the α cell) between two cells α and β, we can use (16.38) to obtain

$$\begin{aligned} \langle \underline{W}, S \rangle &= tr \sum_{\alpha,\beta} \int_{\partial(\alpha\beta)} c \cdot (\underline{\omega}_{\alpha} - \underline{\omega}_{\beta}) \,|\, d\vec{s} \\ &= tr \sum_{\alpha,\beta} c \cdot \Delta a_{\alpha\beta} \\ &= tr \sum_{\alpha,\beta,\gamma} c \cdot (a_{\alpha\beta} + a_{\beta\gamma} + a_{\gamma\alpha}) \end{aligned} \qquad (17.15)$$

where $\Delta a_{\alpha\beta}$ is the change in $a_{\alpha\beta}$ between the vertices at the two ends of the line $\partial(\alpha\beta)$, and where the final summation is taken over the vertex points at which three cells meet.

The most familiar applications involve the abelian group $SO(2)$ of (orientable) two dimensional Riemannian geometry, which is the same as the $U(1)$ group of electromagnetism. For an abelian group the compatibility condition (3.8) implies that on the vertices where three cells meet, e.g at the point of intersection of the three boundary lines $\partial(\alpha\beta)$, $\partial(\beta\gamma)$, $\partial(\gamma\alpha)$, we shall have

$$\exp(a_{\alpha\beta} + a_{\beta\gamma} + a_{\gamma\alpha}) = \mathbb{1} \tag{17.16}$$

This implies that we can write

$$a_{\alpha\beta} + a_{\beta\gamma} + a_{\gamma\alpha} = N^i p_i \tag{17.17}$$

and hence

$$\langle \underline{W}, S\rangle = (\mathrm{tr}\; c \cdot p_i) \sum_{\alpha,\beta,\gamma} N^i \tag{17.18}$$

where p_1, p_2, ... are basic periodic elements in $\mathcal{a}$ (in the sense that they map exponentially onto $\mathbb{1}$ but are not multiples of smaller elements with this property) and where N_1, N_2, ... are integers. In the case of the 1-parameter group $SO(2)$ there is only one basic period p and a single integer N therefore suffices: the fundamental representation matrices can be taken to be given by

$$S \leftrightarrow \begin{pmatrix} \cos\theta & \sin\theta \\ -\sin\theta & \cos\theta \end{pmatrix} = \exp\begin{pmatrix} 0 & \theta \\ -\theta & 0 \end{pmatrix} \tag{17.19}$$

with

$$p \leftrightarrow \begin{pmatrix} 0 & 2\pi \\ -2\pi & 0 \end{pmatrix} \tag{17.20}$$

preserving a metric with matrix form

$$g \leftrightarrow \begin{pmatrix} 1 & 0 \\ 0 & 1 \end{pmatrix} \leftrightarrow \vec{g} \tag{17.21}$$

In this case there is only one independent Casimir operator which may be taken to be

$$c = \vec{g} \cdot \underline{\varepsilon} \longleftrightarrow \begin{pmatrix} 0 & 1 \\ -1 & 0 \end{pmatrix} \Rightarrow \mathrm{tr}\; c \cdot p = 4\pi \tag{17.22}$$

where $\underline{\varepsilon}$ denotes the 2-dimensional alternating tensor, so that (17.18) gives finally

$$\int_S \varepsilon^m{}_l\, \underline{\Omega}^l{}_m \,|\, d\vec{S} = 4\pi\chi \tag{17.23}$$

for an integer χ given by the formula

$$\chi = \sum_{\alpha,\beta,\gamma} N \tag{17.24}$$

which may be considered to be a definition of the Euler numbers if the relevant principle bundle over $\mathcal{M}$ is the orthogonal reduction of the natural tangent frame bundle (In general the Euler number is defined in terms of fields of unit vectors rather than orthogonal frames, but on an orientable 2-dimensional manifold a unit vector obviously specifies a frame and vice versa). This result may be cast in a more familiar form using the fact that in this Riemannian case the 2-dimensional curvature satisfies

$$\varepsilon^{\ell}{}_{m}\,\underline{\Omega}^{m}{}_{\ell} = \varepsilon^{\ell}{}_{m}\,\underline{R}^{m}{}_{\ell} = \Lambda\,\underline{\varepsilon} \qquad (17.25)$$

(where we recall that Λ denotes the Ricci scalar) so that setting $d\vec{S} = \vec{\varepsilon}\,dS$ one obtains the original Gauss-Bonnet theorem in the form

$$\int \Lambda\, dS = 4\pi\chi \qquad (17.26)$$

(note that Λ is twice the ordinary Gaussian curvature). In the standard example of the ordinary 2-sphere

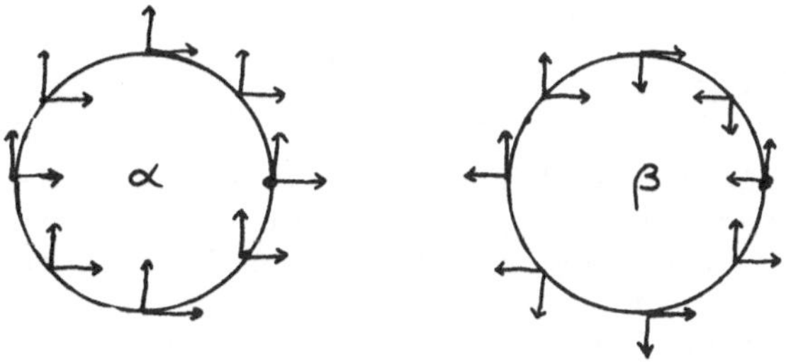

one can easily see that the Euler number (17.24) is equal to 2, e.g by considering two patches covering the north and south hemispheres with a single vertex on which $a_{\alpha\beta}$ (but not its exponential) is discontinuous.

A currently more fashionable example than the classical Gauss-Bonnet theorem is provided by the theory of magnetic monopoles, which is based on the interpretation of the Maxwellian 4-potential form $\underline{A}$ as specifying a guage connection in a principle bundle with the phase-transformation group $U(1)$ (which is isomorphic to $SO(2)$) with fundamental representation

$$S \leftrightarrow (e^{i\theta}) = \exp(i\theta) \qquad (17.27)$$

so that the single period element is

$$p \leftrightarrow (2\pi i) \qquad (17.28)$$

and the Casimir operator may be taken to be just the identity operator (a non vanishing example of the general class 17.3):

$$c \leftrightarrow (1) \quad \Longrightarrow \quad \mathrm{tr}\; c\cdot p = 2\pi i$$

The covariant operator on a complex scalar field Φ with charge e is given by $D = \partial + \underline{\omega}$ with

$$\underline{\omega} = -ie\underline{A} \qquad (17.31)$$

so that the corresponding curvature-form is

$$\underline{\Omega} = -ie\partial\wedge\underline{A} = -ie\underline{F} \qquad (17.32)$$

(since the commutativity of the group gives $\underline{A}\wedge\underline{A}=0$ in this case) where $\underline{F}$ is the ordinary Maxwell 2-form. The analogue of the Gauss Bonnet theorem that is therefore obtained from (17.18) is thus

$$e\,M(\underline{F},S) = 2\pi\chi \qquad (17.33)$$

where the magnetic monopole moment associated with the surface is defined by

$$M(\underline{F},S) = \int_S \underline{F}\,|\,d\vec{S} \qquad (17.33)$$

and where χ is defined as before by (17.24). In this electromagnetic case however the integer χ need not be the same as the ordinary Euler number because the phase bundle structure need not have anything to do with the canonical tangent frame bundle structure: indeed the most obvious possibility is to take the trivial structure so that the principle phase bundle is just $\mathcal{M}\times U(1)$, reducing to $S\times U(1)$ over the two-surface in question, which gives $\chi = 0$. The possibility of different non-trivial phase bundle structures, giving rise to arbitrary integral multiples of $2\pi/e$ for the value of M, was first envisaged by Dirac (originally in the more complicated context of complex spinor fields as described in section 20, for which semi-integral multiples are also possible).

19. The Cartan frame formalism in pseudo-Riemannian geometry.

The existence of a pseudo -Riemannian structure, as specified by the existence of a tangent space metric with co-ordinate components $g_{\mu\nu}$ over an n- dimensionnal base manifold $\mathcal{M}$, may be interpreted as determining and determined by a reduction of the principle tangent frame bundle over $\mathcal{M}$ and of the associated family of tensor bundles to a related family of bundles with the same manifold structure and projection, but with bundle group given no longer by the full linear group $GL(n)$ but merely by the relevant orthogonal subgroup $O(n-p,p)$ preserving the pseudo metric g [For a general manifold $\mathcal{M}$ such a group reduction will not always be possible which implies that a pseudo-metric of given (e.g Lorentzian) signature may not exist . However a positive definite metric (and hence a reduction from $GL(n)$ to $O(n)$) can always be constructed by using the fact that smooth positive definite symmetric tensor fields with support limited to single co-ordinate patches can be made to add to give a globally non degenerate symmetric tensor field that will still be positive definite).

Given the simultaneous presence of the associated families of $GL(n)$ and $O(n-p,p)$ bundles one can immediately construct a corresponding family of associated $GL(n) \times O(n-p,p)$ bundles in the obvious way by taking $\tilde{g}$ to be $O(n-p,p)$ in the language of section 11, i.e. by requireing that this extended family include the original $GL(n)$ and $O(n-p,p)$ families as the special cases for which respectively the $O(n-p,p)$ part or the $GL(n)$ part of the action is the trivial identity operation. This extended family evidently includes bundles of bitensors with mixed $GL(n)$ and $O(n-p,p)$ indices, and among these bitensors bundles is a particularly fundamental one that posesses a canonically defined section, which we shall refer to as the cotangent frame field $\vec{\underline{\theta}}$ say, that is defined as follows. Any tangent vector $\vec{\xi}$ at a point $x \in \mathcal{M}$ can be interpreped simultaneously as a member of the natural $GL(n)$ tangent bundle and also as a member of the bundle with the reduced group $O(n-p,p)$ represented over the fundamental n-dimensional representations space $\mathcal{U}$ say. From the first point of view any local co-ordinate guage will determine a corresponding set of components ξ^μ, while from the second point of view any local guage will determine (in terms of any fixed basis for $\mathcal{U}$) a corresponding alternative set of components ξ^m say for $\vec{\xi}$. The two systems of components will evidently be related by a linear map

$$\theta : \xi^\mu \longmapsto \xi^m = \theta^m{}_\mu \xi^\mu \qquad (18.1)$$

which can obviously be thought of as a bitensioral-field, with one co-variant $GL(n)$ index and one contravariant $O(n-p,p)$ index. In terms of the two kinds of component the pseudo-metric parenthesis of two tangent vectors $\vec{\xi}$ and $\vec{\eta}$ can be expressed in the equivalent forms

$$(\vec{\xi}, \vec{\eta}) = \xi^m g_{mn} \eta^n = \xi^\mu g_{\mu\nu} \eta^\nu \qquad (18.2)$$

which evidently implies that we have

$$\theta^m{}_\mu \theta_m{}^\nu = \delta^\nu_\mu \qquad (18.3)$$

i.e the inverse mapping to θ has bitensorial components given by simply raising and lowering the indices using the respective metric components. Permuting the order of multiplication, we see that this is equivalent to the condition that the local basis of one-forms $\underline{\theta}^m$ that is determined as a function of the choice of gauge by specifying their co-ordinate components to be $\theta^m{}_\mu$, must have the pseudo-orthonormality property

$$(\underline{\theta}^m, \underline{\theta}^n) = g^{mn} \qquad (18.4)$$

in which it is to be emphasised that the components g^{mn} (unlike the $g^{\mu\nu}$)are fixed, independently of position and gauge in $\mathcal{M}$, simply by the once and for all choice of basis in $\mathcal{U}$. It is to be remarked that most discussions of this subject start by defining the $\theta^m{}_\mu$ to be the components of a local basis of one forms, an approach which has the disadvantage of obscuring the fact that they can also be interpreted more fundamentally as components of a globally well defined bitensor field $\vec{\underline{\theta}}$).

The fundamental bitensor $\vec{\underline{\theta}}$ can obviously be used to map ordinary $GL(n)$ tensor fields onto $O(n-p,p)$ tensor fields and vice versa. Whenever one is given an ordinary linear connection Γ say on the $GL(n)$ bundles, it is possible to use $\vec{\underline{\theta}}$ to determine a canonically associated connection on the $O(n-p,p)$ bundles and thus also on the $GL(n) \times O(n-p,p)$ bundles by requiring that the θ mappings should commute with covariant differentiation, which will evidently hold in general if and only if

$$D\vec{\underline{\theta}} = 0 \qquad (18.5)$$

In terms of components this condition is equivalent to

$$\partial_\mu \theta^m{}_\nu - \Gamma_\mu{}^\rho{}_\nu \theta^m{}_\rho + \omega_\mu{}^m{}_n \theta^n{}_\nu \qquad (18.6)$$

which can be solved with the aid of (18.3) to give the fundamental $O(n-p,p)$ components of the connection (in the local guage under consideration) in the form

$$\omega_\mu{}^m{}_n = \theta_n{}^\nu (\Gamma_\mu{}^\rho{}_\nu \theta^m{}_\rho - \partial_\mu \theta^m{}_\nu) \qquad (18.7)$$

Conversely if the $O(n-p,p)$ connection is known in advance one can solve to obtain the $GL(n)$ connection components in the form

$$\Gamma_\mu{}^\rho{}_\nu = \theta_m{}^\rho (\omega_\mu{}^m{}_n \theta^n{}_\nu + \partial_\mu \theta^m{}_\nu) \qquad (18.8)$$

It is even easier to find the relation between the $GL(n)$ components and the $O(n-p,p)$ components of the curvature since (by (11.14)) the condition (18.6) evidently implies

$$(\text{rep}\,\underline{\Omega})\,\vec{\underline{\theta}} = 0 \qquad (18.9)$$

which is equivalent in terms of components to

$$-R_{\mu\nu}{}^\sigma{}_\rho \theta^m{}_\sigma + \Omega_{\mu\nu}{}^m{}_n \theta^n{}_\rho = 0 \qquad (18.10)$$

so that we obtain

$$\Omega_{\mu\nu}{}^{m}{}_{n} = R_{\mu\nu}{}^{\sigma}{}_{\rho}\, \theta^{m}{}_{\sigma}\, \theta_{n}{}^{\rho} \tag{18.12}$$

or equivalently

$$R_{\mu\nu}{}^{\sigma}{}_{\rho} = \Omega_{\mu\nu}{}^{m}{}_{n}\, \theta_{m}{}^{\sigma}\, \theta^{n}{}_{\rho} \tag{18.13}$$

i.e the two kinds of curvature tensor are related directly by the θ mapping.

The latter formula (18.13) provides what is often in practise the most rapid method of evaluating the Riemann tensor components in a space endowed with a known orthonormal (or pseudo orthonormal) basis of 1-forms $\underset{\sim}{\theta}^{m}$, as is the case if the metric is given in terms of an expression of the form

$$ds^{2} = g_{mn}(\theta^{m}{}_{\mu}\, dx^{\mu})(\theta^{n}{}_{\nu}\, dx^{\nu}) \tag{18.14}$$

with fixed g_{mn}. The standard Cartan technique is based on the fact that (18.6) obviously implies that we shall have

$$D \wedge \underset{\sim}{\vec{\theta}} = 0 \tag{18.15}$$

which is equivalent by (11.18) to

$$\tilde{D} \wedge \underset{\sim}{\vec{\theta}} = \underset{\sim}{\vec{\Theta}} \cdot \underset{\sim}{\vec{\theta}} \tag{18.16}$$

Thus writing in the $O(n-p, p)$ indices (but not the $GL(n)$ indices) we obtain

$$\partial \wedge \underset{\sim}{\theta}^{m} + \underset{\sim}{\omega}^{m}{}_{n} \wedge \underset{\sim}{\theta}^{n} = \underset{\sim}{\Theta}^{m} \tag{18.17}$$

where the torsion 2-forms $\underset{\sim}{\Theta}^{m}$ are given in full by

$$\Theta_{\mu\nu}{}^{m} = \Theta_{\mu\nu}{}^{\rho}\, \theta_{\rho}{}^{m} \tag{18.18}$$

The equation (18.19) has the practical advantage over (18.6) that only exterior differentiation involved, but as a result of the antisymmetrisation it contains less information than (18.6). In order to obtain a unique solution for the connection, the Cartan method consists of solving (18.19) for the $\underset{\sim}{\omega}^{m}{}_{n}$ subject to the additional restriction

$$\underset{\sim}{\omega}_{mn} + \underset{\sim}{\omega}_{nm} = 0 \tag{18.19}$$

which is the relevant form of the condition (13.22) that the group $O(n-p, p)$ should preserve the metric components g_{mn}. Once this has been done the $O(n-p, p)$ representative of the curvature 2-form can be read off from the relevant form of (10.31) which gives

$$\underline{\Omega}^m{}_n = \partial \wedge \underline{\omega}^m{}_n + \underline{\omega}^m{}_\ell \wedge \underline{\omega}^\ell{}_n \qquad (18.21)$$

Thus, using (18.13) one obtains the Riemann tensor via a sequence of operations in which the only differentiations involved are of the technically convenient exterior kind.

19. Hermitean and Unitary-Symplectic Structures.

Just as an ordinary metric (or symplectic) determines a parenthesis, as described in section 13, that is a bilinear scalar product of two vectors in the same fundamental representation space $\mathcal{V}$ say, so also an often has occasion to work with a bracket that is a linear-antilinear scalar product of a vector in $\mathcal{V}$ with a vector in the complex conjugate space $\bar{\mathcal{V}}$, as given by

$$\langle w, v\rangle = w \cdot \beta \cdot v = w^{\dot{a}} \beta_{\dot{a}b} v^b \qquad (19.1)$$

for $v \in \mathcal{V}$, $w \in \bar{\mathcal{V}}$, and where as previously the dot an index serves to indicate that it obeys transformation law (2.14) rather than the fundamental law (2.10). By analogy with the ordinary index lowering operation (13.3) we can use the sesquimetric tensor $\beta \in \bar{\mathcal{V}}_* \otimes \mathcal{V}_*$ that is defined by (19.1) to construct a dotting-and-lowering operation

$$\beta : \mathcal{V} \longrightarrow \bar{\mathcal{V}}_* \qquad (19.2)$$

given by

$$\left.\begin{aligned} v &\mapsto {}^{+}\bar{v} = v \cdot \bar{\beta} \\ v^a &\mapsto v_{\dot{a}} = v^b \bar{\beta}_{b\dot{a}} \end{aligned}\right\} \qquad (19.3)$$

and a corresponding undotting and lowering operation

$$\beta : \bar{\mathcal{V}} \rightarrow \mathcal{V}_* \qquad (19.3)$$

given by

$$\left.\begin{aligned} \bar{v} &\rightarrow {}^{+}v = \bar{v} \cdot \beta \\ \bar{v}^{\dot{a}} &\mapsto {}^{+}v_a = \bar{v}^{\dot{b}} \beta_{\dot{b}a} \end{aligned}\right\} \qquad (19.4)$$

where ${}^{+}v$ is the Dirac adjoint of v so that the bracket $\langle w, v\rangle$ is expressible by

$$\langle w, v\rangle = {}^{+}\bar{w} \cdot v = w_a v^a \qquad (19.5)$$

which includes as a special case

$$\langle \bar{v}, v \rangle = {}^{+}v \cdot v = \bar{v}_a v^a \qquad (19.7)$$

Provided β is non-degenerate, so that there is an inverse β^{-1} given by

$$\beta_{\dot{a}c} \beta^{-1 c \dot{b}} = \delta^{\dot{b}}{}_{\dot{a}} \qquad (19.8)$$

we can proceed (in analogy with section 13) to construct an <u>inverse</u> dotting-and-raising operation by contraction (in <u>front</u> instead of behind) with the <u>contravariant</u> version $\vec{\beta}$ of the sesquimetric whose components are given by

$$\beta^{a\dot{b}} = \bar{\beta}^{-1\, \dot{b}a} \qquad (19.9)$$

this being consistent with the application of the raising operation to the covariant form β itself. Thus e.g for a covector $\pi \in V_*$ we have

$$\left.\begin{aligned} \pi &\mapsto {}^{+}\bar{\pi} = \beta^{-T} \cdot \pi \\ \pi_a &\mapsto \pi^{\dot{a}} = \bar{\beta}^{\dot{a}b} \pi_b \end{aligned}\right\} \qquad (19.10)$$

In all the practical applications with which I am familiar β satisfies the hermeticity condition

$$\beta_{\dot{a}b} = \bar{\beta}_{b\dot{a}} \qquad (19.11)$$

and when this is satisfied I shall refer to β as a <u>hermetric</u>. The maximal group leaving such a hermetric invariant in n (complex) dimensions is the unitary group $U(n-p,p)$ where $(n-2p)$ is the invariant signature. If λ_K form an (n^2 -dimensional) basis for the Lie algebra of such a group, then the relation analogous to (13.25) giving the algebra metric as defined by (12.18) in terms of the traces of the products of the fundamental representation matrices $\lambda_K{}^a{}_{\dot{b}}$, is

$$\gamma_{KL} = -(n-1)\, \mathrm{tr}\, \lambda_K \cdot \lambda_L \qquad (19.12)$$

The <u>same</u> algebra generates also the <u>special</u> subgroup $SU(n-p,p)$ as characterised by the additional restriction that the determinant of the transformation matrices be unity.

One may now ask when the dotting and raising and lowering operations specified in terms of β (whether hermitian or not) as above are <u>compatible</u> with ordinary raising and lowering using an <u>independently specified</u> metric or symplectic g (as described in section 13) in the sense that the diagram

$$\begin{array}{ccccc} & \overset{q}{\nearrow} & \mathcal{V}_* & \overset{\beta^{-1}}{\searrow} & \\ \mathcal{V} & & & & \bar{\mathcal{V}} \\ & \underset{\bar\beta}{\searrow} & \bar{\mathcal{V}}_* & \underset{\bar q^{-1}}{\nearrow} & \end{array} \qquad (19.13)$$

is commutative, i.e ordinary lowering followed by dotting-and-raising (and then complex conjugation) gives the same result as dotting-and-lowering followed by (complex conjugation and) ordinary raising. It can be seen that this compatibility condition will be satisfied if and only if we have

$$q_{ac}\,\beta^{-1c\dot b} = \bar\beta_{a\dot c}\,\bar q^{-1\dot c\dot b} \qquad (19.14)$$

If this condition is satisfied, then starting from a given element $v \in \mathcal{V}$ with components v^a we can unambiguously corresponding sets of alternative components v_a, $v^{\dot a}$, $v_{\dot a}$ and their complex conjugates (note for example that $\bar v^a$ not in general the same as $\bar v^{\dot a}$) by successive operations carried out in arbitrary order, and in particular the condition (19.13) itself may thus be expressed as

$$q_a{}^{\dot b} = \bar\beta_a{}^{\dot b}$$

$$= C^{\dot b}{}_a \qquad (19.15)$$

where we have for convenience introduced a new mixed tensor

$$C = \beta^{-T}\cdot q^T \qquad (19.16)$$

which evidently, by (19.13), satisfies

$$C^{-1} = \bar C \qquad (19.17)$$

This means that the corresponding antilinear automorphism

$$C : \mathcal{V} \to \mathcal{V} \qquad (19.18)$$

given by

$$\left.\begin{array}{l} v \mapsto \dagger(\underline{v}) = \overline{(\dagger v)} \\ v^a \mapsto \bar v^a = \bar C^a{}_{\dot b}\,\bar v^{\dot b} \end{array}\right\} \qquad (19.19)$$

which we shall refer to as the charge conjugation mapping, is involutary in the sense that its square is the unit operation.

In the most common situations where there are consistent ordinary and dotted raising and lowering operations the sesquimetric β satisfies the hermiticity condition(19.10) while the parenthetic satisfies the simplecticity condition.

$$q_{ab} + q_{ba} = 0 \tag{19.20}$$

The group simultaneously preserving the standard canonical form of the simplectic tensor q and the hermetric β is describable as a unitary simplectic group, being correspondingly denoted by $USp(n-p,p)$ according to the signature. (As for an ordinary simplectic group $Sp(n)$ such a group is necessarily special in the sense that the representation matrices must have unit determinent). The application we have in mind in setting up the above formalism is to the 4-component Dirac spinors to be described in section 22, for which the relevant unitary simplectic group is $USp(2,2)$, which contains the covering $Sp(2)\mathbb{C} \equiv SL(2)\mathbb{C}$ of the proper orthochronous Lorentz group (though not the Dirac covering spin(4) of the full Lorentz group) as a subgroup. The group $USp(2,2)$ is itself a subgroup of the twistor group $SU(2,2)$ (see Penrose and Mac Callum 1972) which preserves only the Dirac spinor hermetric.

Now the analogue of (19.19) for covectors is

$$C : \mathcal{V}_* \longrightarrow \bar{\mathcal{V}}_*$$

$$v_a \longmapsto \bar{v}_a = \bar{v}_{\dot{a}} \bar{C}^{\dot{a}}{}_b \tag{19.21}$$

When the conditions (19.10) and (19.19) are satisfied we shall have

$$\bar{C}^{\dot{a}}{}_b = - C^{\dot{a}}{}_b \tag{19.22}$$

and under these circumstances we obtain the general contraction switching rule

$$v_a \psi^a = - v_{\dot{a}} \psi^{\dot{a}} \tag{19.23}$$

From (19.10) by itself one has

$$v_a \psi^a = v^{\dot{a}} \psi_{\dot{a}} \tag{19.24}$$

and from (19.19) by itself one has

$$v_a \psi^a = - v^a \psi_a \tag{19.25}$$

20. Charged two-spinor Structures.

One of the most natural ways of setting up a theory of 2-spinors over a 4-dimensional base $\mathcal{M}$ is to start with a family of associated $GL(4) \times GL(2)\mathbb{C}$ bundles, in the manner described in section 11, with

$$\tilde{\mathcal{G}} = GL(2)\mathbb{C} \tag{20.1}$$

this being the group of homogeneous linear transformations of a fundamental complex 2-dimensional spinor representation space $\mathcal{U}$. The $GL(2)\mathbb{C}$ structure is related to the ordinary $GL(4)$ tangent space structure over $\mathcal{M}$ by the existence of a non-degenerate fundamental section $\underline{\sigma}$ in the associated bundle of hermitian $\bar{\mathcal{U}} \otimes \mathcal{U}$ valued 1-forms. A charged 2-spinor structure may be characterised by the existence not just of the fundamental section whose components $\sigma_\mu{}^{\dot{A}B}$ satisfy the hermiticity condition

$$\sigma_\mu{}^{\dot{A}B} = \bar{\sigma}_\mu{}^{B\dot{A}} \tag{20.2}$$

but also of a metric 2-spinor section with components $g_{\dot{A}B\dot{C}D}$ in an associated 4-th order 2-spinor bundle subject to the symmetry requirements

$$g_{\dot{A}B\dot{C}D} = g_{\dot{A}[B|\dot{C}|D]} = \bar{g}_{B\dot{A}D\dot{C}} \tag{20.3}$$

The hermiticity property (20.2) entails that any tangent vector $\vec{\xi}$ to $\mathcal{M}$ has an associated 2-nd order spinor $\vec{\xi}\cdot\underline{\sigma}$ with components $\xi^{\dot{A}B}$ given by

$$\xi^{\dot{A}B} = \xi^\mu \sigma_\mu{}^{\dot{A}B} \tag{20.4}$$

which will itself be hermetean, i.e

$$\xi^{\dot{A}B} = \bar{\xi}^{B\dot{A}} \tag{20.5}$$

provided $\vec{\xi}$ is an ordinary real vector.
The conditions (20.3) imply that the metric 2-spinor can be used to define a corresponding real pseudo-Riemannian metric over $\mathcal{M}$ according to the specification

$$ds^2 = g_{\dot{A}B\dot{C}D}\, dx^{\dot{A}B} dx^{\dot{C}D} \tag{20.6}$$

which is equivalent to taking

$$g_{\mu\nu} = g_{\dot{A}B\dot{C}D}\, \sigma_\mu{}^{\dot{A}B} \sigma_\nu{}^{\dot{C}D} \tag{20.7}$$

The symmetry conditions (20.3) are such that — like the measure tensor field with components $\varepsilon_{\mu\nu\rho\sigma}$ on an (orientable) 4-manifold — the metric spinor is unique to within an arbitrary real scale factor and by conformal scale adjustments one can arrange to cover $\mathcal{M}$ by local gauge patches such that its components reduce to a canonical form given by $g_{1\dot{2}2\dot{1}} = g_{2\dot{1}1\dot{2}} = \pm 1$, $g_{1\dot{1}2\dot{2}} = g_{2\dot{2}1\dot{1}} = \mp 1$ (the other components being always zero) in which the overall sign $\pm$ is fixed globally (independently of the local patch) in order that the section be well behaved. In terms of such a canonical gauge the metric formula (20.6) may be replaced by the more suggestive (though not fully covariant) equivalent expression

$$ds^2 = \mp \left| d\vec{x} \cdot \underline{\sigma} \right| \qquad (20.8)$$

(where the bars indicate the determinant of the 2-dimensional component matrix within, and the sign depends on the (globally constant) signature of g . The existence of the metric 2-spinor section is evidently equivalent to the condition that the structure group $GL(2)\mathbb{C}$ can be reduced to the subgroup of transformations that leave the canonical form of the metric 2-spinor, or equivalently the expression (20.8), invariant. Since an arbitrary 2-spinor gauge transformation

$$\iota^A \longmapsto S^A{}_B \iota^B \qquad \iota \in \mathcal{U} \qquad (20.9)$$

where $S^A{}_B$ are the fundamental representation components of $S \in GL(2)\mathbb{C}$, leads to the corresponding transformation

$$ds^2 \longmapsto \| S \|^2 ds^2 \qquad (20.10)$$

corresponding to a length rescaling factor given by the modulus of the determinant of the fundamental representation matrix of the transformation, we see that the invariance condition characterising the required subgroup is expressible by

$$\| S \|^2 = 1 \qquad (20.11)$$

Thus one sees that the structure group (20.1) may be replaced by the reduced transformation group

$$\tilde{g} = \left(Sp(2)\mathbb{C} \times U(1) \right) / Z_2 \qquad (20.12)$$

this being the subgroup characterised by (20.11). (we recall that the symplectic group $Sp(2)\mathbb{C}$ is more widely albeit less usefully known as $SL(2)\mathbb{C}$. The factoring in (20.12) is performed by identifying the product of an $Sp(2)\mathbb{C}$ and a $U(1)$ element with the product of their negatives.

The condition that the fundamental section $\underline{\sigma}$ be non-degenerate is equivalent to the condition that there exist a dual basis of tangent vector valued covariant two spinor matrices $\vec{\sigma}$ satisfying a similar hermicity condition

$$\sigma^{\mu}{}_{\dot{A}B} = \bar{\sigma}^{\mu}{}_{B\dot{A}} \tag{20.13}$$

the 1-forms and such that their contraction gives the appropriate unit tensor according to the

$$\sigma^{\mu}{}_{\dot{A}B}\ \sigma_{\nu}{}^{\dot{A}B} = \mp\ \delta^{\mu}{}_{\nu} \tag{20.14}$$

or equivalently

$$\sigma^{\mu}{}_{\dot{A}B}\ \sigma_{\mu}{}^{\dot{C}D} = \mp\ \delta^{\dot{C}}{}_{\dot{A}}\ \delta^{D}{}_{B} \tag{20.15}$$

(where the sign $\pm$ is chosen in conformity with that appearing in (20.3) for later convenience) This dual element can be used to solve the relation (20.7) to obtain

$$g_{\dot{A}B\dot{C}D} = g_{\mu\nu}\ \sigma^{\mu}{}_{\dot{A}B}\ \sigma^{\nu}{}_{\dot{C}D} = (\vec{\sigma}_{\dot{A}B}, \vec{\sigma}_{\dot{C}D}) \tag{20.16}$$

If we think of $\vec{\sigma}$ as determining a tetrad of real vectors $\vec{\xi}$, $\vec{\eta}$, $\vec{\zeta}$, $\vec{\tau}$ or equivalently a triad of (what will turn out to be null) covectors

$$\ell_{\mu} = \frac{1}{\sqrt{2}}(\tau_{\mu}+\zeta_{\mu}), \quad n_{\mu} = \frac{1}{\sqrt{2}}(\tau_{\mu}-\zeta_{\mu}), \quad m_{\mu} = \frac{1}{\sqrt{2}}(\xi_{\mu}+i\eta_{\mu}) \tag{20.17}$$

of which the last is complex, by the correspondence

$$\underline{\sigma} \leftrightarrow \begin{pmatrix} \underline{\ell} & \underline{\bar{m}} \\ \underline{m} & \underline{n} \end{pmatrix} \tag{20.18}$$

then (20.16) can be interpreted as providing a corresponding multiplication table which in a canonical gauge as specified by (20.3) gives

$$(\underline{m}, \underline{\bar{m}}) = -(\underline{\ell}, \underline{n}) = \pm 1 \tag{20.19}$$

or equivalently

$$(\vec{\xi}, \vec{\xi}) = (\vec{\eta}, \vec{\eta}) = (\vec{\zeta}, \vec{\zeta}) = -(\vec{\tau}, \vec{\tau}) = \pm 1 \tag{20.20}$$

the other combinations being always zero. This shows that the signature of the metric defined by (20.2), (20.3), 20.7) is necessarily Lorentzian and that the tetrad of real vectors specified implicitly by (20.17) and (20.18) is always orthonormal in any canonically normalised gauge.

Since the structure group for a charged 2-spinor structure can be reduced to the form(20.12) a charged 2-spinor connection is correspondingly required to be specified in terms of a connection 1-form with values in the Lie algebra of $(Sp(2)\mathbb{C} \times U(1)/Z_2$ this restriction being equivalent to the requirement that the canonical form of the metric 2-spinor be conserved so that one has

$$D_\mu g_{\dot{A}B\dot{C}D} = 0 \tag{20.21}$$

This means that (in an arbitrary gauge, whose label α will be suppressed) the gauge connection 1-form $\tilde{\omega}$ can be decomposed into the sum of separate $Sp(2)\mathbb{C}$ and $U(1)$ algebra components given in terms of their fundamental representatives by

$$\omega_\mu{}^A{}_B = \Gamma_\mu{}^A{}_B - ie A_\mu \delta^A{}_B \tag{20.22}$$

where e is a conventional real charge weight factor and where the $Sp(2)\mathbb{C}$ generator is characterised by the property of being trace free, i.e.

$$\Gamma_\mu{}^A{}_A = 0 \tag{20.23}$$

The gauge 1-form $\underline{A}$ appearing as the coefficient of the trace term is required to be effectively real in the sense that

$$A_\mu - \bar{A}_\mu = \frac{2i}{e} \partial_\mu \ln s \tag{20.24}$$

where s is a real conformal scale factor measuring the dilatation of the gauge actually used with respect to a canonical gauge as characterised by (20.8)(for which one would have $\ln s = 0$ so that the second term of (20.22) would then be a genuine $U(1)$ generator).

In order that the connection be compatible with the pseudo-Riemannian property (14.1) the 2-spinor connection is required to satisfy the furthe condition

$$D_\mu \sigma_\nu{}^{\dot{A}B} = 0 \tag{20.25}$$

which, taken in conjonction with (20.13) leads to the desired result

$$D_\mu g_{\nu\rho} = 0 \tag{20.26}$$

Now as in the derivation of (18.11) one sees from (11.14) that (20.25) implies

$$(\text{rep}\,\underline{\Omega})\,\underline{\sigma} = 0 \tag{20.27}$$

which written out in full is equivalent to

$$-R_{\mu\nu}{}^\sigma{}_\rho \sigma_\sigma{}^{\dot{A}B} + \bar{\Omega}_{\mu\nu}{}^{\dot{A}}{}_{\dot{C}} \sigma_\rho{}^{\dot{C}B} + \Omega_{\mu\nu}{}^B{}_C \sigma_\rho{}^{CB} = 0 \tag{20.28}$$

If by analogy with (20. 4) we now make the decomposition

$$\Omega_{\mu\nu}{}^{A}{}_{\cdot} = \mathcal{R}_{\mu\nu}{}^{A}{}_{B} - ie\, F_{\mu\nu}\, \delta^{A}{}_{B} \tag{20·29}$$

with

$$\mathcal{R}_{\mu\nu}{}^{A}{}_{A} = 0 \tag{20·30}$$

and

$$F_{\mu\nu} - \bar{F}_{\mu\nu} = 0 \tag{20·31}$$

(the reality of F (unlike A) being gauge independent in consequence of the fact that Ω (unlike ω) is a well behaved section) then by suitably contracting (20.28) with $\vec{\sigma}$ we can obtain

$$R_{\mu\nu}{}^{\tau}{}_{\rho} = \mp\, \mathcal{R}_{\mu\nu}{}^{B}{}_{C}\, \bar{\sigma}^{\tau}{}_{B\dot{A}}\, \sigma_{\rho}{}^{\dot{A}C} \tag{20·32}$$

and also

$$\mathcal{R}_{\mu\nu}{}^{B}{}_{C} = \mp\tfrac{1}{2}\, R_{\mu\nu}{}^{\sigma}{}_{\rho}\, \bar{\sigma}_{\sigma}{}^{B\dot{A}}\, \sigma^{\rho}{}_{\dot{A}C} \tag{20·33}$$

with the sign $\pm$ still fixed as in (20.8), which shows how the ordinary tangent space curvature completely determines and is determined by the corresponding $Sp(2)\mathbb{C}$ part. The independent $U(1)$ part of the curvature is obtainable directly (in consequence of the abelian nature of $U(1)$) from the familiar Maxwellian formula

$$F_{\mu\nu} = 2\, \partial_{[\mu} A_{\nu]} \tag{20·34}$$

whose $Sp(2)\mathbb{C}$ analogue is

$$\mathcal{R}_{\mu\nu}{}^{A}{}_{B} = 2\, \partial_{[\mu} \Gamma_{\nu]}{}^{A}{}_{B} + 2\, \Gamma_{[\mu|}{}^{A}{}_{C}\, \Gamma_{|\nu]}{}^{C}{}_{B} \tag{20·35}$$

Although it is seldom useful to do so(a better tactic is described in the next section) one can evaluate the $Sp(2)\mathbb{C}$ part of the connection in terms of the tangent space connection by writing out (20.25) in full and contracting with $\vec{\sigma}$ to obtain

$$\Gamma_{\mu}{}^{B}{}_{C} = \pm\, \tfrac{1}{2}\, \sigma^{\nu}{}_{\dot{A}D}\, \nabla_{\mu}\, \sigma_{\nu}{}^{\dot{A}B} \tag{20·36}$$

We complete this section by remarking that using the special algebraic tricks (see e.g. Pirani, 1965) that are available in two dimensions one can go on to establish a useful analogue of (20.24) to the effect that

$$g_{A\dot{E}C\dot{H}}\, g_{B\dot{F}D\dot{G}} - g_{A\dot{E}D\dot{G}}\, g_{B\dot{F}C\dot{H}} = \eta_{\mu\nu\rho\sigma}\, \sigma^{\mu}{}_{A\dot{E}}\, \sigma^{\nu}{}_{B\dot{F}}\, \sigma^{\rho}{}_{C\dot{G}}\, \sigma^{\sigma}{}_{D\dot{H}} \qquad (20.37)$$

where

$$\eta_{\mu\nu\rho\sigma} = \eta_{KL}\, \lambda^{K}{}_{\mu\nu}\, \lambda^{L}{}_{\rho\sigma} = i\, \varepsilon_{\mu\nu\rho\sigma} \qquad (20.38)$$

are components of the tensor introduced in (15.11) for the purposes of defining the involutive adjoint. To establish (20.37) one must verify that the components $\eta_{\mu\nu\rho\sigma}$ are pure imaginary (which is easy) and then that it is totally antisymmetric (which is obvious only for the first and last pairs) and finally that it is correctly normalised. The choice of sign in (20.22) fixes an orientation on the base manifold, which shows that a charged 2-spinor structure is possible only when $\mathcal{M}$ is orientable.

21. Neutral Two-Spinor Structures and the Debever-Newman Penrose Formalism.

The charged 2-spinor structures that have just been defined include as special cases what we shall refer to as neutral spin structures, which are defined by the requirement that the structure group $\mathcal{G}$ can be further restricted to be just the double covering $Sp(2)\mathbb{C}$ of the proper orthochronous Lorentz group $SO(3)\mathbb{C}$ itself. It is in this more specialised sense that the term spin-structure has been used by workers such as Geroch (1968, 1970) who showed that the existence of a neutral 2-spinor-structure implies that the topology of $\mathcal{M}$ must be such that it admits a global orthonormal frame field. (The topological requirements for the existence of a charged spinor structure are slightly less severe though nevertheless still fairly restrictive). As has recently been remarked by Hawking and Pope (1978), the observation that electrically charged fermions such as electrons have charges given by integral — as opposed to half integral — multiples of the same basic unit as bosons suggests that the charged 2-spinor structures relevant to the description of electromagnetically coupled fermions must indeed be capable of being restricted to a Majorama 2-spinor structure (so that the theorem of Geroch applies). The fact that there exist not merely integrally charged fermions such as the electron but also completely neutral fermions such as neutrinos, as well as charged (pseudo-)scalar mesons would seem to imply that the $(Sp(2)\mathbb{C} \times U(1))/Z_2$ structure is constituted as a trivial combination of separately existing $Sp(2)\mathbb{C}$ (neutral 2-spinor) and $U(1)$ (charge) structures. This does not of course imply that the $Sp(2)\mathbb{C}$ and $U(1)$ bundle structures taken separately should necessarily be topologically trivial.

One can proceed as in the previous section to express the more specialised condition of existence of a Majorana 2-spinor structure as the requirement that the 1-form field characterised by (20.3) should exist simultaneously with a spin simplectic field with local components ε_{AB} satisfying

$$\varepsilon_{AB} = \varepsilon_{[AB]} \tag{21.1}$$

This antisymmetry condition suffices to ensure the existence of preferred systems in which the only non zero components are $\varepsilon_{12} = 1$ and $\varepsilon_{21} = -1$ and hence to ensure that the bundle group $\tilde{\mathcal{G}}$ can be reduced from $GL(2)\mathbb{C}$ to the subgroup $Sp(2)\mathbb{C}$ that preserves this canonical form. Whenever the spin simplectic field exists one can always construct a canonically corresponding metric spinor field by taking

$$g_{\dot{A}B\dot{C}D} = \pm\, \varepsilon_{AC}\, \bar{\varepsilon}_{\dot{B}\dot{D}} \tag{21.2}$$

(where the $\pm$ sign agrees with that introduced in (20.8)) but the converse does not hold: assuming the existence of a global solution of (21.2) it is still not unique, since we can always make the $U(1)$ transformation

$$\varepsilon_{AB} \longmapsto e^{-2i\varphi}\, \varepsilon_{AB} \tag{21.3}$$

(for any globally defined real scalar field φ) without violating (21.1) or (21.2).

Given that it exists the tensor ε_{AB} can be used for lowering and raising of neutral 2-spinor indices in accordance with the general principles described in section 13 (remember raising is performed by contraction in front, and lowering by contraction behind). Many simplifications result from the two dimensional version of (13.18) which gives

$$\varepsilon^{AB}\, \varepsilon_{CD} = 2\, \delta^{A}{}_{[C}\, \delta^{B}{}_{D]} \tag{21.4}$$

By contracting this with the product of two sigma matrices one obtains from (20.2) the relation

$$\delta^{A}_{C}\, \bar{\sigma}_{\mu}{}^{E}{}_{\dot{F}}\, \sigma_{\nu}{}^{\dot{F}}{}_{E} = 2\, \bar{\sigma}_{(\mu|}{}^{A}{}_{\dot{B}}\, \sigma_{|\nu)}{}^{\dot{B}}{}_{C} \tag{21.5}$$

and hence by (20.7) and (21.1)

$$\bar{\sigma}_{\mu}{}^{A}{}_{\dot{B}}\, \sigma^{\nu\dot{B}}{}_{C} + \bar{\sigma}^{\nu A}{}_{\dot{B}}\, \sigma_{\mu}{}^{\dot{B}}{}_{C} = \pm\, \delta^{A}{}_{C}\, \delta^{\nu}{}_{\mu} \tag{21.6}$$

Another important application of (20.4) is to the case of a 2-form with tangent space components $F_{\mu\nu}$ and corresponding spinor components

$$F_{\dot{A}B\dot{C}D} = F_{\mu\nu}\,\sigma^{\nu}{}_{\dot{A}B}\,\sigma^{\mu}{}_{\dot{C}D}$$
$$= F_{[\dot{A}|B|\dot{C}]D} + F_{\dot{C}[B|\dot{A}|D]}$$
$$= \varepsilon_{\dot{A}\dot{C}}\,{}^{+}\Phi_{BD} + \varepsilon_{BD}\,{}^{-}\Phi_{\dot{C}\dot{A}} \tag{21.7}$$

where

$$\left.\begin{aligned} {}^{+}\Phi_{AB} &= {}^{+}\Phi_{BA} = \tfrac{1}{2} F_{\dot{C}A}{}^{\dot{C}}{}_{B} \\ {}^{-}\Phi_{\dot{A}\dot{B}} &= {}^{-}\Phi_{\dot{B}\dot{A}} = \tfrac{1}{2} F_{\dot{A}C\dot{B}}{}^{C} \end{aligned}\right\} \tag{21.8}$$

and where

$$F_{\mu\nu} = \overline{F}_{\mu\nu} \iff {}^{-}\Phi_{\dot{A}\dot{B}} = {}^{+}\overline{\Phi}_{\dot{A}\dot{B}} \tag{21.9}$$

so that there is a 1-1 correspondence between symmetric 2-spinors and real 2-forms. The 2-spinor version is particularly convenient for the discussion of the complex adjoint, since by (20.37) one obtains

$$\star F_{\dot{A}B\dot{C}D} = \varepsilon_{\dot{A}\dot{C}}\,{}^{+}\Phi_{BD} - \varepsilon_{BD}\,{}^{-}\Phi_{\dot{A}\dot{C}} \tag{21.10}$$

and in particular, the self adjointness condition is expressible by

$$F_{\mu\nu} = \star F_{\mu\nu} \iff {}^{-}\Phi_{\dot{A}\dot{B}} = 0 \tag{21.11}$$

Just as we were led to impose the condition (20.21) in the charged case, so also in the case of $Sp(2)\mathbb{C}$ spinors one demands that the connection should satisfy the more restrictive condition

$$D_{\mu}\,\varepsilon_{AB} = 0 \tag{21.12}$$

(of which (20.21) is a consequence) as well of course as the condition (20.25). In this case the gauge connection will still have the form (20.2) in arbitrary gauge but the form $\underline{A}$ will be exact in its entirely (not just its imaginary part) and in a canonical gauge (for which $\varepsilon_{12} = -\varepsilon_{21} = 1$) we shall have more particularly

$$\omega_{\mu}{}^{A}{}_{B} = \Gamma_{\mu}{}^{A}{}_{B}\,, \qquad \Gamma_{\mu[AB]} = 0 \tag{21.13}$$

For any gauge we shall correspondingly have

$$\Omega_{\mu\nu}{}^{A}{}_{B} = \mathcal{R}_{\mu\nu}{}^{A}{}_{B}\,, \qquad \mathcal{R}_{\mu\nu[AB]} = 0 \tag{21.14}$$

It is often technically convenient to introduce a four dimensional representation space $\mathcal{V}$ for the proper orthochronous Lorentz group $SO(3)\mathbb{C}$ and hence of its covering $Sp(2)\mathbb{C}$, which allows us to introduce an- as usual non-degenerate- section of the associated $\mathcal{V}_* \otimes \bar{\mathcal{U}} \otimes \mathcal{U}$ bundle with components $\sigma_m{}^{\dot{A}B}$ that have a constant fixed value in any canonical gauge as specified by the condition that $g_{\dot{A}B\dot{C}D}$ and the associated metric components

$$g_{mn} = \sigma_m{}^{\dot{A}B} \sigma_n{}^{\dot{C}D} g_{\dot{A}B\dot{C}D} \qquad (21{\cdot}15)$$

should have (corresponding) canonically fixed values. By the non-degeneracy condition such a constant section determines a corresponding section of the associated $\mathcal{J} \times \mathcal{V}$ bundle with components $\theta_\mu{}^m$ defined by the relation

$$\sigma_\mu{}^{\dot{A}B} = \theta_\mu{}^m \sigma_m{}^{\dot{A}B} \qquad (21{\cdot}16)$$

which can be identified with the tangent-frame field introduced in section 18, in view of the fact that the $\mathcal{V}$ -transformations induced by $Sp(2)\mathbb{C}$ form a subgroup of those unduced by $O(3,1)$.

The constancy of $\sigma_m{}^{\dot{A}B}$ is expressed by the condition that

$$\partial_\mu \sigma_m{}^{\dot{A}B} = 0 \qquad (21{\cdot}17)$$

in any canonical gauge and

$$-a^\ell{}_m \sigma_m{}^{\dot{A}B} + \bar{a}^{\dot{A}}{}_{\dot{C}} \sigma_m{}^{\dot{C}B} + a^B{}_C \sigma_m{}^{\dot{A}C} = 0 \qquad (21{\cdot}18)$$

Since the gauge connection 1-form $\underline{\omega}$ itself will have have values satisfying (21.18) in a canonical gauge, (21.17) and (21.18) combine to give the fully covariant condition

$$D_\mu \sigma_m{}^{\dot{A}B} = 0 \qquad (21{\cdot}19)$$

The condition (21.18) also shows that the $\mathcal{V}$ and $\mathcal{U}$ representation components of any element a in the $Sp(2)\mathbb{C}$ algebra are related by

$$a^\ell{}_m = -\left(s^\ell{}_m{}^B{}_A\, a^A{}_B + \bar{s}^\ell{}_m{}^{\dot{B}}{}_{\dot{A}}\, a^{\dot{A}}{}_{\dot{B}} \right) \qquad (21{\cdot}20)$$

$$a^A{}_B = -\tfrac{1}{2}\, s^m{}_\ell{}^A{}_B\, a^\ell{}_m \qquad (21{\cdot}21)$$

where we have introduced the spin tensor s defined implicitly by

$$\bar{\sigma}^\ell{}_{B\dot{C}}\, \sigma_m{}^{\dot{C}A} = \pm \left(s^\ell{}_m{}^A{}_B + \tfrac{1}{2} \delta^\ell_m \delta^A_B \right) \qquad (21{\cdot}22)$$

or explicitly by

$$\pm S_{mn}{}^{AB} = \bar{\sigma}_{[m|}{}^{A}{}_{\dot{C}}\, \sigma_{n]}{}^{\dot{C}B}$$

$$= \sigma_{m}{}^{\dot{C}D}\, \sigma_{n}{}^{\dot{E}F}\, \bar{\varepsilon}_{\dot{C}\dot{E}}\, \delta^{A}{}_{(D}\, \delta^{B}{}_{F)} \qquad (21.23)$$

The latter expression demonstrates not only the symmetry property

$$S_{mn}{}^{[AB]} = 0 \qquad (21.24)$$

charateristic of $Sp(2)\mathbb{C}$ generators, but also the self **adjointness** property

$$S_{mn}{}^{AB} = \star S_{mn}{}^{AB} = \tfrac{1}{2}\, \eta_{mn}{}^{k\ell}\, S_{k\ell}{}^{AB} \qquad (21.25)$$

as may be seen by comparison with (21.7) and (21.11). These properties provides a key link between the 2-spinor formalism as it was developed particularly by the work of Newman and Penrose, and the self-adjoint bivector formalism, due particularly to Debever, that was introduced in section 15. To make the bridge we start by introducing a set of complex basis vectors Z_α , $\alpha = 1, 2, 3$ for the $Sp(2)\mathbb{C}$ algebra, treated as a 3-complex dimensional space, so as to be able to express any element a in the form

$$a = a^{\alpha} Z_{\alpha} \qquad (21.26)$$

(which is to be contrasted with the alternative representation (15.41) where a was regarded as belonging to the 6-real ($\equiv$ 3-complex) dimensional subset of a 6-complex ($\equiv$ 12 real) dimensional space). The fundamental representation components are correspondingly expressible by

$$a^{A}{}_{B} = a^{\alpha} Z_{\alpha}{}^{A}{}_{B} \qquad (21.27)$$

where the components $Z_{\alpha}{}^{A}{}_{B}$ are <u>constants</u> satisfying

$$Z_{\alpha}{}^{A}{}_{A} = 0 \qquad (21.28)$$

and also obeying the analogue of (12.31), which implies that they can be thought of as representing a field that will satisfy

$$D_{\mu}\, Z_{\alpha}{}^{A}{}_{B} = 0 \qquad (21.29)$$

We can evidently use these basis elements to express the spin tensor in the form

$$S^{m}{}_{n}{}^{A}{}_{B} = S^{m}{}_{n}{}^{\alpha}\, Z_{\alpha}{}^{A}{}_{B} \qquad (21.30)$$

so that (21.21) becomes

$$a^\alpha = -\tfrac{1}{2}\, S^\ell{}_m{}^\alpha\, a^m{}_\ell \tag{21.31}$$

Comparing this with the analogous expression

$$a^\alpha = -\,{}^+Z^{\alpha\,\ell}{}_m\, a^m{}_\ell \tag{21.32}$$

obtained by translating (15.42) into terms of tetrad components, we see that we can arrange for the components a^α defined here to agree with those defined in section 15 by making the identification

$${}^+Z^{\alpha\,\ell}{}_m = \tfrac{1}{2}\, S^\ell{}_m{}^\alpha \tag{21.33}$$

The Debever formalism consists of the application of a procedure analogous to that of section 18 not to the set of 1-forms but to the set of 2-forms

$${}^+\underline{Z}^\alpha = \tfrac{1}{4}\, S_{mn}{}^\alpha\, \underline{\theta}^m \wedge \underline{\theta}^n \tag{21.34}$$

with components given by

$${}^+Z^\alpha{}_{\mu\nu} = {}^+Z_{mn}\, \theta^m_\mu\, \theta^n_\nu \tag{21.35}$$

One sees from (18.6) and (21.29) that these are components of a field satisfying

$$D_\mu\, {}^+Z^\alpha{}_{\mu\nu} = 0 \tag{21.36}$$

As in the ordinary Cartan formalism one proceeds by taking the exterior product (using (11.18), (12.13)) in the form

$$\partial \wedge {}^+\underline{Z}^\alpha + {}^+@_{\beta\gamma}{}^\alpha\, \underline{\omega}^\gamma \wedge {}^+\underline{Z}^\beta = \wedge(\underline{\vec{\Theta}} \cdot Z^\alpha) \tag{21.37}$$

from which—assuming a known (e.g. zero) torsion $\underline{\vec{\Theta}}$ — one can read out the complex gauge connection components $\underline{\omega}^\gamma$, and hence evaluate the expression

$$\underline{\omega} = \underline{\omega}_\alpha\, Z^\alpha \tag{21.38}$$

(where $\underline{\omega}_\alpha = {}^+\gamma_{\alpha\beta}\, \underline{\omega}^\beta$) for the gauge-connection 1-form itself. The curvature is then obtained as

$$\begin{aligned} \underline{\Omega} &= \underline{\Omega}_\alpha\, Z^\alpha \\ &= (\partial \wedge \underline{\omega}_\alpha) Z^\alpha + \tfrac{1}{2}\, \underline{\omega}^\beta \wedge \underline{\omega}^\gamma\, [Z_\beta, Z_\gamma] \end{aligned} \tag{21.39}$$

which gives

$$\underline{\Omega}_\alpha = \partial \wedge \underline{\omega}_\alpha + \tfrac{1}{2}\, @_{\beta\gamma\alpha}\, \underline{\omega}^\beta \wedge \underline{\omega}^\gamma \tag{21.40}$$

To relate the connection and curvature components so obtained to the corresponding spinorial components, one now uses (21.27) to deduce that in any canonical gauge, as restricted by

$$\varepsilon_{AB} = 2\,\delta^1_{[A}\delta^2_{B]} \tag{21.41}$$

so that (21.13) is satisfied, we shall have

$$\underline{\omega}^\alpha Z_\alpha{}^A{}_B = \underline{\Gamma}^A{}_B \tag{21.42}$$

while (in an arbitrary gauge)

$$\begin{aligned} \underline{\Omega}^\alpha Z_\alpha{}^A{}_B &= \underline{\mathcal{R}}^A{}_B \\ &\doteq -(\underline{\mathcal{R}}^\ell{}_m\, {}^{+}Z^{\alpha\ell}{}_m)\, Z_\alpha{}^A{}_B \end{aligned} \tag{21.43}$$

where the last line is obtained by applying (21.27), (21.30) and (21.33). Comparing this with (15.42) we see that it implies

$$\underline{\Omega}_\alpha = \underline{\mathcal{R}}_\alpha \tag{21.44}$$

where $\underline{\mathcal{R}}_\alpha$ is as given by (15.58).

The standard choice of bases for the D.N.P. formalism is made at the Newman-Penrose end by taking

$$\underline{\theta}^1 = \mp \underline{n}\,, \quad \underline{\theta}^2 = \mp \underline{\ell}\,, \quad \underline{\bar{\theta}}^3 = \underline{\theta}^4 = \pm \underline{m} \tag{21.45}$$

where $\underline{n}$, $\underline{\ell}$, $\underline{m}$ are the (complex) null vectors given by (20.18) Since the metric will then be given by

$$\mp g_{mn}\, \underline{\theta}^m \underline{\theta}^n = 2\,\underline{\theta}^1\underline{\theta}^2 - 2\,\underline{\theta}^3\underline{\theta}^4 \tag{21.46}$$

this is equivalent to taking

$$\theta_1{}^\mu = \ell^\mu\,, \quad \theta_2{}^\mu = n^\mu\,, \quad \theta_3{}^\mu = \bar{\theta}_4{}^\mu = m^\mu \tag{21.47}$$

The standard Debever choice (at least in recent work) for the base components $Z^\alpha{}_{AB}$ is given by

$$Z^1{}_{AB} = \delta^1_A\delta^1_B\,, \quad Z^2{}_{AB} = \delta^1_A\delta^2_B + \delta^2_A\delta^1_B\,, \quad Z^3{}_{AB} = \delta^2_A\delta^2_B \tag{21.48}$$

This choice ensures as a consequence of (21.42) that the Debever connection components $\omega_{\mu\alpha}$ are identical with the corresponding Newman-Penrose spin-coefficients $\Gamma_{a AB}$ in accordance with the tabulation

Γ_{mAB}	1	2	3	4	$m \diagup \alpha$
(1,1)	κ	τ	σ	ρ	1
(1,2), (2,1)	ε	γ	β	α	2
(2,2)	π	ν	μ	λ	3
$(A,B) \diagup m$	1	2	3	4	$\omega_{m\alpha}$

(21.49)

in which the coefficients have been given individual Greek labels in accordance with the standard Newman-Penrose usage.

To evaluate these coefficients we first use (21.23), (21.30), (21.33) to obtain

$$^+\underline{Z}^1 = \mp \underline{\theta}^1 \wedge \underline{\theta}^3 \ , \quad ^+\underline{Z}^2 = \mp(\underline{\theta}^1 \wedge \underline{\theta}^2 - \underline{\theta}^3 \wedge \underline{\theta}^4) \ , \quad ^+\underline{Z}^3 = \mp \underline{\theta}^4 \wedge \underline{\theta}^2 \tag{21.50}$$

from (21.48). By (13.25) or (15.36) the complex algebra metric is

$$^+\gamma_{\alpha\beta} = -2\, Z_\alpha{}^{AB} Z_{\beta AB} = {}^+Z_\alpha{}^{\mu\nu}\, {}^+Z_{\beta\mu\nu}$$

$$= \delta^{(1}_\alpha \delta^{3)}_\beta - \tfrac{1}{4}\, \delta^2_\alpha \delta^2_\beta \tag{21.51}$$

In this three dimensional case the structure constants are the components of the ordinary alternating tensor i.e.

$$^+@_{\alpha\beta\gamma} = \varepsilon_{\alpha\beta\gamma} \tag{21.52}$$

whose non-zero values are equal in magnitude to 1/4, this being the square root of the determinant

$$|^+\gamma| = 1/16 \tag{21.53}$$

When the torsion is zero the equations (21.37) reduce to the explicit form

$$\left.\begin{aligned}
\mp\partial\wedge{}^+\underline{Z}^1 &= (2\gamma-\mu)\underline{\theta}^1\wedge\underline{\theta}^2\wedge\underline{\theta}^3 + (\pi-2\alpha)\underline{\theta}^1\wedge\underline{\theta}^3\wedge\underline{\theta}^4 + \nu\,\underline{\theta}^2\wedge\underline{\theta}^3\wedge\underline{\theta}^4 - \lambda\,\underline{\theta}^1\wedge\underline{\theta}^2\wedge\underline{\theta}^4 \\
\mp\partial\wedge{}^+\underline{Z}^2 &= -2\tau\,\underline{\theta}^1\wedge\underline{\theta}^2\wedge\underline{\theta}^3 + 2\rho\,\underline{\theta}^1\wedge\underline{\theta}^3\wedge\underline{\theta}^4 - 2\mu\,\underline{\theta}^2\wedge\underline{\theta}^3\wedge\underline{\theta}^4 + 2\pi\,\underline{\theta}^1\wedge\underline{\theta}^2\wedge\underline{\theta}^4 \\
\mp\partial\wedge{}^+\underline{Z}^3 &= \sigma\,\underline{\theta}^1\wedge\underline{\theta}^2\wedge\underline{\theta}^3 - \kappa\,\underline{\theta}^1\wedge\underline{\theta}^3\wedge\underline{\theta}^4 + (2\beta-\tau)\underline{\theta}^2\wedge\underline{\theta}^3\wedge\underline{\theta}^4 + (\rho-2\varepsilon)\underline{\theta}^1\wedge\underline{\theta}^2\wedge\underline{\theta}^4
\end{aligned}\right\} \tag{21.54}$$

where the left hand sides are evaluated by substituting from (21.50) for known values of the $\underline{\theta}^m$ 1-forms. The complete set of $\Gamma_{m\dot{A}B}$ components can evidently be read out directly from the coefficients appearing on the right hand side of (21·54). Once the connection coefficients have been worked out it is straight forward to evaluate the curvature form $\underline{\mathcal{R}}_\alpha$ using (21.40) and (21.44), and hence by (15.58) to obtain the corresponding conformal and trace free Ricci components $C_{\alpha\beta}$ and $E_{\alpha\dot{\beta}}$ which have Newman-Penrose labels given by

$$C_{\alpha\beta} \leftrightarrow \begin{pmatrix} \Psi_0 & \Psi_1 & \Psi_2 \\ \Psi_1 & \Psi_2 & \Psi_3 \\ \Psi_2 & \Psi_3 & \Psi_4 \end{pmatrix} \qquad E_{\alpha\dot{\beta}} \leftrightarrow \begin{pmatrix} \Phi_{00} & \Phi_{01} & \Phi_{02} \\ \Phi_{10} & \Phi_{11} & \Phi_{12} \\ \Phi_{20} & \Phi_{21} & \Phi_{22} \end{pmatrix} \qquad (21\cdot 55)$$

For further details the reader is referred to the original works of Newman, Penrose and Debever and their colleagues, and particularly to the excellent Review by Pirani (1965). These authors all work with a metric signature (+---) which corresponds to choosing the value - and + repectively for the $\pm$ and $\mp$ signs appearing in our sections 20 and 21 .

22. Dirac and Majorana Spinors.

The original motivation for the use of four- (instead of two-) dimensional spaces was provided by the consideration that the 4-spinors admit representation of groups covering the full Lorentz group $O(3,1)$ not just the proper orthochronous component (that we have seen to be equivalent to $SO(3)\mathbb{C}$). Forty years after their original introduction into physics by Dirac, this consideration seems much less important in view of the fact that modern theories treat parity and time reversal invariance as at most approximate symmetries valid exactly only in idealised limiting cases (such as models limited to the description of "pure" gravitation and electromagnetism). The reason why 4-spinor representations remains as fashionable as ever (even for the description of weak interactions) may more convincingly be explained as due to the technical advantages resulting from the fact that (unlike 2-spinors) the 4-spinors form the fundamental representation space, in the manner described in section 19, for a canonical unitary structure in the charged case and a canonical unitary-symplectic structure in the neutral case.

The fundamental 4-spinor representation space may be constructed from the fundamental 2-spinor representation space $\mathcal{U}$ by taking the direct (not tensor) product $\mathcal{W} = \mathcal{U}_* \oplus \mathcal{U}$ so that to any 4-spinor with components ψ^α say their corresponds a pair of 2-spinors with components φ_A and $\chi^{\dot{A}}$ say. The class of canonical 2-spinor bases evidently determines a corresponding class of canonical Weyl 4-spinor bases in which the two kinds of components are identified according to the correspondence

$$\psi^{a} \longleftrightarrow \begin{pmatrix}\psi^{1}\\ \psi^{2}\\ \psi^{3}\\ \psi^{4}\end{pmatrix} = \begin{pmatrix}\varphi_{1}\\ \varphi_{2}\\ \chi^{\dot 1}\\ \chi^{\dot 2}\end{pmatrix} \longleftrightarrow \begin{pmatrix}\varphi_{A}\\ \chi^{\dot A}\end{pmatrix} \tag{22.1}$$

(This has the same form as the correspondence defining twistors (see Penrose and Mc Callum 1970) though the latter have quite different interpretation and invariance properties). On this space there is a natural hermitian bracket structure defined by

$$\langle \bar\psi, \psi\rangle = \bar\varphi_{\dot A}\,\chi^{\dot A} + \bar\chi^{A}\varphi_{A} = \bar\psi\cdot\beta\cdot\psi \tag{22.2}$$

where the hermetric β , with components $\beta_{\dot a b}$ satisfying (19.10) is given in a Weyl base by the correspondence

$$\beta_{\dot a b} \longleftrightarrow \begin{pmatrix} 0 & \delta^{\dot A}{}_{\dot B}\\ \delta^{A}{}_{B} & 0\end{pmatrix} \tag{22.3}$$

One can check that this hermetric has signature ++-- so that it is left invariant by an action of the transformation group $SU(2,2)$ which contains $GL(2)\mathbb{C}$ as the subgroup specified by the condition that the mixed spinor γ_5 should also be left invariant, where (in accordance with the Streater-Wightman conventions that we use throughout this section) γ_5 is a 4-spinor with components $\gamma_5{}^{a}{}_{b}$ given by the correspondence

$$\gamma_5{}^{a}{}_{b} \longleftrightarrow \begin{pmatrix} -i\,\delta_{A}{}^{B} & 0\\ 0 & i\,\delta^{\dot A}{}_{\dot B}\end{pmatrix} \tag{22.4}$$

so that the matrices $\tfrac{1}{2}(\mathbb{1} + i\gamma_5)$ and $\tfrac{1}{2}(\mathbb{1} - i\gamma_5)$ are the projection operators mapping the 4-spinor representation space $\mathcal{W}$ onto $\mathcal{U}_*$ and $\bar{\mathcal{U}}$ respectively.

A set of ordinary "gamma matrices" γ_μ (μ = 1,2,3,4) with mixed components $\gamma_\mu{}^{a}{}_{b}$ may be defined in a Weyl frame by

$$\gamma_\mu{}^{a}{}_{b} \longleftrightarrow \begin{pmatrix} 0 & i\sqrt{2}\,\bar\sigma_{\mu\dot A B}\\ i\sqrt{2}\,\sigma_{\mu}{}^{\dot A B} & \end{pmatrix} \tag{22.5}$$

It can be seen (by (20.14)) that the products of these matrices satisfy

$$\gamma_{(\mu}\cdot\gamma_{\nu)} = \pm g_{\mu\nu}\mathbb{1} \;,\quad \gamma_{(\mu}\cdot\gamma_{5)} = 0 \;,\quad \gamma_5\cdot\gamma_5 = -\mathbb{1} \tag{22.6}$$

where, as in sections 20 and 21, the sign ± has the upper (+) value if the signature is (+++-) and the lower(-) value if it is (---+).

In the manner described in section 19 one can use the metric β for a dot-lowering operation so as to be able to write the bracket (22.6) in the form

$$\langle\bar{\psi},\psi\rangle = {}^{+}\psi\cdot\psi = \bar{\psi}_a\psi^a \tag{22.7}$$

where the Dirac adjoint ${}^{+}\psi$ is defined by

$$({}^{+}\psi)_a = \bar{\psi}_a = \bar{\psi}^{\dot{b}}\beta_{\dot{b}a} \tag{22.8}$$

The hermiticity of β can be expressed equivently in component or matrix notation by

$$\beta = \beta^{+} \iff \beta_{\dot{a}b} = \bar{\beta}_{b\dot{a}} \tag{22.9}$$

It can be seen that the γ_5 spinor has a corresponding hermiticity property

$$\beta\gamma_5 = \gamma_5^{+}\beta^{+} \iff \gamma_{5\dot{a}b} = \bar{\gamma}_{5b\dot{a}} \tag{22.10}$$

while the 4-vectorial $\underline{\gamma}$ -spinors have the anti-hermiticity property

$$\beta\gamma_\mu = -\gamma_\mu^{+}\beta^{+} \iff \gamma_{\mu\dot{a}b} = -\bar{\gamma}_{\mu b\dot{a}} \tag{22.11}$$

The general 4-spinor structure as set up so far possesses the full invariance of the charged 2-spinor group, including the phase transformations:

$$\varphi_A \mapsto e^{-i\varphi}\varphi_A \Rightarrow \psi^a \mapsto e^{-i\varphi}\psi^a \tag{22.12}$$

In the more specialised <u>neutral</u> case for which the 2-spinor symplectic ε_{AB} is given, one can go on to set up a corresponding symplectic q with components q_{ab} given in terms of a Weyl 4-spinor frame by

$$q_{ab} \longleftrightarrow \begin{pmatrix} \varepsilon^{AB} & 0 \\ 0 & -\bar{\varepsilon}_{\dot{A}\dot{B}} \end{pmatrix} \tag{22.13}$$

and obviously satisfying the antisymmetry condition

$$q = -q^{T} \quad \Leftrightarrow \quad q_{ab} = -q_{ba} \tag{22.14}$$

One can check that the compatibility condition (19.13) is in fact satisfied so that the symplectic q can be used consistently with the hermetric β for raising and lowering of 4-spinor indices. The corresponding charge conjugation operator as defined by (19.15) has components $C^{\dot{a}}{}_{b}$ given in a general frame by

$$C^{\dot{a}}{}_{b} = \beta^{\dot{a}}{}_{b} \tag{22.15}$$

or more explicitly, in a Weyl frame by

$$C^{\dot{a}}{}_{b} \leftrightarrow \begin{pmatrix} 0 & \bar{\varepsilon}_{\dot{A}\dot{B}} \\ -\varepsilon^{AB} & 0 \end{pmatrix} \tag{22.16}$$

Although the Weyl frames used so far are very satisfactory for most purposes, it is sometimes convenient to make a transformation (mixing the φ_{A} and $\chi^{\dot{A}}$ components) to a new kind of base known as a Majorana frame which is characterised by the requirement that the charge conjugation matrix should have the form

$$C^{\dot{a}}{}_{b} \leftrightarrow \left(\delta^{\dot{a}}{}_{b}\right) \tag{22.17}$$

A spinor is said to be Majorana or more briefly "real" if its components in a Majorana frame are all real. The property of being Majorana can be expressed in a frame invariant manner e.g. for a first order spinor ψ as the requirement

$$\bar{\psi} = C\psi \quad \Leftrightarrow \quad \bar{\psi}^{a} = \psi^{a} \tag{22.18}$$

It can be seen that the skew-hermitean vectorial γ matrices we have introduced do in fact satisfy the Majorana ("reality") condition

$$\bar{\gamma}_{\mu} = C\gamma_{\mu}C^{-1} \quad \Leftrightarrow \quad \gamma_{\mu}{}^{a}{}_{b} = \bar{\gamma}_{\mu}{}^{a}{}_{b} \tag{22.19}$$

whereas for the hermitean matrix γ_{5} one has the anti-Majorana ("imaginarity") property

$$\bar{\gamma}_{5} = -C\gamma_{5}C^{-1} \quad \Leftrightarrow \quad \gamma_{5}{}^{a}{}_{b} = -\bar{\gamma}_{5}{}^{a}{}_{b} \tag{22.20}$$

while a similar anti-Majorana property

$$\bar{\beta} = -C^{T}\beta\bar{C} \quad \Leftrightarrow \quad \beta_{\dot{a}b} = -\bar{\beta}_{\dot{a}b} \tag{22.21}$$

holds for the fundamental hermetric itself.

By the defining relation (21.23) one can see that the antisymmetrised product of gamma tensors will be given in a Weyl frame by

$$\gamma_{[\mu|}{}^{a}{}_{c}\,\gamma_{|\nu]}{}^{c}{}_{b} \longleftrightarrow \begin{pmatrix} \mp 2\, s_{\mu\nu\,A}{}^{B} & \\ 0 & \mp 2\, \bar{s}_{\mu\nu}{}^{\dot{A}}{}_{\dot{B}} \end{pmatrix} \tag{22.21}$$

so that from the property (21.25) one can deduce that the gamma tensors will satisfy the useful relations

$$\gamma_5 \cdot \gamma_{[\mu} \cdot \gamma_{\nu]} = \tfrac{1}{2}\, \varepsilon_{\mu\nu\rho\sigma}\, \gamma^{\rho} \cdot \gamma^{\sigma} \tag{22.22}$$

which has as corrollories

$$\gamma_5 = \tfrac{1}{4!}\, \varepsilon_{\mu\nu\rho\sigma}\, \gamma^{\mu} \cdot \gamma^{\nu} \cdot \gamma^{\rho} \cdot \gamma^{\sigma} \tag{22.23}$$

and the product formula

$$\gamma_{\mu} \cdot \gamma_{\nu} = \pm\, g_{\mu\nu}\, \mathbb{1} - \tfrac{1}{2}\, \varepsilon_{\mu\nu\rho\sigma}\, \gamma^{\rho} \cdot \gamma^{\sigma} \cdot \gamma^{5} \tag{22.24}$$

From the way the 4-spinors have been constructed it is evident that a 2-spinor connection will induce an associated 4-spinor connection given in a Weyl frame by

$$\omega_{\mu}{}^{a}{}_{b} \longleftrightarrow \begin{pmatrix} -\omega_{\mu}{}^{B}{}_{A} & 0 \\ 0 & \bar{\omega}{}^{\dot{A}}{}_{\dot{B}} \end{pmatrix} \tag{22.25}$$

while similarly for the curvature

$$\Omega_{\mu\nu}{}^{a}{}_{b} \longleftrightarrow \begin{pmatrix} -\Omega_{\mu\nu}{}^{B}{}_{A} & 0 \\ 0 & \bar{\Omega}_{\mu}{}^{\dot{A}}{}_{B} \end{pmatrix} \tag{22.26}$$

so that using (20.31) and (20.35) we obtain

$$\Omega_{\mu\nu}{}^{a}{}_{b} = \pm \tfrac{1}{4}\, R_{\mu\nu\rho\sigma}\, \gamma^{\rho a}{}_{c}\, \gamma^{\sigma c}{}_{b} - ie\, F_{\mu\nu}\, \delta^{a}{}_{b} \tag{22.27}$$

It is evident that we shall have

$$\left.\begin{aligned} D_{\mu}\, \beta_{ab} = 0 \qquad & D_{\mu}\, \gamma_5{}^{a}{}_{b} = 0 \\ & D_{\mu}\, \gamma_{\nu}{}^{a}{}_{b} = 0 \end{aligned}\right\} \tag{22.28}$$

and in the neutral case also

$$D_{\mu}\, q_{ab} = 0 \tag{22.29}$$

As in the two spinor case one can use a basis satisfying (18.6) to construct a field whose components $\gamma_{m}{}^{a}{}_{b}$ will form a set γ_m "gamma matrices" which will be fixed once and for all for any gauge restricted to a given canonical (e.g Weyl or Majorana) type. One sees that the relations (21.20), (21.21) relating the fundamental $SO(3,1)$ and $Sp(2)\,\mathbb{C}$ representatives of an element a of the neutral 2-spinor algebra can be translated into terms of 4-spinors in the form

$$a_{mn} = \pm \tfrac{1}{2} \gamma_{[m\,l}{}^{a}{}_{c} \gamma_{|n]}{}^{c}{}_{b} a^{b}{}_{a} \tag{22.30}$$

$$a^{a}{}_{b} = \pm \tfrac{1}{4} \gamma_{m}{}^{a}{}_{c} \gamma_{n}{}^{c}{}_{b} a^{mn} \tag{22.31}$$

It is to be remembered that whereas the tetrad representative satisfies the antisymmetry condition

$$a_{mn} = - a_{nm} \tag{22.32}$$

the 4-spinor representative satisfies the symmetry condition

$$a_{ab} = a_{ba} \tag{22.33}$$

as well as the antihermiticity condition

$$a_{\dot{a}b} = - \bar{a}_{b\dot{a}} \tag{22.34}$$

This correspondence can be extended to an arbitrary 2-form $\underline{F}$, which may always be throught of as the covariant form of the space-time representative of an $SO(3,1)$ generator, so that we obtain

$$F_{\mu\nu} = \tfrac{1}{4} \, tr \; \gamma_{[\mu} \cdot \gamma_{\nu]} \cdot \Phi \tag{22.35}$$

where Φ is a field with components $\Phi^{a}{}_{b}$ given in terms of the quantities ${}^{+}\Phi_{AB}$ introduced in (21.8) by

$$\Phi^{a}{}_{b} \leftrightarrow \begin{pmatrix} {}^{+}\Phi_{A}{}^{B} & 0 \\ 0 & {}^{-}\Phi^{\dot{A}}{}_{\dot{B}} \end{pmatrix} \tag{22.36}$$

(One can see from (22.17) that $\underline{F}$ will be real if Φ is Majorana) It is evident from(22.21) that if Φ is the 2nd order 4-spinor corresponding to the 2-form $\underline{F}$ in this way, then $\gamma_5 \cdot \Phi$ will be the 4-spinor analogously corresponding to $*F$ (so that $\underline{F}$ is self dual if $(\mathbb{1} - i\gamma_5)\,\Phi$ vanishes).

An interesting application of the foregoing formalism arises in the Kerr and Kerr Newman black hole solutions of the vaccuum Einstein and Einstein-Maxwell field equations in which the Dirac equation

$$\gamma^{\mu} D_{\mu} \psi = m \psi \tag{22.37}$$

has been shown to be soluble by a (rather elaborate) separation of variables method (Chandrasekhar 1976, Page, 1976, Guren 1977). It has recently been demonstrated (Carter and McLenaghan, 1978) that the Chandrasekhar separation constant can be accounted for as an eigen-value of a certain(appropriately self adjoint) operator

$$L = i\gamma_{\mu} \cdot (f^{\mu\nu} \gamma_5 D_{\nu} + \tfrac{1}{3} * f^{\mu\nu}{}_{;\nu} \mathbb{1}) \tag{22.38}$$

where $\underline{f}$ is a two-form satisfying the conditions

$$f_{\mu(\nu;\rho)} = 0 \tag{22.39}$$

and

$$F_{\rho[\mu} f^{\rho}{}_{\nu]} = 0 \tag{22.40}$$

which turn out to be just what is required to ensure that L commutes with the Dirac operator $\gamma^{\mu} D_{\mu}$, i.e

$$[L, \gamma^{\mu} D_{\mu}] = 0 \tag{22.41}$$

(Such a commutation condition ensures the existence of a corresponding conserved current which may be explicitly constructed using prescription given by Carter 1977). The existence of the form $\underline{f}$ was brought to light in the Kerr solutions by the work of Penrose and Floyd (Penrose (1975), Floyd (1975)) who showed that it could be used to represent the generalised total squared angular momentum constant for classical orbits (that had previously been discovered by Carter1968) in the form $p_{\mu} f^{\mu}{}_{\nu} f^{\nu}{}_{\rho} p^{\rho}$ where $\underline{p}$ is the classical momentum covector. ($\underline{f}$ is in fact proportional to the Debever vector $^{+}\underline{\underline{Z}}_{2}$ as defined by (21.50) with respect to the type-D null eigenvectors)

An analogous but more obvious kind of operator (see e.g Kosmann 1969) can be constructed whenever one has an ordinary space-time symmetry generated by a Killing vector $\vec{k}$ satisfying

$$\vec{k} \pounds g_{\mu\nu} \equiv 2 k_{(\mu;\nu)} = 0 \tag{22.42}$$

$$\vec{k} \pounds F_{\mu\nu} = 0 \tag{22.43}$$

The presence of such a symmetry permits the construction of an operator K satisfying

$$[K, \gamma^{\mu} D_{\mu}] = 0 \tag{22.44}$$

according to the specification

$$K = i k^{\mu} D_{\mu} - \tfrac{1}{4} i \gamma^{\mu} . \gamma^{\nu} k_{\mu;\nu} + c \mathbb{1} \tag{22.45}$$

where the scalar c satisfies

$$\partial_{\mu} c = e k^{\nu} F_{\nu\mu} \tag{22.46}$$

(which is automatically integrable in consequence of (22.43)). In the Kerr-Newmann solutions there are of course two such operators corresponding to energy and axial angular momentum.

It has been shown during the course of the present summer school (McLenaghan and Spindel 1978) that the most general conceivable (appropriately self adjoint) first order linear operator commuting with the Dirac operator consists of a linear combination of operators of the forms (22.45) and (22.38) together with an operator of a third kind, with the form

$$M = y^{\mu\nu\rho} \gamma_\nu \gamma_\rho D_\mu + \frac{3}{4} {}^*y^\mu{}_{;\mu} \gamma_5 \tag{22.47}$$

which will satisfy

$$[M, \gamma^\mu D_\mu] = 0 \tag{22.48}$$

whenever $\underline{y}$ is a <u>three-form</u> satisfying the third order Yano equation ((22.42) and (22.39) being respectively the first and second order Yano equations) namely

$$y_{\mu\nu(\rho;\sigma)} = 0 \tag{22.49}$$

together with the supplementary electromagnetic condition

$$F_{\rho[\mu} y^\rho{}_{\nu\sigma]} = 0 \tag{22.50}$$

the dual vector ${}^*y$ being given in accordance with (13.16) by

$${}^*y^\mu = -\frac{1}{3!} \varepsilon^{\mu\nu\rho\sigma} y_{\nu\rho\sigma} \tag{22.51}$$

An example of this third kind unfortunately does <u>not</u> arise in the Kerr solutions.

Acknowledgments

I wish to thank G.F.R. Ellis, C.W. Misner, E.T. Newman, R. Penrose, and D.W. Sciama who originally introduced me to the topics described here, as well as the many colleagues, particularly G. Gibbons, S.W. Hawking and M. Perry, who have helped me in following recent developments. The actual preparation of the present course was greatly influenced by detailed and enlightenning discussions with J.Madore, R.G.McLenaghan and J.Thierry-Mieg.

Bibliography

The following list contains texts that were directly helpful to me in preparing this course, or that I would recommend from personal knowledge for following up particular points. It is in no way definitive or exhaustive and many possible alternatives are available.

E.M. Corson: Introduction to tensors Spinors and Relativistic Wave Equations, Blackie, Glasgow, 1973.

Y. Choquet-Bruhat, C. Morette-Dewitt, M.Bleik-Dillard: Physical Mathematics - Analysis on Manifolds, North Holland, Amsterdam, 1977.

B.S. Dewitt: Les Houches Lectures, in Relativity, Groups and Topology, Gordon and Breach, New York 1964.

GF.R. Ellis: Cargese Summer School Lectures, in General Relativity and Cosmology, Academic Press 1971.

W. Greub, S. Halperin, R. Vanstone: Connections, Curvature and Cohomology, Vol II, Academic Press, New York 1973.

S.W. Hawking, G.F.R. Ellis: The large Scale Structure of Space-Time, Cambridge U.P., 1973.

N.J. Hicks: Notes on Differential Geometry, Van Nostrand Reinhold Co, London, 1971.

C. Itzykson, J.B. Zuber: Electrodynamique et Theorie Quantique des Champs, Service de Physique Théorique, Centre d'Etudes Nucléaires de Saclay, 91190 Gif sur Yvette, 1978.

S. Kobyashi ,K. Nomizu: Foundations of Differential Geometry, Vol II, Interscience, New York, 1969.

A. Lichnerowicz: Les Houches Lectures, in Relativity, Groups and Topology, Gordon and Breach, New York 1964.

J.W. Milnor, J.D. Stasheff: Characteristic Classes, Princeton U.P., New Jersey, 1974.

C.W. Misner, K.S. Thorne, J.A. Wheeler: Gravitation, Freeman, San Francisco, 1973.

F.A.E. Pirani: Brandeis Summer School Lectures, in Lectures on Theoretical Physics, Prentice Hall, New Jersey ,1965.

J.A. Schouten: Ricci Calculus, Springer Verlag, Heidelberg, 1954.

R.F. Streater and A.S. Wightman, PCT, Spin and Statistics, and All That, Benjamin, New York, 1964.

A. Trautman: Brandeis Summer School Lectures, in Lectures on Theoretical Physics, Prentice Hall, New Jersey, 1965.

A. Trautman, On the Structure of the Einstein-Cartan Equations, Istituto Nazionale di Alta Matematica, Symposia Mathematica, Vol XII, 1973.

M. Veltman: Quantum Theory of Gravity, in Proceedings of 1975, Les Houches Summer School, ed R.Balian, J.Zinn-Justin, North Holland 1976.

REFERENCES

J. Thierry-Mieg (1978) Geometrical Reinterpretation of the Fadeev Popov, Ghosts and BRS transformation, preprint, Obs. de Paris, 92190 Meudon.

C.J. Isham (1978), Proc.Roy.Soc. A362, 383.

R. Debever (1964) Cahier de Physique 168, 303.

M. Cahen, R. Debever, L. Defrise (1976) J. Math. and Mechanics, 16, 761.

R. Debever (1976) Bull. CI. Sci. Acad. Roy. Bel. LXII, 662.

R. Penrose (1960) Ann. Phys. (New York) 10, 171.

E.T. Newman and R. Penrose (1962) J. Math. Phys. 3, 566.

R. Penrose (1968) in Battelle Rencontre, ed. C. Dewitt and J.A. Weeler, Benjamin, New York.

R. Geroch (1968) J. Math. Phys. 9, 1739.
(1970) " " 21, 343.

S.W. Hawking and N.J. Pope (1978) preprint, D.A.M.T.P., Cambridge.

S. Chandrasekkar (1976) Proc. Roy. Soc. A 349, 571.

D.N. Page (1976) Phys. Rev. D 14, 1509.

R. Guren (1977) Proc.Roy. Soc.A356, 465.

B. Carter (1977) Phys. Rev. D16, 3395.

Y. Kosman (1972) Ann. di Mat. Pura ed Appl. IV 91, 317.

B. Carter and R. Mc Lenaghan (1978), preprint, Observatoire de Paris, Meudon.

R. Mc Lenaghan and Ph. Spindel (1978) preprint, Dept de mathematique, U.L.B.Bruxelles.

R. Penrose and MAH McCallum (1972), Physics Reports, 6, 241.

B. Carter (1968) Phys. Rev. 174, 1559.

R. Penrose (1963), Ann.New York Acad. Sci. 224, 125.

Part II
Quantum Gravity

EUCLIDEAN QUANTUM GRAVITY

Stephen W. Hawking

University of Cambridge, D.A.M.T.P.

Silver Street, Cambridge, England

1. INTRODUCTION

In these lectures I am going to describe an approach to Quantum Gravity using path integrals in the Euclidean regime i.e. over positive definite metrics. (Strictly speaking, Riemannian would be more appropriate but it has the wrong connotations). The motivation for this is the belief that the topological properties of the gravitational fields play an essential role in Quantum Theory. Attempts to quantize gravity ignoring the topological possibilities and simply drawing Feynman diagrams corresponding to perturbations around flat space have not been very successful: there seem to be an infinite sequence of undetermined renormalization parameters. The situation is slightly better with supergravity theories; the undetermined renormalization parameters seem to come in only at the third and higher loops around flat space but perturbations around metrics that are topologically non-trivial introduce undetermined parameters even at the one loop level [1] [27] as I shall show later on.

It seems to me that the fault lies not with the pure gravity or supergravity theories themselves but with the uncritical application of perturbation theory to them. In classical general relativity we have found that the perturbation theory has only a limited range of validity. One cannot describe a black hole as a perturbation around flat space. Yet this is what writing down a string of Feynman diagrams amounts to. On a technical level, the failure of perturbation theory can be traced to the fact that the "free" quadratic part of the action for general relativity does not bound the higher "interaction" terms unlike in Yang-Mills theory or Q.E.D. The free action would bound the interaction terms if one added quadratic terms in the curvature to the action. However such additions **alter** the nature of the theory and lead to fourth order equations, tachyons and ghosts, though I shall mention a way in which they can be used as conformal gauge fixing terms.

I must admit that I do not have an answer to the breakdown of the perturbation expansion but, like the man who looked for his key under the lamp-post because that was the only place in which he had any chance of finding it, I feel that if there is an answer, it must involve the topological structure of their gravitational field. I use the path integral approach because that seems to be the only way to handle topological questions. When one does path integrals for non-gravitational fields in flat spacetime one normally performs a Wick rotation of the time axis by replacing t by - it. This converts Minkowski space with Lorentzian metric (signature -+++) into Euclidean space (signature ++++). One reason for doing this is that it makes the path integral better defined. The path integral for (say) a scalar field ϕ in Minkowski space has the form

$$\int D[\phi] \exp (iI[\phi]) \tag{1.1}$$

where $D[\phi]$ is a measure on the space of all field configurations ϕ and $I[\phi]$ is the action of the field ϕ. For real fields on real Minkowski space this integral oscillates and does not converge. However, if one performs a Wick rotation to Euclidean space the path integral becomes

$$\int D[\phi] \exp (-\hat{I}[\phi]) \tag{1.2}$$

where $\hat{I}[\phi] = -iI[\phi]$ is the "Euclidean" action of the field ϕ and is real for fields ϕ that are real on the Euclidean section of complexified spacetime. Since I is usually positive definite, the path integral over all such fields will tend to converge. One can then analytically continue the resultant expression back to Minkowski space. This analytic continuation automatically incorporates the concepts of positive frequency and time ordering. For instance the Feynman propagator

$$\langle 0|T\phi(x)\ \phi(y)|0\rangle \tag{1.3}$$

is positive frequency in $t = x^o - y^o$, i.e. holomorphic in the lower half plane for Re $t > 0$ and negative frequency i.e. holomorphic in the upper half plane for Re $t < 0$. It is therefore holomorphic on the Euclidean space obtained by rotating the time axis 90° clockwise in the complex plane. In fact one could take the attitude that quantum theory and indeed the whole of physics is really defined in the Euclidean region and that it is simply a consequence of our perception that we interpret it in the Lorentzian regime.

I feel that one should adopt a similar Euclidean approach in quantum gravity and supergravity. Of course one cannot simply replace the time coordinates by imaginary quantities because there is no preferred set of time coordinates in general relativity. Instead I think one should perform the path integrals over all positive definite metrics, most of which will not admit a section on which the metric is real and Lorentzian, and then analytically continue the result of the path integral, if necessary. In order to restrict the path integral to positive definite metrics and to exclude integration over metrics with Lorentzian or ultra hyperbolic signatures, one should probably integrate not over the components of the metric g_{ab} but over the components $e^a{}_m$ of a tetrad. This can be regarded as the square root of the metric

$$g_{ab} = e^m{}_a e_{bm} \tag{1.4}$$

spacetime indices a, b ... are raised or lowered by the metric g^{ab} and g_{ab} and tetrad indices m, n... are raised or lowered by the Euclidean metric δ^{mn} and δ_{mn}. Thus the group of tetrad rotations is $SO(4) = \frac{SU(2) \times SU(2)}{Z_2}$ rather than Lorentzian group SO(1,3) that one normally uses.

The metric g_{ab} defined by equation 1.4 for arbitrary real $e_a{}^m$ will be positive semidefinite and will be degenerate where $\det(e_a{}^m) = 0$. I think that one must include such degenerate metrics in the path integral. They provide a way in which one can pass continuously from one spacetime topology to another. I shall return to this point later on.

Tetrads are also essential for dealing with fermion fields. I shall use two component spinors but in a positive definite metric rather than the usual Lorentzian one. In the Lorentzian case one has unprimed spinors λ_A which transform under SL(2,C) and primed spinors $\mu_{A'}$ which transform under the complex conjugate group $\overline{SL(2,C)}$. The complex conjugate operation takes unprimed spinors to primed spinors and vice versa i.e. $\bar{\lambda}_{A'}$ is a primed spinor. When one analytically continues to a positive definite metric, the primed and unprimed spinors transform under independent groups SU(2) and $\widetilde{SU}(2)$. The complex conjugate operation now takes unprimed to unprimed and primed to primed and either raises the index or lowers it with a minus sign. Quantities that were complex are conjugates of each other in the Lorentzian case, like the Weyl spinor Ψ_{ABCD} and $\bar{\Psi}_{A'B'C'D'}$, are analytically continued to the Euclidean region as independent fields ψ_{ABCD} and $\tilde{\psi}_{A'B'C'D'}$. Thus it is possible to have a metric in which $\psi_{ABCD} \neq 0$ but $\tilde{\psi}_{A'B'C'D'} = 0$. Such a metric is conformally self dual i.e. $C_{abcd} = {}^*C_{abcd} = \frac{1}{2} \varepsilon_{abef} C^{ef}{}_{cd}$. If the Ricci spinor $\Phi_{ABC'D'}$ and the curvature scalar Λ are also 0, the metric is self-dual, $R_{abcd} = {}^*R_{abcd}$.

A Dirac 4 spinor ψ can be represented by two 2 component spinors in a column vector

$$\psi = \begin{pmatrix} \lambda_A \\ \bar{\mu}^{A'} \end{pmatrix} \qquad (1.5)$$

The adjoint field $\bar{\psi}$ is represented by a row vector

$$\bar{\psi} = (\mu_A, \bar{\lambda}^{A'}) \qquad (1.6)$$

when one goes to Euclidean space the 4 spinor and its adjoint become independent fields ψ and $\tilde{\psi}$ which are represented by four independent 2 component spinors λ_A, $\tilde{\lambda}^{A'}$, $\tilde{\mu}_{A'}$ and μ^A. A Majorana 4 spinor ψ in Lorentzian space is represented by the column vector

$\psi = \begin{pmatrix} \rho_A \\ \bar{\rho}^{A'} \end{pmatrix}$. When one goes to the Euclidean region one has two independent 2 components spinors ρ_A and $\tilde{\rho}^{A'}$. Thus, contrary to what is often stated, one can deal with Majorana spinors in the Euclidean regime.

The action for the gravitational field is usually taken to be

$$I = \frac{1}{16\pi G} \int R(-g)^{\frac{1}{2}} d^4x \qquad (1.7)$$

I shall use units in which $G = c = \hbar = 1$. When one performs the Wick rotation to the Euclidean region the volume element $(-g)^{\frac{1}{2}}$ becomes $-i(g)^{\frac{1}{2}}$. Thus the Euclidean action would be

$$\hat{I} = -\frac{1}{16\pi} \int R(g)^{\frac{1}{2}} d^4x \qquad (1.8)$$

However this action contains second derivatives of the metric which have to be removed by integration by parts to give an action which is quadratic in first derivatives of the metric, as is required by the path integral approach. One therefore gets a surface term in the action which is often ignored but which turns out to be very important [2, 3]

$$\hat{I} = -\frac{1}{16\pi} \int R(g)^{\frac{1}{2}} d^4x - \frac{1}{8\pi} \int K(h)^{\frac{1}{2}} d^3x + C \qquad (1.9)$$

where h is the induced metric and K is the trace of the second fundamental form of the boundary of the region over which the action is being evaluated. The term C is an arbitrary constant which may depend on the induced metric h on the boundary but not on the metric g in the interior of the region. When the boundary metric h is such that the boundary can be embedded in flat space, it is natural to choose C so that the action is zero when g is the flat metric. However not all boundary metrics h can be embedded even locally in flat space. Also the existence of such a rigid boundary surface is not very physical. I shall therefore be describing ways in which one can eliminate the boundary term and deal only with compact manifolds.

I shall be talking mainly about the Euclidean approach to pure gravity but these ideas can also be extended to supergravity. For instance one could probably regard superspace S as some sort of fibre bundle over the spacetime manifold M with a Grassmannian fibre with coordinates $\Theta_A, \tilde{\Theta}^{A'}$ which represents the Euclidean version of a Majorana spinor. The first, and so far the most successful, application of Euclidean methods was to the thermal properties of black holes. Since much of this work has already been published I shall describe it only briefly. I shall then describe some as yet unpublished work on the gravitational vacuum and shall end up with the volume canonical ensemble which enables one to treat the foamlike structure of spacetime.

2. THERMAL PROPERTIES

The partition function $Z[\beta]$ for the thermal canonical ensemble for a scalar field ϕ in flat spacetime at temperature $T = \beta^{-1}$ is defined as

$$Z[\beta] = \Sigma_n \langle \phi_n \mid \exp(-\beta H) | \phi_n \rangle \qquad (2.1)$$

where $|\phi_n\rangle$ is a complete orthonormal basis of states for the field ϕ. One can also represent Z as a path integral

$$Z[\beta] = \int D[\phi] \exp(-\hat{I}[\phi]) \qquad (2.2)$$

where the path integral is taken over all field configurations ϕ that are real in Euclidean space and are periodic with period β in the Euclidean time coordinate (i.e. they are periodic in imaginary Minkowski time). One can represent the partition function for higher spin boson fields in a similar way but, in the case of fields with gauge degrees of freedom, one has to include the Fadeev-Popov ghosts to subtract out the unphysical degrees of freedom. One can also treat fermion fields in a similar way though in this case the fields have to be antiperiodic i.e. they reverse sign when the Euclidean time coordinate increases by β.

It would be natural to define the partition function for the thermal canonical ensemble for gravity in an analogous manner.

In order to avoid the infra-red problems caused by having an infinite volume of thermal gravitons with an infinite mass, one would like to enclose the system in (say) a spherical box of radius r_o with perfectly reflecting walls. This of course is rather unphysical becuase one cannot make a rigid perfectly reflecting box even for electromagnetic radiation let alone for gravitational radiation. It leads to problems near the walls of the box but I shall ignore these for the present anyway. The world tube of the spherical box identified with period β in the Euclidean time coordinate defines a 3-dimensional manifold ∂M with topology $S^2 \times S^1$ with a positive definite metric h which is the product of the standard metric of a 2-sphere of radius r_o and 1-dimensional metric on a circle of circumference β. The partition function $Z[\beta]$ for the gravitational field at temperature $T = \beta^{-1}$ in the spherical box is defined by a path integral over all metrics g on all manifolds M which have ∂M as their boundary and which induce the given metric h on ∂M.

One expects, or at least one hopes, that the dominant contribution to the path integral will come from metrics near a background metric g_o that extremises the action i.e. is a solution of the classical field equations with the given boundary conditions.

One can expand the action in a Taylor series about the back ground metric g_o

$$\hat{I}[g] = \hat{I}[g_o] + I_2[g_o, \bar{g}] + \text{higher order terms} \tag{2.3}$$

where $g = g_o + \bar{g}$ and I_2 is quadratic in the perturbations $\bar{g}$. Then

$$\log Z = - \hat{I}[g_o] + \log \int D[\bar{g}] \exp(-I_2) \tag{2.4}$$
$$+ \text{higher order terms}$$

One can regard the first term as the contribution to log Z of the background metric and the second, "one loop" term as the contribution of thermal gravitons on the background. The measure $D[\bar{g}]$ for the one loop term can be expressed as

$$D[\bar{g}] = \Pi_n \mu da_n \pi^{-\frac{1}{2}} \tag{2.5}$$

where μ is a normalization quantity or regulator and a_n are the coefficients in the expansion of the perturbation $\bar{g}$ in terms of eigenfunctions of the second order elliptic operator A which determines I_2 i.e.

$$\bar{g} = \Sigma a_n \phi_n \tag{2.6}$$

where $A\phi_n = \lambda_n \phi_n$ and (2.7)

$$I_2 = \int \bar{g}\, A\, \bar{g}(g_o)^{\frac{1}{2}} d^4x \tag{2.8}$$

The one loop term L is formally [4]

$$L = \frac{\det(\mu^{-2}C)}{(\det(\mu^{-2}(A + B)))^{\frac{1}{2}}} \tag{2.9}$$

where B is the gauge fixing operator and C is the ghost operator. The number $N(\lambda)$ of eigenvalues of an operator has an asymptotic expansion of the form

$$N(\lambda) = \sum_{n=o}^{\infty} P_n \lambda^{2-n} \tag{2.10}$$

P_o is proportional to the 4-volume of the background metric times the number of components of the field on which the operator acts. The higher coefficients P_n are polynomials in the metric, the curvature and its covariant derivatives of degree 2n in derivatives of the metric. The one loop term therefore diverges badly. The various regularization schemes all amount to dividing out by the distribution of eigenvalues given by the P_o and P_1 terms. In general however, except in flat space and a few other special background metrics, the net P_2 term will be non-zero. This means that there will be some finite number (not necessarily an integer) of extra eigenvalues whose dimensionality is not cancelled out. Because each extra eigenvalue has dimensions $(\text{length})^{-2}$, one has to divide it by the normalization quantity μ^2. Thus the one loop term is μ dependent. One cannot absorb this dependence in a redefinition of the coupling constant in the original action but, since it is proportional to the Euler number χ one might absorb it in a topological term $k\chi$ where k was a scale dependent topological coupling constant. I shall return to this later.

One background metric which satisfies the boundary conditions for the canonical ensemble is the flat metric on periodically identified Euclidean space i.e. on the manifold M with topology $S^1 \times R^3$. With an appropriate choice of the constant C, the action for this background metric is zero and so it makes no contribution to log Z. The one loop term representing the quadratic fluctuations around the flat space background metric gives the standard result for thermal radiation with two helicity states

$$\log Z = \frac{16\pi^3 r_o^3}{135\beta^3} \tag{2.11}$$

this can be interpreted as the partition function of thermal gravitons in flat space. Higher loops, if they had any meaning which is doubtful, would represent the effects of interactions between these gravitons.

The Schwarzschild metric is another solution with the given boundary conditions. The metric is normally expressed as

$$ds^2 = -(1 - \frac{2M}{r})\,dt^2 + (1 - \frac{2M}{r})^{-1}\,dr^2 + r^2 d\Omega^2 \tag{2.12}$$

Putting $t = -i\tau$ converts this into a positive definite metric for $r > 2M$. There is an apparent singularity at $r = 2M$ but this is like the apparent singularity at the origin of polar coordinates as can be seen by defining a new radial coordinate $x = 4M(1 - 2Mr^{-1})^{\frac{1}{2}}$. Then the metric becomes

$$ds^2 = (\frac{x}{4M})^2\,d\tau^2 + (\frac{r^2}{4M^2})^2\,dx^2 + r^2 d\Omega^2 \tag{2.13}$$

This will be regular at $x = 0$, $r = 2M$ if τ is regarded as an angular variable and is identified with period $8\pi M$ (I am using units in which $G = 1$). The manifold defined by $x \geq 0$, $0 \leq \tau \leq 8\pi M$ is called the Euclidean section of the Schwarzschild solution. On it the metric is positive definite, asymptotically flat and non-singular (the curvature singularity at $r = 0$ does not lie on the Euclidean section).

The Scwarzschild metric will satisfy the boundary conditions if

$$M = \frac{\beta}{8\pi}\,.$$

Because it has R = 0, the action will come entirely from the surface term. This gives [3]

$$\hat{I} = 4\pi M^2 = \frac{\beta^2}{16\pi} \tag{2.14}$$

Thus the Schwarzschild background metric will give a contribution of

$$\frac{-\beta^2}{16\pi}$$

to log Z. By its definition

$$Z = \Sigma \exp(-\beta E_n) \tag{2.15}$$

where E_n is the energy of the nth state of the gravitational field. Thus

$$\langle E \rangle = \frac{\Sigma E_n \exp(-\beta E_n)}{\Sigma \exp(-\beta E_n)} = -\frac{\partial}{\partial\beta} \log Z \tag{2.16}$$

Inserting the contribution $\frac{-\beta^2}{16\pi}$ to log Z from the action of the background Schwarzschild metric one obtains

$$\langle E \rangle = \frac{\beta}{8\pi} = M \tag{2.17}$$

This is what one would expect. However one can also compute the entropy defined as

$$S = -\Sigma p_n \log p_n = \beta \langle E \rangle + \log Z \tag{2.18}$$

where $p_n = Z^{-1} \exp(-\beta E_n)$ is the probability of being in the nth state. Applied to the Schwarzschild background metric this gives

$$S = 4\pi M^2 = \tfrac{1}{4}A$$

where A is the area of the event horizon. This relation between intrinsic gravitational entropy and the area of event horizons is quite general. It arises from a combination of the facts that the gravitational action is not scale invariant (which is what makes it nonrenormalizable) and the fact that the Euclidean section of the Schwarzschild solution has the topology $S^2 \times R^2$ which is different from periodically identified flat space which has topology $S^1 \times R^3$ [5]. There is no analogue in QCD or other lower spin theories

The one loop term contains a μ dependent term of the form

$$\frac{106\chi}{45} \log (\mu\beta) \text{ where } \chi = 2$$

is the Euler number of the Euclidean section of the Schwarzschild solution. This arises from

$$\frac{106\chi}{45}$$

extra eigenvalues whose dimensionality is not cancelled by the regularization procedure. Three of these extra eigenvalues are zero and correspond to translations of the black hole inside the box. They give a contribution of $\log (\mu^3\beta^3 V)$ to log Z. On the other hand a non relativistic particle of mass M in a box of volume V at temperature β^{-1} would have a partition function proportional to $V(M\beta^{-1})^{3/2}$. Thus it seems that to reproduce this one should take μ proportional to β^{-1}.

The eigenfunctions of the operator A + B can be divided into two classes [4]. First there are the eigenfunctions corresponding to metric perturbations $\bar{g}$ which are proportional to the background metric g_0. These represent conformal perturbations. The eigenvalues are all negative for perturbations around flat space or around any metric with R = 0. One therefore has to rotate the contour of integration of the conformal factor to lie parallel to the imaginary axis. The remaining eigenfunctions are traceless and correspond to nonconformal perturbations of the metric. At flat space they are all positive and therefore one adopts the rule that one integrates over real conformal equivalence classes of metrics. However for perturbations around the Schwarzschild background metric, one of these nonconformal eigenvalues is negative [6]. Since it appears in the one loop term under a square root sign, it contributes a factor i to the partition function Z which is therefore imaginary. One might have expected some such pathology in Z because the canonical ensemble for gravity breaks down due to its attractive nature.

One way of saying this is to express Z as the Laplace transform of the density of states N(E) where N(E)dE is the number of states of the gravitational field in the box with energies between E and E + dE:

$$Z[\beta] = \int_0^\infty N(E) \exp(-\beta E) dE \tag{2.19}$$

Thus N(E) is the inverse Laplace transform

$$N(E) = \frac{1}{2\pi i} \int_{-i\infty}^{i\infty} Z[\beta] \exp(\beta E) d\beta \tag{2.20}$$

However if one substitutes for Z the estimate

$$\exp\left(\frac{-\beta^2}{16\pi}\right)$$

arising from the action of the background metric, the integral does not converge. To make it converge one has to rotate the contour of the β integration in 2.20 from the imaginary axis. The fact that Z is imaginary then leads to a real N(E) as one would expect because the micro canonical ensemble for gravity is well-defined: a black hole in a box can be in stable equilibrium with thermal radiation if the total energy in the box is fixed but it is not stable if the temperature of the box is fixed because black holes have negative specific heat [7].

One can also consider ensembles where, in addition to temperature, one has chemical potentials for the angular momentum about some axis and the electric charge. The partition function would be given by a path integral over all fields which on some boundary surface were the same at the points (t, r, Θ, ϕ) and $(t + i\beta, r, \Theta, \phi + i\Omega\beta)$ in a gauge in which $A_o = \Phi$ where Ω is the angular velocity and Φ is the electrostatic potential. In keeping with the Euclidean approach Ω and Φ should be taken to be imaginary and then the partition function should be analytically continued back to real values of Ω and Φ.

The stationary phase metric in the path integral for the partition function will be periodically identified Euclidean space in rotating coordinates and the Euclidean section of the Kerr-Newman solution with imaginary angular momentum and electric charge (imaginary electric charge produces a real field F_{ab} in the Euclidean regime). A similar treatment to that for the Schwarzschild solution shows that the Kerr-Newman solution also has intrinsic quantum gravitational entropy equal to one quarter of the area of the event horizon.

3. THE GRAVITATIONAL VACUUM

In Yang-Mills theory one can describe the vacuum state with $F_{ab} = 0$ on a spacelike surface by potentials A_a which are pure gauge i.e. they are of the form

$$A_a = \Lambda^{-1}\partial_a\Lambda \tag{3.1}$$

where Λ is an element of the gauge group G. If one fixes Λ to be the unit element at infinity or equivalently if one compactifies the spacelike surface by adding a point at infinity to make it a 3-sphere, one has a map from S^3 into G. In the simple case that G is SU(2), it is also a 3-sphere topologically and so these maps can be divided into homotopy equivalence classes characterized by an integer n which is the degree of the mapping. One therefore has a degenerate family of vacuum states described by an integer n [8]. The transition amplitude to tunnelling from an initial vacuum n_1 to a final vacuum n_2 is given by a path integral over all Yang-Mills fields on Euclidean space which die away at large distance and which have a Pontryagin number $n_1 - n_2$. The action of such fields is bounded below by

$$\frac{1}{8\pi^2}\, n_1 - n_2$$

so that the tunnelling is suppressed.

One can perform a similar analysis for the gravitational vacuum. The zero field configuration, flat space, on a spacelike surface can be described by different tetrad fields which can be regarded as maps of the 3-sphere (the spacelike surface compactified by the addition of a point at infinity) into SO(4), the group of tetrad rotations. In fact since one wants to consider fermion fields as well one should consider spin frames rather than just ordinary tetrads. These correspond to maps of the 3-sphere into SU(2) x SU(2), the covering group of SO(4). Since each SU(2) is a 3-sphere topologically, these maps are divided into homtopy classes characterized by two integers (n, m) which are the degrees of the mapping of the 3-sphere on the two factors. One thus has a double infinity of degenerate gravitational vacua. The transition amplitude to tunnel from an initial vacuum (n_1, m_1) to a final vacuum (n_2, m_2) will be given by a path integral over all asymptotically Euclidean metrics on parallelizable manifolds with Euler number $\chi = (n_2 - n_1) + (m_2 - m_1) + 1$
and Hirzebruch signature $\tau = \frac{2}{3}\left[(n_2 - n_1) - (m_2 - m_1)\right]$

Asymptotically Euclidean means that the metric approaches the standard Euclidean metric on R^4 outside some compact region in which the topology will differ from that of R^4. Parallelizable means that one can define continuous tetrad and spin frame fields on the manifold and the conditions on χ and τ ensure that these fields can interpolate between the fields on the initial and final surfaces. In fact the condition that the manifolds admit a spin structure i.e. allow spinors to be defined consistently, implies that the manifold is parallelizable [9, 10]. In the asymptotically Euclidean case it also implies that τ is a multiple of 16. Thus the gravitational vacua are divided into classes which cannot tunnel into each other but live in completely separate universes.

The Euler number χ and the signature τ can be expressed as integrals of the curvature

$$\chi = \frac{1}{128\pi^2} \int R_{abcd} R^{efgh} \varepsilon^{ab}{}_{ef} \varepsilon^{cd}{}_{gh} (g)^{\frac{1}{2}} d^4x \quad + \text{ surface terms} \tag{3.2}$$

$$\tau = \frac{1}{96\pi^2} \int R_{abcd} R^{ab}{}_{ef} \varepsilon^{cdef} (g)^{\frac{1}{2}} d^4x + \text{ surface terms} \tag{3.3}$$

The surface terms are rather complicated in general but for an asymptotically Euclidean metric there is a contribution of one to χ for each asymptotically Euclidean region and a zero contribution to τ.

The Euler number χ is given by

$$\chi = B_o - B_1 + B_2 - B_3 + B_4 \tag{3.4}$$

The pth Betti number B_p is the number of independent closed p-surfaces that are not boundaries of some p + 1 surface. In a compact manifold B_p is equal to the number of square intergrable harmonic p forms. For a compact manifold, $B_p = B_{4-p}$ and $B_o = B_4 = 1$. If the manifold is simply connected $B_1 = B_3 = 0$ so $\chi \geq 2$.

It seems to me that one should probably restrict attention to simply connected manifolds. In this case χ and τ classify compact manifolds that admit spinor structure up to homotopy. It is conjectured (by Poincaré) that they classify the manifolds up to homeomorphisms. It can be shown that there is no possible classification scheme for non-simply connected 4 manifolds.

The B_2 harmonic 2 forms (Maxwell fields) can be divided into B_2^+ self-dual and B_2^- anti-self-dual 2 forms. Then $\tau = B_2^+ - B_2^-$. It is also equal to $8(n^+ - n^-)$ where n^+ and n^- are the numbers of zero modes of the massless Dirac equations with positive helicidies respectively.

In Yang-Mills theory it is convenient to compactify Euclidean space R^4 by adding a point at infinity to convert the manifold into S^4. The original flat metric can be recovered from the metric on S^4 by a conformal transformation that sends the added point to infinity. A similar procedure can be adopted in the gravitational case: one can conformally compactify an asymptotically Euclidean manifold by adding a point at infinity though in this case the resultant compact manifold need not be S^4 but could be any manifold that admitted a spinor structure. In fact one could regard this as a precise definition of what is meant by asymptotically Euclidean analogous to Penrose's definition of asymptotic flatness for Lorentzian manifolds [11]. What one does is start with a compact manifold with a smooth positive definite metric, pick a point z and send it to infinity by a conformal transformation. One obtains an asymptotically Euclidean metric with Euler number one less than that of the compact manifold but with the same signature.

The transition amplitude from the initial to the final vacuum can be expressed as a path integral over all asymptotically Euclidean metrics on the (simply connected) manifold M with the given values of χ and τ

$$Z = \langle 0_- | 0_+ \rangle = \int D[e] \exp(-\hat{I}[g]) \tag{3.5}$$

The path integral can be decomposed into an integral over conformal factors Ω followed by an integral over conformal equivalence classes $\{e^m_a\}$ of tetrads. Under a conformal transformation

$$\tilde{e}^m_a = \Omega e^m_a, \quad \tilde{g}_{ab} = \Omega^2 g_{ab} \tag{3.6}$$

the action becomes

$$\hat{I}[\Omega e] = -\frac{1}{16\pi} \int (\Omega^2 R + 6\Omega_{;a}\Omega_{;b} g^{ab})(g)^{\frac{1}{2}} d^4x$$

$$-\frac{1}{8\pi} \int \Omega^2 K (h)^{\frac{1}{2}} d^3x + C \tag{3.7}$$

The surface term is evaluated over a large sphere in asymptotically Euclidean space and the constant C is chosen to subtract out the flat space value of the surface term. To perform the path integral over the conformal factor one first finds a conformal factor ω which is equal to one on the boundary surface and which is such that the metric $g^*_{ab} = \omega^2 g_{ab}$ has $R^* = 0$ everywhere. This can be thought of as a choice of conformal gauge. The action of the metric g^* is given entirely by the surface term. One then integrates over conformal factors Ω of the form $1 + y$ about the metric g^* where y vanishes on the boundary. Because the kinetic term $(\nabla y)^2$ appears with a minus sign, one has to integrate over imaginary y [4]. This gives

$$(\det(\mu^{-2}\Delta))^{-\frac{1}{2}} \exp(-\hat{I}[g^*]) = Y[\{e^m_a\}] \tag{3.8}$$

where $\Delta = -\Box + \frac{1}{6}R$ is the conformally invariant scalar operator. The transition amplitude Z is then given by an integral over all conformal equivalence classes of tetrads.

$$Z = \int D[\{e\}]\, Y[\{e\}] \tag{3.9}$$

One can express this procedure in terms of the conformally compactified manifold $\tilde{M}$ and metric $\tilde{g}$. Pick a point z in $\tilde{M}$ and send it to infinity by the conformal transformation

$$\omega(x) = 4\pi^2 \Delta^{-1}(x,z) \tag{3.10}$$

where $\Delta^{-1}_{\sim}$ is the Green's function on the compact manifold in the metric $\tilde{g}$. The metric $g^* = \omega^2\tilde{g}$ then is asymptotically Euclidean with $R^* = 0$.

Let (λ_n, ϕ_n) be the eigenvalues and eigenfunctions of Δ on $(\tilde{M},\tilde{g})$. The eigenfunction ϕ_o corresponding to the lowest eigenvalue λ_o will have the same sign everywhere. One can therefore use it as the conformal factor in a regular conformal transformation $g \to g'$ that makes R' have everywhere the same sign as λ_o. From this it follows that if the lowest eigenvalue λ_o is positive, the Green's function $\Delta^{-1}(x,z)$ does not pass through zero anywhere. Therefore the metric g^* is non-singular. On the other hand, if λ_o is negative, $\Delta^{-1}(x,z)$ passes through zero and g^* is singular. I shall give an interpretation of this shortly.

The action of the metric g^* (omitting the C term) is

$$\hat{I}[g^*] = 6\pi^3\Delta^{-1}(z,z) \tag{3.11}$$

This of course is infinite because the Green's function diverges at z but so is the action if the surface term

$$-\frac{1}{8\pi}\int K(h)^{\frac{1}{2}}d^3x$$

is evaluated on a 3-sphere of infinite radius. To obtain a finite value for the action one has to make an (infinite) subtraction of the flat space value of the surface term. In the conformally compactified procedure this corresponds to regularizing $\Delta^{-1}(z,z)$. One obtains the correct result by a conformally invariant dimensional regularization procedure of adding extra flat dimensions. This is equivalent to zeta function regularization [12, 13] plus a correction term $\tilde{R}/288\pi^2$.

<u>The Positive Action Conjecture</u> [4, 5, 6] asserts that the action of any regular asymptotically Euclidean metric with R = 0 is positive or zero, being zero if and only if the metric is flat. In conformally compactified terms, the conjecture is that of $\Delta^{-1}(z,z)$ is greater than or equal to zero for all $(\tilde{M},\tilde{g})$ that do not have negative or zero eigen values for Δ. It is the analogue in one higher dimension of the Positive Energy Conjecture in ordinary general relativity (this has now been proved) [14].

One can represent

$$\Delta^{-1}(z,z) = \Sigma\lambda_n^{-1}\ \phi_n(z)\phi_n(z) \tag{3.12}$$

The regularization makes it zero for the standard metric on S^4. As one deforms the metric away from the spherical metric, the lowest eigenvalue λ_0 decreases and $\Delta^{-1}(z,z)$ becomes positive. When the metric is deformed so far that λ_0 passes through zero. $\Delta^{-1}(z,z)$ passes through infinity and becomes negative and the conformal factor $\Delta^{-1}(x,z)$ passes through zero.

To interpret this physically it is helpful to compare it with the initial value problem for classical general relativity which can be formulated in a similar way. One starts with a compact 3-dimensional manifold $\tilde{S}$ with a smooth metric $\tilde{h}$. One picks a point z and sends it to infinity with a conformal factor $\omega(x) = 4\pi\Delta^{-1}(x,z)$ where Δ^{-1} is now the Green's function for the 3-dimensional conformally invariant operator. The metric $h^* = \omega^4\tilde{h}$ then is asymptotically flat and has $R^* = 0$. It therefore satisfies the constraint equation for the time symmetric initial value problem (one can also deal with the non-symmetric initial value problem in a similar way). The ADM mass is given by the regularized value of $8\pi\Delta^{-1}(z,z)$.

The standard metric on the 3-sphere gives the initial data for flat space. As one deforms the metric $\tilde{h}$ away from the spherical metric one obtains initial data for a time symmetric imploding-exploding gravitational wave wave which will have positive energy or mass. As one increases the strength of the wave one reaches a critical state at which the energy in the wave is so great that it curls up the initial surface and cuts itself off from infinity. This corresponds to an eigenvalue of Δ passing through zero. One then passes to a new class of initial data, that for black holes with an apparent event horizon [15]. Such data can be obtained from a pair $(\tilde{S}, \tilde{h})$ with no negative eigenvalues by sending two points, z_1 and z_2, to infinity by the conformal factor $\omega(x) = 4\pi\Delta^{-1}(x,z_1) + 4\pi\Delta^{-1}(x,z_2)$.

It might seem reasonable to adopt a similar procedure in quantum gravity. Instead of simply sending one point z of the compact manifold $\tilde{M}$ to infinity to obtain an asymptotically Euclidean metric, one might send n points $z_1, z_2 \ldots z_n$ to infinity to obtain a metric with n asymptotically Euclidean regions. One can do this using the conformal factor

$$\omega(x) = 4\pi^2 \int \Delta^{-1}(x,y)\, J(y)(g)^{\frac{1}{2}} d^4y \tag{3.13}$$

where the "infinity current" J(y) is defined by

$$J(y) = \Sigma\delta(y,z_n) \tag{3.14}$$

The action of the metric $g^* = \omega^2\tilde{g}$ which has $R^* = 0$ is

$$\begin{aligned}\hat{I}[g^*] &= 6\pi^3 \int J(x)\Delta^{-1}(x,y)J(y)(g)^{\frac{1}{2}}(g)^{\frac{1}{2}}d^4xd^4y \\ &= 6\pi^3 J\Delta^{-1}J\end{aligned} \tag{3.15}$$

The vacuum to vacuum amplitude, or rather the vacua to vacua amplitude, corresponding to the manifold $\tilde{M}$ with n points removed is then

$$Z_n = \int D[\{e\}]\ (\det\Delta)^{-\frac{1}{2}} \exp(-6\pi^3 J\Delta^{-1}J) \tag{3.16}$$

integrated over all conformal equivalence classes {e} of tetrad fields. One can now sum this amplitude over all numbers, strengths and positions of the infinity points or, equivalently, over all infinity currents J. This produces a factor of $(\det \Delta^{-1})^{-\frac{1}{2}}$ which exactly cancels the factor of $(\det \Delta)^{-\frac{1}{2}}$ arising from the integral over conformal factors.

There remains the integral over conformal equivalence classes of tetrad fields. The components e^m_a of the tetrad at a point x form a 16 dimensional vector space. Identifying the tetrad with components e^m_a with the tetrad Ωe^m_a for any nonzero Ω, i.e. taking conformal equivalence classes, reduces one to a compact 15 dimensional projective space of conformal tetrads which is factored to a 9 dimensional compact space of conformal metrics by the SO(4) group of tetrad rotations. Because the space of conformal tetrads at each point is compact, one can adopt a measure in which it has unit volume. The path integral over conformal equivalence classes of tetrad fields then gives one, so one has

$$\Sigma Z_n = 1 \tag{3.17}$$

One might think that one should also sum over all possible topologies of the compact manifold $\tilde{M}$. However I think that one is already doing this effectively by integrating over all conformal tetrad fields. There will in general be 3 surfaces where $\det(e^m_a) = 0$ and the metric is degenerate. In general the null vector of the matrix e^m_a will not lie in the surface $\det(e) = 0$. This allows one to introduce new coordinates or, equivalently, to go to another manifold with a different topology on which the tetrad field and the metric are nondegenerate. There will in general be 2 surfaces on which the null vector of e^m_a lies in the surface $\det(e) = 0$. The scalar curvature R maybe singular there but it will still be intergrable so the action can still be defined. Thus integrating over all conformal tetrad fields or, equivalently, over all conformal positive semi-definite metrics effectively incorporates a sum over all manifold topologies.

This remarkable result was obtained without any regularization because the divergences in the path integration over conformal factors exactly cancelled the divergences in the action of the asymptotically Euclidean metrics which are normally removed by subtracting out the flat space value of the surface term. One might regard equation 3.17 as a statement of unitarity for gravity when one includes all topological possibilities. However, it is a bit formal. What one would like to do is to use this approach to calculate physical amplitudes and probabilities. I must confess that I have not yet found a way of doing this, though I do have some vague ideas. It seems that most conformal metrics (and all metrics with spinor structure and nonzero τ) have negative or zero eigenvalues of Δ. This means that g^*, the stationary phase metric under conformal transformations, will contain regions which are closed off from infinity. However, there will be other metrics in the conformal equivalence class in which these regions are not closed off. Thus they should have some physical effects though with reduced probabilities because they are not connected in the stationary phase metric. What these effects are could be is a matter of speculation but they might correspond to virtual black holes which could appear, swallow a particle, spit out another particle of a different species with the same charge, momentum and angular momentum and then disappear.

4. ASYMPTOTICALLY LOCALLY EUCLIDEAN METRICS

Under a scale transformation $g \to k^2 g$ (k constant), the action transforms as $\hat{I} \to k^2 \hat{I}$. This implies that the action of any asymptotically Euclidean metric which was a solution of the Einstein equations would have to be zero because it would have to be an extremum of the action under all perturbations including dilations. However the Positive Action Conjecture asserts that any asymptotically Euclidean metric with $R = 0$ has positive or zero action, the action being zero if and only if the metric is flat. Thus, if this conjecture holds, there can be no nontrivial asymptotically Euclidean vacuum gravitational instantons i.e. complete nonsingular solutions of the vacuum field equations. However the positive action conjecture does not exclude the possibility of vacuum instantons which are asymptotically locally Euclidean (I shall abbreviate this by ALE). In other words, outside some compact region they approach the Euclidean metric on flat space identified under some discrete subgroup G of SO(4). The first ALE instanton was found by Eguchi and Hanson [16]. It is asymptotic to Euclidean space with the point x^a identified with the point $-x^a$. It represents the transition between an initial state and its image under TP. Further ALE instantons have been found explicitly by Gibbons and myself [17] and implicitly by Hitchin [18] using Penrose's twistor technique [19]. These correspond to larger discrete subgroups G and their physical interpretation is not yet clear.

All these metrics have self-dual or anti self-dual curvature. This suggests a Generalized Positive Action Conjecture: any asymptotically locally Euclidean metric with $R = 0$ has positive or zero action, the action being zero if and only if the curvature is self-dual or anti self-dual. This conjecture is supported by the fact that one can show that self-dual or anti self-dual metrics are local minima of the action among metrics with $R = 0$. If the generalized conjecture holds, they are global minima of the action among ALE metrics with $R = 0$ just as the self-dual or anti self-dual Yang-Mills instantons are global minima of the Yang-Mills action. One would therefore expect that metrics near them would make the dominant contribution to transition amplitudes determined by these boundary conditions.

A particularly interesting application is to supergravity theories [20]. In a self-dual metric the curvature seen by the primed spinors is zero so there are two independent covariantly constant primed spinors $\iota_{A'}$ and $o_{A'}$ which can be chosen so that

$$\iota_{A'} o^{A'} = 1 \qquad \bar{\iota}_{A'} = o^{A'} \tag{4.1}$$

In an ALE self-dual metric, $B_2^+ = \tau$ and $B_2^- = 0$ so there are τ self-dual harmonic two-forms (Maxwell fields) which are normalizable i.e. L^2 and no L^2 anti self-dual harmonic two-forms. The self-dual harmonic two-forms can be represented by τ symmetric spinors which satisfy the spin-1 equation

$$\nabla^{AA'} \phi^{j}_{AB} = 0 \tag{4.2}$$

Multiplying by one or other of the two covariantly constant spinors one gets 2τ zero modes of the Majorana spin 3/2 equation in the gauge $\gamma^a \psi_a = 0$

$$\nabla^{AA'} \phi^{ij}_{ABA'} = 0 \tag{4.3}$$

where $\phi^{ij}_{ABA'} = \phi^{j}_{AB} \alpha^{i}_{A'}$. None of these modes is a pure gauge transformation i.e. of the form $\nabla_{AA'} \phi_B$. Multiplying by another covariantly constant spinor and symmetrising on the primed indices one obtains 3τ transverse traceless zero modes of the metric perturbation equation in the harmonic gauge

$$\nabla^{AA'} \nabla_{AA'} \phi^{jh}_{BCB'C'} - 2\Psi_{BC}{}^{DE} \phi^{jh}_{DEB'C'} = 0 \tag{4.4}$$

where $\phi^{jh}_{BCB'C'} = \phi^{j}_{BC} \alpha^{h}_{B'C'}$.

Up to three of these modes (depending on the discrete group G) may be pure gauge transformations corresponding to the global rotations of the instanton in asymptotically locally Euclidean space. The relations between the zero modes of different spins are actually global supersymmetry transformations with covariantly constant spinor parameters.

The action of the background self-dual ALE instanton is zero so it makes no contribution to log Z. The one loop term in simple supergravity can be expressed

$$Z = \frac{\det B(\det C)^{\frac{1}{2}}}{(\det E)^{\frac{1}{2}}\ (\det F)^{3/2}} \tag{4.5}$$

where B is the vector ghost operator corresponding to diffeomorphism gauge fixing, C is the spin 3/2 operator (it appears to the power $\frac{1}{2}$ because it is a Majorana field), E is the metric perturbation operator equation 4.4 and F is the spin $\frac{1}{2}$ supersymmetry ghost operator (it appears to the power 3/2 because it is a Majorana field and because when one averages over supersymmetry gauges one introduces an extra factor of $(\det F)^{-\frac{1}{2}}$ [21]). There are also six non-propagating components of the tetrad and six non-propagating components of the auxillary fields [22, 23] but these cancel with the twelve non-propagating tetrad rotation ghosts.

One can use global supersymmetry transformations to relate the non zero modes of the operators B, C, E, F, [20]. The multiplicities are such that the non zero eigenvalues cancel completely in the one loop term 4.5. Thus the infinite dimensional one loop path integral reduces to a finite dimensional integral over zero modes. This cancellation between boson and fermion operators is one of the most attractive features of supergravity theories and raises the hope that one might be able to make some reasonable mathematical sense out of them. It should be pointed out however that in the usual language of Feynman diagrams, the one loop term 4.5 has a logarithmic divergence because there are different numbers of zero modes in the numerator and denominator (the proof that all one loop terms are finite in supergravity applies only in topologically trivial metrics). The one loop term introduces a parameter μ which reflects the measure on the zero modes about which there is, as yet, no agreement.

The 3τ gravitational zero modes in the operator E correspond to global rotation, dilation and other self-dual perturbations of the instanton metric. To deal with them one introduces collective coordinates for the orientation, scale etc. of the instanton. They give rise to a factor of the form

$$\mu^{3\tau}\ \rho^{6\tau-1} d\rho$$

where ρ is some length scale of the instanton. The 2τ spin 3/2 zero modes in the operator C occur in the numerator in equation 4.5 and make the vacuum to vacuum amplitude Z zero in the absence of sources. To deal with this one adds a term

$$\int (\phi_{ABA'}\, \eta^{ABA'} + \tilde{\phi}_{AA'B'}\tilde{\eta}^{AA'B'})(g)^{\frac{1}{2}}d^4x \tag{4.6}$$

where the Majorana spin 3/2 field is represented by the Euclidean spinors $\phi_{ABA'}$ and $\tilde{\phi}_{AA'B'}$ and the source current is represented by η and $\tilde{\eta}$. The vacuum to vacuum amplitude $Z[\eta,\tilde{\eta}]$ in the prescence of the sources is then proportional to

$$\prod_n \int \phi^n_{ABA'}\eta^{ABA'}(g)^{\frac{1}{2}}d^4x \tag{4.7}$$

where $\phi^n_{ABA'}$

are the 2τ positive helicity spin 3/2 zero modes. To the Z given by equation 4.7 one has to add the value of Z of order unity arising from the one loop about flat Euclidean space. One then functionally differentiates with respect to η at 2τ points to get a 2τ point function which violates helicity conservation i.e. it converts τ particles of positive helicity into the same number of negative helicity particles.

The simplest self-dual ALE instanton, the Eguchi-Hanson [16], has $\tau = 1$. One therefore obtains a helicity changing two point function which could be regarded as a spin 3/2 mass term in this metric. However, because the instanton is asymptotic to Euclidean space with the points x^a identified with $-x^a$, one can also interpret this as a 4 point function in Euclidean space which converts two spin 3/2 particles with positive helicity into their TP counterparts, two particles of negative helicity but the same momenta.

The situation will be similar for extended supergravity theories except that there will now be $2\tau N$ spin 3/2 zero modes where N is the number of spin - $\frac{3}{2}$ fields. One will therefore get $2\tau N$ point helicity changing amplitudes. These theories will also contain spin 1 fields and may contain spin $\frac{1}{2}$ and spin 0 fields. There will be no zero modes of the spin $\frac{1}{2}$ and spin 0 fields in a self-dual ALE instanton background metric. There will be τ spin 1 zero modes but these do not appear in the one loop term because they do not arise from a potential. In fact if they are gauge fields (abelian or otherwise) the zero modes will be quantized by the analogue of the Dirac condition for magnetic monopoles.

5. SPACETIME FOAM

I now return to the question of how to treat conformal equivalence classes of metrics on a compact manifold $\hat{M}$ that possess negative or zero eigenvalues of the conformally invariant operator Δ. In these cases the asymptotically Euclidean member g^* of the conformal equivalence class which has $R^* = 0$ and which is the stationary phase point under conformal transformations is singular and contains regions which are cut off by zero conformal factor. On the other hand, one can find a compact metric g' in the conformal equivalence class which has $R' = -4h$ where h is a positive constant. The metric g' can be normalized by requiring that its 4-volume is unity.

The significance of the metrics g' is that they are the stationary phase points under conformal transformations in the volume canonical ensemble that I shall define below. The physical idea is that in quantum gravity one has to sum over metrics of very complicated topology. This "foamlike structure" [24] will be everywhere so that one cannot really think of a spacetime as being asymptotically Euclidean although this is a convenient viewpoint when interpreting amplitudes as S-matrix elements.

To keep things finite it is convenient to consider compact metrics with some given large value of the 4-volume V. This is not to say that spacetime necessarily is compact. It is merely a normalization device like periodic boundary conditions in ordinary quantum mechanics. One would compute the density of certain quantities per unit volume and then take the limit as V tends to infinity.

In order to constrain the 4-volume one adds a term $\Lambda V/8\pi$ to the gravitational action where Λ is a Lagrange multiplier. This is of the same form as the cosmological term but the motivation for its introduction and its actual value are very different: observational evidence indicates that any cosmological term must be so small as to be practically negligible whereas the Lagrange multiplier will turn out to give the dominant contribution when it is of order one in Planck units. One forms the partition function $Z[\Lambda]$ for what I call the volume canonical ensemble [25] by performing a path integral over all compact positive semi-definite metrics

$$Z[\Lambda] = \int D[e] \exp(-\hat{I}[e])$$

$$= \Sigma_n \langle g_n | \exp(-\Lambda V/8\pi) | g_n \rangle \quad (5.1)$$

where the action includes the Λ term. The partition function is the Laplace transform of N(V)dV, the number of states of the gravitational field between V and V + dV,

$$Z[\Lambda] = \int_0^\infty N(V)\ \exp\ (-\Lambda V/8\pi)dV \tag{5.2}$$

Thus N(V) is the inverse Laplace transform

$$N(V) = (1/16\pi^2 i) \int Z[\Lambda]\ \exp\ (\Lambda V/8\pi)d\Lambda \tag{5.3}$$

where the contour is taken parallel to the imaginary Λ axis and to the right of the singularity in $Z[\Lambda]$ at $\Lambda = 0$. This ensures that $N(V) = 0$ for $V \leq 0$.

The path integral 5.1 for Z can be broken up into a path integral over conformal factors Ω in one conformal equivalence class $\{e\}$ of tetrad fields

$$Y\ [\{e\}, \Lambda] = \int D[\Omega]\ \exp\ (-\hat{I}[\Omega e]) \tag{5.4}$$

followed by an integral over conformal equivalence classes

$$Z[\Lambda] = \int D[\{e\}]\ Y\ [\{e\}\ \Lambda] \tag{5.5}$$

Under a conformal transformation $\tilde{e} = \Omega e$, the action (including the Λ term) becomes

$$\hat{I}[\tilde{e}] = -\frac{1}{8\pi} \int (3\Omega\Delta\Omega - \Lambda\Omega^4)(e)d^4x \tag{5.6}$$

Thus $Z[\Lambda]$ can be regarded as the average of $\lambda\phi^4$ theory over all conformal equivalence classes of metrics. The stationary phase point in the integral 5.4 over conformal factors will occur at the metric $-h\Lambda^{-1}g'$ for which $R = 4\Lambda$, $V = h^2\Lambda^{-2}$ and $I = -h^2/8\pi\Lambda$. The quantity h is a function of the conformal equivalence class and will be stationary at those equivalence classes for which g' is a solution of the Einstein equations with a Λ term

$$R_{ab} - \tfrac{1}{2} Rg_{ab} + \Lambda g_{ab} = 0 \tag{5.7}$$

A few solutions are known explicitly on compact manifolds such as S^4, CP^2 and $S^2 \times S^2$. One is really interested however in solutions on very complicated simply-connected manifolds with high Euler number and signature. Although one cannot hope to find such metrics explicitly, one can nevertheless estimate their action. For a solution of 5.7 one has

$$\chi = \frac{1}{32\pi^2} \int (C_{abcd}\, C^{abcd} + \frac{8}{3}\Lambda^2)(e)\, d^4x \tag{5.8}$$

$$\tau = \frac{1}{48\pi^2} \int (C_{abcd}\, {}^*C^{abcd})(e)\, d^4x \tag{5.9}$$

Combining equations 5.8 and 5.9 one has

$$2\chi - 3|\tau| \geq h^2/6\pi^2 \tag{5.10}$$

the equality holding if and only if the Weyl tensor is self dual or anti self-dual.

For solutions of 5.7, h has the lower bound of $-(24)^{\frac{1}{2}}\pi$, the value for the standard metric on S^4. If $\tau \neq 0$ and if the manifold admits a spinor structure, $h \geq 0$. From 5.8 and 5.9 one would expect that for large Euler number $h \sim d\chi^{\frac{1}{2}}$ where $d \leq 2(3)^{\frac{1}{2}}\pi$. This is supported by a number of examples which have been supplied by Nigel Hitchin which indicate that most solutions lie in the range

$$2^{\frac{1}{2}}\pi < d < 2^2\,\pi \tag{5.11}$$

One would hope that one could obtain a value for the partition function $Z[\Lambda]$ by expanding the metric or tetrad in a perturbation series about each solution of the classical field equations. When Λ is small, the action of these solutions will be widely separated an and one should have a good "dilute gas approximation". However when Λ is large, fluctuations around one solution may change the topology and so overlap with fluctuations around another solution, without significantly increasing the action and so being damped. One way of avoiding such an overlap might be to add a gauge fixing term $\alpha \int R^2(e)d^4\chi$ to the action to fix the conformal gauge in which the path integral 5.4 over conformal factors was evaluated. The path integral for Z will then become

$$Z[\Lambda] = \int D[\Omega]\, D[e]\, \det \alpha\Delta(\exp - \tilde{I}[e,\Omega]) \tag{5.12}$$

where

$$\Upsilon[e, \Omega] = -\frac{1}{8\pi}\int (3\Omega\Delta\Omega - \Lambda\Omega^4 - 8\pi\alpha\, R^2)(e)d^4x$$

$$- 16\,\alpha\, h^2[e_o] \qquad (5.13)$$

The det $\alpha\,\Delta$ term in 5.12 is the Fadeev–Popov ghost and the last term in 5.13 is designed to make the gauge-fixing term zero at the solution e_o. The gauge-fixing term $\int R^2(e)d^4x$, unlike the action $\int R(e)d^4x$ would probably not remain bounded if the metric changed topology by passing through a degenerate metric. It might thus confine the perturbation expansion to the topology of the solution about which it was made, and prevent it from overlapping with perturbation expansions centred on other solutions. It would also give rise to a perturbation expansion that was at least formally renormalizable. The fact that it was only a gauge-fixing term and not a part of the action itself might avoid some of the pathologies associated with such terms.

So far no progress has been made with this approach. However one can estimate the ordinary one loop term L about a solution with Euler number χ.

$$L = \left(\frac{\Lambda}{\Lambda_o}\right)^{-\gamma} \qquad (5.14)$$

where γ is the integrated trace anomaly for gravity [26]

$$\gamma = \frac{106\chi}{45} + \frac{73\,h^2}{120\,\pi^2} \qquad (5.15)$$

and Λ_o is related to the normalization constant μ. Thus metrics with Euler number χ make a contribution to Z of the form

$$\left(\frac{\Lambda}{\Lambda_o}\right)^{-\gamma} \exp\,(b\chi\Lambda^{-1}) \qquad (5.16)$$

where $b = d^2/8\pi$. This gives a stationary phase point in the inverse Laplace transform 5.3 for N(V) at $\Lambda = \Lambda_s$ where

$$\Lambda_s = 4\pi V^{-1}\left[\gamma \pm \left(\gamma^2 + \frac{Vb\chi}{2\pi}\right)^{1/2}\right] \tag{5.17}$$

Comparing the contributions from metrics of different χ, one finds that the dominant one comes when

$$-\gamma\chi^{-1} \log \left(\frac{\Lambda_s}{\Lambda_o}\right) + b\,\Lambda_s^{-1} = 0 \tag{5.18}$$

For $\Lambda_o \sim 1$, this is satisfied when $\chi \sim V$ i.e. there is one unit of topology or one gravitational instanton per Planck volume.

REFERENCES

1. M.J. Perry, Nucl.Phys.B., to be published.

2. J. York, Phys.Rev.Lett, 28, 1082, 1972.

3. G.W. Gibbons and S.W. Hawking, Phys.Rev.D15, 2752, 1977.

4. G.W. Gibbons, S.W. Hawking and M.J. Perry, Nucl.Phys.B, to be published.

5. S.W. Hawking, Phys.Rev.D, to be published.

6. D.N.Page, Phys.Rev.D. to be published.

7. S.W. Hawking, Phys.Rev.D13, 191, 1976.

8. R.Jackiw and C. Rebbi, Phys.Lett 67B, 189, 1977.

9. R.P. Geroch, J.Math.Phys 9, 1739, 1968; J.Math.Phys.11, 343, 1970.

10. C.J. Isham, Spinor Fields in Four Dimensional Spacetime, Imperial College preprint, 1978.

11. R. Penrose, Proc.Roy.Soc. A284, 159, 1965.

12. J.S. Dowker and R. Critchley, Phys.Rev.D13, 3224, 1976.

13. S.W. Hawking, Comm.Math.Phys. 55, 133, 1977.

14. S.T. Yau and R. Schoen, "Incompressible Minimal Surfaces, Three Dimensional Manifolds with Non-Negative Scalar Curvature, and the Positive Mass Conjecture in General Relativity.

15. S.W. Hawking, "The Event Horizon" in "Les Astres Occlus" ed. B.S. deWitt and C.M. deWitt, Gordon and Breach, 1973.

16. T. Eguchi and A.J. Hanson, Phys.Lett 74B, 249, 1978.

17. G.W. Gibbons and S.W. Hawking, Gravitational Multi-Instantons, D.A.M.T.P. preprint.

18. N. Hitchin, in preparation.

19. R. Penrose, J.Gen.Rel. and Gravitation, 7, 31, 1976.

20. C.N. Pope and S.W. Hawking, "Symmetry Breaking by Instantons

in Supergravity" D.A.M.T.P. preprint.

21. M.J. Perry, Nucl.Phys.B., to be published.

22. S. Ferrara and P. van Nieuwenhuizen, "The Auxiliary Fields of Supergravity, CERN preprint.

23. K. Stelle and P. West, "Minimal Auxiliary Fields for Supergravity", Imperial College preprint.

24. J.A. Wheeler in "Relativity Groups and Topology", proceedings of the Les Houches Summer School, 1963, ed by B.S. deWitt and C.M. deWitt, Gordon and Breach, New York, 1964.

25. S.W. Hawking, "Spacetime Foam" D.A.M.T.P. preprint.

26. G.W. Gibbons and M.J. Perry "Quantizing Gravitational Instantons, D.A.M.T.P. preprint.

27. M.J. Duff, Abstracts of Contributed Papers for GR VIII Conferences, Waterloo, Ontario (1977)

QUANTUM FIELD THEORY RENORMALIZATION IN CURVED SPACE-TIME

David G. Boulware

Physics Department, FM-15

University of Washington, Seattle, WA 98195, USA

In these few lectures, I shall present a brief outline of a charged scalar field propagating in given background electromagnetic field and given space-time. The theory will be expressed in terms of functional integrals over the scalar field and the general problem of renormalization discussed. The amplitudes which arise as a result of doing the functional integrals may be evaluated in terms of the Green's function, or propagator, of the scalar field, which, in turn, may be expressed in terms of the Schwinger-DeWitt proper time formalism. When the various amplitudes are calculated, the final proper time integral will diverge unless the Green's function is modified or regulated in some way. I shall compare various algorithms for regularizing the theory - point splitting, Pauli-Villars and dimensional continuation - and show their essential equivalence. The renormalization process will be discussed in some detail along with the definition of the renormalized parameters, the ambiguities associated with the parameters and the uniqueness or lack thereof of the process. Finally, I shall discuss the trace anomaly using dimensional regularization and Pauli-Villars regularization and indicate the origin of the anomaly in the two methods as well as showing its uniqueness and independence of the regularization method.

The material presented here is a compendium of many authors' work presented from my own point of view. I shall lean heavily on the work of Schwinger [1] and DeWitt [2,3] who long ago laid the foundation for any work in this field. I have also used the work of Brown [4] and Brown and Cassidy [5,6] who were the first to do the dimensional continuation properly in curved space-times. There has been a large amount of work by many authors on the general topic: Zel'dovich and Starobinsky [7], Parker and Fulling [8], Candelas and Raine [9], Dowker and Critchley [10,11], Davies, Duff and Unruh [12], Christensen [13], Hawking [14], Bernard and Duncan [15], Davies, Fulling, Christensen and Bunch [16], Christensen [17], Christensen and Fulling [18], Adler et al. and Wald's correction [19-21], Wald [22], and Bunch and Davies [23,24]. The trace anomaly has also been extensively discussed in the literature. In addition to the general references Deser, Duff and Isham [25] first explicitly pointed out the gravitational anomaly and Duff [26] and Tsao [27] have calculated the anomaly in a number of cases. Bunch [28,29] and Davies [30] have recently written on the subject, the latter paper provides an excellent discussion of several aspects of renormalization. The anomaly was first discovered for an interacting scalar field by Callan, Coleman and Gross [31] and was important in discussions of the renormalization group Callan [32].

For a discussion of the functional integral and other general techniques used here, the works of Feynman and Hibbs [33], Iliopoulous et al. [34], Abers and Lee [35], and Popov [36] are suggested.

We write the quantum field theory in terms of a functional or Feynman path integral over possible classical fields,

$$\langle +1-\rangle^{g,A,U,\eta} = \int [d\phi] \exp i\int dx\, \mathcal{L}(\phi;g,A,U,\eta;x) \equiv \exp i\, W[g,A,U,\eta]$$

where

$$\mathcal{L}(\phi;g,A,U,\eta;x) \equiv -\frac{1}{2}[\Pi_\mu\phi]^T(x) g^{\mu\nu}(x)\sqrt{-g(x)}\,[\Pi_\nu\phi](x) \tag{1}$$

$$-\sqrt{-g(x)}\left\{\frac{1}{2}\phi(x)[\mu^2+U(x)+\xi R(x)]\phi(x) + (\lambda/8)[(\phi\phi)(x)]^2\right\} + \eta(x)\phi(x),$$

and

$$\Pi_\mu\phi\,(x) \equiv \left(\partial_\mu - ieqA_\mu(x)\right)\phi(x) \quad .$$

and the field ϕ is a real field with two components. The matrix q is Hermitian and anti-symmetric with $q^2 = 1$ and $A_\mu(x)$ is the vector potential for the electromagnetic field. The space-time metric tensor, $g(x)$, is taken to have signature (-1,1,1,1). The scalar potential, U, is a 2x2 matrix which will be used to generate matrix elements of the operator $\phi_a(x)\phi_b(x)$ where a and b indicate the field components. The scalar density, η, acts as a source of the field ϕ and is used to generate matrix elements of ϕ. Since ϕ has more than one component, we will encounter traces over the components, the notation tr will be used to denote such traces, e.g.

$$\mathrm{tr}\phi\phi \equiv \sum_a \phi_a\phi_b \equiv (\phi\phi)$$

and

$$\mathrm{tr}U\phi\phi \equiv \sum_{a,b} U_{ab}\phi_b\phi_a \equiv (\phi U\phi) \quad .$$

The amplitude is presented as the <+|-> matrix element with no designation other than "±". The "±" refers to the initial (-) or final (+) state and, in principle, any states may be used for the initial and final states; however, to do so requires that the functions ϕ over which the integration is done be restricted to those functions which satisfy boundary conditions appropriate to the states. In general, doing the integration thusly is quite difficult (the state of the art for performing functional integrals is not highly developed) and in practice one only attempts to calculate the ground state or some other special state. Usually, the state is (assumed to be) a cyclic state so that any state may be constructed by suitable application of the operator ϕ. In curved space-times, particularly those with non Minkowskian topologies or without asymptotically static regions, the specification of the appropriate state is non-trivial and there is a considerable literature on the choice of the "correct vacuum." I shall discuss these problems further below and just comment that if there is a question of the "correct vacuum," the problem is not being phrased properly. Either there is a unique (or degenerate) state with lowest energy or there is a physical question: what is the state of the system? In the latter case, one can always define the "vacuum" for modes with wavelengths short compared to the inverse curvature. It is the modes with long wavelengths which cause trouble; one cannot tell, locally, whether the field is propagating independently of the background geometry and hence constitutes radiation or whether the field is tied to the geometry and is an induced polarization field. This physical ambiguity makes a unique vacuum definition impossible. One must then ask, "What is the appropriate state in which to calculate?" There is no general answer, one must analyze each case.

In a functional integral formalism, the matrix element of the operator $\phi(x)$ is given by

$$\begin{aligned} \langle +|\phi(x)|-\rangle &= \int [d\phi]\Big(\exp[i\int dx\mathcal{L}]\Big)\,\phi(x) \\ &= \Big(\delta/i\delta\eta(x)\Big)\,\exp iW[g,A,U,\eta] \\ &= \langle +|-\rangle\Big(\delta W[g,A,U,\eta]/\delta\eta(x)\Big) \quad , \end{aligned} \tag{2}$$

hence the average of ϕ, $<\phi>$, is given by

$$<\phi(x)> \equiv <+|\phi(x)|->/<+|-> = \delta W[g,A,U,\eta]/\delta\eta(x) \quad . \qquad (3)$$

In discussing the renormalization of the theory, it is convenient to rewrite W as a functional of $<\phi>$ instead of η. For a given matrix element, each source, η, will yield a definite expectation value, $<\phi>$; we assume the mapping to be one to one. (The one to one mapping may not always exist; there is now an extensive literature dealing with such phenomena: Coleman [37], Polyakov [38], Gibbons and Hawking [40] and Gibbons and Perry [41]. We shall not pursue these questions here.) Then, we may change variables using a Legendre transformation (see Lee and Zinn-Justin [42] and Jona-Lasinio [43]),

$$W[g,A,U,\eta] \equiv \left(\int (dx)\eta(x)<\phi(x)>\right) - \Gamma[<\phi>,g,A,U] \quad , \qquad (4)$$

which defines the functional Γ. If we now take the variational derivative with respect to η, we find

$$<\phi(x)> = <\phi(x)> + \int \frac{\delta<\phi(y)>}{\delta\eta(x)}\left\{\eta(y) - \frac{\delta\Gamma[<\phi>,g,A,U]}{\delta<\phi(y)>}\right\} \qquad (5)$$

hence, under the assumption that $\delta<\phi>/\delta\eta$ has no zeros, the expectation value $<\phi>$ must satisfy the equation

$$\frac{\delta\Gamma[<\phi>,g,A,U]}{\delta<\phi(y)>} = \eta(y) \quad , \qquad (6)$$

which is the field equation for the expectation value $<\phi>$. In order to proceed, the functional Γ must be calculated. It invariably consists of two terms: i) a local term which is constructed from $<\phi>$, A(x), g(x) and U(x) and their derivatives, integrated over all space-time, and ii) a nonlocal term which intrinsically includes fields from different space-time points. The nonlocal term includes the effects of virtual particle exchange and particle production. The local term is, formally, divergent and consists of terms which could be included in an action such as that given in Eq. (1).

The original functional integral is invariant under both electromagnetic gauge transformations and coordinate transformations. Under gauge transformations,

$$\phi(x) \to \phi'(x) = \exp\bigl(ieq\lambda(x)\bigr)\,\phi(x)$$

$$A_\mu(x) \to A'_\mu(x) = A_\mu(x) + \partial_\mu\lambda(x) \qquad (7)$$

$$\eta(x) \to \eta'(x) = \eta(x)\exp\bigl(-ieq\lambda(x)\bigr)$$

$$U(x) \to U'(x) = \exp\bigl(\,ieq\lambda(x)\,\bigr)U(x)\exp\bigl(-ieq\lambda(x)\bigr) \quad ,$$

and the Jacobian of the transformation is

$$(\partial\phi/\partial\phi') = \mathrm{Det}[\exp(ieq\lambda)] = 1$$

where the Fredholm determinant, $(\partial\phi/\partial\phi')$ is calculated by

$$\delta\lambda\ \ell n(\partial\phi/\partial\phi') = \mathrm{Tr}(\partial\phi'/\partial\phi)\delta\lambda(\partial\phi/\partial\phi')$$

$$= \mathrm{tr}\int dxdx'\delta(x-x')\exp\bigl(ieq\lambda(x)\bigr)\ -ieq\delta\lambda(x)$$

$$\times\ \exp\bigl(-ieq\lambda(x)\bigr)\delta(x'-x) \qquad (8)$$

$$= \int dx\delta(0)[-ie\delta\lambda(x)\mathrm{tr}q]$$

or

$$(\partial\phi/\partial\phi') = \exp\left[-ie\int dx\delta(0)\mathrm{tr}q\lambda(x)\right]$$

and, in time honored tradition, $(\delta(0)\mathrm{tr}q)$ is taken to be zero. Thus, the integral is independent of λ and, assuming that the state specification is also gauge invariant,

$$W\left[g,A+\partial\lambda,(\exp\ ieq\lambda)U\bigl(\exp(-ieq\lambda)\bigr),\eta\exp(-ieq\lambda)\right]$$

$$= W[g,A,U,\eta]$$

or

$$\Gamma\left[(\exp\ ieq\lambda)\langle\phi\rangle,g,A+\partial\lambda,(\exp\ ieq\lambda)U\bigl(\exp(-ieq\lambda)\bigr)\right]$$

$$= \Gamma[\langle\phi\rangle,g,A,U] \quad . \qquad (9)$$

The separation of Γ into local and nonlocal terms need not be gauge invariant; however, since the sum is gauge invariant, the local terms may be taken to be gauge invariant and any residual non-gauge invariant terms including with the nonlocal terms so that the sum is gauge invariant. (For example, in flat space,

$$-\int dx A_\mu A^\mu(x) + \int dx dx' \Big\{ A^\mu(x) \langle x| \frac{\mu^2}{-\partial^2+\mu^2} |x'\rangle A_\mu(x')$$

$$+ \partial_\mu A^\mu(x) \langle x| \frac{1}{-\partial^2+\mu^2} |x'\rangle \partial^\nu A_\nu(x') \Big\}$$

is gauge invariant and the sum of a local and a nonlocal term, neither of which is separately gauge invariant. However, the sum may be rewritten as

$$\int dx' A^\mu(x) \langle x| \big(\eta_{\mu\nu}\partial^2 - \partial_\mu\partial_\nu\big)/\big(-\partial^2+\mu^2\big) |x'\rangle A^\nu(x')$$

$$= \frac{1}{2}\int dx dx' F^{\lambda\sigma}(x) \langle x| \Big(1/(-\partial^2+\mu^2)\Big) |x'\rangle F_{\lambda\sigma}(x)$$

which is gauge invariant and nonlocal. See also [30].) Also, the theory is invariant under coordinate transformations, $x^\mu = \xi^\mu(y)$

$$\phi(x) \rightarrow \phi'(y) = \phi\big(\xi(y)\big)$$

$$g_{\mu\nu}(x) \rightarrow g'_{\alpha\beta}(y) = \big(\partial\xi^\mu/\partial y^\alpha\big)\big(\partial\xi^\nu/\partial y^\beta\big) g_{\mu\nu}\big(\xi(y)\big)$$

$$A_\mu(x) \rightarrow A'_\alpha(y) = \big(\partial\xi^\mu/\partial y^\alpha\big) A_\mu\big(\xi(y)\big) \tag{10}$$

$$\eta(x) \rightarrow \eta'(y) = \big(\partial\xi/\partial y\big)\eta\big(\xi(y)\big)$$

$$U(x) \rightarrow U'(x) = U\big(\xi(y)\big)$$

which leaves the action, $\int dx \mathcal{L}$, invariant.

The invariance of the functional integral is impossible to analyze. One is led to consider integrals such as

$$\int dx dx' \delta(x-x')(\partial/\partial x^\sigma)\delta(x-x') f^\sigma(x) = \,?$$

which does not exist and, if taken to be non zero, would have to be renormalized in any case. No matter how the Jacobian is defined, it is independent of ϕ and yields only a (divergent) multiplicative factor which we drop. We thus find,

$$\Gamma[\langle\phi'\rangle, A', g'] = \Gamma[\langle\phi\rangle, A, g] \tag{11}$$

and, again, we may take the local and nonlocal terms to be separately invariant. The local terms must have the dimensions of action, and are related to the formal divergences which appear when quantum corrections are calculated. If a quantum correction is finite, in flat space it always appears as a momentum integral of the schematic form

$$\int d^{4\pi}p(1/p^2 + pk\ldots)^\alpha \sim F(k)$$

and F(k) has dimension $k^{4n-2\alpha}$ with $4n-2\alpha < 0$. However, if the correction is divergent, the same argument holds, but the integral is divergent for large p and the dimension of the (formal) integral is k^p $p \geq 0$. Thus, the term F has divergences, polynomials in momenta, and polynomials times logarithms of momenta and masses. The local terms are the polynomial terms and there is always a coefficient of dimension mass to some power. In our case, the only such local gauge and coordinate invariant term that can be constructed is

$$\begin{aligned}\Gamma^{loc}[\langle\phi\rangle, A, g, U] = \int dx\sqrt{-g}\Big\{&Z_1^{-1}\frac{1}{2}(\Pi_\mu\langle\phi\rangle)^T g^{\mu\nu}(\Pi_\nu\langle\phi\rangle) \\ &+ Z_1^{-1}\mu^2\frac{1}{2}(\langle\phi\rangle\langle\phi\rangle) + Z_1^{-1}\frac{1}{2}\langle\phi\rangle U\langle\phi\rangle + \xi Z_1^{-1}\frac{1}{2}R(\langle\phi\rangle\langle\phi\rangle) \\ &+ \frac{1}{8}\lambda Z_1^{-2}(\langle\phi\rangle\langle\phi\rangle)^2 + \frac{1}{4}\delta Z_3^{-1} g^{\mu\lambda}g^{\nu\sigma}F_{\mu\nu}F_{\lambda\sigma} \\ &- \delta(1/16\pi G)R + \delta\alpha R^2 + \delta\beta C^{\mu\nu\lambda\sigma}C_{\mu\nu\lambda\sigma} + \delta\gamma G\Big\}\end{aligned} \tag{12}$$

where

$$G \equiv R^{\mu\nu\lambda\sigma}R_{\mu\nu\lambda\sigma} - 4R^{\mu\nu}R_{\mu\nu} + R^2 \quad ,$$

and R is the curvature scalar, $\tilde{C}$ the Weyl tensor and $\tilde{F}$ the electro-magnetic field tensor. Furthermore, the coefficients which appear in front of these terms are divergent.

We now recite the standard litany: the quantities which we calculate to be divergent cannot be determined by the theory. These include the masses of the particles, the self coupling, the couplings to the external fields, A and g, and terms which involve only the external fields. The external fields are presumed to be classical manifestations of underlying quantum degrees of freedom. The appearance of F;F or purely gravitational terms are contributions to the self interactions of those fields which are induced by the presence of the quantum field, ϕ. As a result, those fields must themselves be renormalized; terms must be added to $\int dx \mathcal{L}$ of the same form as those which appear in Eq. (11) but with arbitrary coefficients. In order to calculate W, the theory must be regulated in some way so that the coefficients in the local term will be finite. The redefinitions,

$$\langle\phi(x)\rangle = Z_1^{\frac{1}{2}}\langle\tilde{\phi}(x)\rangle \ , \qquad \eta(x) = \tilde{\eta}(x)Z_1^{-\frac{1}{2}} \ ,$$

$$e_o = Z_3^{-\frac{1}{2}}e \ , \qquad A_\mu(x) = Z_3^{-\frac{1}{2}}\tilde{A}_\mu(x) \ ,$$

and

$$U(x) = Z_1\tilde{U}(x) \ , \tag{13}$$

along with addition of the local terms in g and A result the following finite form,

$$\Gamma^{loc}_{ren}[\phi,g,A,U] = \int dx\sqrt{-g}\left\{\frac{1}{2}(\Pi_\mu\phi)^T g^{\mu\nu}(\Pi_\mu\phi) + \frac{1}{2}\mu^2(\phi\phi) + \frac{1}{2}\phi U\phi + \frac{1}{2}R(\phi\phi) + \frac{\lambda}{8}(\phi\phi)^2 + \frac{1}{4}g^{\mu\lambda}g^{\nu\sigma}F_{\mu\nu}F_{\lambda\sigma} - \Lambda - (1/16\Pi G)R + \alpha R^2 + \beta C^{\mu\nu\lambda\sigma}C_{\mu\nu\lambda\sigma} + \gamma G\right\} \tag{14}$$

where the tilde's have been dropped and the values of the renormalized parameters, e,μ,λ etc. must be chosen to fit experiment; the original parameters must be chosen so the renormalized parameters are finite and have the appropriate values. As a result, the original parameters depend upon the regularization and are divergent in the limit of vanishing regularization. Of course, not all the divergences are in the local term. In higher orders of perturbation, the nonlocal terms are also divergent; however, the fundamental proof of renormalizability, Hepp [44] and Zimmerman [45], consists of establishing that the renormalization of the local terms also renormalizes the nonlocal terms.

As has been discussed by DeWitt [2,3], the form for Γ^{loc} which is given in Eq. (13) is valid as an effective action for <g>, <A> etc. even when gravitation, electromagnetism or any other interactions are included, provided only that the theory be renormalizable. (Gravitation is, of course, not renormalizable in the absence of the terms quadratic in the curvature. In presence of the quadratic terms, it is renormalizable, Stelle [46].) If there are no external sources for the gravitational field, the action must be an extremum under variations of <g>, holding <φ> and <A> fixed. The nonlocal terms should be small for fields varying slowly on elementary particle scales, hence a good approximation will simply be the contribution from the local terms:

$$\frac{2}{\sqrt{-g}}\frac{\delta(-\Gamma^{loc}_{ren})}{\delta g_{\mu\nu}(x)} = \Big\{(\Pi^{\mu}\phi)^{T}(\Pi^{\nu}\phi) - \frac{1}{2}g^{\mu\nu}\Big[(\Pi_{\lambda}\phi)^{T}(\Pi_{\lambda}\phi)$$

$$+ \mu^{2}(\phi\phi) + \phi U\phi + \frac{\lambda}{4}(\phi\phi)^{2}\Big] + \xi\Big[G^{\mu\nu} + g^{\mu\nu}\nabla^{2} - \nabla^{\mu}\nabla^{\nu}\Big](\phi\phi)\Big\}$$

$$+ F^{\mu\lambda}F^{\nu}{}_{\lambda} - \frac{1}{4}g^{\mu\nu}F^{\lambda\sigma}F_{\lambda\sigma} + \Lambda g^{\mu\nu} - (1/8\pi G)G^{\mu\nu}$$

$$+ \alpha\Big[4R^{\mu\nu}R - g^{\mu\nu}R^{2} + 4\,g^{\mu\nu}\nabla^{2} - \nabla^{\mu}\nabla^{\nu}\,R\Big]$$

$$+ \beta\Big[8R^{\mu\lambda}R^{\nu}{}_{\lambda} - 2g^{\mu\nu}R^{\lambda\sigma}R_{\lambda\sigma} - \frac{8}{3}R^{\mu\nu}R + \frac{2}{3}g^{\mu\nu}R^{2}$$

$$+ 4\Big[\nabla^{2}R^{\mu\nu} + g^{\mu\nu}\nabla_{\lambda}\nabla_{\sigma}R^{\lambda\sigma} - \nabla_{\lambda}\nabla^{\mu}R^{\mu\lambda} - \nabla_{\lambda}\nabla^{\nu}R^{\mu\lambda}$$

$$- \frac{2}{3}\left(g^{\mu\nu}\nabla^{2} - \nabla^{\mu}\nabla^{\nu}\right)R\Big]\Big] \qquad (15)$$

and we identify the term in curly brackets as the stress enery of the quantum fields. The other terms are ambiguous: at first glance at least some of the terms arose from the quantum field φ, however, their value is independent of φ. In fact, they are self interactions induced by the existence of φ but not involving φ.

This equation is analogous to the equation which may be obtained by requiring that the variation of Γ^{loc} with respect to A vanish,

$$0 = \frac{\delta\Gamma^{loc}}{\delta A_{\mu}(x)} = \partial_{\nu}\mathcal{F}^{\mu\nu}(x) - \sqrt{-g}\,g^{\mu\nu}\phi eq\left(\frac{1}{i}\,\partial_{\nu} - eqA_{\nu}\right)\phi \qquad (16)$$

which we recognize as Maxwell's equations with the first term having a particle contribution from φ but being the electromagnetic field term and second term as the current associated with the charged field φ.

In both equations, the nonlocal terms of Γ yield additional terms in the equations which reflect virtual particle creation and annihilation and vacuum polarization effects. The local field terms here are to be evaluated at the expectation value of φ and, since the solution obtained by dropping the nonlocal terms includes all tree graphs (Boulware and Brown [47]), will include all particle creation, propagation and annihilation effects in the tree approximation.

The field equation for the metric, Eq. (15), is an excellent phenomenological equation for slowly varying fields. However, it possesses undesirable properties and nonphysical solutions. First, it is a fourth order equation. As a result, if we consider a Cauchy surface and use the equation to find how the geometry develops away from the surface, we find that (subject to the constraints which we shall not discuss) we must specify the geometry of the surface <u>and</u> the first <u>three</u> derivatives of the geometry normal to the surface. Another manifestation of the same phenomenon is that there are additional modes besides the gravitational wave modes. These modes, in the case of small perturbations around flat space, have negative energies, negative probabilities and/or propagate faster than light. This situation is in many respects analogous to that of the radiation reaction force acting upon a charged particle. The phenomenological radiation reaction force $(2/3)\alpha\ddot{\vec{v}}$ produces runaway solutions which have extremely rapid variation and which are not observed. They are the result of a renormalization process (the "bare mass" of the charged particle is negative) and the neglect of the internal degrees of freedom of the charge (the nonlocal terms in our case, Coleman [48]). To specify a physical solution to the charge motion, one must either independently specify the initial acceleration <u>or</u> require that the charge not "runaway"; the latter is, of course, correct because only then will the motion be slowly varying so that the approximation of the radiation reaction by the local term is valid. Similarly, we have neglected the nonlocal terms in Γ which, for rapidly varying fields, are comparable to the local terms, hence the correct procedure is to impose as a boundary condition the requirement that the solution be slowly varying. This will eliminate the extra solutions. For a different point of view, see Wald [22].

If the nonlocal terms are included and we attempt to consider more rapidly varying fields, the analysis becomes much more difficult. However, one may speculate that since there is a renormalization group, it is possible to find effective coupling constants which are functions of the scale of variation of the gravitational field (see Weinberg [49]). Then, if these coupling constants α and β vanish as the field becomes rapidly varying, then the conceptual problems discussed above would also vanish. In that case, however, it is not clear that theory is formally renormalizable, because, in perturbation theory around flat space, it is the nonvanishing of α and β at high invariant masses which yields renormalizability.

There is, however, another possibility for dealing with the

renormalizability problem. As has been shown by Weinberg [50] and elaborated by Deser and myself [51], if the underlying quantum theory were to produce a graviton (a zero mass, helicity ±2 particle with nonvanishing coupling to $\mathcal{T}^{\mu\nu}$) as a dynamical effect of other fundamental particles (see Adler [52]), then the gauge restrictions on the self coupling of the graviton force the theory to exhibit general relativity in the slow variation limit. Then, these phenomenological terms we have been discussing will appear as contributions to a theory which is only valid in the slow variation limit.

We now turn to the problem of calculating the functional integral given in Eq. (1). The standard way to approach the problem is to look for extrema of the classical action which appears in the exponent. The extrema are solutions to the classical equation

$$\left\{-(\partial_\nu - ieqA_\nu)g^{\mu\nu}\sqrt{-g}\,(\partial_\mu - ieqA_\mu) + \sqrt{-g}\,(\mu^2 + R) + U\right\}\phi + \frac{1}{2}\lambda\sqrt{-g}\,\bar{\phi}(\bar{\phi}\bar{\phi}) = \eta . \tag{17}$$

There may be more than one solution to Eq. (17); if so, a variety of effects may appear which we shall not discuss but which may be quite important. These effects are associated with degenerate vacua, with instabilities, and with topologically stable field configurations such as magnetic monopoles in non-Abelian gauge theories. There is every reason to believe that a topologically rich theory such as general relativity will have a variety of interesting effects (Refs. [39-41]); however, given a solution to Eq. (17), we expect the renormalization procedure to be formally the same. If there is more than one solution, the functional integral should, in general, be a sum of terms arising from the neighborhood of each extremum. We shall restrict our attention to one particular solution.

If we proceed formally, ignoring the divergences, we may write $\phi \to \bar{\phi} + \phi$ and

$$\mathcal{L}(\bar{\phi} + \phi) = \mathcal{L}(\bar{\phi}) + \sum_{n=2} \frac{1}{n!}\left(\phi \frac{\delta}{\delta\,\bar{\phi}}\right)^n \mathcal{L}(\bar{\phi}) \tag{18}$$

and the functional integral may be written as

$$\int[d\phi]\exp i\int dx \left\{\mathcal{L}(\bar{\phi}) + \frac{1}{2}\phi_\alpha(x)\left(\delta^2\mathcal{L}(\bar{\phi})/\delta\bar{\phi}_\alpha\delta\phi_\beta\right)\phi_\beta(x)\right.$$

$$\times \exp(-i\lambda)\int dx\sqrt{-g}\left[\frac{1}{2}(\phi\phi)(\phi\bar{\phi}) + \frac{1}{8}(\phi\phi)^2\right] \tag{19}$$

In general, the integral cannot be done, even formally, except by expanding the second exponential. Then, the integral may be viewed as an average of the second exponential with the first exponential providing the measure. However, the standard virial expansion may be used here,

$$\langle \exp(iV) \rangle = \exp \sum_{n=1}^{\infty} \left[\langle (iV)^n \rangle c/n! \right] \tag{20}$$

where the subscript c denotes the connected part: in a graphical expansion, only connected graphs are to be included. Thus, the functional integral to be calculated is

$$\int [d\phi] \left[\exp(i/2\hbar) \int \phi \left(\delta^2 \mathcal{L}/\delta\phi\delta\phi\right) \phi \right] f(\phi) \equiv \int d\mu f(\phi)$$

$$= f\left(\hbar\delta/i\delta\psi\right) \int [d\phi] \exp(i/\hbar) \int \left[\psi\phi - \frac{1}{2}\,\phi G^{-1}\phi \right] \Bigg|_{\psi=0} \tag{21}$$

where we have introduced the dimensional constant, h, and written

$$\langle x|G^{-1}|x'\rangle \equiv -\left(\delta/\delta\bar{\phi}(x)\right)\left(\delta/\delta\bar{\phi}(x')\right) \int dy\, \mathcal{L}(\bar{\phi};g,A,U;y) \,.$$

The inverse Green's function operator, G^{-1}, is a function of the metric, g, the external vector potential, A, the potential U, and $\bar{\phi}$, the extermal value of the field. The Gaussian integral may be done by translating ϕ by $G\psi$ and rescaling ϕ by the operator $\sqrt{-g\hbar G}$ to obtain

$$\int d\mu f(\phi) = f\left(\hbar\delta/i\delta\psi\right) \exp(i/2\hbar)(\psi G\psi) \Big|_{\psi=0} \det G^{\frac{1}{2}} \,. \tag{22}$$

The factor $\det G^{\frac{1}{2}} = \det^{\frac{1}{2}} G$ arises from the Jacobian of the transformation,

$$\phi = \sqrt{-i\hbar G}\,\chi$$

$$\mathrm{Det}(\delta\phi/\delta\chi) = \mathrm{Det}(-i\hbar G^{\frac{1}{2}}) = (\mathrm{const.})\det^{\frac{1}{2}} G \,. \tag{23}$$

The factor $\det^{\frac{1}{2}} G$ is independent of $\hbar$ and is the entire contribution for a non interacting theory. Before discussing the one-loop, $\det^{\frac{1}{2}} G$, factor, we shall give a brief discussion of the higher loops and how they must be handled.

As indicated above, the function f which appears in Eq. (22) is restricted to the connected graphs when we are calculating the matrix element $\langle + | - \rangle$. In the expansion, one may express any term as a graph, e.g.

where the lines indicate propagators G, and the $\bar{\phi}$ indicates an explicit factor of $\bar{\phi}$ integrated with the vertex. The graphs shown here are the two-loop graphs; they are of order $\hbar$ just as the n-loop graph is of order $\hbar^{n-1}$ (Boulware and Brown [47]). The second graph has a single propagator connecting two vertices; it is said to be single particle reducible. The single particle reducible graphs may be dropped and $\bar{\phi}$ replaced by $\langle\phi\rangle$, Ref. [42,43]. Thus, in calculating the effective action in higher orders, we are instructed to calculate Γ by including only connected graphs, only single particle irreducible graphs, i.e. ones with no single propagators, $G(g,A,U,\langle\phi\rangle)$, connecting two vertices, and evaluating at $\langle\phi\rangle$ instead of $\bar{\phi}$.

As has been well known for twenty-five years, terms with more than one loop are complicated. Each such term has subgraphs with divergences which must be renormalized and, in general, there are several ways to isolate the divergent subgraphs. This overlapping divergence problem is technically complicated but poses no question of principle, and no features special to the curved background space-time case.

We now discuss the $\det^{\frac{1}{2}} G$ factor. There are a number of subtleties. First we note that since G is the inverse of G^{-1}, if there are homogeneous solutions to the equation,

$$(G^{-1})\psi = 0 \tag{24}$$

then the determinant of G will be infinite. Zeros of G^{-1} may occur for either of two reasons:

i) If the differential operator does determine the solutions, boundary conditions must be imposed. If the boundary conditions are not completely specified, the G^{-1} may have zeros corresponding to the ambiguity of the boundary conditions. The boundary conditions which must be imposed to determine G and its determinant are part of the state specification. In flat space with suitable electromagnetic fields, the usual vacuum state is selected by imposing the positive frequency boundary condition, or, equivalently, giving μ^2 a negative imaginary part.

ii) If the function $\bar{\phi}$ which is a solution to

$$\frac{\delta\mathcal{L}}{\delta\phi}(\bar{\phi}) = 0$$

is one of a family of solutions, $\bar{\phi}(x;\alpha_1 \ldots \alpha_n)$, then $\partial\phi/\partial\alpha_i(x;\alpha_1 \ldots \alpha_n)$ must be a solution to the homogeneous eq. (21). However, in that case a careful analysis (Coleman [37]) shows that the correct form, instead of

$$\int [d\phi] \exp i \int \mathcal{L} = \left(\exp i \int \mathcal{L}(\bar{\phi})\right) \mathrm{Det}^{\frac{1}{2}} G$$

is

$$\int [d\phi] \exp i \int \mathcal{L} = \left(\exp i \int dddx\, \mathcal{L}(\bar{\phi}(\alpha))\right) \mathrm{Det}' G^{\frac{1}{2}} \tag{25}$$

where the det leaves out the zero eigenvalues of G^{-1} associated with the α degeneracy. However, these complications do not affect the renormalization problem.

We briefly discuss the effect of the ambiguity in G due to the state specification or choice of boundary conditions. First, we note that, given any Green's function, $G^{(1)}$, any other Green's function, $G^{(2)}$ may be written as

$$G^{(2)}(x,x') = G^{(1)}(x,x') + i \sum_{i,j} \phi_i(x) \rho_{ij} \phi_j(x') \tag{26}$$

where $\phi_i(x)$ is a solution to the homogeneous equation without regard to boundary conditions. The ϕ_i and ρ are chosen so that $G^{(1)}$ satisfies its boundary conditions. Then, to calculate $\ell n\, \mathrm{Det}(G^{(2)-1}/G^{(1)-1})$ where $G^{(i)-1}$ means the Fredholm determinant of the operator G^{-1} subject to the boundary conditions of $G^{(i)}$, we use

$$\delta\, \ell n\, \mathrm{Det}(G^{(2)-1}/G^{(1)-1}) = \mathrm{Tr}[G^{(2)} - G^{(1)}]\delta G^{-1} \tag{27}$$

$$= i\rho_{ij} \int (\phi_j \delta G^{-1} \phi_i) \quad . \tag{28}$$

However, the functions ϕ_i are themselves dependent upon the parameters being varied in G. If we want the first order variation only, this is the answer. For example, the contribution to $<T^{\mu\nu}>$ is

$$<T^{\mu\nu}(x)> = <T^{\mu\nu}(x)>^{(1)} + \frac{2\delta}{i\delta g_{\mu\nu}(x)} \frac{1}{2} \ell n\, \mathrm{Det}(G^{(2)}/G^{(1)})$$

$$= <T^{\mu\nu}(x)>^{(1)} - \rho_{ij} \int \left(\phi_j \frac{\delta G^{-1}}{\delta g_{\mu\nu}(x)} \phi_i\right) \quad .$$

$$= <T^{\mu\nu}(x)>^{(1)} + \rho_{ij} T^{\mu\nu}(\phi_j, \phi_i; x) \tag{29}$$

where $T^{\mu\nu}(\phi_j;\phi_i;x)$ denotes the classical stress-energy tensor of the ϕ dependent terms in Eq. (15). The changed Green's function may reflect the presence of extra particles and this term is the extra stress energy associated with the particles.

To calculate the extra terms in w we note that any solution, ϕ_i, may be expressed in terms of its value in the neighborhood of the boundary of the region and $G^{(1)}$

$$\phi_i(x)\Big|_{x\,\varepsilon V} = \int_{\partial V} d\sigma'_\mu\, G^{(1)}(x,x')(\overleftrightarrow{\partial^\mu - ieqA^\mu})\phi_i(x')$$
$$\equiv G^{(1)}\cdot\chi_i = -\chi_i\cdot G^{(1)} \quad , \qquad (30)$$

where χ_i indicates the boundary data for the solution ϕ_i. Thus, if we restrict the variation to the interior, we obtain

$$\delta\,\frac{1}{2}\,\ell n\,\mathrm{Det}(G^{(2)-1}/G^{(1)-1}) = \frac{1}{2}\,\rho_{ij}\int \phi_j\delta G^{-1}\phi_i$$
$$= -\,\frac{1}{2}\,\rho_{ij}\chi_j\cdot\delta G^{(1)}\cdot\chi_i$$

or

$$\frac{1}{2}\,\ell n\,\mathrm{Det}(G^{(2)-1}/G^{(1)-1}) = -\,\frac{1}{2}\,\rho_{ij}(\chi_j\cdot G^{(1)}\cdot\chi_i) \quad . \qquad (31)$$

The additional term will in general not vanish. It reflects different particle content in the initial and/or final states and therefore the amplitude will change. Physical quantities such as stress energy and electromagnetic current will also reflect the initial and final states through these terms.

We assume that there are no zeros of G^{-1} and take G as the solution analytic in the lower half μ^2 plane. This defines the vacuum state in Minkowski space-time and in several other cases as well. It is not, however, the appropriate definition to study the Hawking radiation because the final state must be one in which the stress energy along the event horizon is finite. This requires the presence of the Hawking radiation and that the Green's functions be analytic in the Kruskal coordinate, $u = \exp 1/4[r^*-t)/4M]$, and hence that the Green's function be periodic with period $i8\pi M$ in $t_{\overline{2}}r^*$ which is inconsistent with the requirement of analyticity in μ^2.

We then need $G(x,x) \equiv \langle x|G|x\rangle$

$$= \langle x|\,\frac{1}{G^{-1}-i\varepsilon}\,|x\rangle \ ; \qquad (32)$$

however, G^{-1} is a bi-density and it is more convenient to deal with a bi-density of order 1/2, hence, we define

$$\langle x|G^{-1}|x'\rangle \equiv (-g(x))^{\frac{1}{4}}\langle x|\mathcal{H}|x'\rangle\ (-g(x'))^{\frac{1}{4}} \tag{33}$$

where,

$$\langle x|\mathcal{H}|x'\rangle = -\ (-g(x))^{-\frac{1}{4}}\Pi_\mu g^{\mu\nu}\sqrt{-g}\ \Pi_\nu \delta(x-x')(-g(x'))^{-\frac{1}{4}}$$

$$+\ [\mu^2 + U(x) + \xi R(x)]\delta(x-x')$$

and

$$U(x) \equiv U(x) + \frac{1}{2}\lambda[1(\langle\phi\rangle\langle\phi\rangle)(x) + 2\langle\phi\rangle(x)\langle\phi\rangle(x)] \quad .$$

The Green's function, G, may then be written as

$$G(x,x') = (-g(x))^{-\frac{1}{4}}\langle x|1/(\mathcal{H}-i\varepsilon)]|x'\rangle(-g(x'))^{-\frac{1}{4}}$$

$$= i(-g(x))^{-\frac{1}{4}}\int_0^\infty ds\langle x|\exp[-is\mathcal{H}]|x'\rangle(-g(x'))^{-\frac{1}{4}} \tag{34}$$

and

$$\delta\ \ln \mathrm{Det}^{\frac{1}{2}}G = -\frac{1}{2}\,\mathrm{Tr}\ G\delta G^{-1}$$

$$= -\ i\frac{1}{2}\int_0^\infty ds\int dx dx'(-g(x))^{-\frac{1}{4}}\langle x|\exp[-is\mathcal{H}]|x'\rangle(-g(x'))^{-\frac{1}{4}}$$

$$\delta[(-g(x'))^{\frac{1}{4}}\langle x'|\mathcal{H}|x\rangle(-g(x))^{\frac{1}{4}}] \tag{35}$$

$$= -\ i\frac{1}{2}\,\delta\int dx\left\{\int_0^\infty ds(i/s)\mathrm{tr}\langle x|\exp[-is\mathcal{H}]|x\rangle - i\delta(0)\ln\sqrt{-g(x)}\right\} \ .$$

The variations are arbitrary, hence

$$\ln \mathrm{Det}^{\frac{1}{2}}G = i\,\frac{1}{2}\left\{\int dx\left[\int_0^\infty (ds/is)\mathrm{tr}\langle x|\exp[-is\mathcal{H}]|x'\rangle\right.\right.$$

$$\left.\left. +\ 2i\delta(0)\ \ln\sqrt{-g(x)}\right]\right\} \quad . \tag{36}$$

The second, $\delta(0)\ln\sqrt{-g}$, term is peculiar. At first sight it would seem to be required to assure the invariance of $\mathrm{Det}^{\frac{1}{2}}G$ under coordinate transformations, but, because $\mathcal{H}$ is a bi-density of weight ½, the first term alone is invariant under the coordinate transformations. We have here the echo of our earlier discussion of the measure: we should not have integrated over scalar fields but rather over scalar densities of order ½. In that case the determinant would have been $\det^{\frac{1}{2}}\mathcal{H}^{-1}$ and the second term, which is not coordinate invariant would not have appeared. We therefore drop the extra term and obtain the result that, in one-loop order,

$$\langle +|-\rangle = \exp i\left\{\int dx\,\mathcal{L}(\bar{\phi};g,A,U;x) + \frac{1}{2}\int_0^\infty (ds/is)\mathrm{Tr}\,\exp[-is\mathcal{H}]\right\}, \quad (37)$$

where

$$\mathrm{Tr}\,\exp[-is\mathcal{H}] \equiv \int dx\ \mathrm{tr}\langle x|\exp[-is\mathcal{H}]|x\rangle \quad .$$

We must now analyze the extra term, which is being extensively discussed in Leonard Parker's lectures. I shall not attempt to repeat or improve upon his analysis, the result of which in n dimensions, is that the function $\langle x|\exp[-is\mathcal{H}]|x'\rangle$ may be written as

$$\langle x|\exp(-is\mathcal{H})|x'\rangle = i(4\pi is)^{-n/2}D^{\frac{1}{2}}(x,x')\exp i\left(\sigma(x,x')/2s\right)$$

$$\exp(-is\mu^2)\exp\left(ieq\int_{x'}^{x} dy^\lambda A_\lambda(y)\right)F(x,x';is) \quad (38)$$

where σ is one-half the square of the geodesic distance between x and x', D is the bi-density determinant, $D(x,x') = -\det(-\partial^2\sigma(x,x')\partial x\partial x')$, and the factor $\exp ieq\int_{x'}^{x} dy^\lambda A_\lambda(y)$ is integrated along the geodesic connecting x' and x and yields the correct gauge transformation properties for $\langle x|\exp[-is\mathcal{H}]|x'\rangle$. The function F is usually taken to be analytic in s around s = 0 so that it may be expanded in a power series, although, if there are surfaces of discontinuity, F has additional essential singularities at s = 0 which modify the analysis. To calculate F, we need the effect of $\mathcal{H}$ on $\langle x|$, the calculation of which involves

$$\left(\vec{\nabla} - ieq\,\vec{A}(x)\right)\left[\exp ieq\int_{x'}^{x} dy^\lambda A_\lambda(x)\right]$$

$$\equiv \left[\exp ieq\int_{x'}^{x} dy^\lambda A_\lambda(y)\right]\left(\vec{\nabla} + ieq\,\vec{f}(x,x')\right) \quad (39)$$

where

$$\vec{f}(x,x') \equiv \vec{\nabla}\int_{x'}^{x} dy^\lambda A_\lambda(y) - \vec{A}(x)$$

is gauge invariant, vanishes as $x \to x'$ and is orthogonal to $\vec{\nabla}\sigma$, the tangent at x to the geodesic, because the line integral is taken along the geodesic, $y(\lambda')$, $y(0) = x'$, $y(\lambda) = x$,

$$\vec{\nabla}\sigma\cdot\vec{f}(x,x') = \lambda\frac{d}{d\lambda}\int_0^\lambda d\lambda'\left(\frac{\vec{\nabla}\sigma}{\lambda'}\cdot A(y(\lambda'))\right) - \vec{\nabla}\sigma\cdot\vec{A}(x) = 0.$$

To first order in x - x',

$$f^\mu(x,x') = \frac{1}{2}F^\mu{}_\lambda(x)\nabla\sigma^\lambda(x,x') + O(\sigma) \quad . \quad (40)$$

Then, the function F must satisfy the equation,

$$i \frac{\partial}{\partial s} F(x,x';is) - \frac{i}{s} (\nabla\sigma)\cdot\nabla F(x,x';is)$$

$$= \Big\{\eta R(x) + \frac{\lambda}{2} [2\bar{\phi}\bar{\phi} + (\bar{\phi})^2](x) + e^2\vec{f}\cdot\vec{f}(x,x') - ieq(\vec{\nabla}\cdot f)(x,x')$$

$$- 2ieq\, \vec{f}(x,x')\cdot\vec{\nabla} - \Delta^{-\frac{1}{2}}\nabla^2\Delta^{\frac{1}{2}}\Big\} F(x,x';is) \qquad (41)$$

where $\Delta(x,x') \equiv D(x,x')/[(-g(x))(-g(x'))]^{\frac{1}{2}}$ is a bi-scalar. The function F must be used in Eq. (37) to obtain

$$w = \int dx\, \mathcal{L}(<\bar{\phi}>) + \frac{1}{2}\int dx \int_0^\infty \frac{ds}{(is)} <x|e^{-is\mathcal{H}}|x>$$

$$\equiv \int dx\, \mathcal{L}(<\phi>) + w^{(1)}[<\phi>,g,A,U]$$

or

$$w^{(1)} = \frac{1}{2}\int_0^\infty \frac{ids}{(is)} \frac{e^{-is\mu^2}}{(4\pi is)^{n/2}} \int dx\sqrt{-g}\, \mathrm{tr}\, F(x,x';is) \qquad (42)$$

where the trace is over the charge indices.

Because of the factor $s^{-(1+n/2)}$, the integral is divergent. The first few terms in the power series do not converge, hence to identify the divergent terms, we must find the coefficients

$$F(x,x';is) = \sum_{n=0} (is)^n f_n(x,x') \qquad (43)$$

which we insert in eq. (41) to find

$$\Big\{n + \nabla\sigma\cdot\Delta\Big\} f_n(x,x')$$

$$= -\Big\{-(\vec{\nabla} + ieq\vec{f}(x,x') + \vec{\nabla}\ell n\Delta^{\frac{1}{2}})^2 + \breve{U}(x) + \xi R(x)\Big\} f_{n-1}(x,x') \ . \qquad (44)$$

Thus, since $f_0 = 1$,

$$f_1(x,x) = - [\tilde{U}(x) + (\xi - 1/6)R] \qquad (45)$$

because $\vec{f}(x,x) = 0 = \vec{\nabla}\cdot\vec{f}(x,x)$ and the (1/6)R term is calculated in Leonard Parker's lecture from $\nabla^2\Delta^{\frac{1}{2}}(x,x')\big|_{x=x'} = (1/6)R$. To calculate $f_2(x,x)$, $\vec{\nabla}\nabla\vec{f}_1(x,x')\big|_{x'=x}$ is required and may be obtained by using

$$(\vec{\nabla}\vec{\nabla}\sigma)^{\mu\nu}(x,x')\Big|_{x'=x} = g^{\mu\nu}(x)\ \vec{\nabla}\vec{\nabla}(1+\nabla\sigma\cdot\vec{\nabla})f_1(x,x')\Big|_{x'=x} = 3\vec{\nabla}\vec{\nabla}f_1(x,x')\Big|_{x'=x}$$

or

$$\nabla^\mu\nabla^\nu f_1(x,x')\Big|_{x=x'} = -\frac{1}{3}\left\{\frac{e^2}{2} F^{\mu\lambda}F^{\nu}{}_{\lambda} + \frac{ieq}{4}\left(\nabla^\mu j^\nu + \nabla^\nu j^\mu\right)\right.$$

$$\left. - \frac{1}{3} ieq(F^{\lambda\mu}R_{\lambda}{}^{\nu} + F^{\lambda\mu}R_{\lambda}{}^{\mu})\right\} + \frac{1}{3}\left(\frac{1}{4} - \xi\right)\nabla^\mu\nabla^\nu R + \frac{1}{60}\nabla^2 R^{\mu\nu}$$

$$- \frac{1}{45} R^{\mu\lambda}R^{\nu}{}_{\lambda} + \frac{1}{90} R^{\mu}{}_{\alpha}{}^{\nu}{}_{\beta}R^{\alpha\beta} + \frac{1}{90} R^{\mu\alpha\beta\gamma}R^{\nu}{}_{\alpha\beta\gamma} - \frac{1}{3}\nabla^\mu\nabla^\nu\tilde{U} \tag{46}$$

where

$$j^\mu = \nabla_\nu F^{\mu\nu} \quad .$$

The purely gravitational terms are calculated as described in Leonard Parker's lectures or as given by Brown [41]. The eq terms are calculated by expanding $\vec{f}(x,x')$ around $x = x'$ and using Riemann normal coordinates. In any case, these terms do not contribute to f_2 because their traces with $g^{\mu\nu}$ and over the internal indices separately vanish.

The result for $f_2(x,x)$ then becomes using eqs. (44) and (46),

$$f_2(x,x) = \frac{1}{2}[\tilde{U} + (\xi - 1/6)R]^2 - \frac{e^2}{12} F^{\lambda\sigma}F_{\lambda\sigma} - \frac{1}{6}\nabla^2\tilde{U}$$

$$+ \frac{1}{6}\left(\frac{1}{5} - \xi\right)\nabla^2 R + \frac{1}{180}[R_{\mu\nu\lambda\sigma}R^{\mu\nu\lambda\sigma} - R_{\mu\nu}R^{\mu\nu}] \tag{47}$$

and we have,

$$F(x,x;is) \equiv 1 + (is)f_1(x,x) + (is)^2 f_2(x,x) + \tilde{\tilde{F}}(x,x;is)$$

$$\equiv 1 + (is)f_1(x,x) + \tilde{F}(x,x;is) \tag{48}$$

which define $\tilde{\tilde{F}}$ and $\tilde{F}$ as the function F with, respectively, the first 3 and 2 terms of the power series in (is) subtracted.

Then the one-loop contribution to the effective action is

$$w = w^{(1)} + w^{Div} \tag{49}$$

where

$$w^{(1)} \equiv \frac{1}{2}\int dx\sqrt{-g}\int_0^\infty \frac{dsi}{(is)^3(4\pi)^2} e^{-is\mu^2} \mathrm{tr}\tilde{\tilde{F}}(x,x;is) + \text{finite local terms}$$

and

$$w^{Div} \equiv \int dx\sqrt{-g}\int_0^\infty \frac{dsi}{is(4\pi is)^{n/2}}\, e^{-is\mu^2}\left[1 + isf_1(x,x) + (is)^2 f_2(x,x)\right]$$

The theory does not exist. It must be renormalized. First, one must modify Eq. (49) so that it is finite. There are three commonly used methods of doing this: Pauli-Villars regularization, point splitting, and dimensional continuation (Hawking [14] and others have proposed s function renormalization which is closely related to dimensional regularization and is discussed in Leonard Parker's lecture). In each case there are parameters of which the modification of the effective action is a function. Further, there is a limit of the parameters in which the modified w formally returns to the divergent expression given in Eq. (49). Since the first term involving $\tilde{F}$ is finite, coordinate invariant and gauge invariant by itself, that term should not change as we regulate, calculate w, renormalize to define the divergent parts given by the remaining (f_0, f_1 and f_2) terms in w. The $\tilde{F}$ term contains all the nonlocal contributions due the virtual or real production of particles and is dependent upon A,g,U and $\langle\bar{\phi}\rangle$ not just at x but everywhere; as a result, it is in general incalculable. If the future state, $\langle +|$, is the state into which the past state, $|-\rangle$, develops in the presence of the sources η, A, g, and U, $w^{(1)}$ is real and the only effect of the one-loop corrections is to produce a phase for the amplitude $\langle +|-\rangle$. The dependence upon sources does, however, yield all the dynamical information in the theory.

Pauli-Villars regulation consists of adding additional fields with high mass. These fields contribute additional terms to w, which becomes

$$w = \frac{1}{2}\int dx\sqrt{-g}\int_0^\infty dm^2\rho(m^2)\int_0^\infty (dis/is)(4\pi is)^{-n/2} e^{-im^2 s} F(x,x;is) \tag{50}$$

and the weight function, $\rho(m^2)$, is chosen so that

$$\int_0^\infty dm^2\rho(m^2)(m^2)^p = 0 \ , \quad p = 0,\ 1,\ 2 \quad . \tag{51}$$

The general results we shall discuss are independent of the specific form of ρ, but we shall use

$$\rho(m^2) = \delta(m^2-\mu^2) - \delta(m^2-M^2) - (M^2-\mu^2)\delta'(m^2-M^2)$$
$$- \frac{1}{2}(M^2-\mu^2)^2\delta''(m^2-M^2) \quad , \tag{52}$$

to obtain an explicit form for the divergent terms in w,

$$W^{Div} = \frac{1}{2}\int \frac{dx\sqrt{-g}}{(4\pi)^2}(tr1)\left\{\frac{1}{2}\mu^4 \ell n\, m^2/\mu^2 + \frac{1}{4}(m^2-\mu^2)(m^2-3\mu^2)\right\}$$

$$+ \frac{1}{2}\int \frac{dx\sqrt{-g}\, trf_1(x,x)}{(4\pi)^2}\left\{\mu^2 \ell n\, \mu^2/m^2 + \frac{1}{2}\frac{(m^2-\mu^2)(m^2+\mu^2)}{m^2}\right\} \quad (53)$$

$$+ \frac{1}{2}\int \frac{dx\sqrt{-g}\, trf_2(x,x)}{(4\pi)^2}\left\{\ell n(m^2/\mu^2) - \frac{(m^2-\mu^2)}{2m^4}\right\}$$

Dimensional regulation consists of evaluating the effective action in n rather than 4 dimensions. We note that f_0, f_1 and f_2 contain no explicit dependence on n, only the power of is reflects the number of dimensions. Of course, the contracted tensors, e.g. $F^{\lambda\sigma}F_{\lambda\sigma}$, will have a number of components which depends on n but there is no explicit dependence. Then, the integrals which appear in the divergent terms are of the form,

$$\int_0^\infty \frac{dsi(is)^p}{(is)(4\pi is)^{n/2}}e^{-is\mu^2} \equiv Ip(n) \quad p = 0,1,2 \quad (54)$$

and we regulate by continuing in the parameter n/2 until the integral does converge, yielding

$$Ip(n) = \frac{\Gamma(p-n/2)}{(4\pi)^{n/2}}(\mu^2)^{n/2-p} \quad (55)$$

which has a pole at $n = 4$ for $p \le 2$. Thus, the divergent terms become

$$W^{Div} = \frac{1}{2}\int dx\sqrt{-g}\,(tr1)\frac{\Gamma(-n/2)}{(4\pi)^{n/2}}(\mu^2)^{n/2}(K^2)^{2-n/2}$$

$$+ \frac{1}{2}\int dx\sqrt{-g}\, tr\, f_1(x,x)\frac{\Gamma(1-n/2)}{(4\pi)^{n/2}}(\mu^2)^{n/2-1}(K^2)^{2-n/2}$$

$$+ \frac{1}{2}\int dx\sqrt{-g}\, tr\, f_2(x,x)\frac{\Gamma(2-n/2)}{(4\pi)^{n/2}}(\mu^2)^{n/2-2}(K^2)^{2-n/2} \quad (56)$$

which is of exactly the same form as the Pauli-Villars regularization. Further, these terms are of exactly the same form as the local divergent terms in the effective action, Eq. (12), and may be eliminated by the renormalization procedure described there. The dependence on K is to assure that W^{Div} have the units of action (the integral, $\int dx$, is taken over 4 dimensions) and reflects the necessity of introducing a mass scale in the process of renormalization. The divergences themselves have no dependence upon K which is an arbitrary finite parameter. As $n \to 4$,

$$W^{Div} \underset{n\to 4}{\sim} \frac{1}{2} \int dx\sqrt{-g}\,(\mathrm{tr}1)\, \frac{\mu^4}{(4\pi)^2} \left[\frac{1}{(4-n)} + \frac{1}{2}\left[\ln(4\pi K/\mu)^2 + \psi(1) - 3/2\right]\right]$$

$$+ \frac{1}{2} \int dx\sqrt{-g}\,\mathrm{tr} f_1(x,x)\, \frac{\mu^2}{(4\pi)^2} \left[-\frac{2}{4-n} - \ln(4\pi K/\mu)^2 + \psi(1) - 1/2\right]$$

$$+ \frac{1}{2} \int dx\sqrt{-g}\,\mathrm{tr}\, \frac{f_2(x,x)}{(4\pi)^2} \left[\frac{2}{4-n} + \ln(4\pi K/\mu)^2 + \psi(1)\right] \quad . \qquad (57)$$

The structures have exactly the same form as with Pauli-Villars regularization. The finite terms are, however, different and we have dependence upon the finite parameter K. The divergent terms are absorbed into the local effective action of Eq. (13) and we have an arbitrary amount of finite terms to associate with $W^{(1)}$. We shall show that the correct choice is to include no finite terms if $\mu^2 \neq 0$ so that

$$W^{(1)} = \frac{1}{2} \int_0^\infty (dsi/is)(4\pi is)^{-2} \exp[-is\mu^2] \int dx\sqrt{-g}\,\mathrm{tr}\tilde{\tilde{F}}(x,x;is) \ . \qquad (58)$$

This simple result is due to a felicitous choice of finite part. We have not yet justified it.

Point splitting is generally the most complicated way; however, it is also sometimes the only way. If a particular calculational procedure does not allow dimensional continuation or Pauli-Villars regularization, point splitting may be the only regularization procedure available. To do point splitting, we calculate with $x \neq x'$,

$$\mathcal{L}^{(1)}(x,x') = \frac{1}{2} \int_0^\infty (ds/is)\,\mathrm{tr}\langle x|\exp[-is\mathcal{H}]|x'\rangle \qquad (59)$$

and, formally,

$$W^{(1)} = \int dx \mathcal{L}^{(1)}(x,x) \quad ,$$

which is divergent. Several complications arise. This is not gauge invariant, because of the gauge factor,

$$\exp\left[ieq \int_{x'}^{x} dy^\lambda A_\lambda(y)\right] \quad ,$$

which must be dropped to assure gauge invariance (Schwinger [53]). Further, the result is not a density at x or any other point, it is a bi-density of weight 1/2 at x and at x'. We make it a bi-scalar by dividing by $[(-g(x))(-g(x'))]^{\frac{1}{4}}$ and multiply by $(-g(\bar{x}))^{\frac{1}{2}}$ where $\bar{x}$ is the midpoint of the geodesic connecting x' and x so that the result is density of weight 1 at $\bar{x}$ and a bi-scalar at x and x'. Thus, the finite integral to be done is

$$\mathcal{L}^{(1)} = \sqrt{-g(\bar{x})}\,\frac{1}{2}\int_0^\infty (dsi/is)(4\pi is)^{-2}\Delta^{\frac{1}{2}}(x,x')\exp i\left(\tfrac{1}{4}\,\sigma/s\right)\exp(-is\mu^2) \times \mathrm{tr}\tilde{\tilde{F}}(x,x';is) \tag{60}$$

$$\underset{\sigma\to 0}{\to} \sqrt{-g(x)}\,\frac{1}{2}\int_0^\infty (dsi/is)(4\pi is)^{-2}\exp(-is\mu^2)\mathrm{tr}\tilde{\tilde{F}}(x,x;is)$$

and the divergent terms are

$$W^{\mathrm{Div}} = \int d\bar{x}\sqrt{-g(\bar{x})}\,\mathrm{tr}\sum_{p=0}^{2}\Delta^{\frac{1}{2}}f_p(x,x') \times \frac{1}{2}\int_0^\infty (dsi/is)(4\pi is)^{-2}(is)^p\exp\left[(i\sigma/4s) - is\mu^2\right]$$

$$= \int d\bar{x}\sqrt{-g(\bar{x})}\sum_{p=0}^{2}(4\pi)^{-2}\mathrm{tr}\Delta^{\frac{1}{2}}f_p(x,x')\left[(\sqrt{\sigma/2\mu})^{p-2}K_{2-p}(\sqrt{\sigma\mu^2})\right] . \tag{61}$$

The divergent terms are the expansion of $\Delta^{\frac{1}{2}}f_0$ through order $(\nabla\sigma)^4$, $\Delta^{\frac{1}{2}}f_1$ through order $(\nabla\sigma)^2$ and $\Delta^{\frac{1}{2}}f_2$ at x = x'. The result of a fairly tedious calculation is that

$$\Delta^{\frac{1}{2}}(x,x') = 1 + \sigma e_\alpha e_\beta\,\frac{1}{6}\,R^{\alpha\beta} + \sigma^2 e_\alpha e_\beta e_\gamma e_\delta\,\frac{1}{6}\left\{\frac{1}{12}R^{\alpha\beta}R^{\gamma\delta} + \frac{1}{20}\nabla^\alpha\nabla^\beta R^{\gamma\delta} + \frac{1}{15}R^{\alpha\lambda\beta\sigma}R^{\gamma}{}_{\lambda}{}^{\delta}{}_{\sigma}\right\} + O(\sigma^3) \tag{62}$$

and

$$\Delta^{\frac{1}{2}}(x,x')f_1(x,x') = f_1 + \sigma e_\alpha e_\beta\left\{-\frac{e^2}{12}F^{\alpha\lambda}F^{\beta}{}_{\lambda} + \frac{1}{90}\left(R^{\alpha\lambda\sigma\tau}R^{\beta}{}_{\lambda\sigma\tau} + R^{\alpha\lambda\beta\sigma}R_{\lambda\sigma} - 2R^{\alpha\lambda}R^{\beta}{}_{\lambda}\right) - \frac{1}{6}R^{\alpha\beta}[\tilde{U} + (\xi-1/6)R] + \frac{1}{60}\nabla^2R^{\alpha\beta} - \frac{1}{12}\nabla^\alpha\nabla^\beta[\tilde{U} + (\xi-1/10)R]\right\} + O(\sigma^2)$$

where e_α is the tangent vector at $\bar{x}$ of the geodesic connecting x and x' and all functions except σ are evaluated at $\bar{x}$. Then,

$$w^{Div} = \int dx\sqrt{-g(x)}\,(4\pi)^{-2}\Big\{ \sum_{p=0}^{2} \mathrm{tr} f_p(x,x)\Big[(2\mu/\sqrt{\sigma})^{2-p} K_{2-p}(\sqrt{\sigma\mu^2})\Big]$$

$$+ e_\alpha e_\beta \frac{1}{6} R^{\alpha\beta}(2\mu)^2 K_2(\sqrt{\sigma\mu^2})\mathrm{tr}1$$

$$+ e_\alpha e_\beta e_\gamma e_\delta \Big[\frac{1}{72} R^{\alpha\beta}R^{\gamma\delta} + \frac{1}{120} \nabla^\alpha\nabla^\beta R^{\gamma\delta}$$

$$+ \frac{1}{90} R^{\alpha\lambda\beta\sigma}R^{\gamma}{}_{\lambda}{}^{\delta}{}_{\sigma}\Big] 4\mu^2\sigma K_2(\sqrt{\sigma\mu^2})\mathrm{tr}1$$

$$+ e_\alpha e_\beta \mathrm{tr}\Big\{- \frac{e^2}{12} F^{\alpha\lambda}F^{\beta}{}_{\lambda} + \frac{1}{90}\Big(R^{\alpha\lambda\sigma\tau}R^{\beta}{}_{\lambda\sigma\tau} + R^{\alpha\lambda\beta\sigma}R_{\lambda\sigma}$$

$$- 2R^{\alpha\lambda}R^{\beta}{}_{\lambda}\Big) - \frac{1}{6} R^{\alpha\beta}[\tilde{U} + (\xi-1/6)R] + \frac{1}{60} \nabla^2 R^{\alpha\beta}$$

$$- \frac{1}{12} \nabla^\alpha\nabla^\beta[\tilde{U} + (\xi-1/10)R]\Big\}(2\mu\sqrt{\sigma})K_1(\sqrt{\sigma\mu^2}) \quad . \qquad (63)$$

The first terms correspond exactly to the terms we found using dimensional and Pauli-Villars regularization. They are divergent as $\sigma \to 0$, and the dependence upon $\langle\phi\rangle$ etc. is exactly the same. The remaining terms are different and arise from the expansion of $\Delta^{\frac{1}{2}}(x,x)f_p(x,x')$ in x and x' around $\bar{x}$. The new terms are of the form $\vec{e}\vec{e}/\sigma$, $\vec{e}\vec{e}\vec{e}\vec{e}$ or $\vec{e}\vec{e}$ and are dependent upon the direction $\vec{e}$ (since no explicit g's appear). The action cannot depend upon an explicit direction, hence these terms must be dropped. Despite our intentions of dropping the terms, they must be calculated because if the point splitting method is used to regulate, say, a mode sum, the terms which arise must be identified and removed. For a more complete review of this method, including the renormalization of $\langle T^{\mu\nu}\rangle$ see S. M. Christensen [54].

In each method divergences have appeared in the coefficients of $\int dx\sqrt{-g}\, f_p(x,x)$, $p \le 2$. If these terms were finite, we could identify them as, respectively, an effective cosmological constant term, mass term for $\langle\phi\rangle$, $R(\langle\phi\rangle\langle\phi\rangle)$ interaction, Einstein-action, $\langle\phi\rangle$ self coupling, Maxwell action, and gravitational self-interactions involving the square of the Riemann tensor. We are forced to include them: we add terms to the action to cancel these terms, thereby leaving the term $w^{(1)}$ as the one-loop contribution. As we shall see, by doing precisely this, we leave the mass of the $\langle\phi\rangle$ field unchanged, because in flat space w has no term proportional to $(\phi)^2$, hence the propagation of ϕ is changed only by interaction with the gravitational field, the electromagnetic field or with $\langle\phi\rangle$ itself. The $\lambda(\langle\phi\rangle\langle\phi\rangle)^2$ interaction is, with the choice of $w^{(1)}$, fixed so that λ is the coupling constant at zero momentum transfer. Because, in our model, the gravitational and electromagnetic fields have no dynamics, we have no criteria to determine the coefficients for the terms in the $R(\phi\phi)$ coupling; in Einstein's theory, R is, essentially, the trace of the matter stress tensor, hence this term is effectively a matter self interaction term. The major significance of the term is to maintain conformal invariance for $\xi = 1/6$. If ξ is equal to 1/6, there is no renormalization of ξ in the one-loop approximation. There will, presumably, be no renormalization in any loop approximation because such a term would produce an additional divergent contribution to the trace of the stress tensor; however, if dimensional regularization is used, the trace can have no such contributions. At first sight it is remarkable that there is no wave function renormalization of the ϕ field. However, there is no wave function renormalization to first order because the theory has only a ϕ^4 non derivative coupling. Theories with ϕ^3 coupling or derivative couplings have wave function renormalization in first order.

To determine the renormalized parameters in Eq. (14), we must investigate physical processes. The expectation value of ϕ must satisfy

$$\eta(x) = \frac{\delta\Gamma}{\delta\langle\phi(x)\rangle} = \left\{-\nabla^2 + \mu^2 + \xi R + U + \frac{\lambda}{2}(\langle\phi\rangle\langle\phi\rangle)\right\}\langle\phi\rangle - \frac{w^{(1)}}{\delta\langle\phi(x)\rangle}\,. \tag{64}$$

But, since $\langle\phi\rangle$ only enters $\mathcal{H}$ through $\tilde{U}$

$$\left(\delta w^{(1)}/\delta\langle\phi(x)\rangle\right) = \int dy\,\mathrm{tr}\left(\delta\tilde{U}(y)/\delta\langle\phi(x)\rangle\right)\frac{1}{2}\int_0^\infty (ds/is)\left(\delta/\delta\tilde{U}(y)\right)\left(\mathrm{Tr}\,e^{-is\mathcal{H}}\right)$$

$$+ \text{ local terms} \tag{65}$$

$$= -\frac{\lambda}{2}\int_0^\infty (ds/is)\left\{\langle\phi(x)\rangle \mathrm{tr}\left[\widetilde{\widetilde{\langle x|\exp[-is\mathcal{H}]|x\rangle is}}\right]\right.$$

$$\left. + 2\left[\widetilde{\widetilde{\langle x|\exp[-is\mathcal{H}]|x\rangle}}\langle\phi(x)\rangle is\right]\right\}$$

$$+ \text{ local terms,}$$

but the subtraction of terms through order $(is)^2$ of isF corresponds to the subtraction of terms through order (is) in F, hence the result is

$$\left(\delta W^{(1)}/\delta\langle\bar{\phi}(x)\rangle\right) = -\frac{\lambda}{2}\sqrt{-g(x)}\int_0^\infty dsi(4\pi is)^{-2}\exp(-is\mu^2)\Big\{\langle\bar{\phi}(x)\rangle\mathrm{tr}\tilde{F}(x,x;is)$$

$$+ 2\tilde{F}(x,x;is)\langle\bar{\phi}(x)\rangle\Big\} + \text{local terms.} \tag{66}$$

In flat space-time, with vanishing sources, $\tilde{F}$ vanishes. But, in that limit, we want the mass which appears in Eq. (64) to be the correct renormalized mass and we must require that the local terms have no term linear in $\langle\bar{\phi}\rangle$. If we consider terms of higher order in $\langle\bar{\phi}\rangle$ or with non vanishing sources, we must calculate

$$\int_0^\infty \frac{dsi}{(4\pi is)^2} e^{-is\mu^2}\tilde{F}(x,x;is)$$

but this is precisely equal to

$$-\frac{\delta W^{(1)}}{\delta U(x)} = \langle\phi(x)\phi(x)\rangle$$

and has the graphical expansion

(a) (b) (c)

where the prime represents the replacement of F by $\tilde{F}$. There is no replacement in the third term (the subtractions were at most linear in the external potentials). The first, (a), graph is independent of the sources; it is just the 1 in F and is dropped. The second, (b), graph contains the linear response to U and A. The effects of U and A from points not at x is unchanged by the subtraction procedure. The only question is what local term should be included. If F(x,x;is) is calculated in flat space-time, in the linear response approximation, the result is the standard Feynman diagram analysis

$$F(x,x;is) \simeq -is\int\frac{dq}{(2\pi)^4}e^{iqx}\int dx\,\exp\left(-isq^2\alpha(1-\alpha)\right)\tilde{U}(q)$$

and the replacement of F by $\tilde{F}$ yields,

$$F(x,x;is) \simeq -is\int\frac{dq}{(2\pi)^4}e^{iqx}\int_0^1 d\alpha\left[\exp\left(-isq^2\alpha(1-\alpha)\right) - 1\right]\tilde{U}(q).$$

Then the renormalized Feynman diagram yields

$$\int dq(2\pi)^{-4}(\exp iqx)(4\pi)^{-2}\ell n\left[\mu^2/(\mu^2 + \alpha(1-\alpha)q^2)\right]\tilde{U}(q)$$

for the response linear in U. The addition of finite local terms would effectively change the μ in the numerator of the logarithm to K, some unknown mass, but for the amplitude with no additional terms, the response vanishes at zero momentum transfer, q^2. But, since $\tilde{U}$ contains a term proportional to $\langle\phi\rangle^2$ which contributes to scattering which is momentum dependent, we must define the coupling

constant λ in Eq. (64) by the scattering at some value. The standard value, for theoreticians, is the analytic continuation to zero momentum transfer, hence the $\delta w^{(1)}/\delta\langle\bar{\phi}\rangle$ contribution should vanish there, as it does with the choice $\tilde{\tilde{F}}$ in $w^{(1)}$ and no additional local terms. Exactly similar arguments apply to the dependence upon $A^\mu(F^{\lambda\sigma})$, but for the case of space-time curvature, the situation is more subtle. To evaluate the effect at a particular point, x, we choose Riemann normal coordinates around x. Then, the metric only deviates by a small amount in the neighborhood of x. We expand in this deviation and find an exactly similar form, with the coefficient of $R\langle\bar{\phi}\rangle$ with equations of motion being determined to be precisely $\xi R\langle\bar{\phi}\rangle$ for the case of constant curvature, if the renormalized amplitude is taken to be precisely given by $\int \tilde{\tilde{F}}$ with no additional local terms.

These considerations allow us to determine all the terms dependent upon $\langle\bar{\phi}\rangle$. To determine the local terms independent of $\langle\bar{\phi}\rangle$, we must consider variations of the other quantities. The dependence upon A_μ is exactly analogous, and the given amplitude yields, for weak fields, the usual conserved electromagnetic current which responds to field gauge invariantly with vanishing linear response for a homogeneous field.

The identification of the response to the curvature is again more subtle and again requires the use of Riemann normal coordinates. The stress tensor is given by

$$\langle\mathcal{T}^{\mu\nu}(x)\rangle = \frac{2\delta W}{\delta g_{\mu\nu}(x)} = \sqrt{-g}\Big\{[\Pi^\mu\bar{\phi}]^T[\Pi^\nu\bar{\phi}] - \frac{1}{2}g^{\mu\nu}\Big([\Pi^\lambda\bar{\phi}]^T[\Pi_\lambda\bar{\phi}] + \mu^2(\bar{\phi}\bar{\phi}) + \bar{\phi}U\bar{\phi} + \frac{\lambda}{8}(\bar{\phi}\bar{\phi})^2\Big) + 2\xi G^{\mu\nu}(\bar{\phi}\bar{\phi}) + \xi(\nabla^2 g^{\mu\nu} - \nabla^\mu\nabla^\mu)(\bar{\phi}\bar{\phi})\Big\} + \frac{2\delta\, w^{(1)}}{\delta g_{\mu\nu}(x)} \qquad (67)$$

where the first term is the stress energy tensor of the classical field $<\bar{\phi}>$. If the reduction formula is used to create excitations in the initial and/or final states, this term gives the stress energy associated with these excitations with the interactions being calculated in the tree approximation. The second, one-loop, contribution yields a variety of effects. The external fields will induce some stress, e.g. the Casimir effect. If the state is not the vacuum but has $<\phi> = 0$, the stress energy of the excitations will appear in the second term. The stress energy associated with the one-loop approximation to the interactions of the particles will also appear in $\delta w^{(1)}/\delta g_{\mu\nu}$.

To find the one-loop contribution, we require the variation of $w^{(1)}$ with respect to $g_{\mu\nu}$ for which we require

$$\left(\delta/\delta g_{\mu\nu}(x)\right)\int dy(\bar{\psi}\mathcal{H}\psi) = -\left\{\frac{1}{2}\left[(\Pi^{\mu}\bar{\psi})^T(\Pi^{\nu}\psi) + (\Pi^{\nu}\bar{\psi})^T(\Pi^{\mu}\psi)\right]\right.$$

$$- \frac{1}{2} g^{\mu\nu}(\Pi^{\lambda}\bar{\psi})^T(\Pi_{\lambda}\psi) + \frac{1}{4} g^{\mu\nu}\left[\bar{\psi}(-\Pi^2\psi) + (-\Pi^2\bar{\psi})^T\psi\right]$$

$$\left. + \xi[R^{\mu\nu} + g^{\mu\nu}\nabla^2 - \nabla^{\mu}\nabla^{\nu}](\bar{\psi}\psi)\right\}$$

$$= - \left\{\frac{1}{2}\left[(\Pi^{\mu}\bar{\psi})^T(\Pi^{\nu}\psi) + (\Pi^{\nu}\bar{\psi})^T(\Pi^{\mu}\psi)\right] - \frac{1}{2} g^{\mu\nu}(\Pi^{\lambda}\bar{\psi})^T(\Pi_{\lambda}\psi)\right.$$

$$+ \frac{1}{4} g^{\mu\nu}[\bar{\psi}(\mathcal{H}\psi) + (\mathcal{H}\bar{\psi})^T\psi]$$

$$\left. - \frac{1}{2} g^{\mu\nu}\ \bar{\psi}(\mu^2+\tilde{U})\psi + \xi[G^{\mu\nu}+g^{\mu\nu}\nabla^2-\nabla^{\mu}\nabla^{\nu}](\bar{\psi}\psi)\right\} \tag{68}$$

where Π^{μ} denotes the covariant derivative acting on a scalar density of weight 1/2,

$$\Pi^{\mu}\psi = (-g)^{\frac{1}{4}}(\partial^{\mu}-ieqA^{\mu})\ \frac{1}{(-g)^{\frac{1}{4}}}\ \psi \quad .$$

Then,

$$\left(\delta/\delta g_{\mu\nu}(x)\right)\mathrm{Tr}[\exp(-is\mathcal{H})]$$

$$= i\,\mathrm{str}\left\{\Pi^{\{\mu}\Pi^{\nu'\}}<x|e^{-is\mathcal{H}}|x'>\Big|_{x=x'} - \frac{1}{2} g^{\mu\nu}\Pi^{\lambda}\Pi'_{\lambda}<x|e^{-is\mathcal{H}}|x'>\Big|_{x=x'}\right.$$

$$+ \frac{1}{2} g^{\mu\nu} i\ \frac{\partial}{\partial s}\ <x|e^{-is\mathcal{H}}|x> - \frac{1}{2} g^{\mu\nu}\left(\mu^2+U(x)\right)<x|e^{-is\mathcal{H}}|x>$$

$$+ \xi[G^{\mu\nu} + g^{\mu\nu}\nabla^2 - \nabla^\mu\nabla^\nu]\langle x|e^{-is\mathcal{H}}|x\rangle\Big\} \quad . \tag{69}$$

However,

$$\Pi^{\{\mu}\Pi^{\nu'\}}\langle x|e^{-is\mathcal{H}}|x'\rangle\Big|_{x=x'}$$

$$= \sqrt{-g}\Big\{(g^{\mu\nu}/2is) - \frac{1}{6}R^{\mu\nu} + \nabla^\mu\nabla^{\nu'}\Big\}i[\exp(-is\mu^2)](4\pi is)^{-2}F(x,x'is)\Big|_{x=x'}$$

$$= \sqrt{-g}\,i[\exp(-is\mu^2)](4\pi is)^{-2}\Big\{\Big((g^{\mu\nu}/2is) - \frac{1}{6}R^{\mu\nu}\Big)F(x,x;is)$$

$$+ \nabla^\mu\nabla'^\nu F(x,x';is)\Big|_{x=x'}\Big\} \quad , \tag{70}$$

thus,

$$\big(\delta/\delta g_{\mu\nu}(x)\big)\ \mathrm{Tr}\ e^{-is\mathcal{H}}$$

$$= i[\exp(-is\mu^2)](4\pi is)^{-2}\sqrt{-g}\,\mathrm{tr}\Big\{-\frac{1}{2}(g^{\mu\nu}/2is)F(x,x'is)is$$

$$-\frac{1}{2}g^{\mu\nu}\big(\mu^2 + \tilde{U}(x)\big)F(x,x;is)is$$

$$+ [(\xi-1/6)G^{\mu\nu} + \xi(g^{\mu\nu}\nabla^2 - \nabla^\mu\nabla^\nu)]F(x,x;is)is$$

$$-\frac{1}{2}g^{\mu\nu}(s\partial/\partial s) - 2 - is\mu^2\big)F(x,x;is)$$

$$+ (\nabla^\mu\nabla'^\nu - \frac{1}{2}g^{\mu\nu}\nabla^\lambda\nabla'_\lambda)F(x,x';is)\Big|_{x=x'} is\Big\} \tag{71}$$

but we recall that $w^{(1)}$ is renormalized. What appears is not F, but $\tilde{\tilde{F}}$; hence, the first three terms of the power series expansion of the curly brackets must be subtracted leaving

$$\langle \mathcal{J}^{\mu\nu}(x)\rangle^{(1)} = \int_0^\infty \frac{ds}{is}\big(\delta/\delta g_{\mu\nu}(x)\big)\widetilde{\widetilde{(\mathrm{Tr}e^{-is\mathcal{H}})}}$$

$$= \sqrt{-g}\int_0^\infty ds\,i[\exp(-is\mu^2)](4\pi is)^{-2}\mathrm{tr}\Big\{-\frac{1}{4}g^{\mu\nu}\big(\tilde{\tilde{F}}(x,x;is)/is\big)$$

$$- \frac{1}{2} g^{\mu\nu}\left(\mu^2 + U(x)\right) \tilde{F}(x,x;is)$$

$$+ [(\xi-1/6)G^{\mu\nu} + \xi(g^{\mu\nu}\nabla^2 - \nabla^\mu\nabla^\nu)]\tilde{F}(x,x;is)$$

$$+ (\nabla^\mu\nabla'^\nu - \frac{1}{2} g^{\mu\nu}\nabla^\lambda\nabla'_\lambda)\tilde{F}(x,x';is)\Big|_{x=x'} \Big\} + \frac{1}{2} g^{\mu\nu}(4\pi)^{-2} f_2(x,x) \tag{72}$$

where the last, local, term arises from the renormalization of the $(s\partial/\partial s) - 2 - is\mu^2)F$ term to $(s\partial/\partial s) - 2 - is\mu^2)F$ and doing the integral. This last term is local. It is also not conserved; only the full expression is conserved. Also, we have still to determine the coefficients of the purely gravitational counter terms. To do this, we observe that if we either consider small deviations from flat space or, equivalently, use Riemann normal coordinates around x, the deviations from $\sim -\partial^2 + \mu^2$ is of the form $(\partial\partial h)$. Also, the power series of F in (is) is, essentially a power series in $(is\partial^2)$. As a result, all the terms of order $\partial^4 h(\nabla^2 R)$ or of order $(\partial^2 h)^2(R\cdot R)$ will occur in the order $(is)^2$ term in F or in the order (is) term in $\nabla\nabla F$ or $G^{\mu\nu}F$. Upon inspection of the expression for $<\mathcal{T}^{\mu\nu}>^{(1)}$, Eq. (72), we see that only the f_2 in the local term and the $\mu^2 F$ term can be of the appropriate order, but these two terms cancel leaving,

$$<\mathcal{T}^{\mu\nu}>^{(1)} = \sqrt{-g} \int_0^\infty ds\ i[\exp(-is\mu^2)](4\pi is)^{-2}T^{\mu\nu}(x,x;is)$$

where

$$\tilde{T}^{\mu\nu}(x,x;is) \equiv \mathrm{tr}\Big\{ \nabla^\mu\nabla'^\nu - \frac{1}{2} g^{\mu\nu}\nabla^\lambda\nabla'_\lambda\ F(x,x';is)$$

$$+ [(\xi-1/6)G^{\mu\nu} + \xi(g^{\mu\nu}\nabla^2 - \nabla^\mu\nabla^\nu) - \frac{1}{2} g^{\mu\nu}U(x)]F(x,x;is)$$

$$- \frac{1}{2} g^{\mu\nu}\left((2is)^{-1} + \mu^2\right)F(x,x;is)\Big\} \quad . \tag{73}$$

This expression, after all sources are set equal to zero, has leading terms for slow variation of the curvature as R^3 or ∂^2R^2 or ∂^4R. Any finite admixture of the renormalized quantities R^2 or $C^{\mu\nu\lambda\sigma}C_{\mu\nu\lambda\sigma}$ will introduce local terms which are more rapidly varying.

Of course, one may wish to define the R^2 and $C^{\mu\nu\lambda\sigma}C_{\mu\nu\lambda\sigma}$

coupling constants in some other way; in that case, one will move part of the local contact terms into $w^{(1)}$ and identify the new $w^{(1)}$ as the matter action. The resultant new stress tensor will be locally more sensitive to low and slowly varying curvature but will presumably have some other desirable feature. Further, we note that our procedure does not work if the scalar field mass vanishes. In case, there is no factor of $\exp(-is\mu^2)$ which by itself assures convergence for large s. The convergence must be provided by some other parameter and there may be some phvsical question which arises in the guise of an infrared singularity. This infrared problem has nothing to do with the short distance singularities which we have been discussing; however, the absence of a natural mass scale forces us to introduce a new mass scale which can have no physical significance. In the weak field approximation our considerations would yield expressions of the form

$$w \sim \int dpR(p)[\ln(p^2/K^2)]R(p) + \alpha(K)R^2 \equiv w^{(1)} + \alpha(K)R^2$$

where K^2 is the momentum transfer at which $\alpha(K)$ is defined. The full actiion must be independent of K, leading to a renormalization group [Weinberg [49]). Of course, we had this freedom before: we could have defined the R^2 coupling constant in terms of some momentum transfer other than zero and shifted terms from $w^{(1)}$ to $w^{(0)}$, but for $\mu^2 \neq 0$, the zero momentum transfer limit is a natural one at which to define the coupling constant for fields which are slowly varying with respect to the Compton wave length, $1/\mu$.

The conservation of $\langle\mathcal{T}^{\mu\nu}\rangle$ may be checked by considering a variation of the form

$$\delta g_{\mu\nu} = \nabla_\mu \xi_\nu + \nabla_\nu \xi_\mu \quad ,$$

which yields the variation

$$\delta\left\{\int dx\,\mathcal{L}(\langle\phi\rangle,g,A,\eta,U) + w^{(1)}\right\}$$

$$= -\int (\nabla_\nu \langle\mathcal{T}^{\mu\nu}\rangle)\xi_\nu dx$$

$$= -\int dx\xi_\nu(x)\sqrt{-g}\left\{+ F^\mu{}_\lambda(\langle\phi\rangle eq \frac{1}{i}\Pi^\lambda\langle\phi\rangle) - \langle\phi\rangle[\Pi^\mu_; U]\langle\phi\rangle\right.$$

$$- \eta(\Pi^\mu\langle\phi\rangle) - (\Pi^\mu\langle\phi\rangle)^T\Big(\Big[-\Pi^2 + \mu^2 + \xi R + U$$

$$\left. + \frac{\lambda}{2}(\langle\bar{\phi}\rangle\langle\phi\rangle)\Big]\langle\phi\rangle - \eta\Big)\right\} - \int dx\xi_\mu(x)\nabla_\nu \frac{2\delta w^{(1)}}{\delta g_{\mu\nu}(x)} \quad . \tag{74}$$

The term in square brackets is just the equation of motion for $<\phi>$. In the absence of the quantum corrections, it vanishes and the stress tensor is conserved modulo the effects of the sources which may add energy and momentum through the forces on currents, $F^{\mu}{}_{\lambda}j^{\lambda}$, the action of the scalar potential, U, and the production of excitations by η. In the presence of the quantum corrections, the field equation is modified and

$$[-\Pi^2 + \mu^2 + \xi R + U + \frac{\lambda}{2}(<\phi><\phi>)]<\phi> - \eta - \frac{\delta w^{(1)}}{\delta <\phi>(x)} = 0 \quad .$$

Thus, we find that

$$\nabla_{\nu} < \mathcal{J}^{\mu\nu} > = F^{\mu}{}_{\lambda} < j^{\lambda} > - \mathrm{tr}[\Pi^{\mu}_{,}U] < \phi(x)\phi(x) > \sqrt{-g}$$

$$- \eta(\Pi^{\mu}<\phi>) + \nabla_{\nu}\frac{2\delta w^{(1)}}{\delta g_{\mu\nu}(x)} + (\Pi^{\mu}<\phi>)\frac{w^{(1)}}{\delta <\phi>}$$

$$- F^{\mu}{}_{\lambda}\frac{\delta w^{(1)}}{\delta A_{\lambda}(x)} + \mathrm{tr}(\Pi^{\mu}U(x))\frac{\delta w^{(1)}}{\delta U(x)} \quad . \tag{75}$$

Now $<\phi>$ appears in $w^{(1)}$ only through $\tilde{U}$, hence

$$\Pi^{\mu}<\phi>\frac{\delta}{\delta <\phi>} = (\Pi^{\mu}<\phi>)\frac{\delta\tilde{U}}{\delta <\phi>}\frac{\delta}{\delta\tilde{U}} = \left(\Pi^{\mu}(\tilde{U}-U)\right)\frac{\delta}{\delta\tilde{U}}$$

and the sum of the terms involving $w^{(1)}$ is

$$\nabla_{\nu}\frac{2\delta w^{(1)}}{\delta g_{\mu\nu}(x)} - \overset{\mu}{F}(x)\lambda\frac{\delta w^{(1)}}{\delta A_{\lambda}(x)} + \mathrm{tr}\Pi^{\mu}\tilde{U}(x)\frac{\delta w^{(1)}}{\delta\tilde{U}(x)} \tag{76}$$

which vanishes because

$$\delta_{\xi}\int\bar{\psi}\mathcal{H}\psi = \int dx\xi_{\mu}\Big\{F^{\mu}{}_{\lambda}\Big(\bar{\psi}\ eq\ \frac{1}{i}\overleftrightarrow{\Pi}_{\lambda}\psi\Big) - \psi(\Pi^{\mu}\tilde{U})\psi$$

$$- (\Pi^{\mu}\psi)^{T}(\mathcal{H}\psi) - (\mathcal{H}\psi)^{T}(\Pi^{\mu}\psi) + \frac{1}{2}\nabla^{\mu}\Big[\bar{\psi}(\mathcal{H}\psi) + (\mathcal{H}\bar{\psi})^{T}\psi\Big]\Big\}$$

$$= \int dx\xi_{\mu}\Big(F^{\mu}{}_{\lambda}\frac{\delta}{\delta A_{\lambda}(x)} - \mathrm{tr}(\Pi^{\mu}\tilde{U})\frac{\delta}{\delta\tilde{U}(x)}\Big)\int(\bar{\psi}\mathcal{H}\psi)$$

$$- \int dx\xi_{\mu}\Big\{(\Pi^{\mu}\bar{\psi})^{T}(\mathcal{H}\psi) + (\mathcal{H}\bar{\psi})^{T}(\Pi^{\mu}\psi)$$

$$- \frac{1}{2} \nabla^\mu \left[\bar{\psi}(\mathcal{H}\psi) + (\mathcal{H}\bar{\psi})\psi \right] \Big\} \quad , \tag{77}$$

hence

$$\begin{aligned} \delta_\xi \mathrm{Tre}^{-is\mathcal{H}} &= \int dx \xi_\mu(x) \left[F^\mu{}_\lambda \frac{\delta}{\delta A_\lambda(x)} - \mathrm{tr}(\Pi^\mu \tilde{U}) \frac{\delta}{\delta \tilde{U}(x)} \right] (\mathrm{Tre}^{-is\mathcal{H}}) \\ &+ \int dx \xi_\mu(x) \mathrm{tr} \Big\{ i \frac{\partial}{\partial s} \langle x | e^{-is\mathcal{H}} \Pi^\mu | x \rangle + i \frac{\partial}{\partial s} \langle x | \Pi^\mu e^{-is\mathcal{H}} | x \rangle \\ &- \nabla^\mu i \frac{\partial}{\partial s} \langle x | e^{-is\mathcal{H}} | x \rangle \Big\} (is) \\ &= \int dx \xi_\mu(x) \left[F^\mu{}_\lambda \frac{\delta}{\delta A_\lambda(x)} - \mathrm{tr}(\Pi^\mu \tilde{U}) \frac{\delta}{\delta \tilde{U}(x)} \right] \mathrm{Tre}^{-is\mathcal{H}} \end{aligned} \tag{78}$$

which, when used in the expression for $w^{(1)}$, implies that the expression involving $w^{(1)}$ in Eq. (75) vanishes and that

$$\begin{aligned} \nabla_\nu \left(\langle \mathcal{T}^{\mu\nu} \rangle + g^{\mu\nu} \eta \langle \phi \rangle \right) = F^{\mu\lambda} \langle j_\lambda \rangle - \mathrm{tr}[\Pi^\mu_; U] \sqrt{-g}\, \langle \phi(x)\phi(x) \rangle \\ + (\Pi^\mu \eta)^T \langle \phi \rangle \end{aligned} \tag{79}$$

which is the generalized nonconservation law due to the presence of the sources.

We have only verified the relation formally. To present a complete proof is tedious; it is presented by Brown [4] for a non-interacting scalar field.

We now turn to the trace anomaly. We look at the classical expression for $\mathcal{T}^{\mu\nu}(\langle\phi\rangle)$ and find that, for $\xi = 1/6$,

$$\begin{aligned} g_{\mu\nu} T^{\mu\nu}(\langle\phi\rangle) &= \Big\{ - [\Pi_\lambda \langle\phi\rangle]^T [\Pi^\lambda \langle\phi\rangle] \\ &- 2\langle\phi\rangle(\mu^2 + U)\langle\phi\rangle - \frac{\lambda}{2} (\langle\phi\rangle\langle\phi\rangle)^2 - \frac{1}{6} R(\langle\phi\rangle\langle\phi\rangle) \\ &+ \frac{1}{2} \nabla^2 (\langle\phi\rangle\langle\phi\rangle) \Big\} \\ &= - \langle\phi\rangle \left[-\Pi^2 + \mu^2 + U + \frac{1}{6} R + \frac{\lambda}{2} (\langle\phi\rangle\langle\phi\rangle) \right] \langle\phi\rangle - \langle\phi\rangle(\mu^2 + U)\langle\phi\rangle \; . \end{aligned} \tag{80}$$

The first term vanishes due to the equation of motion and the second term is the trace which vanishes if μ^2 and U are zero.

If we calculate the trace of the full stress tensor, we find that

$$g_{\mu\nu}(x)\langle\mathcal{T}^{\mu\nu}\rangle = \Big\{- \langle\phi\rangle(\mu^2 + U)\langle\phi\rangle$$

$$- \langle\phi\rangle\Big[- \Pi^2 + \mu^2 + U + \frac{1}{6} R + \frac{\lambda}{2} (\langle\phi\rangle\langle\phi\rangle)\Big]\langle\phi\rangle\Big\}$$

$$+ \frac{2g_{\mu\nu}(x)}{\sqrt{-g(x)}} \frac{\delta w^{(1)}}{\delta g_{\mu\nu}(x)} \tag{81}$$

which, using the one-loop modified equation of motion yields

$$\langle\mathcal{T}^{\mu}_{\mu}\rangle = \Big\{ -\langle\phi\rangle(\mu^2 + U)\langle\phi\rangle\Big\} - \eta\langle\phi\rangle$$

$$+ \frac{1}{\sqrt{-g(x)}} \Big[2g_{\mu\nu} \frac{\delta w^{(1)}}{\delta g_{\mu\nu}} - \langle\phi\rangle \frac{\delta w^{(1)}}{\delta\langle\phi\rangle}\Big] \quad . \tag{82}$$

Hence, we need the change of the one-loop correction, $w^{(1)}$, under the variation

$$\delta g_{\mu\nu} = 2\delta\lambda(x)g_{\mu\nu}$$

$$\delta\langle\phi\rangle = - \delta\lambda(x)\langle\phi\rangle$$

and, under this variation,

$$\frac{\delta}{\delta\lambda(x)} \int(\bar{\psi}\mathcal{H}\psi) = \Big\{2(\Pi^{\lambda}\bar{\psi})^{T}(\Pi_{\lambda}\psi) - 2\bar{\psi}(-\Pi^2\psi)$$

$$- 2(- \Pi^2\bar{\psi})\psi - 2\bar{\psi}(\tilde{U} - U)\psi - \frac{1}{3} R(\bar{\psi}\psi) - \nabla^2(\bar{\psi}\psi)\Big\}$$

$$= \Big\{- \bar{\psi}(\mathcal{H}\psi) - (\mathcal{H}\bar{\psi})\psi + 2\bar{\psi}(\mu^2 + U)\psi\Big\} \quad . \tag{83}$$

Thus,

$$\left\{2g_{\mu\nu}(x)\frac{\delta}{\delta g_{\mu\nu}(x)} - \langle\phi\rangle\frac{\delta}{\delta\langle\phi(x)\rangle}\right\}\mathrm{Tr}e^{-is\mathcal{H}} \equiv \frac{\delta}{\delta\lambda(x)}\mathrm{Tr}e^{-is\mathcal{H}}$$

$$= -\,i\,\mathrm{str}\left\{2(\mu^2+U)|x\rangle e^{-is\mathcal{H}}|x\rangle - 2i\frac{\partial}{\partial s}\langle x|e^{-is\mathcal{H}}|x\rangle\right\}$$

$$= 2\mathrm{tr}(\mu^2+U)\frac{\delta}{\delta U(x)}\mathrm{Tr}e^{-is\mathcal{H}} - 2is\frac{\partial}{\partial is}\mathrm{tr}\langle x|e^{-is\mathcal{H}}|x\rangle\ . \tag{84}$$

However, what appears in $w^{(1)}$ is not $\mathrm{Tr}e^{-is\mathcal{H}}$ but

$$\widetilde{\widetilde{\mathrm{Tr}e^{-is\mathcal{H}}}} \equiv \frac{i\,e^{-is\mu^2}}{(4\pi is)^2}\int\widetilde{\widetilde{F}}\sqrt{-g}\,dx$$

and

$$\frac{\delta}{\delta\lambda(x)}\widetilde{\widetilde{\mathrm{Tr}e^{-is\mathcal{H}}}} = 2\mathrm{tr}(\mu^2+U)\frac{\delta}{\delta U(x)}(\widetilde{\widetilde{\mathrm{Tr}e^{-is\mathcal{H}}}})$$

$$-\,2\frac{i\,e^{-is\mu^2}}{(4\pi is)^2}\mathrm{tr}\left\{is\frac{\partial}{\partial is} - 2 - is\mu^2\right\}\widetilde{F}(x,x;is) \tag{85}$$

$$= 2\mathrm{tr}(\mu^2+U)\frac{\delta}{\delta U(x)}(\widetilde{\widetilde{\mathrm{Tr}e^{-is\mathcal{H}}}}) - 2is\frac{\partial}{\partial is}\left[\frac{i\,e^{-is\mu^2}}{(4\pi is)^2}\mathrm{tr}\,\widetilde{F}(x,x;is)\right]$$

or

$$\frac{\delta w^{(1)}}{\delta\lambda(x)} = 2\mathrm{tr}(\mu^2+U)\frac{\delta w^{(1)}}{\delta U(x)} - \int_0^\infty\frac{dsi}{is}\,is\frac{\partial}{\partial is}\left[\frac{e^{-is\mu^2}}{(4\pi)^2}\mathrm{tr}\,\widetilde{F}(x,x;is)\right]$$

$$= \mathrm{tr}(\mu^2+U)\frac{\delta w^{(1)}}{\delta U(x)} + \frac{f_2(x,x)\sqrt{-g}}{(4\pi)^2}$$

or

$$\langle\mathcal{T}^{\mu}_{\mu}(x)\rangle = \left\{-\,\mathrm{tr}\big(\mu^2+U(x)\big)\langle\phi(x)\phi(x)\rangle\sqrt{-g} - \eta(x)\langle\phi(x)\rangle + \frac{\sqrt{-g}}{4\pi)^2}\mathrm{tr}f_2(x,x)\right\} \tag{86}$$

and the trace has an anomaly. The f_2 does not appear in the formal arguments. It is independent of μ^2 and the source U.

The anomaly was first discovered in flat space and is intimately related to the renormalization group (Callen [32]). For $\xi = 1/6$, the anomaly is

$$<T^\mu_\mu(x)>^{\text{anomalous}} = \frac{1}{(4\pi)^2}\left\{\left[U^2(x) + \frac{\lambda}{2}\left[(\mathrm{tr}U)(x)(<\phi><\phi>)(x)\right.\right.\right.$$

$$+ 2<\phi>(x)U(x)<\phi>$$

$$\left.\left. + \lambda^2 5(<\phi><\phi>)^2(x)\right] - \frac{e^2}{6}F^{\lambda\alpha}F_{\lambda\sigma}(x) - \frac{1}{6}\nabla^2\left[\mathrm{tr}U + \lambda 2(<\phi><\phi>)\right]$$

$$\left. + \frac{1}{90}\nabla^2 R + \frac{1}{90}\left[R_{\mu\nu\lambda\sigma}R^{\mu\nu\lambda\sigma} - R^{\mu\nu}R_{\mu\nu}\right]\right\} \quad . \qquad (87)$$

The terms involving U are of no interest except for the purpose of generating such quantities as $<T\left(T^\mu_\mu(x)(\phi(y)\phi(y))\right)>$. The $\lambda^2(<\phi><\phi>)^2$ term indicates an anomaly of the form $T^\mu_\mu \sim \lambda^2(\phi\phi)^2$ so that the self interaction term is effectively not scale invariant. This is associated with a renormalization of the effective dimension of the operator, ϕ, which is important in considerations involving the renormalization group. Similar comments apply to the $F^{\lambda\sigma}F_{\lambda\sigma}$ term. The curvature terms are exactly those found by many other authors (Duff [26], Tsao [27] and Brown [4]), except for a factor of two which arises because we have two scalar fields instead of one.

There are several ways to understand the origin of the anomaly. One is the renormalization of the effective scale of ϕ for which the reader is referred to the original references [32]. From the perspective of the renormalization procedures used here, we may follow the work of Deser, Duff and Isham [25] and of Brown [4] in the dimensional regularization and observe that if the unrenormalized stress tensor is calculated, we obtain, formally, in n dimensions,

$$<\mathcal{T}^{\mu\nu}(x)>^{(1)} = \int_0^\infty 2\,\frac{\delta w^{(1)}}{\delta g_{\mu\nu}(x)}$$

$$= \frac{\delta}{\delta g_{\mu\nu}(x)}\int_0^\infty \frac{dsi}{is}\,\frac{e^{-i\mu^2 s}}{(4\pi is)^{n/2}}\int F(y,y;is)dy\sqrt{-g}\,(y)$$

$$= \int_0^\infty dsi\left\{\frac{e^{-is\mu^2}}{(4\pi is)^{n/2}}\,T^{\mu\nu}(x,x;is)\sqrt{-g}\,(x)\right.$$

$$- \frac{\partial}{\partial is} \left[\frac{e^{-is\mu^2}}{(4\pi is)^{n/2}} \frac{1}{2} g^{\mu\nu}(x) F(x,x;is) \right] \Bigg\} \tag{88}$$

where

$$T^{\mu\nu}(x,x;is) = \mathrm{tr}\Big\{ \Big(\nabla^\mu \nabla'^\nu - \frac{1}{2} g^{\mu\nu}(x) \nabla^\lambda \nabla'_\lambda \Big) F(x,x';is) \Big|_{x=x'}$$

$$+ [(\xi - 1/6) G^{\mu\nu} + \xi (g^{\mu\nu} \nabla^2 - \nabla^\mu \nabla^\nu)] F(x,x;is)$$

$$- \frac{1}{2} g^{\mu\nu}(x) \big(\mu^2 + U(x) \big) F(x,x;is) - \frac{1}{4} \frac{g_{\mu\nu}}{is} F(x,x;is)$$

where F, and the derivatives of F which appear here do not contain any explicit dependence on the number of dimensions, n. Of course, the number of dimensions does appear implicitly in the number of components and the number of indices summed over. As a result, if we consider n sufficiently small (< 0), the integral of the derivative term vanishes and the result is

$$< \mathcal{T}^{\mu\nu} >^{(1)} = \int_0^\infty \frac{ds\ i}{(4\pi is)^{n/2}} e^{-is\mu^2} T^{\mu\nu}(x,x;is) \tag{89}$$

and we may analytically continue in the parameter n back to the neighborhood of $n = 4$. To do this, we observe that $T^{\mu\nu}(x,x;is)$ may be expanded in a power series in is starting with $(is)^{-1}$ and that the terms of order $(is)^p$, $p \geq 2$ are convergent. Further, we recall that

$$\int_0^\infty dsi \frac{(is)^p}{(4\pi is)^{n/2}} e^{-is\mu^2} = \frac{\Gamma(p-n/2+1)}{(4\pi)^{n/2}} (\mu^2)^{n/2-p-1}$$

has a pole at $n = 4$. Thus, the s integral may be written as

$$<T^{\mu\nu}>^{(1)} = \frac{1}{n-4} [A^{\mu\nu} - g^{\mu\nu} B] + \alpha^{\mu\nu}(n)$$

where $A^{\mu\nu}$ has no term proportional to $g^{\mu\nu}$. Also, by a similar argument

$$<\phi(x)\phi(x)>^{(1)} = \frac{1}{n-4} C + \beta(n) \quad .$$

Formally, for the unrenormalized theory, with $U = 0$,

$$T_\lambda^\lambda = -\mu^2(\phi\phi) + (n-4)(\text{other terms})$$

hence,

$$0 = \langle T\rangle^{(1)} - \mu^2\langle(\phi\phi)\rangle$$

$$= \frac{1}{n-4}\,[g_{\mu\nu}A^{\mu\nu} - nB \quad C] + g_{\mu\nu\alpha}{}^{\mu\nu}(n) + \mu^2\beta(n) \ .$$

There can be no pole at n = 4, since everything vanishes, hence

$$g_{\mu\nu}A^{\mu\nu} = 4B - C \tag{90}$$

and

$$g_{\mu\nu}\alpha^{\mu\nu} + \mu^2\beta - B = 0 \qquad . \tag{91}$$

However, modulo finite renormalization terms, $\alpha^{\mu\nu}$ and β are respectively the renormalized stress tensor and $(\phi\phi)$ matrix elements. We may define the matrix of the fully renormalized matrix element $\langle(\phi\phi)\rangle$ and $\langle T^{\mu\nu}\rangle$ so as to absorb all the terms in β proportional to μ^2. However, the terms independent of μ^2 cannot be so absorbed. The coefficient of $1/(n-4)$ in the term independent of μ^2 in $\delta w^{(1)}/\delta g_{\mu\nu}$ is simply

$$2\,\frac{\delta}{\delta g_{\mu\nu}}\,\lim_{n\to 4}\,\frac{\Gamma(2-n/2)\,(n-4)}{(4\pi)^2}\int dx\,\sqrt{-g}\,f_2(x,x)(\mu^2)^{n/2-2}$$

$$= -2\,\frac{\delta}{\delta g_{\mu\nu}}\int dx\,\frac{\sqrt{-g}\,f_2(x,x)}{(4\pi)^2} = \left[A_2^{\mu\nu} - g^{\mu\nu}\,\frac{f_2(x,x)}{(4\pi)^2}\right]\sqrt{-g} \ . \tag{92}$$

Thus,

$$B = f_2(x,x)/(4\pi)^2 + O(\mu^2) \qquad , \tag{93}$$

and we can only eliminate the anomaly by adding some local terms to $\langle T^{\mu\nu}\rangle$. These terms must be local, yield a conserved $\langle T^{\mu\nu}\rangle$, be consistent with the coupling constant identifications we made above, and modify the trace so as to eliminate the anomaly. No such terms exist. We set U equal to zero to find:

(1) A $\sqrt{-g}\,\lambda^2(\langle\phi\rangle\langle\phi\rangle)^2$ term would be conserved (the $\langle\phi\rangle$ equation would be modified) and could eliminate that contribution to the anomaly, but at the cost of changing the experimental identification of λ. The anomaly would reappear in higher orders.

(2) The stress tensor contribution from $F^{\lambda\sigma}F_{\lambda\sigma}\sqrt{-g}$, the only possible term quadratic in F, has vanishing trace hence cannot affect the anomaly.

(3) There is no local action term the variation of which yields a stress tensor with a trace $\nabla^2(\langle\phi\rangle\langle\phi\rangle)(x)$, except $R(\langle\phi\rangle\langle\phi\rangle)$ which also produces an extra $R(\langle\phi\rangle\langle\phi\rangle)$ term in the trace.

(4) The $\nabla^2 R$ term may be eliminated by adding a term to the effective action of the form $\int dx\sqrt{-g}\,R^2(x)$ which modifies $\langle T^{\mu\nu}\rangle$ by a term of the form

$$- R(4R^{\mu\nu} - g^{\mu\nu}R) - 4(g^{\mu\nu}\nabla^2 - \nabla^\mu\nabla^\nu)R$$

which is local, conserved, and has trace $-12\nabla^2 R$. We may eliminate this term of the trace anomaly by introducing an action term which did not appear and allowing the stress tensor to have local terms proportional to $\nabla\nabla R$ and R^2. If we could fully eliminate the anomaly by such a device, it would perhaps be worth the price; but the remainder of the anomaly cannot be eliminated and we would be forced to change our identification of the matter stress energy as the part which has minimal local dependence on the curvature.

(5) No local counter term regular at vanishing curvature can change the $R^{\mu\nu\lambda\sigma}R_{\mu\nu\lambda\sigma} - R^{\mu\nu}R_{\mu\nu}$ term. The variation of $\int dx\sqrt{-g}\,C^{\mu\nu\lambda\sigma}C_{\mu\nu\lambda\sigma}$ has vanishing trace and the variation of the Euler-Poincaré invariant is, of course, zero. It has been suggested by M. Brown [55] that counter terms of the form, for example,

$$\int dx\sqrt{-g}\,[R^{\mu\nu\lambda\sigma}R_{\mu\nu\lambda\sigma} - R^{\mu\nu}R_{\mu\nu}] \times \ln[C^{-\frac{1}{4}}(\nabla^2 - 1/6\; R)C^{\frac{1}{4}}] \qquad (94)$$

with

$$C \equiv C^{\mu\nu\lambda\sigma}C_{\mu\nu\lambda\sigma}$$

may be used to exactly cancel the gravitational anomaly. The renormalized amplitude that we have used is a functional of the metric all of whose derivatives with respect to the metric exist at flat space. This ansatz introduces terms so that the new amplitude has flat space as a singular point. It is true that one must go to the third derivative to find the singularity but one would expect the terms to introduce singularities in graviton scattering amplitudes.

If Pauli-Villars regularization is used, the result is, of course, the same but the origin of the anomaly appears somewhat differently. The regulated effective action is

$$\frac{1}{2}\int_0^\infty \frac{ds}{is}\int_0^\infty dm^2\rho(m^2)\ e^{-ism^2}\mathrm{Tr}e^{-is(\mathcal{H}-\mu^2)} \tag{95}$$

and we recall that

$$2g_{\mu\nu}(x)\frac{\delta}{\delta g_{\mu\nu}(x)}\mathrm{Tr}e^{-is\mathcal{H}} = -\frac{e^{-is\mu^2}}{(4\pi is)^2}\,2\left[is\frac{\partial}{\partial is}-2\right]\mathrm{tr}F(x,x;is)\sqrt{-g}$$

if we set $\widetilde{U} = 0$, eliminating the self interactions, hence

$$2g_{\mu\nu}(x)\frac{\delta}{\delta g_{\mu\nu}(x)}\mathrm{tr}\int dy\sqrt{-g}\,f_p(y,y) = (4-2p)\sqrt{-g}\,\mathrm{tr}\,f_p(x,x) \tag{96}$$

and the renormalized effective action must satisfy

$$2g_{\mu\nu}(x)\frac{\delta W^{ren}}{\delta g_{\mu\nu}(x)} = \lim_{M^2\to\infty} 2g_{\mu\nu}(x)\frac{\delta}{\delta g_{\mu\nu}(x)}\left\{\int_0^\infty dm^2\rho(m^2)\frac{dis\,e^{-ism^2}}{is(4\pi is)^2}\left[\mathrm{Tr}F - \sum_{p=0}^{2}(is)^p\mathrm{Tr}f_p\right]\right\} = \sqrt{-g}\,\langle T^\mu_\mu\rangle^{ren} \tag{97}$$

$$= \lim_{M^2\to\infty}\sqrt{-g}\int_0^\infty dm^2\rho(m^2)\frac{dis}{(4\pi)^2}e^{-ism^2}\left\{-\frac{\partial}{\partial is}\frac{\mathrm{tr}F(x,x;is)}{(is)^2} - \frac{2\mathrm{tr}f_0}{(is)^3} - \frac{\mathrm{tr}f_1(x,x)}{(is)^2}\right\}$$

$$= \sqrt{-g}\int_0^\infty dm^2\rho(m^2)\frac{dis\,e^{-ism^2}}{(4\pi)^2(is)^2}\left\{-m^2\mathrm{tr}F(x,x;is) - \frac{2\mathrm{tr}f_0}{(is)} - \mathrm{tr}f_1(s,s)\right\} .$$

However, the renormalized matrix element $\langle\phi\phi\rangle$ is

$$<\phi\phi(x)>^{ren} = \lim_{M^2\to\infty} \int_0^\infty dm^2 \frac{dis}{(4\pi is)^2} e^{-ism^2} \rho(m^2)\Big\{trF(x,x;is)$$

$$- trf_0 - istrf_1(x,x)\Big\} \qquad (98)$$

hence

$$<T^\mu_\mu(x)>^{ren} + \mu^2<(\phi\phi)(x)>^{ren}$$

$$= \lim_{M^2\to\infty} \Big\{\int_0^\infty dm^2 \frac{\rho(m^2)}{(4\pi)^2} \frac{dis}{(is)^2} e^{-ism^2}\Big[- (m^2-\mu^2)trF(x,x;is)$$

$$-\left(\frac{2}{is} + \mu^2\right) trf_0 - (1 + \mu^2 is)trf_1(x,x)\Big] \quad . \qquad (99)$$

Because of the factor $(m^2 - \mu^2)$ the first term has no contribution from $m^2 = \mu^2$, and it vanishes if the $F \sim (is)^3$ because the (is) integral goes as $(1/m^2)^2$. Thus, we need only look at the terms through order $(is)^2$. The f_0 and isf_1 terms are exactly canceled by the explicit terms which appear, leaving

$$<T^\mu_\mu(x)>^{ren} + \mu^2<\phi\phi(x)>^{ren} = + \frac{1}{(4\pi)^2} trf_2(x,x)I$$

$$I = - \lim_{M^2\to\infty} \int_0^\infty dis\ dm^2(m^2 - \mu^2)\rho(m^2)e^{-ism^2}$$

$$= - \lim_{M^2\to\infty} \int_0^\infty dm^2\left(\frac{m^2-\mu^2}{m^2}\right)\rho(m^2) = 1 \quad . \qquad (100)$$

This result arises because the formal result

$$<T^\mu_\mu> = - \mu^2<\phi\phi>$$

becomes

$$<T^\mu_\mu> = - \int_0^\infty dm^2\rho(m^2)m^2<\phi\phi>(m^2)$$

and we could renormalize each m term so that

$$<\phi\phi>(m^2) \underset{m^2\sim\infty}{\sim} 1/m^2\ f^2/(4\pi)^2 \quad .$$

This term vanishes as $m^2 \to \infty$ and does not contribute to $\langle\phi\phi\rangle^{ren}$, but

$$m^2\langle\phi\phi\rangle(m^2) \underset{m^2\to\infty}{\longrightarrow} \frac{f_2}{(4\pi)^2}$$

yielding the anomaly.

REFERENCES

1. J. Schwinger, Phys. Rev. 82, 914 (1951).
2. B. S. DeWitt, Dynamical Theory of Groups and Fields (Gordon and Breach, New York, 1965).
3. B. S. DeWitt, Physics Reports 19C, 295 (1975).
4. L. S. Brown, Phys. Rev. D15, 1469 (1976).
5. L. S. Brown and J. P. Cassidy, Phys. Rev. D15, 2810 (1976).
6. L. S. Brown and J. P. Cassidy, Phys. Rev. D16, 1712 (1977).
7. Y. B. Zel'dovich and A. A. Starobinsky, Zh Eskp. Teor. Fiz. 61, 2161 (1971)[Eng. Trans. Sov. Phys. JETP 34, 1159 (1972].
8. L. Parker and S. A. Fulling, Phys. Rev. D9, 341 (1974).
9. P. Candelas and D. J. Raine, Phys. Rev. D12, 965 (1975).
10. J. S. Dowker and R. Chritchley, Phys. Rev. D13, 3224 (1976).
11. J. S. Dowker and R. Chritchley, Phys. Rev. D16, 3390 (1977).
12. P.C.W. Davies, M. J. Duff, and W. Unruh, Phys. Rev. D13, 2720 (1976).
13. S. M. Christensen, Phys. Rev. D14, 2490 (1976).
14. S. Hawking, Comm. Math. Phys. 55, 133 (1977).
15. D. Bernard and A. Duncan, Ann. Phys. (N. Y.) 107, 201 (1977).
16. P.C.W. Davies, S. A. Fulling, S. M. Christensen, and T. S. Bunch, Ann. Phys. (N. Y.) 109, 108 (1977).
17. S. M. Christensen, Phys. Rev. D14, 2490 (1976).
18. S. M. Christensen and S. A. Fulling, Phys. Rev. D15, 2088 (1977).
19. S. L. Adler, J. Lieberman, and Y. J. Ng, Ann. Phys. (N. Y.) 106, 279 (1977).
20. S. L. Adler and J. Lieberman, Ann. Phys., to be published.
21. R. M. Wald, Ann. Phys. 110, 472 (1978).
22. R. M. Wald, Comm. Math. Phys. 54, 1 (1977).
23. T. S. Bunch and P.C.W. Davies, Proc. Roy. Soc. A360, 117 (1978).
24. T. S. Bunch and P.C.W. Davies, J. Phys. A. 11, 603 (1978).
25. S. Deser, M. J. Duff and C. Isham, Nucl. Phys. B111, 45 (1976).
26. M. J. Duff, Nucl. Phys. B125, 334 (1977).
27. H. S. Tsao, Phys. Lett. 68B, 79 (1977).
28. T. S. Bunch, "Regularization of the Stress Tensor of Quantum Fields in Curved Space-Times," University of London, King's College Ph.D. Thesis, unpublished. (1977)

29. T. S. Bunch, "On Renormalization of the Quantum Stress Tensor in Curved Space Time by Dimensional Regularization", University of Wisconsin-Milwaukee preprint, unpublished.
30. P.C.W. Davies, "Quantum Fields in Curved Space," Einstein Centenary volume of Gen. Rel. and Grav., to be published.
31. C. G. Callan, S. Coleman, and R. Jackiw, Ann. Phys. (N. Y.) 59, 42 (1970).
32. C. G. Callan, Phys. Rev. D2, 1541 (1971).
33. R. P. Feynman and R. R. Hibbs, Quantum Mechanics and Path Integrals, (McGraw Hill, Inc., New York, 1965).
34. J. Iliopoulous, C. Itzakson, and A. Martin, Rev. Mod. Phys. 47, 165 (1975).
35. E. Abers and B. W. Lee, Physics Reports 9, 1 (1973).
36. V. N. Popov, "Functional Integrals in Quantum Field Theory", CERN preprint TH2424, unpublished.
37. S. Coleman, "The Uses of Instantons", Lectures delivered at the 1977 International School of Subnuclear Physics, Ettore Majorana.
38. A. M. Polyakov, Nucl. Phys. B121, 429 (1977).
39. G. W. Gibbons and S. Hawking, "Gravitational Multi-Instantons", D.A.M.P.T. preprint, unpublished.
40. C. N. Pope and S. Hawking, "Symmetry Breaking by Instantons in Supergravity", D.A.M.P.T. preprint, unpublished.
41. G. W. Gibbons and M. J. Perry, "Quantizing Gravitational Instantons", D.A.M.P.T. preprint, unpublished.
42. B. W. Lee and J. Zinn-Justin, Phys. Rev. D5, 3121, 3137, 3155 (1972).
43. G. Jona-Lasinio, Nuovo Cimento 34, 1790 (1964).
44. K. Hepp, Comm. Math. Phys. 1, 95 (1965).
45. W. Zimmerman, in Brandeis University Summer Institute in Theoretical Physics, M. Chretien ed., MIT Press, Cambridge, Mass. (1970).
46. K. Stelle, Phys. Rev. D16, 953 (1977).
47. D. G. Boulware and L. S. Brown, Phys. Rev. 172, 1628 (1968).
48. S. Coleman, "Classical Electron Theory From a Modern Standpoint", Rand Corporation Report RM-2820-PR (1961).
49. S. Weinberg, "Ultraviolet Divergences in Quantum Theories of Gravitation", Einstein Centenary volume, Gravitational Theories Since Einstein, S. W. Hawking and W. Israel eds., Cambridge University Press, to be published.
50. S. Weinberg, Phys. Rev. 138, 988 (1965).
51. D. G. Boulware and S. Deser, Ann. Phys. (N. Y.) 89, 193 (1975).
52. S. Adler, J. Lieberman, Y. J. Ng, and H. S. Tsao, Phys. Rev. D14, 359 (1976).
53. J. Schwinger, Phys. Rev. Lett. 3, 296 (1959).
54. S. M. Christensen, Phys. Rev. D17, 946 (1978).
55. M. Brown, "Actions and Anomalies", University of Texas preprint, unpublished.

ASPECTS OF QUANTUM FIELD THEORY IN CURVED SPACE-TIME: EFFECTIVE ACTION AND ENERGY-MOMENTUM TENSOR *

Leonard Parker

Dept. of Physics, University of Wisconsin-Milwaukee

Milwaukee, Wisconsin 53201

1. INTRODUCTION

The theory of quantized fields in curved spacetime has reached a high level of development, and a number of important physical consequences have been predicted. By treating the metric of the gravitational field classically, one avoids the nonrenormalizability problems of quantized gravity, but nevertheless retains a wide domain of applicability. I have already given a recent review of quantized fields in curved spacetime [L. Parker, 1977], in which the creation of elementary particles by strong gravitational fields (as in cosmology and near black holes) was emphasized. The present lectures will emphasize material which was not covered in the previous review.

In particular, the proper-time formalism will be introduced, and used to calculate the effective action, and the so-called anomalous trace of the energy-momentum tensor of a quantized free scalar field in an arbitrary curved spacetime. The trace anomaly turns out to be a coefficient in the proper-time expansion of the propagator of the quantized field. Therefore, the derivation of the expression for that coefficient in terms of the Riemann tensor is given in detail. The formally infinite expression for the effective action is made well-defined or regularized by the method of ζ-function regularization. The use of dimensional regularization for that purpose is also outlined, and the result compared with that of the ζ-function approach. As an application, the anomalous trace of the energy-momentum tensor is used to obtain the expectation values of the energy density and pressure of

* Work supported by the National Science Foundation(PHY77-07111).

quantized fields obeying conformally invariant field equations in Robertson-Walker universes. The trace anomaly is also used to show that no creation of real particles obeying conformally invariant field equations occurs in Robertson-Walker expansions. In the final section, the proper-time representation of the propagator is again considered, and path-integral representations of the propagator of a scalar field are given for arbitrary coupling to the scalar curvature.

Although most of this material is in the nature of a review, some of the results given or methods used are new. For example, the one parameter family of path integral representations of the propagator given in the final section is new, and serves to tie together previously known special cases. The relation between the trace anomaly and particle creation in Robertson-Walker universes given in Section 6 is new. The derivations of the effective action in Eqs. (5.34) through (5.43), and of the trace anomaly in Eqs. (5.48) through (5.56), are new. The way Riemann normal coordinates are used in Section 3 to calculate the coefficients in the proper-time expansion of the propagator in terms of the Riemann tensor is new, although related to other work. It seems simpler and more straightforward than the method of working in general coordinates, and requires less mathematical background than some of the related methods involving geodesic coordinates which have been applied to the heat equation.

2. PROPER-TIME FORMALISM

Consider a scalar field ϕ with the Lagrangian

$$\mathcal{L} = -\frac{1}{2} g^{\mu\nu}\partial_\mu\phi\partial_\nu\phi - \frac{1}{2}\xi R\phi^2 - \frac{1}{2}m^2\phi^2 , \tag{2.1}$$

where R is the scalar curvature of the spacetime, ξ is an arbitrary real number, and m is the mass. [We use units with $\hbar = c = 1$, metric signature (-+++), and the conventions of Misner, Thorne, and Wheeler (1973).] The field equation is

$$-g^{\mu\nu}\nabla_\mu\nabla_\nu\phi + (m^2 + \xi R)\phi = 0 , \tag{2.2}$$

where ∇_μ denotes the covariant derivative. The Green function, G(x,x'), by definition is a solution of the equation

$$(-\nabla^\mu\nabla_\mu + \xi R + m^2)G(x,x') = [-g(x)]^{-\frac{1}{2}}\delta(x-x') . \tag{2.3}$$

The factor $[-g]^{-\frac{1}{2}}$ is introduced so that G will be a scalar at x and at x' (i.e., a biscalar), rather than a scalar density.

The problem of finding G can be reduced to that of finding the propagation amplitude of a fictitious particle moving in a curved space in accordance with a Schrödinger equation. One does that by writing G(x,x') as the matrix element of an abstract operator G in a Hilbert space (Schwinger 1951, DeWitt 1965, 1975):

$$G(x,x') \equiv \langle x|G|x'\rangle \, . \tag{2.4}$$

Let

$$H(x) \equiv -\nabla^{\mu}\nabla_{\mu} + \xi R + m^2 \tag{2.5}$$

(with $\nabla^{\mu}\nabla_{\mu}$ acting on a scalar), and

$$\langle x|H|x'\rangle \equiv H(x)\delta(x-x')[-g(x')]^{-\frac{1}{2}} \, . \tag{2.6}$$

The completeness relation in this abstract Hilbert space is

$$1 = \int d^4x \, \sqrt{-g(x)} \, |x\rangle\langle x| \, , \tag{2.7}$$

and the orthonormality relation is

$$\langle x|x'\rangle = [-g(x)]^{-\frac{1}{2}}\delta(x-x') \, . \tag{2.8}$$

(Our notation differs from that of DeWitt in that we make use of scalars or biscalars rather than scalar densities.) The equation for G(x,x') can now be written as

$$HG = 1 \, . \tag{2.9}$$

[Taking a matrix element of Eq. (2.9) and using the completeness relation gives $\int d^4x' \sqrt{-g(x')} \langle x|H|x'\rangle\langle x'|G|x''\rangle = \langle x|x''\rangle = [-g(x)]^{-\frac{1}{2}}\delta(x-x'')$, and hence $H(x)G(x,x') = [-g(x)]^{-\frac{1}{2}}\delta(x-x')$, which is Eq. (2.3).] The operator H is clearly Hermitian with respect to the scalar product in the Hilbert space,

$$\langle\psi|H|\phi\rangle = \int d^4x \, \sqrt{-g(x)} \, \langle\psi|x\rangle H(x)\langle x|\phi\rangle \, , \tag{2.10}$$

as follows by integration by parts (where $|\psi\rangle$, $|\phi\rangle$ are elements of the Hilbert space represented by square integrable functions).

In the proper-time formalism one writes

$$G = H^{-1} = \int_0^\infty i ds\, e^{-isH} , \tag{2.11}$$

or

$$G(x,x') = \int_0^\infty i ds\, \langle x|e^{-isH}|x'\rangle , \tag{2.12}$$

where it is understood that $H \to H - i\varepsilon$, with ε a small positive quantity (which is taken to zero when such a limit becomes well defined). Let

$$|x,s\rangle \equiv e^{isH}|x\rangle \tag{2.13}$$

and (using Hermiticity of H),

$$\psi(x,s) \equiv \langle x,s|\psi\rangle = \langle x|e^{-isH}|\psi\rangle = e^{-isH(x)}\langle x|\psi\rangle . \tag{2.14}$$

Thus

$$i \frac{\partial}{\partial s} \psi(x,s) = H(x)\psi(x,s) . \tag{2.15}$$

From the definition of H(x) in Eq. (2.5), one sees that Eq. (2.15) has the form of a non-relativistic Schrödinger equation, expressed in curvilinear spatial coordinates, for a fictitious particle of mass 1/2 moving in a 4-dimensional space (coordinates x^μ) of indefinite metric under the potential $\xi R + m^2$. The parameter s plays the role of a "time" (it is of dimension length squared). It does not change when the "space" coordinates x^μ are transformed, and is therefore called the proper-time.

In particular, the integrand $\langle x,s|x',0\rangle$ of G(x,x') in Eq. (2.12) satisfies the Schrödinger equation

$$i \frac{\partial}{\partial s} \langle x,s|x',0\rangle = [-\nabla_\alpha\nabla^\alpha + \xi R + m^2]\langle x,s|x',0\rangle \tag{2.16}$$

subject to the boundary condition

$$\lim_{s\to 0} \langle x,s|x',0\rangle = (-g)^{-\frac{1}{2}}\delta(x-x') \ . \tag{2.17}$$

One way to represent the solution for $\langle x,s|x',0\rangle$ is through a Feynman path integral representation, and we will discuss that formulation later (in Section 7). For the present, however, it is convenient to approach the curved space result via curvilinear coordinates in flat space.

In Minkowski space [metric $g_{\mu\nu} = \eta_{\mu\nu} = \mathrm{diag}(-1,1,1,1)$], the well-known solution of Eqs. (2.16) and (2.17) is

$$\langle x,s|x',0\rangle = \frac{i}{(4\pi is)^2}\, e^{-im^2 s}\, \exp\left\{\frac{i}{4s}\,\eta_{\alpha\beta}(x^\alpha - x^{\alpha'})(x^\beta - x^{\beta'})\right\} , \tag{2.18}$$

as is readily checked by direct substitution (the first factor of i appears because $\eta_{oo} = -1$). In n-dimensions the only change is that $(4\pi is)^{-2}$ is replaced by $(4\pi is)^{-n/2}$ in Eq. (2.18). As this equation involves only scalars under transformations of the coordinates, we can write it in curvilinear coordinates in the flat spacetime as

$$\langle x,s|x',0\rangle = \frac{i}{(4\pi is)^2}\, e^{im^2 s}\, \exp\left(\frac{i\tau^2}{4s}\right) , \tag{2.19}$$

where

$$\tau = \int_0^s ds'\, \left(g_{\alpha\beta}\,\frac{dx^\alpha}{ds'}\,\frac{dx^\beta}{ds'}\right)^{\frac{1}{2}} \tag{2.20}$$

is the proper arc length along the geodesic from x' to x. Therefore, in curved spacetime we write

$$\langle x,s|x',0\rangle = \frac{i}{(4\pi is)^2}\, e^{-im^2 s}\, \exp\left(\frac{i\tau^2}{4s}\right) U(x,x';is) , \tag{2.21}$$

where $U(x,x';is) \to 1$ when $R^\alpha{}_{\beta\gamma\delta} \to 0$ globally, and the topology of the spacetime is simple.

In renormalizing ultraviolet divergences we will only be interested in the behavior of $G(x,x')$ in a neighborhood of the point x'. Therefore, it is convenient to work in a special locally (or infinitesimally) inertial coordinate system called Riemann normal coordinates with origin at x'.

3. USE OF RIEMANN NORMAL COORDINATES TO DERIVE PROPER-TIME EXPANSION OF PROPAGATOR

Suppose the geodesics emanating from x' are continuous and do not intersect in a neighborhood of x'. One defines the Riemann normal coordinates y of a point x in that neighborhood by

$$y^\mu = \tau\xi^\mu \ , \tag{3.1}$$

where τ is an affine parameter along the geodesic from x' to x (for non-null geodesics, τ will be taken to be the proper arc length along the geodesic), and

$$\xi^\mu = (dx^\mu/d\tau)_{x'} \tag{3.2}$$

is the tangent to the geodesic at x'. If the spacetime is n-dimensional, then the ξ^μ can be thought of as functions of n-1 "angles" which are constant along a given geodesic, while τ is analogous to a radial coordinate which locates points along the geodesic for given ξ^μ. Along a given geodesic one has $d^2y^\mu/d\tau^2 = 0$, which implies that in these coordinates the affine connections or Christoffel symbols vanish at x':

$$\Gamma^\mu{}_{\nu\lambda}(x') = 0 \ . \tag{3.3}$$

Thus, the y^μ define a locally (or more precisely "infinitesimally") inertial frame. Without loss of generality, we take the metric at x' (i.e., at y = 0) in these coordinates to have the value

$$g_{\mu\nu}(0) = \eta_{\mu\nu} \ . \tag{3.4}$$

It follows from Eqs. (3.1) and (3.2) that for non-null geodesics,

$$\tau^2 = \eta_{\mu\nu}y^\mu y^\nu \ . \tag{3.5}$$

It can be shown (A. Z. Petrov 1969, p. 36) that

$$g_{\mu\nu}(y)y^{\nu} = \eta_{\mu\nu}y^{\nu} \ . \tag{3.6}$$

Equations (3.5) and (3.6) hold in the neighborhood of x' in which the Riemann normal coordinates are valid.

According to Eq. (2.21),

$$\langle x,s|x',0\rangle = \phi U \ , \tag{3.7}$$

with

$$\phi = i(4\pi is)^{-n/2} \exp(-im^2 s) \exp(\frac{i\tau^2}{4s}) \tag{3.8}$$

being a function of τ but not ξ^{μ}. (We are working in n dimensions for generality.) For a function of the form ϕU with ϕ independent of ξ^{μ}, one finds (S. Minakshisundaram and A. Pleijel 1949) that

$$\nabla^{\mu}\nabla_{\mu}(\phi U) = \phi\nabla^{\mu}\nabla_{\mu}U + 2\,\frac{d\phi}{d\tau}\frac{\partial U}{\partial\tau} + [\frac{d^2\phi}{d\tau^2} + \frac{n-1}{\tau}\frac{d\phi}{d\tau} + (\frac{\partial}{\partial\tau}\ln\sqrt{-g})\,\frac{d\phi}{d\tau}]U \ . \tag{3.9}$$

[The derivation of Eq. (3.9) is given in the appendix.] With the above form of ϕ this is

$$\nabla^{\mu}\nabla_{\mu}(\phi U) = \phi\nabla^{\mu}\nabla_{\mu}U + \frac{i\tau}{s}\,\phi\,\frac{\partial U}{\partial\tau} + \phi U\,[\frac{in}{2s} - \frac{\tau^2}{4s} + \frac{i\tau}{s}\frac{\partial}{\partial\tau}\ln\,(-g)^{\frac{1}{4}}] \ . \tag{3.10}$$

Then substituting Eqs. (3.7) and (3.8) and (3.10) into the Schrödinger equation, (2.16), one readily obtains the equation satisfied by U(x,x';is) in Riemann normal coordinates, namely,

$$i\,\frac{\partial U}{\partial s} = -\nabla^{\mu}\nabla_{\mu}U + \xi RU - \frac{i\tau}{s}\,[\frac{\partial U}{\partial\tau} + U\,\frac{\partial}{\partial\tau}\ln\,(-g)^{\frac{1}{4}}] \ . \tag{3.11}$$

Defining (in these coordinates),

$$F(x,x';is) = [-g(x)]^{\frac{1}{4}}U(x,x';is) \ , \tag{3.12}$$

this becomes

$$i \frac{\partial F}{\partial s} = -(-g)^{\frac{1}{4}}\nabla^{\mu}\nabla_{\mu}[(-g)^{-\frac{1}{4}}F] + \xi RF - \frac{i\tau}{s}\frac{\partial F}{\partial \tau} , \qquad (3.13)$$

where $(-g)^{-\frac{1}{4}}F$ is regarded as a scalar in applying $\nabla^{\mu}\nabla_{\mu}$, that is

$$\nabla_{\mu}\nabla^{\mu}[(-g)^{-\frac{1}{4}}F] = (-g)^{-\frac{1}{2}}\partial_{\mu}\left\{\sqrt{-g}\, g^{\mu\nu}\partial_{\nu}[(-g)^{-\frac{1}{4}}F]\right\} . \qquad (3.14)$$

Although it is convenient to continue working in Riemann normal coordinates, we will rewrite Eqs. (3.12) and (3.13) in general coordinates at this point to facilitate comparison with the literature. To write Eq. (3.13) in manifestly covariant form we take F to be a scalar at x and x'. In view of Eq. (3.12), we therefore look for the biscalar which reduces to $(-g(x))^{\frac{1}{4}}$ in the normal coordinates at x'. The quantity

$$(-g(x))^{-\frac{1}{2}}\det[-\frac{\partial}{\partial x^{\mu}}\frac{\partial}{\partial x^{\nu'}}(\tau^2)](-g(x'))^{-\frac{1}{2}} \qquad (3.15)$$

transforms as a scalar at x and at x'. Furthermore, in Riemann coordinates one has $\tau^2 = \eta_{\mu\nu}(y^{\mu}-y^{\mu'})(y^{\nu}-y^{\nu'})$, where for the moment we let $y^{\mu'}$ be the coordinate of the point x'. Thus $\partial_{\mu}\partial_{\nu'}(\tau^2) = -2\eta_{\mu\nu}$, and $[-g(x')]^{-\frac{1}{2}} = 1$ in those coordinates, so that the expression in Eq. (3.15) reduces to $-2^4[-g(x)]^{-\frac{1}{2}}$ in normal coordinates. Therefore, if we let

$$\sigma(x,x') \equiv \frac{1}{2}\tau^2 , \qquad (3.16)$$

the biscalar

$$\Delta(x,x') \equiv -[-g(x)]^{-\frac{1}{2}}\det[-\partial_{\mu}\partial_{\nu'}\sigma(x,x')][-g(x')]^{-\frac{1}{2}} \qquad (3.17)$$

reduces to $[-g(x)]^{-\frac{1}{2}}$ in normal coordinates with origin at x'. (In flat spacetime $\Delta(x,x') = 1$.) Thus, the definition of F in arbitrary coordinates is

$$F(x,x';is) = \Delta^{-\frac{1}{2}}(x,x')\, U(x,x';is) , \qquad (3.18)$$

and Eq. (3.13) can be written in manifestly covariant form as

$$i\,\frac{\partial F}{\partial s} = -\Delta^{-\frac{1}{2}}\nabla^{\mu}\nabla_{\mu}[\Delta^{\frac{1}{2}}F] + \xi RF - \frac{i\tau}{s}\frac{\partial F}{\partial \tau}\,. \tag{3.19}$$

The final term in Eq. (3.19) can be rewritten in a form often found in the literature, by using the relation

$$\tau\,\frac{\partial}{\partial \tau}\,F = g^{\mu\nu}\partial_{\mu}\sigma\partial_{\nu}F\,, \tag{3.20}$$

which is readily derived by means of normal coordinates (i.e., $g^{\mu\nu}\partial_{\mu}(\tau^2/2)\partial_{\nu}F = g^{\mu\nu}\eta_{\mu\lambda}y^{\lambda}\partial_{\nu}F = g^{\mu\nu}g_{\mu\lambda}y^{\lambda}\partial_{\nu}F = y^{\nu}\partial_{\nu}F = \tau\;\partial y^{\nu}/\partial\tau\cdot\partial_{\nu}F = \tau\partial F/\partial\tau$). Thus, in a general coordinate system in n dimensions, Eq. (2.21) takes the form

$$\langle x,s|x',0\rangle = e^{-im^2 s}\,\frac{i}{(4\pi is)^{n/2}}\,\exp\!\left(\frac{i\tau^2}{4s}\right)\Delta^{\frac{1}{2}}(x,x')F(x,x';is)\,, \tag{3.21}$$

where F satisifes Eq. (3.19). In flat spacetime in curvilinear coordinates the biscalar $\Delta(x,x') = 1$ (since it is unity in cartesian coordinates) and F reduces to unity, so that Eq. (2.19) is recovered.

We will need $[\partial^2F(x,x;is)/\partial(is)^2]_{s=0}$, which turns out to be essentially the trace anomaly. Thus, write

$$F(x,x';is) = 1 + isf_1(x,x') + (is)^2f_2(x,x') + \ldots\,, \tag{3.22}$$

where the leading term is determined by the requirement that F reduce to unity in flat spacetime. In Eq. (3.22) we are in general ignoring a term in F which has an essential singularity at $s = 0$ [e.g., like $\exp(-R^2/s^2)$, which vanishes faster than any power of s as $s \to 0$]. A term with an essential singularity at $s = 0$ is often connected with the phenomenon of particle creation and depends on non-local boundary conditions. Such a term will not affect the quantity $[\partial^2F/\partial(is)^2]_{s=0}$, which is of interest here, but may be quite significant for large s. The series of Eq. (3.22) will only be used for small s. The expressions for f_1 and f_2, with $x = x'$, in terms of the Riemann tensor were first calculated by DeWitt (1964, 1975) in general coordinates (in his notation $f_j \to a_j$). The calculation is simplified in Riemann normal coordinates, and because $f_2(x,x)$ is essentially the trace anomaly we carry it out here in detail. Our method of using the known expansion of $g_{\mu\nu}$ about the origin to calculate f_1 and f_2 evidently is new, although mathematicians have used such coordinates in a related context (Minakshisundaram and Pleijel 1949). We do not consider terms which appear when the topology is not simple or when $\langle x,s|x',0\rangle$ is required to vanish on a boundary (such terms are considered in McKean and Singer 1967). Our main results used later are Eqs. (3.31) and (3.58).

To calculate $f_1(x',x')$ and $f_2(x',x')$, let us return to Eq. (3.13) in normal coordinates. Substituting the power series expansion of F and equating the coefficients of equal powers of s gives

$$f_1(x,x') = L(1) - \xi R(x) - \tau\partial f_1(x,x')/\partial\tau \tag{3.23}$$

$$2f_2(x,x') = L(f_1) - \xi R(x)f_1(x,x') - \tau\partial f_2(x,x')/\partial\tau \;, \tag{3.24}$$

where

$$L(f) \equiv (-g(x))^{-\frac{1}{4}}\partial_\mu\left\{\sqrt{-g(x)}\; g^{\mu\nu}(x)\partial_\nu[(-g(x))^{-\frac{1}{4}}f(x,x')]\right\}, \tag{3.25}$$

with $\partial_\mu = \partial/\partial y^\mu$, and x regarded as a function of the normal coordinates y such that $x \to x'$ as $y \to 0$. The coincidence limits of f_1 and f_2 as $y \to 0$ are the key quantities which enter into renormalization. These limits can be calculated directly from Eqs. (3.23), (3.24), and (3.25) using the known Taylor series for $g_{\mu\nu}$ in Riemann normal coordinates (Petrov 1969, p. 36):

$$\begin{aligned} g_{\alpha\beta} = \overset{\circ}{g}_{\alpha\beta} &- \frac{1}{3} R_{\alpha\mu\beta\lambda}y^\mu y^\lambda - \frac{1}{3!} R_{\alpha\gamma\beta\lambda;\mu}y^\lambda y^\mu y^\gamma \\ &+ \frac{1}{5!}\left\{-6R_{\alpha\delta\beta\gamma;\lambda\mu} + \frac{16}{3} R_{\lambda\beta\mu}{}^{\rho}R_{\gamma\alpha\delta\rho}\right\}y^\lambda y^\mu y^\gamma y^\delta \\ &+ O(y^5) \;, \end{aligned} \tag{3.26}$$

where the coefficients in the series are evaluated at y = 0. It follows that

$$g/\overset{\circ}{g} = 1 - \frac{1}{3} R_{\mu\lambda} y^\mu y^\lambda - \frac{1}{3!} R_{\gamma\lambda;\mu} y^\gamma y^\lambda y^\mu$$

$$- \frac{1}{4!} \left(- \frac{4}{3} R_{\mu\lambda} R_{\sigma\varepsilon} + \frac{4}{15} R_{\alpha\mu\lambda}{}^{\beta} R^{\alpha}{}_{\varepsilon\sigma\beta} + \frac{6}{5} R_{\mu\lambda;\sigma\varepsilon}\right) y^\mu y^\lambda y^\sigma y^\varepsilon$$

$$+ O(y^5) \ . \tag{3.27}$$

(The expression for g given in Petrov 1969, p. 44, has several sign errors, the most important being the sign of the 4/3 in the fourth order term.) In the present context, $\overset{\circ}{g}_{\alpha\beta} = \eta_{\alpha\beta}$ and $\overset{\circ}{g} = -1$.

To find $f_1(x',x')$ from Eq. (3.23) we need

$$\lim_{y\to 0} L(1) = \lim_{y\to 0} (-g)^{-\frac{1}{4}} \partial_\mu \left\{\sqrt{-g}\, g^{\mu\nu} \partial_\nu [(-g)^{-\frac{1}{4}}]\right\} \ . \tag{3.28}$$

In Riemann normal coordinates, $\sqrt{-g}\, g^{\mu\nu} = \eta^{\mu\nu} + O(y^2)$, and

$$(-g)^{-\frac{1}{4}} = 1 + \frac{1}{12} R_{\alpha\beta} y^\alpha y^\beta + O(y^3) \ . \tag{3.29}$$

Hence, one finds immediately that

$$\lim_{y\to 0} L(1) = \frac{1}{6} \eta^{\mu\nu} R_{\mu\nu} = \frac{1}{6} R \ , \tag{3.30}$$

where the fact that $g^{\mu\nu} = \eta^{\mu\nu}$ at y = 0 has been used. Thus, noting that $\tau = 0$ at y = 0, the coincidence limit of Eq. (3.23) is

$$f_1(x',x') = (\tfrac{1}{6} - \xi) R(x') \ . \tag{3.31}$$

The calculation of the coincidence limit of f_2 from Eq. (3.24) is somewhat more complicated. In the same way as $\lim_{y\to 0} L(1)$ was found above, one obtains

$$\lim_{y\to 0} L(f_1) = \lim_{y\to 0} \eta^{\mu\nu} \partial_\mu \partial_\nu f_1 + \frac{1}{6} R f_1(x',x') \tag{3.32}$$

Applying $\lim_{y\to 0} \eta^{\mu\nu} \partial_\mu \partial_\nu$ to Eq. (3.23) yields

$$\lim_{y\to 0} \eta^{\mu\nu}\partial_\mu\partial_\nu f_1 = \lim_{y\to 0} \eta^{\mu\nu}\partial_\mu\partial_\nu L(1) - \xi R_{,\mu}{}^{;\mu}$$

$$- \lim_{y\to 0} \eta^{\mu\nu}\partial_\mu\partial_\nu(\tau\partial f_1/\partial\tau) \ , \tag{3.33}$$

where we have used $\eta^{\mu\nu}\partial_\mu\partial_\nu R = R_{,\mu}{}^{;\mu}$ at $y = 0$. Now

$$\lim_{y\to 0} \eta^{\mu\nu}\partial_\mu\partial_\nu(\tau\partial f_1/\partial\tau) = \lim_{y\to 0} \eta^{\mu\nu}\partial_\mu\partial_\nu(y^\lambda\partial_\lambda f_1)$$

$$= 2 \lim_{y\to 0} \eta^{\mu\nu}\partial_\mu\partial_\nu f_1 \ , \tag{3.34}$$

so that Eq. (3.33) becomes

$$3 \lim_{y\to 0} \eta^{\mu\nu}\partial_\mu\partial_\nu f_1 = M - \xi R_{,\mu}{}^{;\mu} \ , \tag{3.35}$$

where we have let

$$M \equiv \lim_{y\to 0} \eta^{\mu\nu}\partial_\mu\partial_\nu L(1) = \lim_{y\to 0} \eta^{\nu\sigma}\partial_\nu\partial_\sigma\left\{(-g)^{-\frac{1}{4}}\partial_\mu[\sqrt{-g}\, g^{\mu\lambda}\partial_\lambda(-g)^{-\frac{1}{4}}]\right\}. \tag{3.36}$$

Substituting Eqs. (3.31) and (3.35) into Eq. (3.32), and using the result in Eq. (3.24), gives for the coincidence limit of f_2,

$$f_2(x',x') = \frac{1}{6}\,(M - \xi R_{,\mu}{}^{;\mu}) + \frac{1}{2}\,(\frac{1}{6} - \xi)^2 R^2 \ . \tag{3.37}$$

To evaluate M we need $(-g)^{-\frac{1}{4}}$ to order y^4 because up to 4 derivatives act on it in Eq. (3.36). From Eq. (3.27), we find

$$(-g)^{-\frac{1}{4}} = 1 + a_{\alpha\beta}y^\alpha y^\beta + a_{\alpha\beta\gamma}y^\alpha y^\beta y^\gamma + a_{\alpha\beta\gamma\delta}y^\alpha y^\beta y^\gamma y^\delta + O(y^5) \ , \tag{3.38}$$

with

$$a_{\alpha\beta} = \frac{1}{12} R_{\alpha\beta} \ , \tag{3.39}$$

$$a_{\alpha\beta\gamma} = \frac{1}{24} R_{\alpha\beta,\gamma} \ , \tag{3.40}$$

$$a_{\alpha\beta\gamma\delta} = \frac{1}{96}\,(\frac{1}{3} R_{\alpha\beta}R_{\gamma\delta} + \frac{4}{15} R_{\rho\alpha\beta}{}^{\lambda}R^{\rho}{}_{\delta\gamma\lambda} + \frac{6}{5} R_{\gamma\delta;\alpha\beta}) \ . \tag{3.41}$$

Hence

$$\partial_\lambda(-g)^{-\frac{1}{4}} = 2a_{\alpha\lambda}y^\alpha + 3a_{(\alpha\beta\lambda)}y^\alpha y^\beta + 4a_{(\alpha\beta\gamma\lambda)}y^\alpha y^\beta y^\gamma + O(y^4) , \tag{3.42}$$

where parentheses indicate symmetrization over the enclosed indices [e.g. $a_{(\alpha\beta)} = \frac{1}{2!}(a_{\alpha\beta}+a_{\beta\alpha})$, etc.]. Because $\partial_\lambda(-g)^{-\frac{1}{4}}$ is already of first order in y, we only need $\sqrt{-g}\, g^{\mu\lambda}$ to second order. Now

$$g^{\mu\nu} = \eta^{\mu\nu} + b^{\mu\nu}{}_{\lambda\sigma}y^\lambda y^\sigma + O(y^3) , \tag{3.43}$$

with

$$b^{\mu\nu}{}_{\lambda\sigma} = \frac{1}{3} R^\mu{}_\sigma{}^\nu{}_\lambda , \tag{3.44}$$

and

$$\sqrt{-g}\, g^{\mu\lambda} = \eta^{\mu\lambda} + c^{\mu\lambda}{}_{\alpha\beta}y^\alpha y^\beta + O(y^3) , \tag{3.45}$$

with

$$c^{\mu\lambda}{}_{\alpha\beta} = b^{\mu\lambda}{}_{\alpha\beta} - 2\eta^{\mu\lambda}a_{\alpha\beta} . \tag{3.46}$$

Now M can be calculated directly. Using Eq. (3.38) to second order, one has

$$M = \lim_{y\to 0} \Big\{\eta^{\nu\sigma}\partial_\nu\partial_\sigma\partial_\mu[\sqrt{-g}\, g^{\mu\lambda}\partial_\lambda(-g)^{-\frac{1}{4}}]$$

$$+ 2a_{\nu\sigma}\eta^{\nu\sigma}\partial_\mu[\sqrt{-g}\, g^{\mu\lambda}\partial_\lambda(-g)^{-\frac{1}{4}}]\Big\} ,$$

$$M = \lim_{y\to 0} \eta^{\nu\sigma}\partial_\nu\partial_\sigma\partial_\mu[\sqrt{-g}\, g^{\mu\lambda}\partial_\lambda(-g)^{-\frac{1}{4}}] + (\tfrac{1}{6} R)^2 , \tag{3.47}$$

where the final term was found from Eq. (3.30). The first term on the right of Eq. (3.47) comes only from the third order term in $\sqrt{-g}\, g^{\mu\lambda}\partial_\lambda(-g)^{-\frac{1}{4}}$, which from Eqs. (3.42) and (3.45) is

$$[4\eta^{\mu\lambda}a_{(\alpha\beta\gamma\lambda)} + 2c^{\mu\lambda}{}_{\alpha\beta}a_{\lambda\gamma}]y^\alpha y^\beta y^\gamma . \tag{3.48}$$

Therefore,

$$M = 24\eta^{(\nu\sigma}\eta^{\mu\lambda)}a_{\nu\sigma\mu\lambda} + 12\eta^{\nu\sigma}c^{\mu\lambda}{}_{(\nu\sigma}a_{\mu)\lambda} + (\tfrac{1}{6}R)^2 \ . \tag{3.49}$$

One has

$$24\eta^{(\nu\sigma}\eta^{\mu\lambda)}a_{\nu\sigma\mu\lambda} = \frac{1}{12}(\eta^{\nu\sigma}\eta^{\mu\lambda} + \eta^{\mu\sigma}\eta^{\nu\lambda} + \eta^{\nu\mu}\eta^{\sigma\lambda})(\tfrac{1}{3}R_{\nu\sigma}R_{\mu\lambda}$$
$$+ \frac{4}{15}R_{\rho\nu\sigma}{}^{\gamma}R^{\rho}{}_{\mu\lambda\gamma} + \frac{6}{5}R_{\mu\lambda;\nu\sigma}) \ . \tag{3.50}$$

With the aid of the identities

$$R_{\nu\rho\sigma\gamma}R^{\nu\gamma\sigma\rho} = \frac{1}{2}R_{\nu\rho\sigma\gamma}R^{\nu\rho\sigma\gamma} \ , \tag{3.51}$$

$$R_{\mu\lambda;}{}^{\mu\lambda} = \frac{1}{2}R_{,\nu}{}^{;\nu} \ , \tag{3.52}$$

$$R^{\nu}{}_{\rho\nu\gamma} = R_{\rho\gamma} \ , \tag{3.53}$$

one readily reduces Eq. (3.50) to

$$24\eta^{(\nu\sigma}\eta^{\mu\lambda)}a_{\nu\sigma\mu\lambda} = \frac{1}{36}R^2 + \frac{1}{5}R_{;\alpha}{}^{\alpha} + \frac{7}{90}R_{\alpha\beta}R^{\alpha\beta}$$
$$+ \frac{1}{30}R_{\alpha\beta\gamma\delta}R^{\alpha\beta\gamma\delta} \ . \tag{3.54}$$

The next term in the expression for M is

$$12\eta^{\nu\sigma}c^{\mu\lambda}{}_{(\nu\sigma}a_{\mu)\lambda} = \eta^{\nu\sigma}b^{\mu\lambda}{}_{(\nu\sigma}R_{\mu)\lambda} - \frac{1}{6}\eta^{\nu\sigma}\eta^{\mu\lambda}R_{(\nu\sigma}R_{\mu)\lambda} \ . \tag{3.55}$$

Symmetrizing over the indices $\nu\sigma\mu$ and inserting the expression for $b^{\mu\lambda}{}_{\nu\sigma}$ from Eq. (3.44) yields

$$12\eta^{\nu\sigma}c^{\mu\lambda}{}_{(\nu\sigma}a_{\mu)\lambda} = \frac{1}{9}\eta^{\nu\sigma}(R^{\mu}{}_{\sigma}{}^{\lambda}{}_{\nu}R_{\mu\lambda} + R^{\mu}{}_{\sigma}{}^{\lambda}{}_{\mu}R_{\nu\lambda} + R^{\mu}{}_{\mu}{}^{\lambda}{}_{\nu}R_{\sigma\lambda})$$
$$- \frac{1}{18}\eta^{\nu\sigma}\eta^{\mu\lambda}(R_{\nu\sigma}R_{\mu\lambda} + R_{\mu\sigma}R_{\nu\lambda} + R_{\nu\mu}R_{\sigma\lambda}) \ ,$$

$$12\eta^{\nu\sigma}c^{\mu\lambda}{}_{(\nu\sigma}a_{\mu)\lambda} = -\frac{1}{18}R^2 - \frac{1}{9}R_{\alpha\beta}R^{\alpha\beta} \ . \tag{3.56}$$

Substituting Eq. (3.54) and Eq. (3.56) into Eq. (3.49) finally gives

$$M = \frac{1}{5} R_{,\alpha}{}^{;\alpha} + \frac{1}{30} R_{\alpha\beta\gamma\delta}R^{\alpha\beta\gamma\delta} - \frac{1}{30} R_{\alpha\beta}R^{\alpha\beta} . \qquad (3.57)$$

Then Eq. (3.37) yields the coincidence limit of f_2 as

$$f_2(x',x') = \frac{1}{6}(\frac{1}{5} - \xi)R_{,\alpha}{}^{;\alpha} + \frac{1}{2}(\frac{1}{6} - \xi)^2 R^2 + \frac{1}{180} R_{\alpha\beta\gamma\delta}R^{\alpha\beta\gamma\delta} - \frac{1}{180} R_{\alpha\beta}R^{\alpha\beta} . \qquad (3.58)$$

Because $f_2(x',x')$ is a scalar, the above expression is already in covariant form. It holds in n-dimensions, and for either metric signature (±1,1,1,1).

4. RELATION OF PROPAGATOR TO GENERALIZED ζ-FUNCTION

To show how $f_2(x,x)$ is related to $\langle T_\mu{}^\mu \rangle$, we will use the ζ-function approach (Dowker and Critchley 1976a 1977a; Hawking 1977). Define

$$G^\nu(x,x') = \int_0^\infty ids \langle x|e^{-isH^\nu}|x'\rangle = \langle x|G^\nu|x'\rangle , \qquad (4.1)$$

where

$$G^\nu = \int_0^\infty ids\ e^{-isH^\nu} = H^{-\nu} = \sum_n \frac{|\phi_n\rangle\langle\phi_n|}{\lambda_n{}^\nu} . \qquad (4.2)$$

The last equality is the spectral representation of $H^{-\nu}$ obtained by applying $H^{-\nu}$ to unity in the form $\sum_n |\phi_n\rangle \langle\phi_n|$, where

$$H|\phi_n\rangle = \lambda_n|\phi_n\rangle . \qquad (4.3)$$

We are assuming that the $|\phi_n\rangle$ are complete, and taking the λ_n as discrete. (If necessary, boundary conditions can be imposed for that purpose and the boundary taken to infinity in the end. The boundary terms evidently do not affect the final result.) The mathematical aspects of the theory appear to be better understood if the metric has the Euclidean or Riemannian signature (+,+,+,+), as $G(x,x')$ then satisfies an elliptic differential equation, and can be uniquely specified by requiring it to vanish asymptotically (if the metric is asymptotically flat). When the form of the metric permits, one can complexify the coordinates and analytically continue until the metric has the Euclidean signature (Hartle and Hawking 1976, Gibbons and Hawking 1977, Chitre and Hartle 1977, Lapedes 1978). We will continue to work with Minkowskian signature, it being understood that the Euclideanized version of the theory can be dealt with in a parallel manner, and that it may sometimes be necessary to analytically continue. We will avoid the question of possible zero eigenvalues of H by assuming that either they do not lead to problems if H is replaced by $H-i\varepsilon$ (with $\varepsilon\to 0$ after physical quantities have been calculated), or that H (and H^{-1}) is redefined by taking its spectral representation with the zero eigenvalues λ_n omitted [the contribution of the zero eigenvalues to the effective action in that case must be considered separately (Gibbons 1977)].

It is evident from Eq. (4.2) that G^ν is a generalization of the Riemann ζ-function, $\zeta_R(\nu) = \sum_{n=1}^{\infty} n^{-\nu}$. One has

$$\mathrm{Tr}\, G^\nu = \sum_n \langle\phi_n|G^\nu|\phi_n\rangle = \sum_n \lambda_n^{-\nu} \equiv \zeta(\nu) \ . \tag{4.4}$$

(This reduces to the Riemann ζ-function when H is essentially the harmonic oscillator Hamiltonian.) This generalized ζ function formed from the eigenvalues of H will converge only for a limited range of ν, but can be defined over a wider range by analytic continuation. A useful integral representation is found as follows. One has

$$\int_0^\infty ids\ (is)^{\nu-1}\ e^{-isH} = H^{-\nu}\int_0^{\pm\infty} ids'\ (is')^{\nu-1}\ e^{-is'} = H^{-\nu}\Gamma(\nu) \ ,$$

where it is understood that $H\to H-i\varepsilon$ and $\mathrm{Re}\nu > 1$ (the integration contour can be deformed to the real axis of the integration variable z = is' for either sign of the eigenvalues of H). Therefore,

$$G^{\nu} = H^{-\nu} = \Gamma(\nu)^{-1} \int_0^{\infty} ids\ (is)^{\nu-1}\ e^{-isH}\ , \tag{4.5}$$

where this can be taken as a definition of G^{ν} even if $\mathrm{Re}\nu < 1$. Taking the trace over states $|x\rangle$ gives

$$\zeta(\nu) = \mathrm{Tr}\ G^{\nu} = \Gamma(\nu)^{-1} \int d^4x \sqrt{-g} \int_0^{\infty} ids(is)^{\nu-1} \langle x|e^{-isH}|x\rangle\ . \tag{4.6}$$

One can make use of this integral representation to deduce the analytic properties of $\zeta(\nu)$, and to find $\zeta(0)$.

The coincidence limits of Eqs. (3.21) and (3.22) give (for small s, and dimension $n = 4$),

$$\langle x|e^{-isH}|x\rangle = e^{-im^2 s} \frac{i}{(4\pi is)^2} [1 + isf_1(x,x)$$

$$+ (is)^2 f_2(x,x) + \ldots]\ , \tag{4.7}$$

where f_1 and f_2 are given by Eqs. (3.31) and (3.58), respectively. Thus,

$$\zeta(\nu) = \Gamma(\nu)^{-1} \int d^4x \sqrt{-g} \left\{ \int_0^{s_o} ids(is)^{\nu-1} \langle x|e^{-isH}|x\rangle \right.$$

$$\left. + \int_{s_o}^{\infty} ids\ (is)^{\nu-1} \langle x|e^{-isH}|x\rangle \right\}\ ,$$

which leads to

$$\zeta(\nu) = \Gamma(\nu)^{-1} \left\{ i(4\pi)^{-2} \int d^4x \sqrt{-g} \left[\frac{(is_o)^{\nu-2}}{\nu-2} + \frac{(is_o)^{\nu-1}}{\nu-1} (f_1 - m^2) \right.\right.$$

$$\left.\left. + \frac{(is_o)^{\nu}}{\nu} (f_2 - m^2 f_1 + \frac{1}{2} m^4) + \ldots \right] + \text{finite term} \right\}\ . \tag{4.8}$$

(We took $\mathrm{Re}\nu > 2$ in doing the integrals, but then analytically continued.) Therefore $\zeta(\nu)$ has a simple pole at $\nu = 2$ with residue $i(4\pi)^{-2}\int d^4x\,\sqrt{-g}$, and a simple pole at $\nu = 1$ with residue $i(4\pi)^{-2}\int d^4x\,\sqrt{-g}\,[f_2(x,x) - m^2]$. For small ν,

$$\Gamma(\nu)^{-1} = \nu + \gamma\nu^2 + O(\nu^3) \,, \tag{4.9}$$

where γ = Euler's constant = 0.577 ... .

Hence, only one term survives as $\nu \to 0$, and

$$\zeta(0) = i(4\pi)^{-2}\int d^4x\,\sqrt{-g}\,[f_2(x,x) - m^2 f_1(x,x) + \frac{1}{2}\,m^4] \,. \tag{4.10}$$

As $\zeta(\nu)$ is analytic near $\nu = 0$, its derivative $\zeta'(0)$ also exists (and will be evaluated later).

One has $\zeta'(\nu) = -\sum_n (\ln \lambda_n)\lambda_n^{-\nu}$, so that formally

$$\zeta'(0) = -\sum_n \ln \lambda_n = -\ln(\prod_n \lambda_n) = -\ln \det H \tag{4.11}$$

Although the product of the eigenvalues appearing in Eq. (4.11) diverges, the derivative of the generalized ζ-function at $\nu = 0$ is well defined through analytic continuation. Therefore, we define det H through the equation

$$\ln \det H \equiv -\zeta'(0) \,. \tag{4.12}$$

If a is a constant, the ζ-function corresponding to the operator aH is

$$\mathrm{Tr}(aH)^{-\nu} = \sum_n (a\lambda_n)^{-\nu} = a^{-\nu}\,\zeta(\nu) \,, \tag{4.13}$$

where $\zeta(\nu)$ is $\mathrm{Tr}\,H^{-\nu}$. Therefore,

$$\begin{aligned}\ln \det aH &= -\frac{d}{d\nu}\left(a^{-\nu}\zeta(\nu)\right)\Big|_{\nu=0} = -\zeta'(0) + \zeta(0)\ln a \\ &= \ln \det H + \zeta(0)\ln a \,.\end{aligned} \tag{4.14}$$

5. THE ANOMALOUS TRACE OF THE ENERGY-MOMENTUM TENSOR

The action of the neutral scalar field is

$$S = \int d^4x \mathcal{L} = -\frac{1}{2}\int d^4x \sqrt{-g}\, \phi(x)H(x)\phi(x) \ , \qquad (5.1)$$

where $\mathcal{L}$ is given by Eq. (2.1), H(x) by Eq. (2.5), and we have performed an integration by parts. For the charged scalar field one would have $\mathcal{L} = -\sqrt{-g}\,\{g^{\mu\nu}\partial_\mu\phi^*\partial_\nu\phi + \xi R|\phi|^2 + m^2|\phi|^2\}$ and $S = -\int d^4x \sqrt{-g}\, \phi^* H\phi$ with H given by Eq. (2.5). The charged field can be related to two real fields ϕ_1 and ϕ_2 described by the neutral theory by writing $\phi = (\phi_1 + i\phi_2)/\sqrt{2}$. For fields of other spin, the action will have a similar form, although for spin $\frac{1}{2}$ H will be of first order. For simplicity, we discuss the scalar case. The expression for the energy-momentum tensor is

$$T^{\mu\nu}(x) = \frac{2}{\sqrt{-g(x)}} \frac{\delta S}{\delta g_{\mu\nu}(x)} \ . \qquad (5.2)$$

For the neutral scalar field that is

$$T^{\mu\nu} = \partial^\mu\phi\partial^\nu\phi - \frac{1}{2}g^{\mu\nu}\partial_\alpha\phi\partial^\alpha\phi - \frac{1}{2}g^{\mu\nu}m^2\phi^2$$

$$+ \xi[g^{\mu\nu}\nabla_\alpha\nabla^\alpha(\phi^2) - \nabla^\mu\nabla^\nu(\phi^2) + G^{\mu\nu}\phi^2] \ , \qquad (5.3)$$

where $G^{\mu\nu} = R^{\mu\nu} - \frac{1}{2}g^{\mu\nu}R$. As a consequence of the field equation (2.2), one has

$$T^\mu{}_\mu = (6\xi-1)\partial_\alpha\phi\partial^\alpha\phi + \xi(6\xi-1)R\phi^2 + (6\xi-2)m^2\phi^2 \ . \qquad (5.4)$$

For example, in the conformally invariant case ($\xi = \frac{1}{6}$, m = 0), this gives

$$T^\mu{}_\mu = 0 \ . \qquad (5.5)$$

By contrast, one finds that the renormalized expectation value of $T_\mu{}^\mu$ does not vanish. It seems that the requirements of general covariance and the local conservation of $\langle T_{\mu\nu}\rangle$ lead to contributions to the renormalized trace which would not be expected from the classical expression. One speaks in general of the contribution to $\langle T_\mu{}^\mu\rangle$, beyond that expected from the classical expression, as the trace anomaly.

The existence of a trace anomaly was first noticed by Capper and Duff (1974,1975) using dimensional regularization. The general form of such an anomaly (without the values of the coefficients) was given by Deser, Duff, and Isham (1976). The evaluation of the trace anomaly in specific cases as well as for an arbitrary metric has been carried out by a number of methods: point-splitting regularization (Davies, Fulling, and Unruh 1976, Christensen 1976, Davies and Fulling 1977, Christensen and Fulling 1977, Adler, Lieberman, and Ng 1977, Wald 1977,1978ab); dimensional regularization (Brown 1977, Brown and Cassidy 1977ab, Tsao 1977, Bunch 1978b); ζ-function regularization (Dowker and Critchley 1976a, 1977a, Hawking 1977); adiabatic regularization (Bunch 1978a, Hu 1978, based on results of Fulling, Parker, and Hu 1974); and Pauli-Villars regularization (Bernard and Duncan 1977, Vilenkin 1978). In this Section, we will obtain the trace anomaly using ζ-function regularization, as it permits a relatively succinct derivation; and an outline of a related method using dimensional regularization will be given. Space will not permit an outline of the point-splitting, Pauli-Villars, and adiabatic regularization methods, which also have important applications, especially if one is interested in the expectation value of $T_{\mu\nu}$ in a given state, rather than a matrix element between in and out states [in that connection the method of Zeldovich and Starobinsky (1972), which is related to Pauli-Villars and to adiabatic regularization, is also useful]. Several recent suggestions have been made as to how one may possibly avoid the appearance of an anomalous trace (Fradkin and Vilkovisky 1978, Brown 1978), but it is too early to comment on them here.

Returning to Eq. (5.2), one can express $\langle T^{\mu\nu}\rangle$ as a functional integral over field configurations:

$$\langle T^{\mu\nu}(x)\rangle = \frac{2}{\sqrt{-g(x)}} \left\langle \frac{\delta S}{\delta g_{\mu\nu}(x)} \right\rangle$$

$$= \frac{2}{\sqrt{-g(x)}} \int d[\phi] \frac{\delta S}{\delta g_{\mu\nu}(x)} e^{iS} \Big/ \int d[\phi] e^{iS}$$

$$= -\frac{2i}{\sqrt{-g(x)}} \frac{\delta}{\delta g_{\mu\nu}(x)} \ln \int d[\phi] e^{iS} = \frac{2}{\sqrt{-g(x)}} \frac{\delta}{\delta g_{\mu\nu}(x)} W , \qquad (5.6)$$

where

$$W \equiv -i \ln \int d[\phi] e^{iS} . \qquad (5.7)$$

As is evident from Eq. (5.6), W plays a role analogous to S and is thus an effective action. When vacua can be defined in distinct asymptotic regions of spacetime with a timelike separation, then $\langle T^{\mu\nu}\rangle$ of Eq. (5.6) represents (by extension of what occurs in flat spacetime),

$$\langle T^{\mu\nu}\rangle = \frac{\langle 0,\text{out}|T^{\mu\nu}|0,\text{in}\rangle}{\langle 0,\text{out}|0,\text{in}\rangle} , \qquad (5.8)$$

and

$$\langle 0,\text{out}|0,\text{in}\rangle = \int d[\phi] e^{iS} = e^{iW} \qquad (5.9)$$

is the vacuum to vacuum, or vacuum persistence amplitude. (One could equally well obtain Eq. (5.6) by means of the Schwinger action principle with the identification (5.9).)

When particle production occurs, the in and out vacuum states will differ by more than a phase factor. To relate the above definition of $\langle T^{\mu\nu}\rangle$, which is a matrix element, to a true expectation value like $\langle 0,\text{in}|T^{\mu\nu}|0,\text{in}\rangle/\langle 0,\text{in}|0,\text{in}\rangle$, one would have to express $|0,\text{out}\rangle$ in terms of $|0,\text{in}\rangle$, which requires solving for the completë S-matrix including the particle production. In cases when in and out vacua cannot be defined, the appropriate interpretation of $\langle T^{\mu\nu}\rangle$ in Eq. (5.6) must be sought by extension from the analogous flat space situation. For example, when the Schwarzschild metric is Euclideanized one finds that the metric is periodic in the imaginary time with period $i\beta$ (Gibbons and Hawking 1977). In analogy with what happens in flat spacetime when one Euclideanizes and imposes a periodicity $i\beta$ in the imaginary time, one can interpret Eq. (5.6) as giving $\langle T^{\mu\nu}\rangle$ for a state in which the black hole is in equilibrium with a heat bath at a temperature proportional to β^{-1}, and

$$Z \equiv \int d[\phi]e^{iS} = e^{iW} \tag{5.10}$$

as the partition function of the system (corresponding to a canonical ensemble).

The effective action of Eq. (5.7) can be expressed in terms of the generalized ζ-function. One expands the field ϕ in terms of the complete orthonormal set $\phi_n(x) = \langle x|\phi_n\rangle$:

$$\phi(x) = \sum_n C_n \phi_n(x) , \tag{5.11}$$

with the C_n real in the present case (if $\phi(x)$ were complex, the measure $d[\phi]$ given below would involve integrations over the real and imaginary parts of C_n, but the results would only be altered by numerical factors corresponding to the doubling of the number of degrees of freedom). Then

$$S = -\frac{1}{2}\int \phi H\phi \sqrt{-g}\, d^4x = -\frac{1}{2}\sum_n C_n^2 \lambda_n , \tag{5.12}$$

where we have used the orthonormality relation

$$\int d^4x \sqrt{-g}\, \phi_n(x)\, \phi_m(x) = \delta_{nm} . \tag{5.13}$$

From Eq. (5.13), $\phi_n(x)$ has dimension ℓ^{-2} (where ℓ is a length), while from Eq. (5.12) the dimension of $\phi(x)$ is ℓ^{-1}, as H has dimension ℓ^{-2} and the action S is dimensionless in these units. Therefore C_n has dimension ℓ. The functional integration over all field configurations $\phi(x)$ can be written as an integral over the coefficients C_n in Eq. (5.11). Thus,

$$d[\phi] = \prod_n (\mu dC_n) , \tag{5.14}$$

where μ is a (complex) normalization constant which must have dimension ℓ^{-1} for $d[\phi]$ and hence Z to be dimensionless. Then

$$\int d[\phi]\exp(iS) = \prod_n \int \mu dC_n \exp(-\frac{1}{2} i C_n^2 \lambda_n)$$

$$= \left\{\prod_n \left[\frac{2\pi\mu^2}{i\lambda_n}\right]\right\}^{\frac{1}{2}} = \left\{\det \left[\frac{iH}{2\pi\mu^2}\right]\right\}^{-\frac{1}{2}} . \qquad (5.15)$$

Therefore, referring to Eq. (5.7) and Eq. (4.14), one has

$$W = \frac{i}{2} \ln \det\left[\frac{iH}{2\pi\mu^2}\right] = -\frac{i}{2} \zeta'(0) - \frac{i}{2} \zeta(0)\ln(-2\pi i\mu^2) . \qquad (5.16)$$

Because of Eq. (4.10) and the expressions for $f_2(x,x)$ and $f_1(x,x)$ in Eqs. (3.31) and (3.58), the terms in the effective action involving $\zeta(0)$ could be absorbed by a finite renormalization of the cosmological constant (the part proportional to $\int d^4x \sqrt{-g}$), of the gravitational constant (the part proportional to $\int d^4x \sqrt{-g}\, R$), and by the addition to the action of a geometrical counterterm involving $\int d^4x \sqrt{-g}\, f_2(x,x)$.

Let us use Eq. (5.16) to find $\langle T_\mu{}^\mu\rangle$ for a massless scalar field. Consider a constant infinitesimal scale transformation of the metric

$$\tilde{g}_{\mu\nu} = ag_{\mu\nu} = (1+\varepsilon)g_{\mu\nu} , \qquad (5.17)$$

with

$$\delta g_{\mu\nu} = \varepsilon g_{\mu\nu} . \qquad (5.18)$$

One has by definition of the variational derivative,

$$W[g_{\mu\nu}+\delta g_{\mu\nu}] = W[g_{\mu\nu}] + \int d^4x\, \delta g_{\mu\nu}(x) \frac{\delta W[g_{\mu\nu}]}{\delta g_{\mu\nu}(x)}$$

$$= W[(1+\varepsilon)g_{\mu\nu}] = W[g_{\mu\nu}] + \varepsilon \frac{\partial}{\partial a} W[ag_{\mu\nu}]\Big|_{a=1} ,$$

or, as a consequence of Eq. (5.18),

$$\int d^4x\, g_{\mu\nu} \frac{\delta W[g_{\mu\nu}]}{\delta g_{\mu\nu}} = \frac{\partial}{\partial a} W[ag_{\mu\nu}]\Big|_{a=1} \qquad (5.19)$$

Therefore, Eq. (5.6) gives

$$\int d^4x \sqrt{-g} \langle T^\mu{}_\mu \rangle = 2 \frac{\partial}{\partial a} W[ag_{\mu\nu}] \bigg|_{a=1} . \tag{5.20}$$

Under the transformation $g_{\mu\nu} \to ag_{\mu\nu}$ one has (with $m = 0$, but ξ arbitrary) that $H \to a^{-1}H$. Then Eq. (4.14) with $a \to a^{-1}$, and Eq. (5.16) yield

$$W[ag_{\mu\nu}] = W[g_{\mu\nu}] - \frac{i}{2} \zeta(0) \ln a , \tag{5.21}$$

where it has been assumed that the normalization μ is not changed by the scale transformation. Hence

$$\int d^4x \sqrt{-g} \langle T_\mu{}^\mu \rangle = -i\zeta(0) = (4\pi)^{-2} \int d^4x \sqrt{-g}\, f_2(x,x) , \tag{5.22}$$

which implies that

$$\langle T_\mu{}^\mu \rangle = (4\pi)^{-2} f_2(x,x) + \text{covariant 4-divergence} , \tag{5.23}$$

where $f_2(x,x)$ is given by Eq. (3.58):

$$f_2(x,x) = \frac{1}{6}\left(\frac{1}{5} - \xi\right) R_{,\alpha}{}^{;\alpha} + \frac{1}{2}\left(\frac{1}{6} - \xi\right)^2 R^2$$
$$+ \frac{1}{180} R_{\alpha\beta\gamma\delta} R^{\alpha\beta\gamma\delta} - \frac{1}{180} R_{\alpha\beta} R^{\alpha\beta} .$$

For the conformally invariant scalar field ($\xi = \frac{1}{6}$, $m = 0$) one has an additional symmetry under a coordinate-dependent scale transformation (conformal transformation), $g_{\mu\nu} \to a(x) g_{\mu\nu}$. Under such a transformation one has (see, for example, Parker 1973)

$$H(x) \to a^{-3/2}(x) H(x) a^{\frac{1}{2}}(x) . \tag{5.24}$$

As will be shown later, this implies that the 4-divergence term in Eq. (5.23) vanishes. Thus, for $\xi = \frac{1}{6}$, $m = 0$, one has

$$\langle T_\mu{}^\mu \rangle = \frac{1}{(4\pi)^2 180} (R_{,\alpha}{}^{;\alpha} + R_{\alpha\beta\gamma\delta}R^{\alpha\beta\gamma\delta} - R_{\alpha\beta}R^{\alpha\beta}) . \qquad (5.25)$$

This result is often expressed in terms of the topological invariant

$$G \equiv R_{\alpha\beta\gamma\delta}R^{\alpha\beta\gamma\delta} - 4R_{\alpha\beta}R^{\alpha\beta} + R^2 \qquad (5.26)$$

which satisfies (in 4-dimensions only)

$$\frac{\delta}{\delta g_{\mu\nu}} \int d^4x \sqrt{-g}\, G = 0 , \qquad (5.27)$$

and the Weyl tensor (in 4-dimensions)

$$C^\alpha{}_{\beta\gamma\delta} = R^\alpha{}_{\beta\gamma\delta} - \frac{1}{2} (\delta^\alpha{}_\gamma R_{\beta\delta} - \delta^\alpha{}_\delta R_{\beta\gamma} - g_{\beta\gamma}R^\alpha{}_\delta + g_{\beta\delta}R^\alpha{}_\gamma)$$
$$+ \frac{1}{6} R(\delta^\alpha{}_\gamma g_{\beta\delta} - \delta^\alpha{}_\delta g_{\beta\gamma}) , \qquad (5.28)$$

which is the conformally invariant tensor formed from $R^\alpha{}_{\beta\gamma\delta}$, $R_{\alpha\beta}$, and R. One has

$$C_{\alpha\beta\gamma\delta}C^{\alpha\beta\gamma\delta} = R_{\alpha\beta\gamma\delta}R^{\alpha\beta\gamma\delta} - 2R_{\alpha\beta}R^{\alpha\beta} + \frac{1}{3} R^2 . \qquad (5.29)$$

Eliminating $R_{\alpha\beta\gamma\delta}R^{\alpha\beta\gamma\delta}$ and $R_{\alpha\beta}R^{\alpha\beta}$ in Eq. (5.25) in favor of G and $C_{\alpha\beta\gamma\delta}C^{\alpha\beta\gamma\delta}$, one finds

$$\langle T_\mu{}^\mu \rangle = (4\pi)^{-2}[\frac{1}{180} R_{,\alpha}{}^{;\alpha} + \frac{1}{120} (C_{\alpha\beta\gamma\delta}C^{\alpha\beta\gamma\delta} - \frac{1}{3} G)] . \qquad (5.30)$$

Note that no R^2 term is present, although such a term could have appeared. The R^2 term is also found to be absent in the corresponding expressions for conformally invariant fields of higher spin.

I will show in Section 6 that the absence of an R^2 term when $\langle T_\mu{}^\mu \rangle$ is expressed in terms of $R_{,\alpha}{}^{;\alpha}$, $C_{\alpha\beta\gamma\delta}C^{\alpha\beta\gamma\delta}$, and G is required by the theorem that no creation of particles governed by conformally invariant field equations occurs in any statically bounded Robertson-Walker expansion of the universe. The original proof of that theorem (L. Parker 1966, 1968, 1969) used only conformal invariance of the field equations. It must therefore be respected by the trace anomaly.

Can one add a term W_c to the effective action to cancel this $\langle T_\mu{}^\mu \rangle$? Such a counterterm W_c would give rise to a contribution to $\langle T_\mu{}^\mu \rangle$ of (for a derivation of this see Eq. (5.46) below),

$$g_{\mu\nu} \langle T_c^{\mu\nu}(x) \rangle = \frac{2}{\sqrt{-g(x)}} g_{\mu\nu}(x) \frac{\delta W_c[g_{\mu\nu}]}{\delta g_{\mu\nu}(x)}$$

$$= \frac{2}{\sqrt{-g(x)}} \left. \frac{\delta W_c[a(x) g_{\mu\nu}]}{\delta a(x)} \right|_{a=1} \tag{5.31}$$

To give terms of the type appearing in Eq. (5.25), it is generally argued that W_c must be formed from $\int d^4x \sqrt{-g}$ of $R_{\alpha\beta\gamma\delta}R^{\alpha\beta\gamma\delta}$, $R_{\alpha\beta}R^{\alpha\beta}$, and R^2, or equivalently of $C_{\alpha\beta\gamma\delta}C^{\alpha\beta\gamma\delta}$, G, and R^2. The topological invariant G makes no contribution to $\delta W_c / \delta a(x)$, while $C_{\alpha\beta\gamma\delta}C^{\alpha\beta\gamma\delta} \times \sqrt{-g}$, being invariant under conformal transformation, also makes no contribution. Finally,

$$\frac{2}{\sqrt{-g(x)}} \frac{\delta}{\delta a(x)} \int d^4x' [-\det(a g_{\mu\nu})]^{\frac{1}{2}} R^2(a g_{\mu\nu}) \Big|_{a=1} = -12 R_{,\alpha}{}^{;\alpha}. \tag{5.32}$$

Thus, W_c could only alter the coefficient of the $R_{,\alpha}{}^{;\alpha}$ term in $\langle T_\mu{}^\mu \rangle$. Therefore, the coefficient of that term in $\langle T_\mu{}^\mu \rangle$ should not be regarded as determined by the theory. However, the remaining terms in $\langle T_\mu{}^\mu \rangle$ cannot be altered by the addition of such a W_c to the effective action. Horowitz and Wald (1978) have given arguments suggesting that the coefficient of the $R_{,\alpha}{}^{;\alpha}$ term in $\langle T_\mu{}^\mu \rangle$ should perhaps be set equal to zero. M. Brown (1978) has suggested the possibility of using some unconventional counterterms in W_c to alter the coefficients of the other terms in the trace anomaly. We will not be able to discuss such matters here.

In summary, for conformally invariant massless fields $\langle T_\mu{}^\mu \rangle$ is of the form

$$\langle T_\mu{}^\mu \rangle = (4\pi)^{-2}(AC_{\alpha\beta\gamma\delta}C^{\alpha\beta\gamma\delta} + BG + CR_{,\alpha}{}^{;\alpha}) , \tag{5.33}$$

where A, B, C have been found by various workers (cited earlier) to be as follows:

spin 0: $A = \frac{1}{120}$, $B = -\frac{1}{360}$, $C = \frac{1}{180}$

spin $\frac{1}{2}$: $A = \frac{1}{40}$, $B = -\frac{11}{720}$, $C = \frac{1}{60}$
(2 component neutrino)

spin 1: $A = \frac{1}{10}$, $B = -\frac{31}{180}$, $C = \frac{1}{15}$ or $-\frac{1}{10}$.
(photon)

In the spin 1 case, the value C = 1/15 seems to come from dimensional regularization, while the value C = -1/10 comes from ζ-function regularization. In any event, the value of C can be altered by adding a counterterm to the action.

Before showing that the trace anomaly for the conformally invariant scalar field ($\xi = \frac{1}{6}$, $m = 0$) is given by Eq. (5.25), I will obtain an explicit expression for the effective action W of Eq. (5.16) (with ξ and m general), in terms of the function $F(x,x;is)$ appearing in Eq. (3.21). The scalar $\Delta(x,x) = 1$, as is readily found by evaluating it in normal coordinates at x. Therefore, Eq. (3.21) gives (in 4-dimensions),

$$\langle x,s|x,0\rangle = \langle x|e^{-isH}|x\rangle = (4\pi is)^{-2} i\; e^{-ism^2} F(x,x;is) . \tag{5.34}$$

Therefore Eq. (4.6) can be written as

$$\zeta(\nu) = \Gamma(\nu)^{-1}(4\pi)^{-2} i\int d^4x \sqrt{-g} \int_0 ids(is)^{\nu-3}F(x,x;is)e^{-ism^2} \tag{5.35}$$

Taking $\mathrm{Re}\,\nu > 2$, we can perform three integrations by parts, dropping the boundary terms at $s = 0$ because they involve (is) to a positive power, and the boundary terms at $s = \infty$ because $m^2 \to m^2 - i\varepsilon$ is understood. Thus, one finds

$$\zeta(\nu) = \frac{-i(4\pi)^{-2}}{\Gamma(\nu+1)(\nu-1)(\nu-2)} \int d^4x \sqrt{-g} \int_0^\infty ids(is)^\nu \frac{\partial^3}{\partial(is)^3} \times [F(x,x;is)e^{-ism^2}] , \qquad (5.36)$$

where $\nu\Gamma(\nu) = \Gamma(\nu+1)$ has been used. As this agrees with the previous expression when $\mathrm{Re}\,\nu > 2$, it is a valid analytic continuation, and is clearly well behaved at $\nu = 0$. For example, at $\nu = 0$ the s-integration can be done, and gives

$$\zeta(0) = \frac{i}{2(4\pi)^2} \int d^4x \sqrt{-g} \frac{\partial^2}{\partial(is)^2} [F(x,x;is)e^{-ism^2}]_{s=0}$$

$$= i(4\pi)^{-2} \int d^4x \sqrt{-g} [f_2(x,x) - m^2 f_1(x,x) + \frac{1}{2} m^4] , \qquad (5.37)$$

in agreement with Eq. (4.10). One obtains an explicit expression for $\zeta'(0)$ by taking $d/d\nu$ of Eq. (5.36) and setting $\nu = 0$. Hence using (see Eq. (4.9)),

$$\frac{d}{d\nu} [\Gamma(\nu+1)^{-1}]_{\nu=0} = \gamma = 0.577 \ldots , \qquad (5.38)$$

one finds

$$\zeta'(0) = -i(4\pi)^{-2}\Big\{(\frac{\gamma}{2} - \frac{1}{2} - \frac{1}{4}) \int d^4x \sqrt{-g} \Big[- \frac{\partial^2}{\partial(is)^2} (F(x,x;is)e^{-im^2 s})\Big]_{s=0}$$

$$+ \frac{1}{2} \int d^4x \sqrt{-g} \int_0^\infty ids(\ln is) \frac{\partial^3}{\partial(is)^3} [F(x,x;is)e^{-ism^2}]\Big\},$$

or

$$\zeta'(0) = \frac{i}{32\pi^2} \Big\{(\gamma - \frac{3}{2}) \int d^4x \sqrt{-g} [f_2(x,x) - m^2 f_1(x,x) + \frac{1}{2} m^2]$$

$$- \int d^4x \sqrt{-g} \int_0^\infty ids(\ln is) \frac{\partial^3}{\partial(is)^3} [F(x,x;is)e^{-ism^2}]\Big\} , \qquad (5.39)$$

which is clearly convergent. Therefore, Eq. (5.16) yields for the effective action,

$$W = \int d^4x \sqrt{-g}\,(64\pi^2)^{-1}\Big\{-\int_0^\infty ids(\ln is)\,\frac{\partial^3}{\partial(is)^3}\,[F(x,x;is)e^{-ism^2}]$$

$$+\,[\gamma-\tfrac{3}{2}+2\ln(-2\pi i\mu^2)][f_2(x,x)-m^2f_1(x,x)+\tfrac{1}{2}m^2]\Big\}, \quad (5.40)$$

where f_1 and f_2 are given in Eqs. (3.31) and (3.58), respectively. This expression for W is clearly convergent. The effective Lagrangian can be read off directly from Eq. (5.40). Because

$$-\int_0^\infty ids\,\frac{\partial^3}{\partial(is)^3}\,[F(x,x;is)e^{-ism^2}] = 2(f_2-m^2f_1+\frac{m^4}{2})\ ,$$

one can write Eq. (5.40) in the form

$$W = \int d^4x\sqrt{-g}\,L_{eff}\ , \quad (5.41)$$

with

$$L_{eff} = (64\pi^2)^{-1}\Big\{-\int_0^\infty ids(\ln isk^2)\,\frac{\partial^3}{\partial(is)^3}\,[F(x,x;is)e^{-ism^2}]$$

$$+\,(\gamma-\tfrac{3}{2})[f_2(x,x)-m^2f_1(x,x)+\frac{m^4}{2}]\Big\}\ , \quad (5.42)$$

where we have defined

$$k^2 \equiv -2\pi i\mu^2\ . \quad (5.43)$$

The coefficient $(\gamma-\frac{3}{2})$ of the last term of L_{eff} can be altered by renormalizing the cosmological and gravitational constants in the Einstein Lagrangian, and by adding terms quadratic in $R_{\alpha\beta\gamma\delta}$ to the Lagrangian. For the case $\xi = \frac{1}{6}$, the effective Lagrangian has been calculated using dimensional regularization by Brown (1977) and Brown and Cassidy (1977a), and agrees with Eq. (5.42) modulo a renormalization of the constant $(\gamma-\frac{3}{2})$ (when $\xi=\frac{1}{6}$, one has $f_1(x,x) = 0$).

Now let me turn to the conformally invariant scalar field ($\xi=\frac{1}{6}$, $m=0$) and show that its trace anomaly is given by Eq. (5.25) with no additional 4-divergence. Consider the change in W under an infinitesimal conformal transformation

$$g_{\mu\nu}(x) \rightarrow a(x)g_{\mu\nu}(x) \equiv \tilde{g}_{\mu\nu}(x) , \tag{5.44}$$

with

$$a(x) = 1 + \varepsilon(x) \quad \text{or} \quad \delta g_{\mu\nu}(x) = \varepsilon(x)g_{\mu\nu}(x) . \tag{5.45}$$

By regarding W as a functional of a(x), with $g_{\mu\nu}(x)$ being a given function, one obtains

$$W[(1+\varepsilon)g_{\mu\nu}] = W[g_{\mu\nu}] + \int d^4x \left.\frac{\delta W[ag_{\mu\nu}]}{\delta a(x)}\right|_{a=1} \varepsilon(x) ,$$

so that (as $\delta g_{\mu\nu} = \varepsilon g_{\mu\nu}$),

$$\frac{\delta W[g_{\mu\nu}]}{\delta g_{\mu\nu}(x)} g_{\mu\nu}(x) = \left.\frac{\delta W[ag_{\mu\nu}]}{\delta a(x)}\right|_{a=1} ,$$

and

$$<T_\mu{}^\mu(x)> = \frac{2}{\sqrt{-g}} \frac{\delta W}{\delta g_{\mu\nu}} g_{\mu\nu} = \frac{2}{\sqrt{-g(x)}} \left.\frac{\delta W[ag_{\mu\nu}]}{\delta a(x)}\right|_{a=1} , \tag{5.46}$$

where $\delta g_{\mu\nu}(x)$ and hence a(x) is not varied on the boundary of integration of W. For the case $\xi = \frac{1}{6}$, $m = 0$, one has $f_1(x,x) = 0$ and

$$\begin{aligned} f_2(x,x) &= \frac{1}{180} (R_{,\alpha}{}^{;\alpha} + R_{\alpha\beta\gamma\delta}R^{\alpha\beta\gamma\delta} - R_{\alpha\beta}R^{\alpha\beta}) \\ &= [\frac{1}{180} R_{,\alpha}{}^{;\alpha} + \frac{1}{120} (C_{\alpha\beta\gamma\delta}C^{\alpha\beta\gamma\delta} - \frac{1}{3} G)] , \end{aligned} \tag{5.47}$$

where Eqs. (5.26) and (5.29) have been used. The quantity $C_{\alpha\beta\gamma\delta}C^{\alpha\beta\gamma\delta} \sqrt{-g}$ is invariant under conformal transformation, while the remaining terms in Eq. (5.47) when integrated over the space-time form a topological invariant. Therefore, when $\xi = \frac{1}{6}$ and $m = 0$, the last term in Eq. (5.40) contributes nothing to $\delta W/\delta a(x)$, if one assumes that the normalization constant μ (or equivalently k) is invariant under conformation transformation.

Thus, for m = 0 and $\xi = \frac{1}{6}$, the only part of W which may vary under a conformal transformation of the metric is

$$W_{(1)} = -(64\pi^2)^{-1}\int d^4x\sqrt{-g(x)}\int_0^\infty ids\ \ln(k^2 is)\ \frac{\partial^3}{\partial(is)^3}F(x,x;is) \tag{5.48}$$

Because of Eq. (5.34) with m = 0, this can be rewritten

$$W_{(1)} = i(64\pi^2)^{-1}\,\mathrm{Tr}\int_0^\infty ids\ \ln(k^2 is)\ \frac{\partial^3}{\partial(is)^3}\left[(4\pi is)^2 e^{-isH}\right] . \tag{5.49}$$

Let $|\tilde{x}\rangle = a^{-1}(x)|x\rangle$, so that $1 = \int d^4x(-\tilde{g}(x))^{\frac{1}{2}}|\tilde{x}\rangle\langle\tilde{x}| = \int d^4x[-g(x)]^{\frac{1}{2}}|x\rangle\langle x|$, and $\widetilde{\mathrm{Tr}}(\ ) = \mathrm{Tr}(\ )$. Then the part of the effective action of interest, evaluated with the metric $\tilde{g}_{\mu\nu}$, will have the same form as Eq. (5.49) but with H replaced by $\tilde{H}$. Here the abstract operator $\tilde{H}$ is defined by

$$\tilde{H}(x)\delta(x-x') = [-\tilde{g}(x')]^{\frac{1}{2}}\langle\tilde{x}|\tilde{H}|\tilde{x}'\rangle = [-g(x')]^{\frac{1}{2}}\langle x|a^{-1}\tilde{H}a|x'\rangle, \tag{5.50}$$

where

$$\tilde{H}(x) = -\tilde{\nabla}^\mu\tilde{\nabla}_\mu + \frac{1}{6}\tilde{R} \tag{5.51}$$

with ~ denoting quantities formed from the metric $\tilde{g}_{\mu\nu}$ of Eq. (5.44), and $\tilde{\nabla}^\mu\tilde{\nabla}_\mu$ acting on a scalar. As was pointed out earlier

$$\tilde{H}(x) = a^{-3/2}(x)\ H(x)a^{\frac{1}{2}}(x) . \tag{5.52}$$

Therefore, using Eq. (2.6), one has $\tilde{H}(x)\delta(x-x') = a^{-3/2}(x)H(x)a^{\frac{1}{2}}(x)\delta(x-x') = a^{-3/2}(x)a^{\frac{1}{2}}(x')H(x)\delta(x-x') = [-g(x')]^{\frac{1}{2}}\langle x|a^{-3/2}Ha^{\frac{1}{2}}|x'\rangle$. Comparing this with Eq. (5.50), one finds that the abstract operators satisfy $\tilde{H} = a^{-\frac{1}{2}}Ha^{-\frac{1}{2}}$. Now

$$\mathrm{Tr}\ e^{-is\tilde{H}} = \mathrm{Tr}\ e^{-isa^{-\frac{1}{2}}Ha^{-\frac{1}{2}}} = \mathrm{Tr}\ e^{-isa^{-1}H} , \tag{5.53}$$

so that

$$\tilde{W}_{(1)} = i(64\pi^2)^{-1}\mathrm{Tr}\int_0^\infty ids\ \ln(k^2 is)\ \frac{\partial^3}{\partial(is)^3}\ [(4\pi is)^2\ e^{-isa^{-1}H}]$$

$$= i(64\pi^2)^{-1}\mathrm{Tr}\int_0^\infty ids'\ln(k^2 is'a)\ \frac{\partial^3}{\partial(is')^3}\ [(4\pi is')^2 e^{-is'H}], \tag{5.54}$$

where the change of variable $s' = sa^{-1}$ has been made (the eigenvalues of a are positive). Hence

$$\tilde{W}_{(1)} = W_{(1)} + i(64\pi^2)^{-1}\int d^4x\ \sqrt{-g(x)}\ \ln a(x)\int_0^\infty ids'$$

$$\times\ \frac{\partial^3}{\partial(is')^3}\ [(4\pi is')^2 \langle x|e^{-is'H}|x\rangle]$$

$$= W_{(1)} + (64\pi^2)^{-1}\int d^4x\ \sqrt{-g(x)}\ \ln a(x)\left[\frac{\partial^2}{\partial(is)^2}\ F(x,x;is)\right]_{s=0}$$

$$= W_{(1)} + (32\pi^2)^{-1}\int d^4x\ \sqrt{-g(x)}\ [\ln a(x)]f_2(x,x) \tag{5.55}$$

Therefore, according to Eq. (5.46) one has

$$\langle T_\mu{}^\mu(x)\rangle = (4\pi)^{-2}\ f_2(x,x)\ , \quad (\xi = \tfrac{1}{6},\ m = 0) \tag{5.56}$$

where f_2 is given by Eq. (5.47), in agreement with the trace anomaly of Eq. (5.25). This derivation of the trace anomaly is simpler than those generally found in the literature.

Finally, consider the question of conservation of the energy-momentum tensor obtained from the effective action by means of Eq. (4.7), i.e.

$$\langle T^{\mu\nu}(x)\rangle = 2(-g(x))^{-\frac{1}{2}}\delta W/\delta g_{\mu\nu}(x)\ . \tag{5.57}$$

As $F(x,x;is)$ is a scalar it is clear from Eq. (5.40) that W is invariant under coordinate transformations. The conservation of $\langle T^{\mu\nu}\rangle$ then follows in the usual way. Under an infinitesimal coordinate transformation

$$x^{\mu'} = x^{\mu} - \varepsilon\xi^{\mu}(x) , \tag{5.58}$$

where ξ^{μ} is a vector field, one has

$$0 = \delta W = \int d^4x \frac{\delta W}{\delta g_{\mu\nu}(x)} \delta g_{\mu\nu}(x)$$

$$= \frac{1}{2}\int d^4x \sqrt{-g(x)} <T^{\mu\nu}(x)>\delta g_{\mu\nu}(x) . \tag{5.59}$$

For this infinitesimal coordinate transformation, one finds that

$$\delta g_{\mu\nu}(x) = \varepsilon \mathcal{L}_{\xi} g_{\mu\nu}(x) = 2\varepsilon\nabla_{(\mu}\xi_{\nu)} , \tag{5.60}$$

where $\nabla_{(\mu}\xi_{\nu)} = \frac{1}{2}(\nabla_{\mu}\xi_{\nu} + \nabla_{\nu}\xi_{\mu})$, and $\mathcal{L}$ denotes the Lie derivative. Thus,

$$0 = \int d^4x \sqrt{-g(x)} <T^{\mu\nu}(x)>\nabla_{\mu}\xi_{\nu} = -\int d^4x \sqrt{-g(x)} (\nabla_{\mu}<T^{\mu\nu}>)\xi_{\nu} , \tag{5.61}$$

where because $<T^{\mu\nu}>\xi_{\nu}$ is a vector, we have used

$$(-g)^{-\frac{1}{2}}\partial_{\mu}(\sqrt{-g} <T^{\mu\nu}>\xi_{\nu}) = \nabla_{\mu}(<T^{\mu\nu}>\xi_{\nu})$$

$$= (\nabla_{\mu}<T^{\mu\nu}>)\xi_{\nu} + <T^{\mu\nu}>\nabla_{\mu}\xi_{\nu} , \tag{5.62}$$

and we have assumed that the vector field ξ_{ν} vanishes on the boundary. As $\xi_{\nu}(x)$ is otherwise arbitrary, it follows that

$$\nabla_{\mu}<T^{\mu\nu}> = 0 . \tag{5.63}$$

Finally, for the purpose of comparison, we close this section with an outline of the derivation of the effective action using the dimensional regularization method based on the proper-time formalism. One starts as before [see Eq. (5.15)] with

$$W = -i \ln \int d[\phi]e^{iS[\phi]} = \frac{i}{2} \ln \det[\frac{iH}{2\pi\mu^2}]$$

$$= \frac{i}{2} \mathrm{Tr} \ln [\frac{iH}{2\pi\mu^2}] . \tag{5.64}$$

Hence

$$\delta W = \frac{i}{2}\,\mathrm{Tr}\, H^{-1}\delta H = \frac{i}{2}\,\mathrm{Tr}\int_0^\infty ids\; e^{-isH}\delta H$$

$$= -\delta\Big[\frac{i}{2}\,\mathrm{Tr}\int_0^\infty \frac{ids}{is}\, e^{-isH}\Big] , \tag{5.65}$$

and one can let

$$W = -\frac{i}{2}\int d^n x\,\sqrt{-g(x)}\int_0^\infty \frac{ids}{is}\,\langle x|e^{-isH}|x\rangle$$

$$= \frac{1}{2}\int d^n x\,\sqrt{-g}\int_0^\infty \frac{ids}{is}\,(4\pi is)^{-n/2}F(x,x;is)e^{-ism^2} , \tag{5.66}$$

where the n-dimensional version of Eq. (5.34) was used. In passing from Eq. (5.64) to (5.65) the scale factor involving μ^2 has disappeared, and the possibility of additional terms in W which do not contribute to δW in Eq. (5.65) has been ignored (evidently μ^2 would be present in such terms). The scale factor will be reintroduced later. [The ζ-function approach made use of a definition of $\det(iH/2\pi\mu^2)$, and thus bypassed those steps.]

To separate out the pole at $n = 4$, one performs 3 integrations by parts, obtaining

$$W = -\frac{1}{2}(4\pi)^{-n/2}\Big(-\frac{n}{2}\Big)^{-1}\Big(-\frac{n}{2}+1\Big)^{-1}\Big(-\frac{n}{2}+2\Big)^{-1}\,\cdot$$

$$\cdot\int d^n x\,\sqrt{-g}\int_0^\infty ids\,(is)^{-\frac{n}{2}+2}\,\frac{\partial^3}{\partial(is)^3}\,[F(x,x;is)e^{-ism^2}] . \tag{5.67}$$

Now one introduces an arbitrary parameter k of dimension mass (or ℓ^{-1}), so as to make $k^2 s$ dimensionless, and writes

$$(is)^{-\frac{n}{2}+2} \rightarrow (isk^2)^{-\frac{n}{2}+2} = e^{(-\frac{n}{2}+2)\ln(isk^2)}$$

$$= 1 + \Big(-\frac{n}{2}+2\Big)\ln(isk^2) + O\Big[\Big(-\frac{n}{2}+2\Big)^2\Big] . \tag{5.68}$$

When this is substituted in W, the leading term can be integrated over s, and one finds

$$W = -(4\pi)^{-\frac{n}{2}}(-\tfrac{n}{2})^{-1}(-\tfrac{n}{2}+1)^{-1}(-\tfrac{n}{2}+2)^{-1}\int d^4x\,\sqrt{-g}\,\cdot$$

$$\cdot\,[f_2(x,x) - m^2 f_1(x,x) + \tfrac{1}{2}m^4]$$

$$-\tfrac{1}{2}(4\pi)^{-\frac{n}{2}}(-\tfrac{n}{2})^{-1}(-\tfrac{n}{2}+1)^{-1}\int d^nx\,\sqrt{-g}\int_0^\infty ids\,\ln(isk^2)\frac{\partial^3}{\partial(is)^3}\,\cdot$$

$$\cdot\,[F(x,x;is)e^{-ism^2}] + O(-\tfrac{n}{2}+2)\ , \tag{5.69}$$

where now W is dimensionless only at $n = 4$. The leading term, which diverges at $n = 4$ can be renormalized by adding counterterms to the action. Then, setting $n = 4$ gives

$$W = (\text{constant})\int d^4x\,\sqrt{-g}\,[f_2(x,x) - m^2 f_1(x,x) + \tfrac{1}{2}m^4]$$

$$-\tfrac{1}{4}(4\pi)^{-2}\int d^4x\cdot\sqrt{-g}\int_0^\infty ids\,\ln(isk^2)\,\frac{\partial^3}{\partial(is)^3}\,\cdot$$

$$\cdot\,[F(x,x;is)e^{-ism^2}]\ , \tag{5.70}$$

which agrees with the effective action of Eqs. (5.41)-(5.43), obtained from the ζ-function method. In the latter method, a factor of $\Gamma^{-1}(\nu)$ avoided the appearance of a divergent term.

6. APPLICATION OF TRACE ANOMALY TO PARTICLE CREATION AND VACUUM POLARIZATION IN ROBERTSON-WALKER UNIVERSES

A conformal Killing vector field ξ^μ by definition satisfies the equation

$$\mathcal{L}_\xi g_{\mu\nu}(x) = \lambda(x)g_{\mu\nu}(x)\ , \tag{6.1}$$

where

$$\mathcal{L}_\xi g_{\mu\nu} = \xi^\alpha\partial_\alpha g_{\mu\nu} + g_{\mu\alpha}\partial_\nu\xi^\alpha + g_{\alpha\nu}\partial_\mu\xi^\alpha = 2\nabla_{(\mu}\xi_{\nu)} \tag{6.2}$$

is the Lie derivative of the metric, and $\lambda(x)$ is a scalar. When a conformal Killing vector field exists and $\nabla_\mu T^{\mu\nu} = 0$, one has

$$\nabla_\mu (T^{\mu\nu}\xi_\nu) = T^{\mu\nu}\nabla_{(\mu}\xi_{\nu)} = \frac{1}{2}\lambda T^{\mu\nu}g_{\mu\nu} \quad . \tag{6.3}$$

(I thank R. Sorkin for pointing out to me this relation between the divergence of $T^{\mu\nu}\xi_\nu$ and the trace $T^\mu{}_\mu$.) In the Robertson-Walker universes, with

$$ds^2 = -dt^2 + a^2(t)h_{ij}(\vec{x})dx^i dx^j \ , \tag{6.4}$$

where $h_{ij}(\vec{x})dx^i dx^j$ is the line element of a 3-space of constant curvature, one has a time-like future directed conformal Killing vector field given by

$$\xi^\mu = a(t)\delta^\mu{}_o \tag{6.5}$$

with

$$\lambda = 2\dot{a}(t) \ . \tag{6.6}$$

Integrating the expectation value of Eq. (6.3) over the 4-volume bounded by constant time hypersurfaces at t_1 and t_2 gives

$$\frac{1}{2}\int d^4x \sqrt{-g}\ \lambda \langle T_\mu{}^\mu \rangle = \int_{t_2} d^3x \sqrt{-g} \langle T^{o\nu}\rangle \xi_\nu - \int_{t_1} d^3x \sqrt{-g} \langle T^{o\nu}\rangle \xi_\nu . \tag{6.7}$$

If $\langle T^{\mu\nu}\rangle$ refers to an expectation value in a state having the symmetry properties of the spacetime, then $\langle T_\mu{}^\mu\rangle$ and $\langle T^{oo}\rangle \equiv \rho$ are functions of t alone, and one has

$$\rho(t_2)a^4(t_2) = \rho(t_1)a^4(t_1) - \int_{t_1}^{t_2} dt\ a^3\dot{a}\langle T_\mu{}^\mu\rangle \ . \tag{6.8}$$

Here ρ is the proper energy-density measured by a comoving detector on the geodesic $\vec{x}$ = constant (in general, $T^{\mu\nu}V_\mu V_\nu$ is the proper-energy density measured by a detector having 4-velocity $V^\mu = dx^\mu/d\tau$, and for a comoving detector in these coordinates

$v^{\mu} = \delta^{\mu}{}_{o}$).

For conformally invariant massless fields one has

$$\langle T_{\mu}{}^{\mu}\rangle = (4\pi)^{-2}(AC_{\alpha\beta\gamma\delta}C^{\alpha\beta\gamma\delta} + BG + CR_{,\alpha}{}^{;\alpha}) , \qquad (6.9)$$

where the values of A, B, C for various spins were listed after Eq. (5.33). In the Robertson-Walker universes one has

$$C_{\alpha\beta\gamma\delta}C^{\alpha\beta\gamma\delta} = 0 , \qquad (6.10)$$

which is equivalent to

$$R_{\alpha\beta\gamma\delta}R^{\alpha\beta\gamma\delta} = 2\, R_{\alpha\beta}R^{\alpha\beta} - \frac{1}{3}R^2 . \qquad (6.11)$$

Therefore,

$$G = R_{\alpha\beta\gamma\delta}R^{\alpha\beta\gamma\delta} - 4\, R_{\alpha\beta}R^{\alpha\beta} + R^2$$

$$= \frac{2}{3}R^2 - 2\, R_{\alpha\beta}R^{\alpha\beta} . \qquad (6.12)$$

The non-vanishing components of $R_{\mu\nu}$ are

$$R_{oo} = -3a^{-1}\ddot{a}, \quad R_{ij} = (a\ddot{a} + 2\dot{a}^2 + 2\varepsilon)h_{ij}(\vec{x}) , \qquad (6.13)$$

where

$$\varepsilon = \begin{cases} +1 \text{ for positive spatial curvature (closed)} \\ \;\;0 \text{ for zero spatial curvature (flat)} \\ -1 \text{ for negative spatial curvature (open).} \end{cases} \qquad (6.14)$$

Thus,

$$R = 6(a^{-1}\ddot{a} + a^{-2}\dot{a}^2 + \varepsilon a^{-2}) , \qquad (6.15)$$

$$R_{,\alpha}{}^{;\alpha} = -\ddot{R} - 3a^{-1}\dot{a}\dot{R}$$

$$= 6(-a^{-1}\ddddot{a} - 3a^{-2}\dot{a}\dddot{a} + 5a^{-3}\dot{a}^2\ddot{a} - a^{-2}\ddot{a}^2 + 2\varepsilon a^{-3}\ddot{a}) , \qquad (6.16)$$

and

$$G = 24\,(a^{-3}\dot{a}^{2}\ddot{a} + \varepsilon a^{-3}\ddot{a}) \; . \tag{6.17}$$

It follows that for conformally invariant massless fields the integrand in the final term in Eq. (6.8) is a time derivative, so that the result depends only on t_1 and t_2, but not on the form of a(t) between those times. In particular, one has

$$\int dt\; a^{3}\dot{a}G = 6(\dot{a}^{4} + 2\varepsilon\dot{a}^{2}) + \text{const.} \; , \tag{6.18}$$

and

$$\int dt\; a^{3}\dot{a}R_{,\alpha}{}^{;\alpha} = 6(-a^{2}\dot{a}\dddot{a} - a\dot{a}^{2}\ddot{a} + \frac{1}{2}a^{2}\ddot{a}^{2} + \frac{3}{2}\dot{a}^{4} + \varepsilon\dot{a}^{2}) + \text{const.} \tag{6.19}$$

Hence

$$\int_{t_1}^{t_2} dt\; a^{3}\dot{a}\,\langle T_{\mu}{}^{\mu}\rangle = g(t_2) - g(t_1) \; , \tag{6.20}$$

where

$$g(t) = \frac{6}{(4\pi)^2}\,[B(\dot{a}^{4} + 2\varepsilon\dot{a}^{2}) + C(-a^{2}\dot{a}\dddot{a} - a\dot{a}^{2}\ddot{a} + \frac{1}{2}a^{2}\ddot{a}^{2} + \frac{3}{2}\dot{a}^{4} + \varepsilon\dot{a}^{2})] \; . \tag{6.21}$$

A statically bounded (or asymptotically static) expansion is one for which a(t) approaches constant values at early and at late times. In such an expansion g(t) vanishes at early and late times, and it follows from Eq. (6.8) that the energy-density present initially is simply red-shifted, with no production of real particles resulting from the expansion. (Real particles produced by the expansion would presumably be left over if the expansion were very gradually halted.) This is consistent with the independent proof (Parker 1966, 1968, 1969) based on the conformal invariance of the field equations, and also brings up to date the considerations based on $\langle T_{\mu}{}^{\mu}\rangle$ in Parker (1975, 1977). Thus, for conformally invariant fields in Robertson-Walker universes, one has $|0,\mathrm{out}\rangle \propto |0,\mathrm{in}\rangle$, so that $\langle T^{\mu\nu}\rangle$, and hence ρ, correspond to expectation values in a given state.

From Eqs. (6.8) and (6.20), one can obtain the explicit expressions for the average energy-density ρ and pressure p of these fields. Thus, one has

$$\rho(t_2)a^4(t_2) + g(t_2) = \rho(t_1)a^4(t_1) + g(t_1) = E \ , \qquad (6.21)$$

where E is a constant which depends on the choice of state. Hence

$$\begin{aligned}\rho(t) &= -a^{-4}(t)g(t) + a^{-4}(t)E \\ &= -\frac{6}{(4\pi)^2}\, a^{-4}[B(\dot{a}^4 + 2\varepsilon\dot{a}^2) + C(-a^2\dot{a}\dddot{a} - a\dot{a}^2\ddot{a} \\ &\qquad + \frac{1}{2}a^2\ddot{a}^2 + \frac{3}{2}\dot{a}^4 + \varepsilon\dot{a}^2)] + Ea^{-4} \ . \qquad (6.22)\end{aligned}$$

Here B and C are the constants appearing in Eq. (6.9), and discussed after Eq. (5.33), while ε is defined in Eq. (6.14). The corresponding pressure is given by

$$p = \frac{1}{3}(\rho + \langle T_\mu{}^\mu\rangle) \ , \qquad (6.23)$$

with $\langle T_\mu{}^\mu\rangle$ obtained from Eqs. (6.9), (6.10), (6.16), and (6.17). These expressions for ρ and p agree with those obtained in a different way by Bunch (1977). The terms involving E in ρ and p behave like a classical gas of massless particles. The other terms vanish during any period when a(t) becomes static, and therefore cannot correspond to the accumulation of real particles created by the expansion. Those terms can be thought of as due to virtual pairs or vacuum polarization. In a static universe, the constant E would correspond to the vacuum energy calculated by a number of workers [Ford (1975, 1976), Dowker and Critchley (1976b, 1977b), Dowker and Altaie (1978), Gibbons (1977)].

For $\xi \neq \frac{1}{6}$, Eq. (5.23) can be written in the form

$$\langle T_\mu{}^\mu \rangle = (4\pi)^{-2}[AC_{\alpha\beta\gamma\delta}C^{\alpha\beta\gamma\delta} + BG + CR_{,\alpha}{}^{;\alpha} + DR^2]$$

$$+ \text{4-divergence} , \qquad (6.24)$$

with

$$A = \frac{1}{120} , \; B = -\frac{1}{360} , \; C = \frac{1}{6}\left(\frac{1}{5} - \xi\right), \; D = \frac{1}{2}\left(\frac{1}{6} - \xi\right)^2 . \qquad (6.25)$$

The term

$$(4\pi)^{-2}\int_{t_1}^{t_2} dt\; a^3\dot{a}R^2 = (4\pi)^{-2}36\int_{t_1}^{t_2} dt\; a^3\dot{a}(a^{-1}\ddot{a} + a^{-2}\dot{a}^2 + \varepsilon a^{-2})^2 , \qquad (6.26)$$

which appears in

$$\int_{t_1}^{t_2} dt\; a^3\dot{a}\langle T_\mu{}^\mu \rangle ,$$

depends on the behavior of a(t) during the interval from t_1 to t_2, and does not vanish for a statically bounded expansion. Thus, it is clear from Eq. (6.8) that the presence of the R^2 term in $\langle T_\mu{}^\mu \rangle$ reflects the fact that real particle creation is occuring during the expansion (or contraction). In that case $|0,\text{out}\rangle$ is not proportional to $|0,\text{in}\rangle$, and $\langle T^{\mu\nu} \rangle$ obtained from the variation of W is not an expectation value in a given state.

7. PATH INTEGRAL REPRESENTATIONS OF THE PROPAGATOR FOR ARBITRARY COUPLING TO THE SCALAR CURVATURE

In the proper-time representation recall that

$$G(x,x') = \int_0^\infty i ds\; \langle x,s|x',0\rangle , \qquad (7.1)$$

where $\langle x,s|x',0\rangle$ satisfies the Schrödinger equation, (2.16), with the boundary condition of Eq. (2.17). In this Section, we consider the Feynman path integral representations of $\langle x,s|x',0\rangle$ and thus of G(x,x'). For comparison with the literature, it is convenient to introduce a dimensionless parameter μ, and replace Eq.

(2.16) by

$$i \frac{\partial}{\partial s} \langle x,s|x',0\rangle = [-\frac{1}{2\mu} \nabla_\alpha \nabla^\alpha + \frac{1}{2\mu} \xi R + m^2]\langle x,s|x',0\rangle. \quad (7.2)$$

[Eq. (2.16) is recovered if $\mu = 1/2$.] This has the form of the Schrödinger equation of a particle moving in a 4-dimensional hypersurface, with $\langle x,s|x',0\rangle$ representing the probability amplitude for the fictitious particle to propagate from x' at $s' = 0$ to x at $s' = s$.

When the 4-space is flat, the classical Lagrangian whose Hamiltonian gives the above Schrödinger equation is

$$L = \frac{1}{2} \mu \, g_{\alpha\beta} \frac{dx^\alpha}{ds} \frac{dx^\beta}{ds} - m^2 , \quad (7.3)$$

and the Feynman path integral solution of Eq. (7.2) with the boundary condition of Eq. (2.17) is

$$\langle x,s|x',0\rangle = \int d[x(s')]\exp[i\int_0^s ds' \, L(x,dx/ds)] \, . \quad (7.4)$$

Here the functional integration is over all paths from x' at $s'=0$ to x at $s'=s$, and $d[x(s')]$ denotes the integration measure. The path integral can be given a more precise meaning in the usual way (Feynman and Hibbs 1969, for an alternative way of defining path integrals see C. DeWitt-Morette 1972, 1974) by dividing the interval $[0,s]$ into $N+1$ equal segments of length ε, with $x^\alpha(n\varepsilon) \equiv x^\alpha_n$ $(n = 0,1,\ldots,N+1)$, $x_o = x'$, $x_{N+1} = x$, and $(N+1)\varepsilon = s$. Then

$$\langle x,s|x',0\rangle = \lim_{N\to\infty} [\frac{1}{i} (\frac{\mu}{2\pi i\varepsilon})^2]^{N+1} \int \prod_{j=1}^{N} \left\{d^4x_j[-g(x_j)]^{\frac{1}{2}}\right\} \cdot$$

$$\cdot \exp\left[i \sum_{\ell=0}^{N} \int_{\ell\varepsilon}^{(\ell+1)\varepsilon} ds' \, L(x,dx/ds')\right] , \quad (7.5)$$

where each integral in the exponential is evaluated along the geodesic path connecting x_ℓ to $x_{\ell+1}$. [In n dimensions each factor

$(\mu/2\pi i\varepsilon)^2$ is replaced by $(\mu/2\pi i\varepsilon)^{n/2}$. The factor 1/i in the normalization is needed because $\eta_{oo} = -1$. The Euclideanized expressions can readily be obtained from those given here.] The various terms in Eq. (7.5) are manifestly invariant under coordinate transformations of the x^μ, and by going to rectangular coordinates ($g_{\mu\nu} = \eta_{\mu\nu}$) it is clear that Eq. (7.5) is the well known flat space expression written in curvilinear coordinates. The corresponding expression for the Green function,

$$G(x,x') = \int_0^\infty ids\ e^{-im^2 s}\int d[x(s')]\exp\left[i\,\frac{1}{2}\,\mu\int_0^s ds' g_{\alpha\beta}\frac{dx^\alpha}{ds'}\frac{dx^\beta}{ds'}\right], \tag{7.6}$$

can be shown to correspond to the usual flat space Feynman propagator (propagating positive frequencies into the future and negative frequences into the past) if the replacement $\exp(-im^2 s) \to \exp(-im^2 s - s^{-1}\delta)$ is made, and the small positive quantity δ is taken to zero at the end of the calculation.

In curved spacetime it can be shown (DeWitt 1957, Cheng 1972, Hartle and Hawking 1976) that the path integral expression of Eq. (7.4) satisfies Eq. (7.2) with $\xi = 1/3$. (DeWitt and Cheng had $\mu = 1$, while Hartle and Hawking had $\mu = 1/2$.) Thus, for example, with $\mu = 1/2$, the Green function satisfying

$$(-\nabla_\mu\nabla^\mu + \frac{1}{3}R + m^2)G(x,x') = [-g(x)]^{-\frac{1}{2}}\delta(x-x') \tag{7.7}$$

is

$$G(x,x') = \int_0^\infty ids\ e^{-im^2 s}\int d[x(s')]\exp\left[i\,\frac{1}{4}\int_0^s ds' g_{\alpha\beta}\frac{dx^\alpha}{ds'}\frac{dx^\beta}{ds'}\right], \tag{7.8}$$

where it is understood that $\exp(-im^2 s) \to \exp(-im^2 s - s^{-1}\delta)$. This gives a natural generalization of the Feynman propagator to curved spacetime, at least when the spacetime is asymptotically flat. However, Eq. (7.7) corresponds neither to the minimal ($\xi = 0$) nor the conformal ($\xi = 1/6$) coupling. The Euclideanized (or Riemannianized) version of Eq. (7.8) arrived at through a different approach, was applied by Hartle and Hawking (1976) to the Schwarzschild black hole (for which R vanishes), and by Chitre and Hartle (1977) to a cosmological metric.

In the appendix of Hartle and Hawking (1976), it was pointed out that if one replaces each factor of

$$\exp[i\int_{\ell\varepsilon}^{(\ell+1)\varepsilon} ds'\ L]$$

in Eq. (7.5) by

$$[g(x_{\ell+1})/g(x_\ell)]^{p/2}\ \exp[i\int_{\ell\varepsilon}^{(\ell+1)\varepsilon} ds'\ L]\ ,$$

then $\langle x,s|x',0\rangle$ would satisfy Eq. (7.2) with $\xi = (1-p)/3$. However, the expression obtained in that way for $\langle x,s|x',0\rangle$ does not transform correctly (i.e. as a scalar) under coordinate transformations. I will show below that a path integral expression for $\langle x,s|x',0\rangle$ satisfying Eq. (7.2) with $\xi = (1-p)/3$ and also having the correct transformation properties, is obtained from Eq. (7.5) by replacing each factor of

$$\exp[i\int_{\ell\varepsilon}^{(\ell+1)\varepsilon} ds'\ L]$$

by

$$[\Delta(x_{\ell+1},x_\ell)]^p\ \exp[i\int_{\ell\varepsilon}^{(\ell+1)\varepsilon} ds'\ L]\ ,$$

where $\Delta(x_{\ell+1},x_\ell)$ is the biscalar defined in Eq. (3.17). The boundary condition of Eq. (2.17),

$$\lim_{s\to 0}\ \langle x,s|x',0\rangle = (-g)^{\frac{1}{2}}\delta(x-x') \qquad (7.9)$$

is also satisfied.

The proposed path integral solution of Eq. (7.2) can be written as

$$\langle x,s|x',0\rangle = \lim_{N\to\infty} \left[\frac{1}{i}\left(\frac{\mu}{2\pi i\varepsilon}\right)^2\right]^{N+1} \int \prod_{j=1}^{N} \left\{d^4x_j[-g(x_j)]^{\frac{1}{2}}\right\} \cdot$$

$$\cdot \exp\left\{\sum_{\ell=0}^{N}\left[i\,\frac{1}{2}\,\mu\int_{\ell\varepsilon}^{(\ell+1)\varepsilon} ds' g_{\alpha\beta}\frac{dx^\alpha}{ds'}\frac{dx^\beta}{ds'} + p\ln\Delta(x_{\ell+1},x_\ell)\right]\right\}, \tag{7.10}$$

with

$$p = 1 - 3\xi\ , \tag{7.11}$$

and $x_o = x'$, $x_{N+1} = x$. One can prove that Eq. (7.10) satisfies Eq. (7.2) as follows. From Eq. (7.10) one has

$$\langle x,s+\varepsilon|x',0\rangle = \frac{1}{i}\left(\frac{\mu}{2\pi i\varepsilon}\right)^2\int d^4x_{N+1}[-g(x_{N+1})]^{\frac{1}{2}}\, e^{-im^2\varepsilon}\ \cdot$$

$$\cdot\ [\Delta(x,x_{N+1})]^p \exp\left[i\,\frac{1}{2}\,\mu\int_{(N+1)\varepsilon}^{(N+2)\varepsilon} ds' g_{\alpha\beta}\frac{dx^\alpha}{ds'}\frac{dx^\beta}{ds'}\right]\langle x_{N+1},s|x',0\rangle, \tag{7.12}$$

where now $x_{N+2} = x$, and lim $N\to\infty$, $\varepsilon\to 0$ is understood. Introduce Riemann normal coordinates with origin at x, and let y^μ be the coordinates of the point x_{N+1}. [Riemann normal coordinates are described in Eqs. (3.1)-(3.6).] It appears to be justified to assume that the coordinates are good out to the point x_{N+1}, as the main contribution to the integration comes from x_{N+1} close to x as $\varepsilon\to 0$. In these coordinates the geodesic from x_{N+1} to x is linear, and one has

$$\int_{(N+1)\varepsilon}^{(N+2)\varepsilon} ds' g_{\alpha\beta}\frac{dx^\alpha}{ds'}\frac{dx^\beta}{ds'} = \varepsilon^{-1}\eta_{\alpha\beta}y^\alpha y^\beta = \varepsilon^{-1}\tau^2\ . \tag{7.13}$$

One also has [see the discussion of Eqs. (3.15)-(3.17)] in these coordinates,

$$\Delta(x,x_{N+1}) = [-g(y)\ ]^{-\frac{1}{2}}\ , \tag{7.14}$$

where y is the coordinate of the point x_{N+1}. Thus, Eq. (7.12) becomes

$$<x,s+\varepsilon|x',0> = \frac{1}{i}\left(\frac{\mu}{2\pi i\varepsilon}\right)^2\int d^4y\ e^{-im^2\varepsilon}[-g(y)]^{\frac{1}{2}(1-p)}\ \cdot$$

$$\cdot\ \exp[i\ \tfrac{1}{2}\ \mu\varepsilon^{-1}\eta_{\alpha\beta}y^\alpha y^\beta]<x_{N+1},s|x',0>\ . \tag{7.15}$$

From Eq. (3.27),

$$[-g(y)]^{\frac{1}{2}(1-p)} = 1-\frac{1}{6}\ (1-p)R_{\mu\lambda}y^\mu y^\lambda + \ldots \tag{7.16}$$

As the origin of these coordinates is at x, we also have

$$<x_{N+1},s|x',0> = <x,s|x',0> + y^\lambda\partial_\lambda<x_{N+1},s|x',0>\Big|_{y=0}$$

$$+\ \frac{1}{2}\ y^\lambda y^\mu\partial_\lambda\partial_\mu<x_{N+1},s|x',0>\Big|_{y=0} + \ \ldots\ . \tag{7.17}$$

Also in these coordinates with origin at x we have

$$<x_{N+1},s|x',0> = <x,s|x',0> + y^\lambda\ \frac{\partial}{\partial y^\lambda}\ <x_{N+1},s|x',0>\Big|_{y=0}$$

$$+\ \frac{1}{2}\ y^\lambda y^\mu\ \frac{\partial}{\partial y^\lambda}\ \frac{\partial}{\partial y^\mu}\ <x_{N+1},s|x',0>\Big|_{y=0} + \ \ldots\ . \tag{7.18}$$

Because of the exponential involving τ^2/ε in Eq. (7.15), the contributions to the integral come mainly from the range in which $y^2 \lesssim \varepsilon$ (assuming $\mu \sim 1$). Therefore, we can extend the range of integration over y from $-\infty$ to ∞ and carry out the integrations explicitly (the terms of interest which are of order ε will come from the terms of order y^2 in the integrand).

The Gaussian integrals yield

$$\int_{-\infty}^{\infty} d^4y\ \exp[i\ \tfrac{1}{2}\ \mu\varepsilon^{-1}\eta_{\alpha\beta}y^\alpha y^\beta] = i\left(\frac{2\pi i\varepsilon}{\mu}\right)^2\ , \tag{7.19}$$

and

$$\int_{-\infty}^{\infty} d^4y\ y^\mu y^\lambda\exp[i\ \tfrac{1}{2}\ \mu\varepsilon^{-1}\eta_{\alpha\beta}y^\alpha y^\beta] = i\left(\frac{2\pi i\varepsilon}{\mu}\right)^2\eta^{\mu\lambda}\left(\frac{i\varepsilon}{\mu}\right)\ . \tag{7.20}$$

The integrals involving odd numbers of y's vanish by symmetry. Thus, Eq. (7.15) becomes

$$\langle x,s+\varepsilon|x',0\rangle = \int \frac{1}{i}\left(\frac{\mu}{2\pi i\varepsilon}\right)^2 d^4y\ e^{-im^2\varepsilon}[1-\frac{1}{6}(1-p)R_{\mu\lambda}y^\mu y^\lambda + \ldots]\ \cdot$$

$$\cdot\ \exp[i\,\frac{1}{2}\,\mu\varepsilon^{-1}\eta_{\alpha\beta}y^\alpha y^\beta][\langle x,s|x',0\rangle + y^\lambda\partial_\lambda\langle x,s|x',0\rangle$$

$$+\ \frac{1}{2}\,y^\lambda y^\mu\partial_\lambda\partial_\mu\langle x,s|x',0\rangle + \ldots]$$

$$= e^{-im^2\varepsilon}\Big\{\langle x,s|x',0\rangle + \frac{i\varepsilon}{\mu}\,\eta^{\mu\lambda}[\frac{1}{2}\,\partial_\mu\partial_\lambda\langle x,s|x',0\rangle$$

$$-\ \frac{1}{6}\,(1-p)R_{\mu\lambda}\langle x,s|x',0\rangle] + O(\varepsilon^2)\Big\}\ . \tag{7.21}$$

Expanding $\langle x,s+\varepsilon|x,0\rangle$ and $\exp(-im^2\varepsilon)$ in powers of ε, and equating the coefficients of equal powers of ε on each side of Eq. (7.21), confirms the normalization, and gives for the terms of order ε the equation

$$\frac{\partial}{\partial s}\langle x,s|x',0\rangle = \frac{i}{\mu}\,[\frac{1}{2}\,\eta^{\mu\lambda}\partial_\mu\partial_\lambda - \frac{1}{6}\,(1-p)\eta^{\mu\lambda}R_{\mu\lambda}]\langle x,s|x',0\rangle$$

$$-\ im^2\langle x,s|x',0\rangle\ , \tag{7.22}$$

where the derivatives are evaluated at the point x (i.e., $y=0$). In explicitly covariant forms valid for arbitrary coordinates, this can be written as

$$i\,\frac{\partial}{\partial s}\langle x,s|x',0\rangle = [-\frac{1}{2\mu}\,\nabla^\mu\nabla_\mu + \frac{1}{6\mu}\,(1-p)R + m^2]\langle x,s|x',0\rangle\ , \tag{7.23}$$

which is Eq. (7.2) with

$$\xi = \frac{1}{3}\,(1-p)\ . \tag{7.24}$$

This completes the proof that Eq. (7.12) satisfies Eq. (7.2) for arbitrary values of ξ.

For convenience, I will abbreviate Eq. (7.12) as

$$\langle x,s|x',0\rangle = e^{-im^2s}\int d[x(s')][\Delta^p]\exp\left[i\,\frac{1}{2}\,\mu\int_0^s ds' g_{\alpha\beta}\,\frac{dx^\alpha}{ds'}\,\frac{dx^\beta}{ds'}\right]\ , \tag{7.25}$$

where $[\Delta^p]$ indicates that a factor of $[\Delta(x_{\ell+1},x_\ell)]^p$ goes with each factor of $\exp[i\,\frac{1}{2}\,\mu\int_{\ell\varepsilon}^{(\ell+1)\varepsilon} ds' g_{\alpha\beta}\dot{x}^\alpha\dot{x}^\beta]$ in the definition of the path integral.

Thus, putting $\mu = 1/2$ in Eq. (7.25), one can write the generalized Feynman propagator satisfying Eq. (2.3), i.e.,

$$(-\nabla^\mu\nabla_\mu + \xi R + m^2)G(x,x') = [-g(x)]^{-\frac{1}{2}}\delta(x-x') , \tag{7.26}$$

in the manifestly covariant path integral form

$$G(x,x') = \int_0^\infty ids\; e^{-im^2 s}\int d[x(s')][\Delta^{(1-3\xi)}] \cdot$$
$$\cdot \exp\left[i\,\frac{1}{4}\int_0^s ds' g_{\alpha\beta}\,\frac{dx^\alpha}{ds'}\,\frac{dx^\beta}{ds'}\right] , \tag{7.27}$$

where it is understood that $\exp(-im^2 s) \to \exp(-im^2 s - s^{-1}\delta)$ with $\delta \to 0^+$ at the end of the calculation. This expression is valid for any value of ξ.

It is surprising that one can find other path integral expressions for the amplitude $\langle x,s|x',0\rangle$ satisfying Eq. (7.2). For example, one has

$$\langle x,s|x',0\rangle = e^{-im^2 s}\int d[x(s')]\exp\left\{i\int_0^s ds'\left[\frac{1}{2}\,\mu g_{\alpha\beta}\,\frac{dx^\alpha}{ds'}\,\frac{dx^\beta}{ds'}\right.\right.$$
$$\left.\left. - \frac{1}{2\mu}\,(\xi - \frac{1}{3})R\right]\right\} . \tag{7.28}$$

To prove that Eq. (7.28) satisfies Eq. (7.2), one writes

$$\langle x,s+\varepsilon|x',0\rangle = e^{-im^2\varepsilon}\,\frac{1}{i}\,(\frac{\mu}{2\pi i\varepsilon})^2\int d^4x_{N+1}[-g(x_{N+1})]^{\frac{1}{2}} \cdot$$
$$\cdot \exp\left\{i\int_{(N+1)\varepsilon}^{(N+2)\varepsilon} ds'\left[\frac{1}{2}\,\mu g_{\alpha\beta}\,\frac{dx^\alpha}{ds'}\,\frac{dx^\beta}{ds'} - \frac{1}{2\mu}\,(\xi - \frac{1}{3})R\right]\right\}\langle x_{N+1},s|x',0\rangle, \tag{7.29}$$

which is analogous to Eq. (7.12). Here $x_{N+2} = x$ and $s = (N+1)\varepsilon$. As before, introduce Riemann normal coordinates y^μ with origin at x.

The geodesic from x_{N+1} to x is given by the equation

$$y^{\mu}(s') = y^{\mu} - \varepsilon^{-1} y^{\mu}[s' - (N+1)\varepsilon] \ , \tag{7.30}$$

where the y^{μ} are the coordinates of point x_{N+1}. Expanding $R(y(s'))$ about $y = 0$ and integrating along the geodesic from x_{N+1} to x, one finds

$$\int_{(N+1)\varepsilon}^{(N+2)\varepsilon} ds' R(y(s')) = \varepsilon[R + \frac{1}{2} y^{\mu} R_{;\mu} + O(y^2)] \ , \tag{7.31}$$

where R and $R_{;\mu}$ are evaluated at the point x. The rest of the proof is analogous to Eqs. (7.15) through Eq. (7.23), but with the present expression, and need not be repeated here. More generally (Parker 1978) one can show in the same way that the following path integral satisfies the Schrodinger equation (7.2):

$$\langle x,s|x',0\rangle = e^{-im^2 s}\int d[x(s')][\Delta^P]\exp i\int_0^s ds'\left\{\frac{1}{2}\mu g_{\alpha\beta}\frac{dx^{\alpha}}{ds'}\frac{dx^{\beta}}{ds'}\right.$$

$$\left. - \frac{1}{2\mu}[\xi + \frac{1}{3}(p-1)]R\right\} \ , \tag{7.32}$$

where now p is arbitrary [Eq. (7.25) corresponds to the choice $p = 1 - 3\xi$, and Eq. (7.28) to $p = 0$]. The amplitude $\langle x,s|x',0\rangle$ given by Eq. (7.32) is independent of the value of p because it satisfies Eq. (7.2) and the boundary condition of Eq. (7.9).

To show that the above boundary condition is satisfied, write Eq. (7.32) as

$$\langle x,s|x',0\rangle = \lim_{N\to\infty}\int\prod_{j=1}^{N}\left\{d^4x_j[-g(x_j)]^{\frac{1}{2}}\right\} \cdot$$

$$\cdot \langle x,s|x_N,N\varepsilon\rangle\langle x_N,N\varepsilon|x_{N-1},(N-1)\varepsilon\rangle \ldots \langle x_1,\varepsilon|x',0\rangle \ , \tag{7.33}$$

with

$$<x_{\ell+1},(\ell+1)\varepsilon|x_\ell,\ell\varepsilon> = \frac{1}{i}\left(\frac{\mu}{2\pi i\varepsilon}\right)^2 [\Delta(x_{\ell+1},x_\ell)]^p e^{-im^2\varepsilon} \cdot$$

$$\cdot \exp\left[i\int_{\ell\varepsilon}^{(\ell+1)\varepsilon} ds'\left(\frac{1}{2}\mu g_{\alpha\beta}\frac{dx^\alpha}{ds'}\frac{dx^\beta}{ds'} - \frac{\lambda}{2\mu}R\right)\right] , \qquad (7.34)$$

and

$$\lambda = \xi + \frac{1}{3}(p-1) . \qquad (7.35)$$

Working in Riemann normal coordinates at x_1 and using the previously given expansions about $y = 0$, one finds for an arbitrary function $f(x)$ that

$$\int d^4x'[-g(x')]^{\frac{1}{2}}<x_1,\varepsilon|x',0>f(x')$$

$$= \frac{1}{i}\left(\frac{\mu}{2\pi i\varepsilon}\right)^2\int d^4y \exp[i\frac{1}{2}\mu\varepsilon^{-1}\eta_{\alpha\beta}y^\alpha y^\beta][1-\frac{1}{6}(1-p)R_{\mu\lambda}y^\mu y^\lambda+\dots]\cdot$$

$$\cdot [1-im^2\varepsilon+\dots][1-i\frac{\varepsilon\lambda}{2\mu}R+\dots][f(x_1)+\varepsilon y^\mu\partial_\mu f(x_1)+\dots]$$

$$= f(x_1) + O(\varepsilon) . \qquad (7.36)$$

Similarly,

$$\int d^4x[-g(x_\ell)]^{\frac{1}{2}}<x_{\ell+1},(\ell+1)\varepsilon|x_\ell,\ell\varepsilon>f(x_\ell) = f(x_{\ell+1}) + O(\varepsilon). \qquad (7.37)$$

Therefore, using Eq. (7.33) and Eq. (7.37) repeatedly, one finds

$$\int d^4x'[-g(x')]^{\frac{1}{2}}<x,s|x',0>f(x') = \lim_{N\to\infty}\{f(x) + O(N\varepsilon)\}$$

$$= f(x) + O(s) , \qquad (7.38)$$

since $s = (N+1)\varepsilon$. Taking the limit $s\to 0$ now yields the boundary condition of Eq. (7.9).

APPENDIX: DERIVATION OF EQ. (3.9)

With ϕ a function of τ alone, consider

$$\nabla_\mu\nabla^\mu(\phi U) = |g|^{-\frac{1}{2}}\partial_\mu[|g|^{\frac{1}{2}}g^{\mu\nu}\partial_\nu(\phi U)]$$

$$= U\left\{g^{\mu\nu}\partial_\mu\partial_\nu\phi + |g|^{-\frac{1}{2}}[\partial_\mu(|g|^{\frac{1}{2}}g^{\mu\nu})]\partial_\nu\phi\right\}$$

$$+ 2g^{\mu\nu}\partial_\mu\phi\partial_\nu U + \phi\nabla_\mu\nabla^\mu U \; . \tag{A.1}$$

In Riemann normal coordinates y^μ, in which Eqs. (3.1)-(3.6) are satisfied, one can express τ and ξ^μ as functions of the y^μ by

$$\tau = (\eta_{\mu\nu}y^\mu y^\nu)^{\frac{1}{2}} \tag{A.2}$$

and

$$\xi^\lambda = y^\lambda/\tau = y^\lambda(\eta_{\mu\nu}y^\mu y^\nu)^{-\frac{1}{2}} \; , \tag{A.3}$$

or

$$\frac{\partial\tau}{\partial y^\mu} = \eta_{\mu\nu}\xi^\nu \; , \tag{A.4}$$

and

$$\frac{\partial\xi^\lambda}{\partial y^\mu} = \tau^{-1}\delta^\lambda{}_\mu - \tau^{-1}\eta_{\mu\nu}\xi^\nu\xi^\lambda \; . \tag{A.5}$$

Hence

$$\frac{\partial}{\partial y^\mu} = \frac{\partial\tau}{\partial y^\mu}\frac{\partial}{\partial\tau} + \frac{\partial\xi^\lambda}{\partial y^\mu}\frac{\partial}{\partial\xi^\lambda}$$

$$= \eta_{\mu\alpha}\xi^\alpha\frac{\partial}{\partial\tau} + \tau^{-1}\frac{\partial}{\partial\xi^\mu} - \tau^{-1}\eta_{\mu\alpha}\xi^\alpha\xi^\lambda\frac{\partial}{\partial\xi^\lambda} \; . \tag{A.6}$$

Because ϕ is taken to be a function of τ alone one has

$$\partial_\mu\phi(\tau) = \eta_{\mu\alpha}\xi^\alpha d\phi/d\tau \; , \tag{A.7}$$

and

$$\partial_\nu\partial_\mu\phi = \eta_{\mu\alpha}\xi^\alpha\eta_{\nu\beta}\xi^\beta d^2\phi/d\tau^2$$

$$+ \tau^{-1}\eta_{\mu\nu}d\phi/d\tau - \tau^{-1}\eta_{\mu\alpha}\eta_{\nu\beta}\xi^\beta\xi^\alpha d\phi/d\tau . \qquad (A.8)$$

As a consequence of Eqs. (3.1), (3.2), and (3.6), one has

$$g^{\mu\nu}\eta_{\mu\alpha}\xi^\alpha\eta_{\nu\beta}\xi^\beta = 1 , \qquad (A.9)$$

and hence

$$g^{\mu\nu}\partial_\mu\partial_\nu\phi = d^2\phi/d\tau^2 - \tau^{-1}d\phi/d\tau + \tau^{-1}g^{\mu\nu}\eta_{\mu\nu}d\phi/d\tau . \qquad (A.10)$$

Next consider the second term in Eq. (A.1). One has

$$\partial_\mu(|g|^{\frac{1}{2}}g^{\mu\nu})\partial_\nu\phi = \partial_\mu(|g|^{\frac{1}{2}}g^{\mu\nu})\eta_{\nu\alpha}\xi^\alpha d\phi/d\tau$$

$$= \partial_\mu(|g|^{\frac{1}{2}}g^{\mu\nu}\eta_{\nu\alpha}\xi^\alpha)d\phi/d\tau$$

$$- |g|^{\frac{1}{2}}g^{\mu\nu}\eta_{\nu\alpha}\partial_\mu\xi^\alpha d\phi/d\tau$$

$$= (\partial_\mu|g|^{\frac{1}{2}})\xi^\mu d\phi/d\tau + |g|^{\frac{1}{2}}\partial_\mu\xi^\mu d\phi/d\tau$$

$$- |g|^{\frac{1}{2}}g^{\mu\nu}\eta_{\mu\nu}\tau^{-1}d\phi/d\tau$$

$$+ |g|^{\frac{1}{2}}g^{\mu\nu}\eta_{\nu\alpha}\tau^{-1}\eta_{\mu\beta}\xi^\beta\xi^\alpha d\phi/d\tau \qquad (A.11)$$

From Eqs. (A.6) and (A.9) one has

$$\xi^\mu\partial_\mu = \partial/\partial\tau , \qquad (A.12)$$

and from Eq. (A.5)

$$\partial_\mu\xi^\mu = \tau^{-1}(n-1) , \qquad (A.13)$$

where n is the dimension of the spacetime.

Therefore Eq. (A.11) yields

$$|g|^{-\frac{1}{2}}\partial_\mu(|g|^{\frac{1}{2}}g^{\mu\nu})\partial_\nu\phi = (\frac{\partial}{\partial\tau}\ln|g|)\frac{d\phi}{d\tau}$$

$$+ n\tau^{-1}d\phi/d\tau$$

$$- g^{\mu\nu}\eta_{\mu\nu}\tau^{-1}d\phi/d\tau \ . \qquad (A.14)$$

Finally, the third term in Eq. (A.1) is

$$2g^{\mu\nu}\partial_\mu\phi\partial_\nu U = 2g^{\mu\nu}\eta_{\mu\alpha}\xi^\alpha\partial_\nu U d\phi/d\tau$$

$$= 2\xi^\nu\partial_\nu U d\phi/d\tau$$

$$= 2\ \partial U/\partial\tau\ d\phi/d\tau \ , \qquad (A.15)$$

where Eq. (3.6) (with τ factored out), and Eq. (A.12) were used. Substituting Eqs. (A.10), (A.14), and (A.15) into Eq. (A.1) gives

$$\nabla_\mu\nabla^\mu(\phi U) = U\left[\frac{d^2\phi}{d\tau^2} + \frac{n-1}{\tau}\frac{d\phi}{d\tau} + (\frac{\partial}{\partial\tau}\ln|g|)\frac{d\phi}{d\tau}\right]$$

$$+ 2\frac{\partial U}{\partial\tau}\frac{d\phi}{d\tau} + \phi\nabla_\mu\nabla^\mu U \ , \qquad (A.16)$$

which is the same as Eq. (3.9). This equation is given, but not derived, in Minakshisundaram and Pleijel (1949).

REFERENCES

Adler, S. L., Lieberman, J., and Ng, Y. J. 1977 Ann. Phys. (N.Y.) 106, 279.

Bernard, C. and Duncan, A. 1977 Ann. Phys. (N.Y.) 107, 201.

Brown, L. S. 1977 Phys. Rev. D15, 1469.

Brown, L. S. and Cassidy, J. P. 1977a Phys. Rev. D15, 2810.

_______. 1977b Phys. Rev. D16, 1712.

Brown, M. 1978 "Actions and Anomalies," preprint, U. of Texas at Austin.

Bunch, T. S. 1977 Ph.D. Thesis, King's College, London (unpublished).

_______. 1978a J. Phys. A11, 603.

_______. 1978b J. Phys. A (to be published).

Bunch, T. S. and Davies, P.C.W. 1977 Proc. Roy. Soc. A357, 381.

Capper, D. M. and Duff, M. J. 1974 Nuovo Cim. A23, 173.

_______. 1975 Phys. Lett. 53A, 361.

Cheng, K. S. 1972 J. Math. Phys. 13, 1723.

Chitre, D. M. and Hartle, J. B. 1977 Phys. Rev. D16, 251.

Christensen, S. M. 1976 Phys. Rev. D14, 2490.

Christensen, S. M. and Fulling, S. A. 1977 Phys. Rev. D15, 2088.

Davies, P.C.W., Fulling, S. A., and Unruh, W. G. 1976 Phys. Rev. D13, 2720.

Davies, P.C.W. and Fulling, S. A. 1977 Proc. Roy. Soc. A354, 59.

Deser, S., Duff, M. J., and Isham, C. J. 1976 Nucl. Phys. B111, 45.

DeWitt, B. S. 1957 Rev. Mod. Phys. 29, 377.

_______. 1964 Dynamical Theory of Groups and Fields (Gordon and Breach, N.Y.).

_______. 1975 Phys. Reports 19C, 295.

DeWitt-Morette, C. 1972 Commun. Math. Phys. 28, 47.

_______. 1974 Commun. Math. Phys. 37, 63.

Dowker, J. S. and Critchley, R. 1976a Phys. Rev. D13, 3224.

_______. 1976b J. Phys. A9, 535.

_______. 1977a Phys. Rev. D16, 3390.

_______. 1977b Phys. Rev. D15, 1484.

Dowker, J. S. and Altaie, M. B. 1978 Phys. Rev. D17, 417.

Feynman, R. P. and Hibbs, A. R. 1969 Quantum Mechanics and Path Integrals (McGraw-Hill, N.Y.).

Ford, L. M. 1975 Phys. Rev. D11, 3370.

_______. 1976 Phys. Rev. D14, 3304.

Fradkin, E. S. and Vilkowisky, G. A. 1978 Phys. Lett. 73B, 209.

Fulling, S. A., Parker, L., and Hu, B. L. 1974 Phys. Rev. D10, 3905.

Gibbons, G. W. and Hawking, S. W. 1977 Phys. Rev. D15, 2752.

Gibbons, G. W. 1977 "On Functional Integrals in Curved Spacetime," preprint (to appear in Commun. Math. Phys.).

Hartle, J. B. and Hawking, S. W. 1976 Phys. Rev. D13, 2188.

Hawking, S. W. 1977 Commun. Math. Phys. 55, 133.

Horowitz, G. and Wald, R. M. 1978 Phys. Rev. D17, 414.

Hu, B. L. 1978 (to be published).

Lapedes, A. S. 1978 Phys. Rev. D17, 2556.

McKean, H. P. and Singer, J. M. 1967 J. Diff. Geom. 1, 43.

Minakshisundaram, S. and Pleijel, A. 1949 Can. J. Math. 1, 242.

Misner, C. W., Thorne, K. S., and Wheeler, J. A. 1973 Gravitation (W. H. Freeman and Company, San Francisco).

Parker, L. 1966 Ph.D. Thesis, Harvard University (unpublished).

_______. 1968 Phys. Rev. Lett. 21, 562.

_______. 1969 Phys. Rev. 183, 1057.

_______. 1973 Phys. Rev. D7, 976.

_______. 1975 "Quantized Fields and Particle Creation in Curved Spacetime," in Relativity, Fields, Strings and Gravity, edited by C. Aragone (Universidad Simon Bolivar, Caracas, Venezuela).

_______. 1977 "The Production of Elementary Particles by Strong Gravitational Fields," pp. 107-227 in Asymptotic Structure of Space-Time, edited by F. P. Esposito and L. Witten (Plenum Press, N.Y.).

_______. 1978 "Path Integrals for a Particle in Curved Space," preprint (University of Wisconsin-Milwaukee).

Petrov, A. Z. 1969 Einstein Spaces (Pergamon Press, N.Y.).

Tsao, H. S. 1977 Phys. Lett. 68B, 79.

Vilenkin, A. 1978 "Pauli-Villars Regularization and Trace Anomalies," preprint, Case Western Reserve Univ., Cleveland, Ohio.

Wald, R. M. 1977 Commun. Math. Phys. 54, 1.

_______. 1978a Ann. Phys. (N.Y.) 110, 472.

_______. 1978b Phys. Rev. D17, 1477.

Zeldovich, Ya. B. and Starobinsky, A. A. 1972 JETP 34, 1159 [Zh. Eksp. Teor. Fiz. 61, 2161 (1971)].

THE FORMAL STRUCTURE OF QUANTUM GRAVITY*

Bryce S. De Witt

Department of Physics
University of Texas
Austin, Texas 78712

INTRODUCTION. NOTATION.

In 1956 Utiyama pointed out that the gravitational field can be regarded as a non-Abelian gauge field. In 1963 Feynman found that in order to construct a quantum perturbation theory for a non-Abelian gauge field he had to introduce new graphical rules not previously encountered in quantum field theory. He showed, in one-loop order, that to preserve unitarity one must add to every standard closed-loop graph another, involving a closed integral-spin fermion loop. In 1966 an explicitly gauge invariant functional-integral algorithm was found which extended Feynman's new rules to all orders (De Witt (1967b)). A short time later it was shown that the algorithm could be obtained by a method of factoring out the gauge group (Fadde'ev and Popov (1967)).

Formally the gravitational field and the Yang-Mills field can be treated identically. In the computation of amplitudes for specific physical processes, however, the two differ by the fact that the Yang-Mills field yields a renormalizable theory while the gravitational field does not.

* Lectures given at the Institut d'Études Scientifiques de Cargèse, Summer 1978. This work was supported in part by a research grant from the National Science Foundation (U.S.), by a travel grant from the North Atlantic Treaty Organization, by a Senior Research Fellowship of the Science Research Council (U.K., Imperial College, London), and by organized research funds of the University of Texas.

Some of the proposals that have been made for dealing with quantum gravity despite its nonrenormalizability will be discussed briefly later. But it must be admitted at the outset that we are dealing with an incomplete theory. The student may take comfort in the fact that every formal statement will be true for all field theories, even those, like supergravity, possessing supergauge groups, provided they are formulated in such a way that the action of the (super)gauge group on the field variables is expressible without use of field equations, and the group operations thus given are closed.

To emphasize the generality of the formalism we shall, most of the time, suppress field symbols such as $g_{\mu\nu}$ for the gravitational field or $A^{\alpha}{}_{\mu}$ for the Yang-Mills field in favour of a generic symbol φ^i . We shall go even further. The index i (or j, k, l, etc.) will be understood to label not only a field component but also a spacetime point x. Thus, in the gravitational case, i will be understood to stand for the set $\{\mu, \nu, x\}$ and, in the Yang-Mills case, for the set $\{\alpha, \mu, x\}$. In a supergauge theory the set i may include spinor indices. When it does, i (or φ^i) is said to be fermionic; otherwise it is bosonic.

The reason for including the continuous label x in the set i is that much of the formalism of quantum field theory is purely combinatorial, with summations over dummy indices being accompanied by integrations over spacetime. In order to avoid having to write a lot of integral signs we lump x and the field indices together and adopt the convention that the repetition of a lower case Latin index implies a combined summation-integration. Correspondingly, a comma followed by a lower case Latin index will denote functional differentiation:

$$A_{,i} \overset{\text{def}}{=} A\frac{\overleftarrow{\delta}}{\delta\varphi^i} \quad . \tag{1}$$

The change in a functional A (of the field φ^i) resulting from an infinitesimal variation $\delta\varphi^i$ is then

$$\delta A = A_{,i}\,\delta\varphi^i . \tag{2}$$

If A, or any of the φ^i , is fermionic one must distinguish left from right differentiation:

$$_{i,}A \overset{\text{def}}{=} \frac{\overrightarrow{\delta}}{\delta\varphi^i}A \quad , \tag{1'}$$

$$\delta A = \delta\varphi^i\,{}_{i,}A \quad . \tag{2'}$$

Evidently

$$ {}_{i,}A = (-1)^{i(A+1)} A_{,i} \tag{3} $$

where we adopt the rule that when an index (such as i) or a dynamical quantity (such as A) appears as an exponent of -1 it is to be understood as assuming the value 0 or 1 according as it is bosonic or fermionic. The summation-integration convention is not to be understood as applying to indices appearing as exponents. Such indices may participate in summation-integrations induced by their appearance twice elsewhere in an expression, but they themselves may not induce summation-integrations. Note that we are here treating all quantities (A, ϕ^i, $\delta\phi^i$ etc.) as supernumbers, i.e. as even or odd elements of an infinite-dimensional Grassmann algebra. Bosonic quantities commute with everything; fermionic quantities anticommute among themselves. In the quantum theory this perfect commutativity or anticommutativity is broken. The corresponding quantities will then be written in boldface.

The following notation will sometimes be convenient for expressing repeated functional differentiation:

$$ \cdots {}_{ij,}A_{,kl}\cdots \overset{\text{def}}{=} \cdots \frac{\overrightarrow{\delta}}{\delta\phi^i}\frac{\overrightarrow{\delta}}{\delta\phi^j} A \frac{\overleftarrow{\delta}}{\delta\phi^k}\frac{\overleftarrow{\delta}}{\delta\phi^l}\cdots \tag{4} $$

Note the particular examples:

$$ \phi^i{}_{,j} = \delta^i{}_j \, , \tag{5} $$

$$ (\phi^i\phi^j)_{,kl} = (-1)^{ij}\delta^i{}_k\delta^j{}_l + \delta^i{}_l\delta^j{}_k \, . \tag{6} $$

If x belongs to the set i and x' belongs to the set j, the "generalized Kronecker delta" $\delta^i{}_j$ includes, as a factor, the spacetime delta function $\delta(x, x')$.

When we are displaying specific details of a given field theory lower case indices from the middle of the Greek alphabet will be used to label tensor components. Co-ordinates in a given local patch, or chart, will be denoted by x^μ , with μ running from 0 to n - 1, n being the dimensionality of spacetime. (With an eye to the ultimate application of methods such as dimensional regularization and the renormalization group, we do not here hold n fixed at 4.) Commas followed by lower-case mid-alphabet Greek indices will denote ordinary differentiation with respect to the co-ordinates.

The following abbreviations will be useful:

$$\delta g_{\mu\nu}/\delta g_{\sigma'\tau'} = \delta_{\mu\nu}{}^{\sigma'\tau'} \overset{\text{def}}{=} \tfrac{1}{2}(\delta_\mu{}^\sigma\delta_\nu{}^\tau + \delta_\mu{}^\tau\delta_\nu{}^\sigma)\,\delta(x,x'), \tag{7}$$

$$\delta A^\alpha{}_\mu/\delta A^{\beta'}{}_{\nu'} = \delta^\alpha{}_{\mu\beta'}{}^{\nu'} \overset{\text{def}}{=} \delta^\alpha{}_\beta\,\delta_\mu{}^\nu\,\delta(x,x'). \tag{8}$$

It is straightforward to verify that

$$\delta_{\mu\nu}{}^{\sigma'\tau'}{}_{;}{}^{\nu} = -\delta_\mu{}^{\sigma'}{}_{;}{}^{\tau'} - \delta_\mu{}^{\tau'}{}_{;}{}^{\sigma'}, \tag{9}$$

$$\delta^\alpha{}_{\mu\beta'}{}^{\nu'}{}_{;}{}^{\mu} = -\delta^\alpha{}_{\beta'}{}_{;}{}^{\nu'}, \tag{10}$$

where $\delta_\mu{}^{\sigma'} \overset{\text{def}}{=} \delta_\mu{}^\sigma\delta(x,x')$, $\delta^\alpha{}_{\beta'} \overset{\text{def}}{=} \delta^\alpha{}_\beta\,\delta(x,x')$, the semicolons denote covariant differentiation, and tensor indices may be lowered and raised by the metric tensor $g_{\mu\nu}$ and its inverse $g^{\mu\nu}$ respectively. (Note that we always leave the semicolons in the lower position regardless of what happens to the indices). The delta functions appearing above are 2-point tensor densities, or bitensor densities, of total weight unity. In eqs. (9) and (10) the apportionment of the weight between the points x and x' is arbitrary; in eqs. (7) and (8) all the weight is at x'. In eqs. (9) and (10) the derivatives on the left are at x, on the right at x'.

In eqs. (8) and (10) lower case Greek indices from the beginning of the alphabet appear. These are associated with the Yang-Mills group. The laws of covariant differentiation of tensors bearing various kinds of indices are determinded as follows: let the symbol T represent a tensor field, with indices suppressed, in which we imagine all the components strung out in a single column. Let T also be coupled to the Yang-Mills field. Then

$$T_{;\mu} \overset{\text{def}}{=} T_{,\mu} + G^\nu{}_\sigma\Gamma^\sigma{}_{\nu\mu}T + G_\alpha A^\alpha{}_\mu T, \tag{11}$$

where

$$\Gamma^\sigma{}_{\nu\mu} \overset{\text{def}}{=} \tfrac{1}{2}g^{\sigma\tau}(g_{\tau\nu,\mu} + g_{\tau\mu,\nu} - g_{\nu\mu,\tau}), \tag{12}$$

and the $G^\mu{}_\nu$ and G_α are respectively the matrix generators of the representations of the linear group and Yang-Mills Lie group to which T corresponds. These generators satisfy

$$[G^\mu{}_\nu, G^\sigma{}_\tau] = \delta^\mu{}_\tau G^\sigma{}_\nu - \delta^\sigma{}_\nu G^\mu{}_\tau, \tag{13}$$

$$[G_\alpha, G_\beta] = G_\gamma f^\gamma{}_{\alpha\beta}, \tag{14}$$

where $f^\gamma{}_{\alpha\beta}$ are the structure constants of the Yang-Mills Lie group. When the suppressed indices on T are restored their positions generally determine the particular representations involved. Thus a Yang-Mills index in the upper position indicates the adjoint representation of the Lie group and one in the lower position the contragredient representation, etc. For physical reasons (positive probability) the Yang-Mills Lie group is required to be compact. Therefore given representations and their contragredient forms are equivalent, and Yang-Mills indices may be lowered (and raised) by the matrix $\gamma_{\alpha\beta}$ (and its inverse $\gamma^{\alpha\beta}$) that connects the adjoint and co-adjoint representations. When the Lie group is simple $\gamma_{\alpha\beta}$ may be taken to be the Kronecker delta, and all Yang-Mills indices may be dropped to the lower position.

We make no attempt here to list the rather complicated additional structures that appear in supergravity theories. (The student should consult the published literature for those details). We remark only that when spinors are present the local frame group and its spin representations must be introduced. The local frame group is completely analogous to the Yang-Mills group and makes a corresponding contribution to the covariant derivative on the right side of eq. (11), with $A^{\alpha}{}_{\mu}$ replaced by the connection components in the local frame and the G_{α} replaced by the generators of the relevant spin representation of the Lorentz group. (See De Witt (1965) for details).

In the next section we shall show how to place the Yang-Mills group and the diffeomorphism group (with which tensor indices are associated) on a common footing. Despite the analogies between the two groups, as displayed for example by the similarity of the second and third terms on the right of eq. (11), the diffeomorphism group is much more complicated than the Yang-Mills group and much less is known about its structure. Moreover, there is a lack of symmetry between the two groups in the fact that when they are combined into a single group (as is necessary when both Yang-Mills and gravitational fields are present) they are united not in a direct product but in a semi-direct product based on the automorphisms of the Yang-Mills group under diffeomorphisms. The same is true for the combined diffeomorphism and local frame groups, when spinor fields are present.

Our list of notational conventions is completed with the following statements and definitions:

$$T_{;\mu\nu} - T_{;\nu\mu} = -(G^{\sigma}{}_{\tau} R^{\tau}{}_{\sigma\mu\nu} + G_{\alpha} F^{\alpha}{}_{\mu\nu}) T \tag{15}$$

$$R^{\tau}{}_{\sigma\mu\nu} \stackrel{\text{def}}{=} \Gamma^{\tau}{}_{\sigma\nu,\mu} - \Gamma^{\tau}{}_{\sigma\mu,\nu} + \Gamma^{\tau}{}_{\mu\rho}\Gamma^{\rho}{}_{\sigma\nu} - \Gamma^{\tau}{}_{\nu\rho}\Gamma^{\rho}{}_{\sigma\mu} \tag{16}$$

$$F^{\alpha}{}_{\mu\nu} \stackrel{\text{def}}{=} A^{\alpha}{}_{\nu,\mu} - A^{\alpha}{}_{\mu,\nu} + f^{\alpha}{}_{\beta\gamma} A^{\beta}{}_{\mu} A^{\gamma}{}_{\nu} , \tag{17}$$

$$R_{\mu\nu} \stackrel{\text{def}}{=} R^{\sigma}{}_{\mu\sigma\nu} , \qquad R \stackrel{\text{def}}{=} R_{\mu}{}^{\mu} . \tag{18}$$

Unless otherwise specified we shall assume that spacetime is globally hyperbolic and complete*. Without loss of generality x^0 may then be assumed to be a global time co-ordinate, in the sense that it defines a foliation of spacetime into smooth complete hypersurfaces x^0 = constant, arranged in a temporal order. These hypersurfaces need not be everywhere spacelike, although if they are noncompact they must be asymptotically spacelike. The signature of the metric tensor will be $- + + \ldots$, and units (when needed) will be chosen to be "absolute", with $\hbar = c = 32\pi G = 1$. The absolute units of length, time and mass respectively are 1.6×10^{-32} cm., 5×10^{-43} sec. and 2×10^{-6} g., which gives an idea of the domains in which quantum gravity becomes relevant.

THE GAUGE GROUP

The gauge group of quantum gravity is the diffeomorphism group and that of Yang-Mills theory is the Yang-Mills group. We begin with the latter.

Elements of the Yang-Mills group are locally parametrized by a set of differentiable scalar functions $\xi^{\alpha}(x)$, with $\xi^{\alpha} = 0$ denoting the identity element. Elements infinitesimally close to the identity are parameterized by infinitesimal scalars. The action of such an element on the Yang-Mills potentials is given by

$$\delta A^{\alpha}{}_{\mu} = -\delta\xi^{\alpha}{}_{,\mu} + f^{\alpha}{}_{\gamma\beta} A^{\beta}{}_{\mu} \delta\xi^{\gamma} = -\delta\xi^{\alpha}{}_{;\mu} , \tag{19}$$

the covariant derivative being determined by noting that $\delta\xi^{\alpha}$ transforms (under inner automorphisms) according to the adjoint representation of the group. It will be convenient to rewrite eq. (19) in the form

$$\delta A^{\alpha}{}_{\mu} = \int Q^{\alpha}{}_{\mu\beta'} \delta\xi^{\beta'} d^n x' , \qquad d^n x' \stackrel{\text{def}}{=} \prod_{\mu=0}^{n-1} dx^{\mu} , \tag{20}$$

* There is some evidence (although hardly overwhelming yet) that quantization suppresses the singularities in spacetime that often develop automatically in classical general relativity.

or, in the generic notation,

$$\delta\phi^i = Q^i{}_\alpha \, \delta\xi^\alpha , \tag{21}$$

where

$$Q^\alpha{}_{\mu\beta'} \overset{\text{def}}{=} -\delta^\alpha{}_{\beta';\mu} . \tag{22}$$

In passing from eq. (20) to the generic form (21) one replaces the labels α, μ, x by the index i and the labels β', x' by the index α, and one understands that repetition of the latter index implies a combined summation-integration.

In quantum gravity the action of the diffeomorphism group can be expressed in identical generic form. The diffeomorphism group is the group of mappings $f: M \to M$ of the spacetime manifold M into itself such that f is one-to-one and both f and f^{-1} are differentiable. In practice one may require f and f^{-1} to be C^∞ and, if the sections $x^0 =$ constant are noncompact, to reduce asymptotically (i.e. "at spatial infinity") to the local identity mapping. Such mappings define a "dragging" of all tensor fields defined on M, and if the mapping is infinitesimally close to the identity the "dragging" may be viewed as a physical displacement of the fields through an infinitesimal vector $\delta\xi$. If all fields, including the metric (i.e. gravitational) field, are displaced by the same amount the physics remains unchanged. It is conventional in physics, therefore, to adopt an opposite viewpoint and to regard an infinitesimal diffeomorphism as leaving the "physical" points of the manifold untouched while dragging all co-ordinate patches (i.e. the complete atlas) through the negative vector $-\delta\xi$. Locally this is expressed by the co-ordinate transformation $x^\mu \to \xi^\mu$ where

$$\xi^\mu = x^\mu + \delta\xi^\mu . \tag{23}$$

Let T be a tensor field and δT its change under dragging through $\delta\xi$. The <u>Lie derivative</u> of T with respect to $\delta\xi$ is defined by

$$\mathcal{L}_{\delta\xi} T \overset{\text{def}}{=} -\delta T . \tag{24}$$

Let p be a point of M and p' the point to which it is dragged under $\delta\xi$. Then the co-ordinates of p in the new co-ordinate system (23) are identical with those of p' in the old co-ordinate system. Moreover, the components of T at p in the new co-ordinate system are identical with those of T + δT at p' in the old co-ordinate system. One has only to cast eq. (24) into component language, therefore, to regard the Lie derivative as expressing the negative of the change in the <u>functional form</u> of the components of T, <u>viewed as functions of the local co-ordinates</u>, under the diffeomorphism. This enables one to compute

$$\pounds_{\delta\xi} T = T_{,\mu}\delta\xi^{\mu} - G^{\nu}{}_{\mu} T \delta\xi^{\mu}{}_{,\nu} \tag{25a}$$

$$= T_{;\mu}\delta\xi^{\mu} - G^{\nu}{}_{\mu} T \delta\xi^{\mu}{}_{;\nu} , \tag{25b}$$

which yields, in particular, the gauge transformation law for the metric tensor:

$$\delta g_{\mu\nu} = -\pounds_{\delta\xi} g_{\mu\nu} = -\delta\xi_{\mu;\nu} - \delta\xi_{\nu;\mu} . \tag{26}$$

Equation (26) may be rewritten

$$\delta g_{\mu\nu} = \int Q_{\mu\nu\sigma'}\delta\xi^{\sigma'} d^n x' , \qquad Q_{\mu\nu\sigma'} \overset{\text{def}}{=} -\delta_{\mu\sigma';\nu} - \delta_{\nu\sigma';\mu} , \tag{28}$$

which, if the labels μ, ν, x are replaced by i and the labels σ', x' by α, takes the generic form (21).

Lower case Greek indices from the first part of the alphabet, as in eq. (21), will from now on be called <u>group</u> indices. If the group indices are allowed to label fermionic as well as bosonic gauge parameters then the generic form (21) holds also for the <u>super</u>gauge transformations of supergravity theories. Although we shall not go into the specific details of such theories we shall, in all that follows, allow for their possible presence.

STRUCTURE CONSTANTS

By invoking the requirement that the commutator of two infinitesimal gauge group operations be itself a group operation (the closure property) one arrives at the functional differential identity

$$Q^{i}{}_{\alpha,j} Q^{j}{}_{\beta} - (-1)^{\alpha\beta} Q^{i}{}_{\beta,j} Q^{j}{}_{\alpha} = Q^{i}{}_{\gamma} c^{\gamma}{}_{\alpha\beta} , \tag{29}$$

where the c's are certain coefficients known as the <u>structure constants</u> of the gauge group. They posses the symmetry

$$c^{\gamma}{}_{\alpha\beta} = -(-1)^{\alpha\beta} c^{\gamma}{}_{\beta\alpha} . \tag{30}$$

The structure constants of the Yang-Mills group may be determined by straightforward computation from eqs. (19), (20) and (22). They are the components of the following 3-point tensor density:

$$c^{\alpha}{}_{\beta'\gamma''} = f^{\alpha}{}_{\beta\gamma}\, \delta(x,x')\, \delta(x,x'') . \tag{31}$$

The weights are at x' and x''.

The structure constants of the diffeomorphism group may be determined by recalling the commutation law for the Lie derivative:

$$[\pounds_X, \pounds_Y]T = \pounds_{[X,Y]}T. \tag{32}$$

Here $[X, Y]$ is the Lie bracket of the vectors X and Y:

$$[X,Y] = \pounds_X Y = -\pounds_Y X. \tag{33}$$

The structure constants are the components of the 3-point tensor density defined by

$$\begin{aligned} \int d^n x' \int d^n x''\, c^\mu{}_{\nu'\sigma''} X^{\nu'} Y^{\sigma''} &= -[X,Y]^\mu \\ &= X^\mu{}_{,\nu} Y^\nu - Y^\mu{}_{,\nu} X^\nu = X^\mu{}_{;\nu} Y^\nu - Y^\mu{}_{;\nu} X^\nu. \end{aligned} \tag{34}$$

Evidently

$$c^\mu{}_{\nu'\sigma''} = \delta^\mu{}_{\nu',\tau}\delta^\tau{}_{\sigma''} - \delta^\mu{}_{\sigma'',\tau}\delta^\tau{}_{\nu'} = \delta^\mu{}_{\nu';\tau}\delta^\tau{}_{\sigma''} - \delta^\mu{}_{\sigma'';\tau}\delta^\tau{}_{\nu'}. \tag{35}$$

The weights are at x' and x''.

The action of the gauge group on the field variables φ^i, expressed by eq. (21), is a realization of the group. This realization is always a faithful one, which implies that $Q^i{}_\alpha X^\alpha = 0$ for all i if and only if $X^\alpha = 0$ for all α. By functionally differentiating eq. (29) with respect to φ^k, multiplying by $Q^k{}_\gamma$, judiciously permuting the indices α, β, γ, adding the results, and invoking the faithfulness of the realization, one obtains the following cyclic identity satisfied by the structure constants:

$$c^\delta{}_{\alpha\epsilon} c^\epsilon{}_{\beta\gamma} + (-1)^{\alpha(\beta+\gamma)} c^\delta{}_{\beta\epsilon} c^\epsilon{}_{\gamma\alpha} + (-1)^{\gamma(\alpha+\beta)} c^\delta{}_{\gamma\epsilon} c^\epsilon{}_{\alpha\beta} = 0. \tag{36}$$

In the case of the Yang-Mills group this identity reduces to the corresponding identity for the constants $f^\alpha{}_{\beta\gamma}$. In the case of the diffeomorphism group it is the Jacobi identity for Lie brackets.

CONFIGURATION SPACE. ORBITS

For each point x in the spacetime manifold M, the field φ (index i suppressed) takes its "value" in a certain finite-dimensional differentiable manifold Φ_x, which may but need not be a vector space or subspace thereof. In pure gravity theory, for example, Φ_x is the subspace of Sym $(T_x^* \otimes T_x^*)$ containing all local symmetric covariant second rank tensors at x having nonvanishing determinant and signature $- + + \ldots$. Here T_x^* is the dual of the tangent space to M at x, and "Sym" denotes the symmetric part of the tensor product $T_x^* \otimes T_x^*$.

The set of all Φ_x with x in M, may be regarded as forming a fiber bundle over M. Each Φ_x is a fiber, and each field φ is a cross section of the bundle.* The bundle may but need not be a simple product bundle.

The set of all cross sections,* i.e. of all field configurations φ may be assembled into a space Φ called the <u>configuration space</u>. Because of differentiability requirements on the field configurations Φ is endowed naturally with a functional differentiable structure and may be viewed as an infinite dimensional differentiable manifold, or, if φ^i includes fermion fields, as a differentiable <u>super</u>manifold (also known as a Z_2-graded manifold. See Kostant (1977)). Since M is never compact (at least in the time direction) the fields φ are usually constrained to obey also special boundary conditions "at infinity". In both Yang-Mills and gravity theory these can be of considerable importance.

Let the gauge group be denoted by G and let ξ be an element of G. Denote by ${}^{\xi}\varphi$ the "point" of Φ to which φ is displaced under the action of ξ. The set of points ${}^{\xi}\varphi$ for all ξ in G is known as the <u>orbit</u> of φ and denoted by Orb.(φ). The set of all orbits can be assembled into a space called the <u>space of orbits</u>, denoted by the quotient symbol Φ/G. Since all fields on a given orbit describe the same physics it is the space of orbits that constitutes the real <u>physical configuration space</u> of the theory. In pure gravity theory Φ is the space, Lor (M), of Lorentzian (also called pseudo-Riemannian) metrics on M, G is the group, Diff (M), of diffeomorphisms of M, and the physical configuration space, Lor (M)/Diff (M), is the space of <u>Lorentzian geometries</u> on M.

Because gauge groups can be "co-ordinatized" by differentiable functions (i.e. the gauge parameters) G, like Φ, can be regarded as an infinite dimensional differentiable manifold (or supermanifold). In Yang-Mills theory G may have a simple product structure inherited from the associated Lie group of the theory (see (31)), or it may itself be a twisted bundle. The diffeomorphism group of gravity theory, by contrast, cannot be viewed as a bundle but has a structure that is much less well understood. Some, but only a little, of its complexity will emerge as we go along.

Since both Φ and G are differentiable (super) manifolds the quotient space Φ/G too is a differentiable (super) manifold, or rather it is a differentiable (super) manifold that may have a boundary.

---

* If the fiber bundle admits no global cross sections the field must be defined by introducing overlapping patches.

To see how a boundary can arise consider a typical, i.e. generic, orbit. Modulo a possible discrete center it is a copy of G, because it provides a realization of G and has the same dimensionality. Not all orbits need have this dimensionality. There is often a class of degenerate orbits having fewer dimensions. These are the orbits that remain invariant under the action of nontrivial continuous subgroups of G. They are the boundary points of Φ/G. To see this think of Φ as being $\mathbb{R}^3$ and G as being the group of rotations about a fixed axis. The orbits are then circles perpendicular to and centered on the axis, and the orbit space is a half-plane whose boundary points correspond to the points on the axis, which remain invariant under the group.

The greater the dimensionality of the subgroup that leaves a given orbit invariant, the smaller the dimensionality of the orbit. Fischer (1970) has shown that if the invariance group has only one dimension then the orbit is an ordinary boundary point of Φ/G. If the invariance group has two dimensions then the orbit lies on a boundary of the boundary and so on. The whole orbit manifold, with its boundary, and its boundaries of boundaries, etc., is known as a stratified manifold.

In gravity theory the boundary orbits are the symmetrical geometries, i.e. those that possess Killing vectors. The boundary structure of Lor (M)/Diff (M) in general depends critically on M. Since there exists no complete classification of n-dimensional manifold ($n > 3$) that can possess globally hyperbolic metric tensors, there exists also no complete classification of possible configuration spaces for the gravitational field. On some spacetimes there may be no Lorentzian geometries possessing Killing vectors. Such spacetimes are called wild. If M is wild Lor (M)/ Diff (M) has no boundary points. For technical reasons it is frequently necessary to regard certain familiar spacetimes as wild. For example, asymptotically flat spacetimes diffeomorphic to $\mathbb{R}^n$ are usually treated as wild. The reason for this is to keep Lorentz transformations distinct from gauge transformations, by requiring the gauge parameters $\delta\xi^\mu$ to vanish at infinity. Flat Minkowski spacetime is then not a boundary point of Lor($\mathbb{R}^n$)/Diff ($\mathbb{R}^n$) because the Poincaré isometries are not regarded as being contained in Diff ($\mathbb{R}^n$).

METRICS ON CONFIGURATION SPACE

It turns out to be both possible and useful to regard Φ and Φ/G not merely as differentiable (super)manifolds but as pseudo-Riemannian (super)manifolds as well. Let $d\varphi^i$ be an infinitesimal displacement in Φ. We may associate with this displacement a (super) arc length ds, given by

$$ds^2 = d\varphi^i \; {}_i\gamma_j \; d\varphi^j \quad , \tag{36}$$

where ${}_i\gamma_j$ are the components of a (super) metric tensor on Φ. The ${}_i\gamma_j$ are functionals of φ having the symmetry

$${}_i\gamma_j = (-1)^{i+j+ij} \; {}_j\gamma_i \tag{37}$$

and forming an invertible continuous matrix. The inverse, denoted by γ^{ij}, satisfies

$${}_i\gamma_k \gamma^{kj} = \delta_i{}^j \; , \qquad \gamma^{ik} {}_k\gamma_j = \delta^i{}_j \; , \qquad \gamma^{ij} = (-1)^{ij} \gamma^{ji} \, . \tag{38}$$

If the metric ${}_i\gamma_j$ is chosen in such a way that the actions of G on Φ are isometries then ${}_i\gamma_j$ induces also a metric on the orbit space Φ/G. One simply defines the distance between neighbouring orbits in Φ/G to be the orthogonal distance between them in Φ. This requires selecting ${}_i\gamma_j$ in such a way that the continuous matrix $(-1)^{\alpha(i+1)} Q^i{}_\alpha \; {}_i\gamma_j \; Q^j{}_\beta$ is nonsingular, on all orbits so that a vector cannot be simultaneously tangent to and orthogonal to any of them.

Every element of G infinitesimally close to the identity generates a vector field $Q^i{}_\alpha \delta\xi^\alpha$ on Φ (see eq. (21)). Each of these fields is a linear combination of the basic vector fields $Q^i{}_\alpha$, and the structure of G is determined by the (super) Lie bracket relations (29) that they satisfy. The condition that G act isometrically on Φ may be translated into the statement that the (super) Lie derivatives of the metric ${}_i\gamma_j$ with respect to the Q's all vanish:

$$0 = {}_i\gamma_j \overleftarrow{\pounds}_{Q_\alpha} = {}_i\gamma_{j,k} Q^k{}_\alpha + (-1)^{\alpha(j+k)} {}_{i,}Q^k{}_\alpha \; {}_k\gamma_j + (-1)^{\alpha j} {}_i\gamma_k \, Q^k{}_{\alpha,j} \, . \tag{39}$$

It is not difficult to verify that eq. (29) is the integrability condition for (39). Equation (39) generally has an infinity of solutions differing nontrivially from one another. It it did not, i.e. if the solution were unique up to a constant factor, this would mean that G acts transitively on Φ and hence that Φ/G is trivial, the theory having no physical content.

In order to understand what eq. (39) says in more familiar terms it is helpful to note that the fields φ^i encountered in practice usually provide <u>linear</u> realizations of their gauge groups. This is certainly true for the Yang-Mills and gravitational fields. What it means is that the functional derivatives $Q^i{}_{\alpha,j}$ are independent of the φ^i and, when regarded as continuous matrices (in i and j), yield a matrix <u>representation</u> of the (graded) Lie algebra associated with the group.

Of course this simplicity is generally lost if the ϕ's are replaced by nonlinear functions of themselves. But it is remarkable that there is usually a "natural" set of field variables of which the Q's are linear functionals. In gravity theory, in fact, there is a family of "natural" fields, namely all tensor densities of the form

$$\mathfrak{g}^{\mu\nu} \stackrel{\text{def}}{=} g^{r} g^{\mu\nu} \quad \text{or} \quad \mathfrak{g}_{\mu\nu} \stackrel{\text{def}}{=} g^{-r} g_{\mu\nu}\,, \quad r \neq 1/n\,, \tag{40}$$

where $g \stackrel{\text{def}}{=} -\det(g_{\mu\nu})$.

Consider now the way in which the Q's themselves change under infinitesimal gauge transformations. Using eq. (29) one finds

$$\begin{aligned}\delta Q^{i}{}_{\alpha} &= Q^{i}{}_{\alpha,j}\,\delta\phi^{j} = Q^{i}{}_{\alpha,j}\,Q^{j}{}_{\beta}\,\delta\xi^{\beta} \\ &= (-1)^{\alpha\beta}\left(Q^{i}{}_{\beta,j}\,Q^{j}{}_{\alpha} - Q^{i}{}_{\gamma}\,c^{\gamma}{}_{\beta\alpha}\right)\delta\xi^{\beta}\,,\end{aligned} \tag{41}$$

which says that $Q^{i}{}_{\alpha}$ is a two-point function that transforms at the point associated with i according to the representation generated by the matrices $(Q^{i}{}_{\alpha,j})$ and at the point associated with α contragrediently to the representation generated by the matrices $(c^{\alpha}{}_{\gamma\beta})$. In gravity theory this says that the function $Q_{\mu\nu\sigma'}$ of eq. (28) transforms like a covariant tensor at x and like a covariant vector density of unit weight at x',* which indeed it does. Equation (41) yields an analogous statement about the transformation law of the two-point function $Q^{\alpha}{}_{\mu\beta'}$ of eq. (22) under the Yang-Mills group.

We are now ready to interpret eq. (39). Under the gauge group the γ's change according to

$$\begin{aligned}\delta\,{}_{i}\gamma_{j} &= {}_{i}\gamma_{j,k}\,\delta\phi^{k} = {}_{i}\gamma_{j,k}\,Q^{k}{}_{\alpha}\,\delta\xi^{\alpha} \\ &= -(-1)^{\alpha j}\left[(-1)^{\alpha k}\,{}_{i,}Q^{k}{}_{\alpha}\,{}_{k}\gamma_{j} + {}_{i}\gamma_{k}\,Q^{k}{}_{\alpha,j}\right]\delta\xi^{\alpha}\,,\end{aligned} \tag{42}$$

which says that the γ's are two-point functions that transform at each point contragrediently to the representation generated by the matrices $(Q^{i}{}_{\alpha,j})$ In gravity theory this implies that when the ϕ^{i} are chosen to be the components of the covariant metric tensor, ${}_{i}\gamma_{j}$ must transform at each point like a symmetric contravariant tensor density of unit weight. Any γ's that transform in this way, and have an inverse γ^{ij}, provide an acceptable metric on Lor (M).

* Under the diffeomorphism group a tensor density of weight w, having p covariant and q contravariant indices, transforms contragrediently to a tensor density of weight 1 - w, having q covariant and p contravariant indices.

Among all such metrics on Lor (M) there is a unique (up to a constant factor) 1-parameter family of them that may be characterized as local. These are given by

$$\gamma^{\mu\nu\sigma'\tau'} = \gamma^{\mu\nu\sigma\tau}\,\delta(x,x') \tag{43}$$

$$\gamma^{\mu\nu\sigma\tau} \overset{\text{def}}{=} \tfrac{1}{2} g^{1/2}\left(g^{\mu\sigma}g^{\nu\tau} + g^{\mu\tau}g^{\nu\sigma} + \lambda g^{\mu\nu}g^{\sigma\tau}\right), \quad \lambda \neq -\tfrac{2}{n} \tag{44}$$

For Yang-Mills theory in flat spacetime the corresponding metric is

$$\gamma_{\alpha}{}^{\mu}{}_{\beta'}{}^{\nu'} = \gamma_{\alpha\beta}\,\eta^{\mu\nu}\,\delta(x,x'), \tag{45}$$

where $\eta^{\mu\nu}$ is the Minkowski metric. The matrices $(-1)^{\alpha(i+1)} Q^{i}{}_{\alpha} {}_{;}\gamma_{ij} Q^{j}{}_{\beta}$ in the two cases are readily calculated to be respectively

$$-2g^{1/2}\left\{\delta_{\mu\nu';\sigma}{}^{\sigma} + R_{\mu}{}^{\sigma}\delta_{\sigma\nu'} - (1+\lambda)\left[\delta(x,x')\right]_{;\mu\nu'}\right\} \tag{46}$$

and

$$-\delta_{\alpha\beta';\mu}{}^{\mu}. \tag{47}$$

The continuous matrix (47) is effectively the negative of the Yang-Mills-invariant Laplace-Beltrami operator. If the Yang-Mills field is untwisted (see the lectures by Avis and Isham in this volume) it is a nonsingular operator having a unique Green's function for each choice of boundary conditions at infinity. If the Yang-Mills field is twisted, however it may have zero eigenvalues, which means that the choice (45) fails to yield a globally valid metric on the orbit manifold. Although this is an important and interesting situation we shall not attempt to deal with it in these lectures.

The continuous Matrix (46) too may become singular. Its structure is simplest when $\lambda = -1$ $(n \neq 2)$; it is then effectively a slightly generalized form of the standard Laplace-Beltrami operator. Considerable evidence exists to indicate that it is nonsingular when the spacetime manifold is diffeomorphic to $\mathbb{R}^n$. But for other topologies it may have zero eigenvalues. Again we exclude this situation from consideration.

When the matrices (46) and (47) are nonsingular, expressions (43) and (45) constitute metrics on the space of fields which define, by orthogonal projection, globally valid nonsingular metrics on the space of orbits. It is possible to develop a theory of geodesics on these configuration spaces. The geometry defined by (45) is flat and the geodesics in the space of Yang-Mills potenitials are trivial. The geometry defined by (43) and (44), on the other hand, is not flat, and the resulting theory of geodesics on the space of metric tensors $g_{\mu\nu}$ is not trivial. It can nevertheless be shown that any pair of points in this space can be connected by a unique geodesic. It can also be shown that if a geodesic intersects one orbit orthogonally then it intersects every orbit in its path orthogonally, and, moreover, traces out a geodesic curve in the space of orbits. Methods for proving these theorems can be found in DeWitt (1967a). Using these theorems together with the fact that a vector in the space of metric tensors cannot be simultaneously parallel and orthogonal to an orbit, one can then prove that any pair of orbits can be connected by a unique geodesic. It should be stated that all of these theorems depend upon the maintenance of fixed boundary conditions (on fields and diffeomorphisms) at infinity.

VOLUME ELEMENTS ON CONFIGURATION SPACE

With a metric defined on the space of fields it is possible to introduce a formal volume element $\mu\, d\phi$ $(d\phi \overset{\text{def}}{=} \prod_i d\phi^i)$ by choosing

$$\mu = \text{const.} \times |\det({}_{;}\gamma_{j})|^{1/2} . \tag{48}$$

This volume element is gauge invariant and can be used to define gauge invariant functional integrals over configuration space. When fermionic fields are present the determinant in eq. (48) is the <u>super</u> determinant (see Nath (1976)), which satisfies the variational law

$$\delta \ln \det({}_{;}\gamma_{j}) = (-1)^{i} \gamma^{ij} \delta_{;j}\gamma_{i} . \tag{49}$$

This law, combined with eq. (39), yields the following equation of "divergenceless flow" that could in principle be used to select a gauge invariant volume element independently of a metric:

$$(-1)^{i(\alpha+1)} (\mu Q^{i}{}_{\alpha})_{,i} = 0 . \tag{50}$$

The delta functions contained in the metrics (43) and (45) give these metrics a block structure that yields simple formal expressions for their determinants. In Yang-Mills theory the determinant is a constant; in gravity theory it is given by

$$\det(\gamma^{\mu\nu\sigma'\tau'}) = \prod_{x} \gamma(x) , \tag{51}$$

where $\gamma(x)$ is the determinant of the $\frac{1}{2}n(n+1) \times \frac{1}{2}n(n+1)$ matrix $\gamma^{\mu\nu\sigma\tau}$. It is not a difficult computation to show that

$$\gamma = (-1)^{n-1}\left(1 + \frac{n\lambda}{2}\right) g^{\frac{1}{4}(n-4)(n+1)} . \tag{52}$$

In a 4-dimensional spacetime γ, and hence $\det(\gamma^{\mu\nu\sigma'\tau'})$, is seen to be a constant, independent of the $g_{\mu\nu}$. The functional μ, in the volume element over the configuration space of gravity theory, may therefore be taken to be a constant. Without loss of generality it may be chosen equal to 1. This will no longer be true in other dimensions, or when other fields are present in addition to the gravitational field, if we stick to the $g_{\mu\nu}$ as the basic field variables. However, we can in principle replace the $g_{\mu\nu}$ by one of the family of variables defined in eq. (40) and choose r so that μ remains constant. In practice, as we shall see later, this is unnecessary. To set μ "effectively" equal to unity it turns out to be necessary only to choose basic fields that transform linearly under the gauge group.

GROUP COORDINATES

The scalar functions $\xi^\alpha(x)$ that parameterize the elements of the Yang-Mills group may be regarded as "coordinates" in the group manifold. In the case of the diffeomorphism group the group coordinates may be taken to be the functions $\xi^\mu(x)$ that define the coordinate transformation $x^\mu \to \xi^\mu$ associated with each diffeomorphism in each coordinate chart or patch. Note that the functions $\xi^\mu(x)$ are neither scalars nor components of vectors, Note also that in both cases the coordinatization of the group cannot generally be achieved without bringing in the whole apparatus of charts, atlases and consistency conditions in the regions of intersection of overlapping charts.

Any group may be regarded as acting on itself through multiplication either on the left or on the right. Every group thus provides a realization of itself, and if it is a gauge group G, possesses a set of functionals $Q^\alpha{}_\beta[\xi]$ analogous to the functionals $Q^i{}_\alpha[\varphi]$ over the configuration space Φ. The $Q^\alpha{}_\beta$ are then defined by

$$[(I+\delta\xi)\xi]^\alpha = \xi^\alpha + Q^\alpha{}_\beta[\xi]\delta\xi^\beta \quad \text{for all } \delta\xi^\alpha \text{ and all } \xi \text{ in } G, \tag{53}$$

where I denotes the identity element of G and $I+\delta\xi$ denotes an element of G whose coordinates differ by infinitesimal amounts $\delta\xi^\alpha$ from those of I. Using the fact that $I^\mu(x) = x^\mu$, and that $(\xi\xi')^\mu(x) = \xi^\mu(\xi'(x))$ for all ξ and ξ' in G, it is not difficult to verify that the $Q^\alpha{}_\beta$ for the diffeomorphism group are given explicitly by

$$Q^\mu{}_{\nu'}[\xi] = \delta^\mu{}_\nu\, \delta(\xi(x), x'). \tag{54}$$

In the case of the Yang-Mills groups the $Q^\alpha{}_\beta$ take the forms

$$Q^\alpha{}_{\beta'}[\xi] = q^\alpha{}_\beta(\xi(x))\,\delta(x, x'), \tag{55}$$

where the $q^\alpha{}_\beta$ are the corresponding quantities for the associated Lie group.

Now let $\delta\xi^\alpha = \xi^\alpha \delta t$ for some fixed ξ^α and consider the curve in G defined by

$$\xi(t) = \lim_{\delta t\to 0} (I+\delta\xi)^{t/\delta t}. \tag{56}$$

Evidently

$$\left.\begin{aligned} \xi(s)\xi(t) &= \xi(t)\xi(s) = \xi(s+t), \\ \xi(0) &= I, \qquad \xi^{-1}(t) = \xi(-t), \end{aligned}\right\} \tag{57}$$

$$d\xi^\alpha(t)/dt = Q^\alpha{}_\beta[\xi(t)]\,\xi^\beta . \tag{58}$$

The points on the curve are seen to constitute a one-parameter Abelian subgroup of G.

If it could be proved that all the elements of G in a neighbourhood N of the identity can be obtained by a process of exponentiation of the form (56) then it would follow that the one-parameter Abelian subgroups completely span the neighbourhood N. A special set of coordinates $\xi_c{}^\alpha$, known as canonical coordinates, could be introduced in N for which the functions $\xi^\alpha(t)$ above take the simple forms

$$\xi_c{}^\alpha(t) = \xi^\alpha t . \tag{59}$$

Let us assume that our coordinates are already canonical, so that we may drop the subscript c. Then we have

$$I^\alpha = 0 , \qquad \xi^{-1\,\alpha} = -\xi^\alpha , \tag{60}$$

$$\xi^\alpha = Q^\alpha{}_\beta[\xi]\,\xi^\beta = Q^{-1\alpha}{}_\beta[\xi]\,\xi^\beta , \tag{61}$$

where $(Q^{-1\alpha}{}_\beta)$ is the (continuous) matrix inverse to $(Q^\alpha{}_\beta)$. By taking note of the fact that the $Q^\alpha{}_\beta$ must satisfy an identity analogous to (29), namely

$$Q^\gamma{}_{\alpha,\delta}\,Q^\delta{}_\beta - (-1)^{\alpha\beta}\,Q^\gamma{}_{\beta,\delta}\,Q^\delta{}_\alpha = Q^\gamma{}_\delta\,c^\delta{}_{\alpha\beta} , \tag{62}$$

we may show that in a canonical coordinate system the $Q^\alpha{}_\beta$ are completely determined by the structure constants.

We begin by rewriting eq. (62) in the equivalent form

$$Q^{-1\alpha}{}_{\beta,\gamma} - (-1)^{\beta\gamma} Q^{-1\alpha}{}_{\gamma,\beta} + (-1)^{\epsilon(\delta+\beta)}\,c^\alpha{}_{\delta\epsilon}\,Q^{-1\delta}{}_\beta\,Q^{-1\epsilon}{}_\gamma = 0 . \tag{63}$$

Multiplying this equation on the right by ξ^γ and using eq. (61) we get

$$Q^{-1\alpha}{}_{\beta,\gamma}\,\xi^\gamma - (-1)^{\beta\gamma} Q^{-1\alpha}{}_{\gamma,\beta}\,\xi^\gamma + c^\alpha{}_{\delta\epsilon}\,\xi^\epsilon\,Q^{-1\delta}{}_\beta = 0. \tag{63}$$

On the other hand, differentiating eq. (61) with respect to ξ^β we find

$$(-1)^{\beta\gamma} Q^{-1\alpha}{}_{\gamma,\beta}\, \xi^\gamma + Q^{-1\alpha}{}_\beta = \delta^\alpha{}_\beta . \tag{65}$$

Addition of eqs. (64) and (65) yields

$$Q^{-1}{}_{,\alpha}\, \xi^\alpha + Q^{-1} - c\cdot\xi\, Q^{-1} = 1 , \tag{66}$$

where "1" denotes the unit matrix (delta function) and

$$Q^{-1}[\xi] \overset{\mathrm{def}}{=} (Q^{-1\alpha}{}_\beta[\xi]), \quad c\cdot\xi \overset{\mathrm{def}}{=} ((-1)^{\beta\gamma} c^\alpha{}_{\gamma\beta}\, \xi^\gamma) = -(c^\alpha{}_{\beta\gamma}\, \xi^\gamma). \tag{67}$$

The solution of eq. (66) satisfying the necessary boundary condition

$$Q^\alpha{}_\beta[I] = \delta^\alpha{}_\beta \tag{68}$$

(see eq. (53)) is

$$Q^{-1}[\xi] = \frac{e^{c\cdot\xi} - 1}{c\cdot\xi} \overset{\mathrm{def}}{=} 1 + \frac{1}{2!} c\cdot\xi + \frac{1}{3!}(c\cdot\xi)^2 + \cdots . \tag{69}$$

The series (69) converges for all values of the ξ^α. For certain values the (continuous) matrix Q^{-1} may have vanishing roots. For these values some of the $Q^\alpha{}_\beta$, and hence the canonical coordinate system itself, become singular. In the case of an untwisted Yang-Mills group it can be shown that the one-parameter Abelian subgroups do span a neighbourhood of the identity. (This is, in fact, a corollary of the corresponding theorem for the associated Lie group.) Indeed they span the entire group - or, rather, that part of the group that is connected to the identity, i.e. the proper group. The whole group can therefore be parameterized by canonical coordinates (supplemented, perhaps, with some discrete labels).

Canonical coordinates for the Yang-Mills group have a periodic, or angular, nature. At a given point x of the spacetime manifold let the $\xi^\alpha(x)$ in eq. (55) increase in magnitude but maintain fixed ratios to one another. Eventually all of the $Q^\alpha{}_\beta$ will become singular at once. One has returned to the identity element of the associated Lie group. By allowing the canonical coordinates to range from $-\infty$ to ∞ one evidently covers the gauge group an infinity of times. Despite the fact that the $Q^\alpha{}_\beta$ become singular for certain values of the ξ's, the canonical coordinates are good in that, no matter what their values, they always define a unique element of the group.

NO CANONICAL COORDINATES FOR THE DIFFEOMORPHISM GROUP

If canonical coordinates could be introduced into the diffeomorphism group one could dispense with the apparatus of charts, atlases, etc. in parameterizing the group. Every diffeomorphism could be characterized by a (finite) vector field just as those infinitesimally close to the identity can be characterized by an infinitesimal vector field. And a vector field has a meaning independent of charts and atlases.

Unfortunately the one-paramater Abelian subgroups of the diffeomorphism group do not span a neighbourhood of the identity. If the dimensionality n of the spacetime manifold M is greater than or equal to 2 there are C^∞ diffeomorphisms arbitrarily close to the identity that cannot be obtained by exponentiation as in eq. (56). The proof, which we now outline, was first given by Freifeld (1968).

It suffices to confine attention to $\mathbb{R}^2$ or, equivalently, to the complex plane $\mathbb{C}$. Let x be a point of $\mathbb{C}$. Instead of breaking x into its real and imaginary parts we may treat x and its complex conjugate x* formally as independent variables. A C^∞ diffeomorphism $\xi : \mathbb{C} \to \mathbb{C}$ is then a one-to-one complex function $\xi(x, x^*)$, of class C^∞ in both x and x*, whose inverse, $x(\xi, \xi^*)$ is C^∞ in ξ and ξ^*.

Let N be a positive integer and α a positive real number. Suppose ξ has the analytic form

$$\xi(x, x^*) = e^{\frac{2\pi i}{N}} x + \alpha x^{N+1} \tag{70}$$

in a finite neighbourhood of the origin (e.g., in a circle of finite radius), and suppose that outside of this neighbourhood ξ changes smoothly (C^∞) to the identity function $\xi(x, x^*) = x$. If N is chosen large and α is chosen small then ξ and all its derivatives may be made uniformly close to those of the identity. We shall show that ξ does not lie on a one-parameter subgroup of C^∞ diffeomorphism $\xi(t) : \mathbb{C} \to \mathbb{C}$ with $\xi(0) = I$.

Suppose we assume that it does lie on such a subgroup. Without loss of generality we may also assume that $\xi(1) = \xi$, and then we have

$$\xi(0, x, x^*) = x\,, \qquad \xi(1, x, x^*) = \xi(x, x^*)\,, \tag{71}$$

as well as

$$\begin{aligned} \xi(s, \xi(t, x, x^*), \xi^*(t, x, x^*)) &= \xi(t, \xi(s, x, x^*), \xi^*(s, x, x^*)) \\ &= \xi(s+t, x, x^*)\,. \end{aligned} \tag{72}$$

Note that the diffeomorphism (70) leaves the origin fixed. Therefore

$$\xi(0, 0, 0) = 0\,, \qquad \xi(1, 0, 0) = 0\,. \tag{73}$$

Define

$$z(t) \overset{\text{def}}{=} \xi(t, 0, 0) . \tag{74}$$

The function $z(t)$ describes a closed curve passing through the origin in the complex plane. Using eqs. (72) and (73) we find

$$\begin{aligned} \xi(z(t), z^*(t)) &= \xi(1, z(t), z^*(t)) \\ &= \xi(1, \xi(t, 0, 0), \xi^*(t, 0, 0)) = \xi(t, \xi(1, 0, 0), \xi^*(1, 0, 0)) \\ &= \xi(t, 0, 0) = z(t) , \end{aligned} \tag{75}$$

which implies that the diffeomorphism (70) leaves every point on this closed curve fixed. But the only curve passing through the origin that (70) leaves fixed is the degenerate curve consisting of the single point x = 0. Therefore every one of the diffeomorphisms $\xi(t)$ must leave the origin fixed:

$$\xi(t, 0, 0) = 0 \quad \text{for all } t. \tag{76}$$

Since ξ and the $\xi(t)$ are C^∞ we may consider their formal Taylor series at the origin. The formal Taylor series for ξ, which is just expression (70), must lie on the one-parameter group of formal Taylor series for the $\xi(t)$, which may be written in the form

$$\xi(t, x, x^*) = \sum_{m,n=0}^{\infty} a_{m,n}(t)\, x^m x^{*n} . \tag{77}$$

Furthermore, these formal Taylor series must satisfy (formally) eqs. (72).

In view of eqs. (71) and (76) it is evident that

$$\left.\begin{aligned} &a_{0,0}(t) = 0 \quad \text{for all } t ; \\ &a_{1,0}(0) = 1 , \quad \text{all other } a_{m,n}(0)\text{'s vanish} ; \\ &a_{1,0}(1) = e^{\frac{2\pi i}{N}} , \quad a_{N+1,0}(1) = \alpha , \quad \text{all other } a_{m,n}(1)\text{'s vanish.} \end{aligned}\right\} \tag{78}$$

Moreover, inserting (77) into (72) with s = t = ½, one finds

$$\begin{aligned} e^{\frac{2\pi i}{N}} x + \alpha x^{N+1} &= \sum_{m,n=0}^{\infty} a_{m,n}(\tfrac{1}{2})[\xi(\tfrac{1}{2}, x, x^*)]^m [\xi^*(\tfrac{1}{2}, x, x^*)]^n \\ &= a_{1,0}(\tfrac{1}{2})[a_{1,0}(\tfrac{1}{2})x + a_{0,1}(\tfrac{1}{2})x^* + \cdots] \\ &\quad + a_{0,1}(\tfrac{1}{2})[a_{1,0}(\tfrac{1}{2})x + a_{0,1}(\tfrac{1}{2})x^* + \cdots] + \cdots , \end{aligned}$$

whence

$$e^{\frac{2\pi i}{N}} = [a_{1,0}(\tfrac{1}{2})]^2 + |a_{0,1}(\tfrac{1}{2})|^2, \tag{79}$$

$$0 = a_{0,1}(\tfrac{1}{2})\, \mathrm{Re}\, a_{1,0}(\tfrac{1}{2}). \tag{80}$$

Suppose $a_{0,1}(\frac{1}{2}) \neq 0$. Then $a_{1,0}(\frac{1}{2})$ must be pure imaginary, and the right hand side of eq. (79) must be a real number, which contradicts the left hand side. Therefore

$$a_{0,1}(\tfrac{1}{2}) = 0, \quad a_{1,0}(\tfrac{1}{2}) = e^{\frac{1}{2}\left(\frac{2\pi i}{N} + 2\pi i K\right)},$$

for some integer K. Repeating this reasoning for $s = t = \frac{1}{4}$, $s = t = \frac{1}{8}$, etc., one obtains, by continuity,

$$\left.\begin{aligned} &a_{0,1}(t) = 0 \quad \text{for all } t, \\ &a_{1,0}(t) = e^{\beta t}, \quad \beta = 2\pi i\left(\tfrac{1}{N} + K\right). \end{aligned}\right\} \tag{81}$$

We now have

$$\xi(t, x, x^*) = e^{\beta t} x + \sum_{m+n\geq 2}{}' a_{m,n}(t)\, x^m x^{*n}. \tag{82}$$

Insertion of this formal series into (72) yields

$$a_{m,n}(s+t) = e^{\beta s} a_{m,n}(t) + e^{(m-n)\beta t} a_{m,n}(s), \quad m+n=2. \tag{83}$$

This functional equation can be solved by differentiating with respect to s and setting s = 0:

$$\left(\frac{d}{dt} - \beta\right) a_{m,n}(t) = \dot{a}_{m,n}(0)\, e^{(m-n)\beta t}, \quad m+n=2. \tag{84}$$

(Here the dot denotes the derivative.) The Green's function for the operator $d/dt - \beta$ appropriate to the boundary conditions (78) is $[\theta(t-t')\theta(t') - \theta(t'-t)\theta(-t')]e^{\beta(t-t')}$ where θ is the step function. Use of this Green's function yields

$$a_{m,n}(t) = \frac{\dot{a}_{m,n}(0)}{(m-n-1)\beta}\left[e^{(m-n)\beta t} - e^{\beta t}\right], \quad m+n=2, \tag{85}$$

which is easily verified to satisfy (83). Now if N is large $a_{m,n}(1)$ $(m+n=2)$ must vanish in virtue of the last of eqs. (78). But the right hand side of (85) does not vanish at t = 1 unless $\dot{a}_{m,n}(0) = 0$. Therefore

$$a_{m,n}(t) = 0 \quad \text{for all } t \text{ when } m+n=2, \tag{86}$$

and hence

$$\xi(t, x, x^*) = e^{\beta t} x + \sum_{m+n\geq 3} a_{m,n}(t)\, x^m x^{*n}. \tag{87}$$

Inserting this series into (72) one gets

$$a_{m,n}(s+t) = e^{\beta s} a_{m,n}(t) + e^{(m-n)\beta t} a_{m,n}(s), \quad m+n=3, \tag{88}$$

which is identical with eq. (83) except that now m + n = 3. The solution is the same as before:

$$a_{m,n}(t) = \frac{\dot{a}_{m,n}(0)}{(m-n-1)\beta}\left[e^{(m-n)\beta t} - e^{\beta t}\right], \quad m+n=3. \tag{89}$$

It is now possible for the factor m−n−1 in the denominator to vanish, in which case this solution is replace by its limit as m−n $\to$ 1:

$$a_{m,n}(t) = \dot{a}_{m,n}(0)\, t\, e^{\beta t}, \qquad m-n=1. \tag{90}$$

Once again, comparing (89) and (90) with the boundary condition $a_{m,n}(1)=0$ (m + n = 3), one must conclude that

$$a_{m,n}(t) = 0 \quad \text{for all } t \text{ when } m+n=3. \tag{91}$$

In fact, continuing in this way one finds

$$a_{m,n}(t) = 0 \text{ for all } t \text{ and all } m,n \text{ with } 2 \leq m+n \leq N. \tag{92}$$

One arrives finally at the case m + n = N + 1, where one obtains

$$a_{N+1,0}(t) = \frac{1}{N\beta}\, \dot{a}_{N+1,0}(0)\, e^{\beta t}\left(e^{N\beta t} - 1\right). \tag{93}$$

This expression vanishes at t = 1, precisely where we don't want it to! According to (78) we must have $a_{N+1,0}(1) = \alpha$. We have thus arrived at a contradiction, Q.E.D.

Since only a small (infinitesimal) neighbourhood of the origin is really involved in the above analysis, it follows that the restriction to $\mathbb{R}^2$ is not essential. For any differentiable manifold of dimension greater than or equal to 2 there exist C^∞ diffeomorphisms arbitrarily close to the identity that do not lie on one-parameter subgroups of C^∞ diffeomorphisms.

GAUGE CONDITIONS

A gauge condition is a set of constraints that picks out a subspace in the configuration space Φ , of codimension equal to the dimension of the gauge group G. The gauge condition is said to be globally valid if this subspace intersects each orbit in precisely one point. Such a subspace exists in the Yang-Mills case only if the gauge group corresponds to an untwisted fiber bundle. For the diffeomorphism group it probably exists if spacetime is diffeomorphic to $\mathbb{R}^n$. We confine our attention to these cases. The subspace may then be regarded as representing the orbit manifold Φ/G. Each orbit is represented by the point at which it intersects the subspace.

To express this idea in equations one may think of the variables φ^i as being replaced by other variables I^A, P^α, where the I^A label individual orbits and are gauge invariant, and the P^α label corresponding points in each orbit. The point on each orbit that is selected by the given gauge condition may be chosen as the origin of the "coordinates" P^α in that orbit. The gauge condition is then simply $P^\alpha = 0$.

It will actually prove convenient to work with the continuum of gauge conditions

$$P^\alpha[\varphi] = \zeta^\alpha, \tag{94}$$

where the ζ^α are constants (i.e., independent of the φ^i) whose values range over some preselected domain. Explicit functional forms for the P^α in terms of the φ^i may be obtained (in principle) as follows. Remembering that each (generic) orbit is a copy of G, choose the P^α to be a set of group coordinates. Since the action of the gauge group on each (generic) orbit mimics its action on itself such P's must be solutions of the functional differential equations *

$$P^\alpha{}_{,i}[\varphi]Q^i{}_\beta[\varphi] = Q^\alpha{}_\beta[P[\varphi]] . \tag{95}$$

The domain over which the ζ^α in eq. (94) range may then be taken to be the full domain of the group coordinates.

Equations (95) do not suffice completely to determine the P^α. Additional conditions are needed to "line up" corresponding points on adjacent orbits. One possible way to do the lining up is as follows. Introduce into the configuration space Φ one of the metrics γ_{ij} previously discussed. Choose a generic orbit and call it the base orbit. Call the identity element on that orbit the base point. Let V be the subspace of Φ generated by the set of all geodesics emanating from the base point in directions orthogonal to the base orbit. As previously noted, these geodesics intersect all orbits in their paths orthogonally. Using the fact that every pair of points in Φ can be connected by a unique geodesic (at least in the Yang-Mills and gravitational cases) and the fact that a geodesic cannot be simultaneously orthogonal to and tangent to an orbit, one can show that V ultimately intersects all orbits. To keep it from intersecting a

* These equations are readily verified to be integrable in virtue of the identities (29) and (62).

given orbit more than once one may terminate each of the generating geodesics as soon as it strikes a boundary point of Φ/G. V is then topologically (but not necessarily metrically) a copy of Φ/G.

To gain an appreciation of some of the metrical situations that can arise think of Φ as being $\mathbb{R}^3$ and G as being the group of screw motions with fixed nonvanishing pitch about some axis. The orbits are then helices and all, including the axis itself, are generic. If $\mathbb{R}^3$ bears the Cartesian metric then the orbit space Φ/G is topologically but not metrically a plane. Note that in this example there exist no surfaces that intersect all orbits orthogonally, although every plane not containing the axis is perpendicular to some orbit at its intersection point and is a surface like V, based on that orbit.

Returning now to the general problem, we may place the identity element on each orbit at the point where the orbit intersects V. If another subspace V' is constructed like V but starting from another point on the base orbit, it too will intersect all the orbits. Because the group operations are isometries of ${}_i\gamma_j$, the P^α will be constants over V'. That is, once the identity points are "lined up" all the other points are automatically lined up too. The gauge condition (94) is therefore globally valid for all ξ^α in the domain G.

Unfortunately in practice it is almost hopelessly difficult to implement constructions like this one, which are guaranteed to yield globally valid gauge conditions. In the present construction, because V is generally orthogonal to none but the base orbit, one is faced with the problem of solving global functional constraints rather than functional differential equations. Even the functional differential equations that one has, namely eqs. (95), are highly nontrivial. In the Yang-Mills and gravitational cases they take respectively the forms

$$[\delta P^\alpha(x)/\delta A^\beta{}_\mu(x')]_{;\mu'} = q^\alpha{}_\beta(P(x))\,\delta(x,x'), \tag{96}$$

$$2[\delta P^\mu(x)/\delta g_{\nu\sigma}(x')]_{;\sigma'} = \delta^\mu{}_\nu\,\delta(P(x),x'), \tag{97}$$

(see eqs. (22), (28), (54) and (55)).

By far the bulk of all work on Yang-Mills theory and quantum gravity has made use of <u>linear</u> gauge conditions, i.e., conditions (94) with P^α taken in the form

$$P^{\alpha}[\varphi] = P^{\alpha}{}_{i}[\varphi_B]\,\phi^{i}\,, \qquad \phi^{i} = \varphi^{i} - \varphi_B^{i} \tag{98}$$

where φ_B^{i} is some fiducial field, often called a background field. To ensure that the subspaces defined by (94) do indeed intersect the orbits uniquely, at least in the vicinity of the background field and with the ζ^{α} close to zero, one often makes use of the orthogonality idea by choosing

$$P^{\alpha}{}_{i}[\varphi_B] = (-1)^{\alpha(j+1)}\, Q^{j}{}_{\alpha}[\varphi_B]\; {}_{j}\gamma_{i}[\varphi_B]\,. \tag{99}$$

For example, if ${}_{i}\gamma_{j}$ has the form (43) then, with the choice (99), the condition $P^{\alpha} = 0$ becomes

$$g_B^{1/2}\,(2\,\phi_{\mu}{}^{\nu} + \lambda\,\delta_{\mu}{}^{\nu}\,\phi_{\sigma}{}^{\sigma})_{;\nu} = 0\,, \qquad \phi_{\mu\nu} = g_{\mu\nu} - g_{B\mu\nu}\,. \tag{100}$$

Here indices are raised and lowered by means of the background metric $g_{B\mu\nu}$ and the covariant derivative is defined in terms of it. With λ set equal to -1 (see the comments following eqs. (46) and (47)) this is a very popular gauge condition in quantum gravity. The corresponding condition in Yang-Mills theory, with ${}_{i}\gamma_{j}$ given by (45), is

$$\phi_{\alpha}{}^{\mu}{}_{;\mu} = 0\,, \qquad \phi^{\alpha}{}_{\mu} = A^{\alpha}{}_{\mu} - A_B{}^{\alpha}{}_{\mu}\,. \tag{101}$$

Here indices are raised and lowered by means of the metrics $\gamma_{\alpha\beta}$ and $\eta_{\mu\nu}$ and the covariant derivative is defined in terms of the background field $A_B{}^{\alpha}{}_{\mu}$. Condition (101) is known as the Lorentz condition.

Linear gauge conditions are extremely convenient in perturbation theory, where the field φ^{i} is treated as if it never gets very far from the background φ_B^{i}. Covariant (with respect to the background) gauge conditions like (100) and (101) are usually the best, but for some purposes noncovariant gauges (e.g., the Coulomb gauge in Yang-Mills theory) are more useful. In non-perturbative studies, however, linear gauge conditions have to be used with great care (see Gribov (1977)). At least five things can go wrong with linear gauge conditions when applied globally: (1) The subspace defined by a linear condition may or may not have a boundary, and if it does this boundary may not coincide with the boundary (if any) of Φ/G. (2) The subspace defined by a linear condition may intersect some orbits more than once. (3) There may be some orbits that it does not intersect at all. (4) Even if it intersects all orbits when the ζ^{α} in eq (94) have certain values, it may not intersect all orbits when the ζ^{α} have other values. This means that there is no natural domain for the ζ^α. (5) When G is "twisted" there are no globally valid gauge conditions at all, linear or otherwise.

If any of the above situations hold, the subspace defined by (94) will not represent Φ/G faithfully. It is possible in some cases to patch things up so that the advantages of linear gauge conditions can be maintained. This has been done in certain global studies in Yang-Mills theory. However, the diffeomorphism group, as we have repeatedly emphasized, is a much more complicated group than the Yang-Mills group and both the difficulties to which it gives rise globally and the opportunities that it presents for technical innovation are almost unknown at the present time. In order to keep all options open we shall first develop the formal theory using foolproof gauge conditions, such as those based on group coordinates, and then make some remarks about how things might go when other gauge conditions are used.

THE ACTION. VERTEX FUNCTIONS. RENORMALIZABILITY.

The dynamical behavior of any field is determined by its action functional S. The action functionals of (pure) Yang-Mills and gravity thories are respectively

$$S_A = -\frac{1}{4}\int F_{\alpha\mu\nu} F^{\alpha\mu\nu}\, d^n x \,, \tag{103}$$

$$S_g = 2\int g^{1/2} R\, d^n x \,. \tag{104}$$

As long as the limits of integration are not specified these integrals must be regarded as purely formal expressions that serve merely to yield the dynamical equations:

$$0 = \delta S_A / \delta A^\alpha{}_\mu \equiv -F_\alpha{}^{\mu\nu}{}_{;\nu} \,, \tag{105}$$

$$0 = \delta S_g / \delta g_{\mu\nu} \equiv -2 g^{1/2} (R^{\mu\nu} - \tfrac{1}{2} g^{\mu\nu} R) \,. \tag{106}$$

For some purposes, however, values need to be assigned to the actions. Integration boundaries must then be specified and, in the case of the gravitational field, a surface integral must be split off from (104) so that the integrand involves derivatives of $g_{\mu\nu}$ of order no higher than the first.

We refer the student to standard references (e.g., Misner, Thorne and Wheeler (1973)) for analyses of the initial value problems associated with eqs. (105) and (106). From these analyses it is readily deduced that in a spacetime of n dimensions the Yang-Mills field has n-2 degrees of freedom per spatial point and the gravitational field has $\frac{1}{2}n(n-3)$.

In the generic notation, eqs. (103) and (104) are written

$$S_{,i} = 0 . \tag{107}$$

Gauge invariance of the theory is guaranteed by the identity

$$S_{,i} Q^i{}_\alpha \equiv 0 , \tag{108}$$

or, more explicitly,

$$F_\alpha{}^{\mu\nu}{}_{;\nu\mu} \equiv 0 , \qquad 4[g^{1/2}(R^{\mu\nu} - \tfrac{1}{2} g^{\mu\nu} R)]_{;\nu} \equiv 0 . \tag{109}$$

The left hand side of eq. (107) transforms linearly under the gauge group and hence the gauge group leaves the field equations intact. This is most easily seen by functionally differentiating eq. (108), which yields

$$\delta S_{,i} \equiv S_{,ij}\delta\varphi^j \equiv S_{,ij} Q^j{}_\alpha \delta\xi^\alpha \equiv -(-1)^{i\alpha} S_{,j} Q^j{}_{\alpha,i} \delta\xi^\alpha . \tag{110}$$

The action functionals (103), (104) may be expanded in functional Taylor series about a background field. In generic notation one writes

$$\left.\begin{aligned} S &= S_B + (S_{,i})_B \phi^i + \tfrac{1}{2!}(S_{,ij})_B \phi^j\phi^i + \tfrac{1}{3!}(S_{,ijk})_B \phi^k\phi^j\phi^i + \cdots \\ \phi^i &= \varphi^i - \varphi_B{}^i . \end{aligned}\right\} \tag{111}$$

If the background fields satisfy the classical field equations then the second term on the right may be omitted.

The functional derivatives $(S_{,i_1 \ldots i_N})_B$ with $N \geq 3$ are known as (bare) <u>vertex functions</u>. In the case of the Yang-Mills field the vertex functions with $N > 4$ vanish and the Taylor series terminates. In the case of the gravitational field the series may or may not terminate depending on what choice is made for the basic field variables. By expressing inverse matrices in terms of minors and determinants, and by examining the number of determinants needed to yield unit total weight for the integrand of (104), one easily verifies that if the basic field variables are taken to be $\mathfrak{g}^{\mu\nu} = g^r g^{\mu\nu}$ and r is chosen to be 5/(4n+2) then the vertex functions with $N > 2n+1$ vanish. Alternatively, if $\mathfrak{g}_{\mu\nu} = g^{-r} g_{\mu\nu}$ are chosen as the basic field variables, with r = 5/6n-2), then the vertex functions with $N > 3n-1$ vanish.* There are three reasons, however, why neither of these choices is useful. First, the vertex functions of gravity theory are exceedingly complicated,

* Both of these choices require $n \neq 2$. (See eqs. (40)).

involving thousands of terms already for N=4. Nobody is going to work out the vertex functions up to maximum order even with the aid of a computer. Second, any imagined advantage in these choices is lost as soon as one tries to introduce dimensional regularization into the quantum theory. A specific choice of field variables has then to be made, and it cannot vary continuously with the dimension. Third, although a series that terminates has an infinite radius of convergence, the range of the variables $\phi^{\mu\nu}$ $(\stackrel{\text{def}}{=} g^{\mu\nu} - g_B{}^{\mu\nu})$ or $\phi_{\mu\nu}$ $(\stackrel{\text{def}}{=} g_{\mu\nu} - g_{B\mu\nu})$ is in fact limited. These variables must avoid regions where the signature of the metric tensor changes.

The third reason is the most important, at least in perturbation theory. As is well known, the Feynman rules are obtained by inserting the expansion (111) into the Feynman functional integral (see the next section) and evaluating the integral as a sum (asymptotic series) of Gaussian integrals, with the ϕ^i ranging from $-\infty$ to ∞. Any constraint on the ϕ^i would make these integrals almost impossible to evaluate, and although one may for some purposes wish to extend the Feynman integrand into nonphysical regions, one never does this by naively removing constraints.

These remarks suggest, in fact, that none of the variables (40) is good to use in perturbation theory. A better choice would be something like

$$\left.\begin{aligned} &\underset{\sim}{\phi} \stackrel{\text{def}}{=} [\ln(\underset{\sim}{g}\,\underset{\sim}{\eta}^{-1})]\underset{\sim}{\eta}\,, \qquad \underset{\sim}{g} = e^{\underset{\sim}{\phi}\,\underset{\sim}{\eta}^{-1}}\underset{\sim}{\eta}\,, \\ &\underset{\sim}{\phi} \stackrel{\text{def}}{=} (\phi_{\mu\nu}),\ \underset{\sim}{g} \stackrel{\text{def}}{=} (g_{\mu\nu}),\ \underset{\sim}{\eta} \stackrel{\text{def}}{=} (\eta_{\mu\nu}),\ \underset{\sim}{\eta}^{-1} \stackrel{\text{def}}{=} (\eta^{\mu\nu}), \end{aligned}\right\} \tag{112}$$

which maintains the signature of the spacetime metric. With these variables the series (113), of course, does not terminate, and one speaks of gravity theory as being a <u>non-polynomial Lagrangian theory</u> (Isham, Strathdee and Salam (1971), (1972)). It will be noted that all such "safe" variables inevitably transform <u>non</u>linearly under the diffeomorphism group.

Regardless of the choice of variables it is not difficult to draw preliminary conclusions about the renormalizability or nonrenormalizability (in perturbation theory) of a given quantum field theory. Although momentum space is not, in an absolute sense, appropriate for use in quantum gravity, conclusions about the high energy behaviour of amplitudes in perturbation theory may be safely drawn with its aid. Consider a Feynman graph with L_e external lines, L_i internal Lines, and V_N Nth-order vertices $(N \geq 3)$. L_e, L_i and the V_N are related by the topological condition

$$L_e + 2L_i = \sum_N N V_N\ . \tag{113}$$

The number of independent closed loops, or momentum integrations, in the graph is given by

$$I = L_i - \sum_N V_N + 1 . \tag{114}$$

In quantum gravity 2 powers of momentum are associated with each vertex, -2 powers with each internal line, and n powers with each momentum integration. The superficial degree of divergence of the graph is therefore

$$D = -2L_i + 2\sum_N V_N + nI = (n-2)I + 2, \tag{115}$$

which, for $n > 2$, increases without limit as the number of independent closed loops increases. This means that for $n > 2$, there is an infinite number of primitive divergences, and, if one attempts to compute order by order, an infinite number of experimentally determined coupling constants is needed to determine the theory. These conclusions are not altered if account is taken of the "ghost" contributions which, as we shall see in the following sections, must be included. The theory is said to be nonrenormalizable.

In Yang-Mills theory, in contrast, only one power of momentum is associated with each 3rd-order vertex, and the 4th-order vertices have no momentum dependence at all. This leads to

$$D = -2L_i + \sum_N (4-N)V_N + nI = 4 + (n-4)I - L_e . \tag{116}$$

For $n > 4$ this theory too is nonrenormalizable, but for $n=4$ and an arbitrary background field there are only four primitive divergences (corresponding to $L = 1,2,3,4$), and the theory is renormalizable. The proof of renormalizability is not trivial and depends crucially on gauge invariance as well as some of the formal developments to be discussed in the following sections. The primitive divergences turn out to be related in virtue of gauge invariance.

The nonrenormalizability of standard quantum gravity has stimulated investigations of alternative theories in which terms of the form $g^{\frac{1}{2}}(\alpha R^2 + \beta R_{\mu\nu} R^{\mu\nu})$ are added to the integrand of expression (104). Such theories generally suffer from "physical ghosts" with negative probabilities, but they do improve the convergence situation. Each vertex now carries 4 powers of momentum and each internal line carries -4 powers. This leads to

$$D = -4L_i + 4\sum_N V_N + nI = (n-4)I + 4 \tag{117}$$

When $n=4$ all diagrams have the same superficial degree of divergence, namely 4. There is an infinity of primitive divergences, but they are all related by gauge invariance, and only three experimental coupling constants are required. The proof of renormalizability has been carried out by Stelle (1977) using methods similar to those applied to Yang-Mills theory.

It will be observed, in virtue of eq. (115), that the same methods should work in the case of standard quantum gravity when n=2, although since the number of degrees of freedom in the field is then negative it is not clear what such a theory means. Weinberg (1979) has studied the asymptotic stability of quantum gravity when $n=2+\epsilon$, $\epsilon \ll 1$, and has given plausibility arguments concerning its relevance for trying to make sense out of the theory when n=4.

In addition to its ultraviolet divergences quantum gravity also possesses infrared divergences. Gravitational field quanta -gravitons- are massless. This fact in itself need not lead to difficulties worse than those encountered in quantum electrodynamics where the divergences are completely understood and are removable by standard methods. Gravitons, however, are coupled to other massless quanta (photons, neutrinos, etc.) as well as to themselves. In Yang-Mills theory as well as in massless electrodynamics such a situation gives rise to infrared divergences of a new type that cannot be removed by standard techniques or argued away on physical grounds. In quantum gravity these new divergences are miraculously absent (Weinberg (1965) and DeWitt (1967c)). It appears therefore that the mysteries of Yang-Mills theory and gravity theory lie at opposite ends of the momentum spectrum. There is an increasing body of evidence that the Yang-Mills field solves its infrared dilemma by adopting a nonstandard behaviour at long wavelengths, which is intimately related to the phenomena of quark confinement and dynamical symmetry breaking. These phenomena may also bear a technical relation to the failure of gauge conditions to be globally valid in Yang-Mills theory (Gribov (1977)). No analogous phenomena are known to exist for gravity, at least when spacetime is diffeomorphic to $\mathbb{R}^n$. The mysteries of gravitation theory thus appear to lie solely at the high end of the momentum spectrum.

THE FEYNMAN FUNCTIONAL INTEGRAL. FACTORING OUT THE GAUGE GROUP.

Consider a transition amplitude of the form $\langle \text{out} \mid \text{in} \rangle$ where the vectors $|\text{in}\rangle$ and $|\text{out}\rangle$ refer to states in which the field is maximally specified (in the quantum mechanical sense, e.g., in terms of complete sets of commuting observables) in regions "in" and "out" respectively. These states need not be "vacuum" states and the regions "in" and "out" need not refer to the infinite past and future respectively. If the background field (which enters naturally in most calculations) has singularities in the past and/or future, $|\text{in}\rangle$ and $|\text{out}\rangle$ may be defined not in terms of observables at all but by some analytic continuation procedure (e.g., to the "Euclidean sector") that removes the singularities. It will be assumed only that the "in" and "out" regions lie respectively to the past and future of the region of dynamical interest.

There are many ways of showing that the amplitude $\langle\text{out}\,|\,\text{in}\rangle$ can be expressed as a formal functional integral:

$$\langle\text{out}|\text{in}\rangle = N\int e^{iS[\varphi]}\mu[\varphi]\,d\varphi\,, \qquad d\varphi \overset{\text{def}}{=} \prod_i d\varphi^i\,. \tag{118}$$

Here N is a normalization constant, $S[\varphi]$ is the classical action functional, $\mu[\varphi]$ is chosen to make the "volume element" $\mu d\varphi$ gauge invariant (see eq. (50)), and the integration is to be extended over all fields φ that satisfy the boundary conditions appropriate to the given "in" and "out" states. We have remarked earlier (and will show later) that μ may be set "effectively" equal to unity if the φ^i are chosen to transform linearly under the gauge group. We shall assume that such a choice has been made and henceforth drop μ from the theory. (It can always be restored if desired.)

Expression (118) was first derived by Feynman (1948) in ordinary quantum mechanics, without gauge groups, and later (1950) applied by him to field theory. The full extension to field theories with gauge groups is the work of many people, and the student is referred to the literature for details.* When the Feynman integral is applied to the gravitational field the only additional comment that needs to be made is that the integration may have to embrace as many topologies as can be reached by analytic continuation from the given background topology.

If any of the fields φ^i in (118) are fermionic the integration with respect to them is to be carried out according to the formal rules for integrating with respect to anticommuting variables that were first introduced by Berezin (1966). These rules are analogous in many ways to the well known rules for ordinary definite integrals from $-\infty$ to ∞ with integrands that vanish asymptotically. For example, integrals of total derivatives vanish, and the position of the zero point may be shifted. On the other hand, with Berezin rules, transformations of variables and evaluation of Gaussian integrals lead to determinants precisely inverse to those of standard theory. When both bosonic and fermionic fields are involved it is the <u>super</u> determinant that appears.

* Useful modern references are Fadde'ev (1969), (1976) and Abers and Lee (1973).

All physical amplitudes can be deduced from expression (118) by examining how $\langle \text{out} \mid \text{in} \rangle$ changes under variations in the action. Physical amplitudes can alternatively be obtained by judicious use of

$$\langle \text{out} | T(A[\underset{\sim}{\varphi}]) | \text{in} \rangle = N \int A[\varphi] e^{iS[\varphi]} d\varphi , \tag{119}$$

where $A[\underset{\sim}{\varphi}]$ is any functional of the field operators $\underset{\sim}{\varphi}^i$, and the "T" symbol removes ambiguities about ordering the $\underset{\sim}{\varphi}^i$ by arranging them chronologically (with appropriate $\pm$ signs thrown in if any of the $\underset{\sim}{\varphi}^i$ are fermionic).

When a gauge group is present the integration in (118) is redundant. Furthermore (119) is generally ambiguous, unless A is gauge invariant, in which case the integration in (119) too is redundant. This is because, owing to the gauge invariance of the classical action, the exponent in the integrands of (118) and (119) remains constant as φ ranges over a group orbit in the configuration space Φ. One can remove this redundancy and/or ambiguity by adopting a gauge condition like (94). The details of the procedure were first given by Fadde'ev and Popov (1967).

Let ξ be an element of the gauge group G, with coordinates ξ^α, and let ${}^\xi\varphi$ be the field to which φ is displaced under the action of ξ. Define

$$\Delta[\zeta,\varphi] \stackrel{\text{def}}{=} \int_G \delta[P[{}^\xi\varphi] - \zeta] \det Q^{-1}[\xi]\, d\xi , \qquad d\xi \stackrel{\text{def}}{=} \prod_\alpha d\xi^\alpha , \tag{120}$$

where $\delta[\ ]$ is the delta functional, $P^\alpha[\varphi]$ are the functionals appearing in eq. (94), Q^{-1} is the inverse of the matrix formed out of the $Q^\alpha{}_\beta$ of eq. (53), and the integration extends over the entire gauge group! We shall assume that the gauge condition (94) is globally valid. The integrand in (120) then "switches on" at only one point in G, namely that point for which ${}^\xi\varphi$ is equal to the unique field φ_ξ lying on the orbit containing φ and picked out by the gauge condition:

$$P^\alpha[\varphi_\xi] = \zeta^\alpha . \tag{121}$$

By building infinitesimal parallelepipeds in the group manifold G and making use of eq. (53) one can verify that the combination $\det Q^{-1}[\xi] d\xi$ appearing in (120) is a right-invariant volume element, satisfying

$$\det Q^{-1}[\xi\xi'] \, d(\xi\xi') = \det Q^{-1}[\xi] \, d\xi \quad \text{for all } \xi' \text{ in } G. \tag{122}$$

The presence of this volume element renders the functional gauge invariant:

$$\Delta[\zeta, {}^{\xi'}\varphi] = \Delta[\zeta, \varphi] \quad \text{for all } \xi' \text{ in } G. \tag{123}$$

Several comments must be made about its use, however. In the case of the diffeomorphism group, with the Q's given by (54), it is easily checked that

$$Q^{-1\mu}{}_{\nu'}[\xi] = \delta^{\mu}{}_{\nu}\, \delta(x, \xi(x')) \frac{\partial(\xi(x'))}{\partial(x')} \,. \tag{124}$$

No one has ever discovered how to evaluate or give a meaning to the determinant of this continuous matrix. The right-invariant volume element of the diffeomorphism group, therefore, can only be defined (and used) purely formally. The same is true for the invariance group of supergravity theory. It should be noted that when the group is a supergauge group, possessing anticommuting as well as commuting coordinates, det Q^{-1} is a superdeterminant, and the integral (120) involves the Berezin rules. (Remark: The delta functional in (120) presents no difficulty. Delta functions of anticommuting variables turn out to be easy to define. They can even be given Fourier representations.)

The gauge invariance of Δ makes it an easy functional to evaluate. One has only to shift ϕ to ϕ_ζ so that the integrand in (120) switches on at the identity element I. All quantities can then be expanded in power series in $\xi^\alpha - I^\alpha$. For example, the argument of the delta functional becomes

$$\begin{aligned} P^\alpha[{}^\xi\phi_\zeta] - \zeta^\alpha &= P^\alpha[\phi_\zeta] - \zeta^\alpha + P^\alpha{}_{,i}[\phi_\zeta]\, Q^i{}_\beta[\phi_\zeta](\xi^\beta - I^\beta) + \cdots \\ &= F^\alpha{}_\beta[\phi_\zeta](\xi^\beta + I^\beta) + \cdots \end{aligned} \tag{125}$$

where

$$F^\alpha{}_\beta[\phi] \overset{\text{def}}{=} P^\alpha{}_{,i}[\phi]\, Q^i{}_\beta[\phi] \,. \tag{126}$$

Similarly, making use of eq. (68), we find

$$Q^{-1\alpha}{}_\beta[\xi] = \delta^\alpha{}_\beta - Q^\alpha{}_{\beta,\gamma}[I](\xi^\gamma - I^\gamma) + \cdots \tag{127}$$

and hence

$$\begin{aligned} \Delta[\zeta, \phi] &= \int_G \delta[F[\phi_\zeta](\xi - I) + \cdots][1 - (-1)^\alpha Q^\alpha{}_{\alpha,\beta}[I](\xi^\beta - I^\beta) + \cdots] d\xi \\ &= (\det F[\phi_\zeta])^{-1} \,, \end{aligned} \tag{128}$$

F being the matrix with elements $F^\alpha{}_\beta$. If any of the group indices is fermionic the determinant is again a superdeterminant. Note that if the P's are constructed according to eq. (95), $F[\phi]$ is identical to the matrix $Q[P[\phi]]$. Although this construction will not be assumed in what follows, we shall, for simplicity and conveneience, assume that the gauge condition (94) is globally valid for all ζ^α lying in the ranges of the functionals $P^\alpha[\phi]$.

The next step is to insert unity into the integrand of (118), in the guise of $(\Delta[\zeta,\varphi])^{-1}\int_G \delta[P[{}^{\xi}\varphi]-\zeta]\det Q^{-1}[\xi]\,d\xi$, and interchange the order of integrations, obtaining

$$\langle \text{out}|\text{in}\rangle = N\int_G \det Q^{-1}[\xi]\,d\xi \int d\varphi\, e^{iS[\varphi]}(\Delta[\zeta,\varphi])^{-1}\delta[P[{}^{\xi}\varphi]-\zeta]. \tag{129}$$

We have assumed a choice of variables for which the volume element $d\varphi$ is gauge invariant ($\mu=1$). $S[\varphi]$ and $\Delta[\zeta,\varphi]$ are also gauge invariant. Therefore a superscript ξ may be affixed to every φ in the integrand of (129) that does not already bear one. But every ${}^{\xi}\varphi$ is then a dummy, and hence all the ξ's may be removed. Making used of (128) one immediately obtains

$$\langle \text{out}|\text{in}\rangle = N'\int e^{iS[\varphi]}\det F[\varphi]\,\delta[P[\varphi]-\zeta], \tag{130}$$

where $F[\varphi_\zeta]$ has been replaced by $F[\varphi]$ in the integrand because of the presence of the delta functional, and where

$$N' \stackrel{\text{def}}{=} N\int_G \det Q^{-1}[\xi]\,d\xi. \tag{131}$$

The gauge group has now been factored out, and its "volume" has been absorbed into the new normalization constant N'. The integration in (130) is restricted to the subspace $P^{\alpha}[\varphi]=\zeta^{\alpha}$.

The technique of confining the fields φ^i to a particular subspace can also be used to remove the ambiguity from the integral (119) when $A[\varphi]$ is not gauge invariant. Strictly speaking, matrix elements are definable only for gauge invariant operators. However, given a non-gauge-invariant operator $A[\underset{\sim}{\varphi}]$, one can construct a gauge invariant operator out of it by the following definition:

$$T(A[\underset{\sim}{\varphi}_\zeta]) \stackrel{\text{def}}{=} T\Big((\Delta[\zeta,\underset{\sim}{\varphi}])^{-1}\int_G A[{}^{\xi}\underset{\sim}{\varphi}]\,\delta[P[{}^{\xi}\underset{\sim}{\varphi}]-\zeta]\det Q^{-1}[\xi]\,d\xi\Big). \tag{132}$$

The chronological ordering symbol is used here so that the non-commutativity (or non-anticommutativity) of $A[{}^{\xi}\underset{\sim}{\varphi}]$ with both $(\Delta[\zeta,\underset{\sim}{\varphi}])^{-1}$ and the delta functional can be effectively ignored. Note that because the gauge group acts linearly on the $\underset{\sim}{\varphi}$'s there is no ambiguity about the symbol ${}^{\xi}\underset{\sim}{\varphi}$. Note, however, that diffeomorphisms in gravity theory can drag the field in very complicated ways. The chronological operation, which orders field operators solely by the value of the coordinate x^0, rearranges the "physical" fields in correspondingly complicated ways as the variable ξ in the integral (132) ranges over the group.

Applying eq. (119) to the operator $T(A[\underset{\sim}{\varphi}_\zeta])$ and following the same reasoning as was used in passing from eq. (129) to eq. (130), one finds

$$\langle \text{out} | T(A[\underset{\sim}{\varphi}_\zeta]) | \text{in} \rangle = N' \int A[\varphi] e^{iS[\varphi]} \det F[\varphi] \delta[P[\varphi] - \zeta] \, d\varphi , \qquad (133)$$

valid for any functional $A[\varphi]$.

AVERAGING OVER GAUGES

It is possible to develop a perturbation theory based on eqs. (130) and (133), but it is usually more convenient to work with a formalism from which the delta functionals have been eliminated. Note that although the parameters ζ^α appear on the right side of (130), the amplitude $\langle \text{out} \,|\, \text{in} \rangle$ is actually independent of them. Therefore nothing changes if we integrate over these parameters, with a weight factor.

In practically all studies of non-Abelian gauge theories to date, Gaussian weight factors of the form $\exp(\frac{1}{2} i \zeta^\alpha {}_\alpha M_\beta \zeta^\beta)$, where M is a nonsingular constant matrix having the symmetry ${}_\alpha M_\beta = (-1)^{\alpha+\beta+\alpha\beta} {}_\beta M_\alpha$, have been used. From a fundamental standpoint a Gaussian weight factor can be used only if the bosonic ζ's can range from $-\infty$ to ∞ without the gauge condition (94) becoming globally invalid — for example, if the P's satisfy eq. (95), with $Q^\alpha{}_\beta$'s based on canonical group coordinates (eq. (69)). This condition, however, is almost universally violated. Indeed, the Gaussian weight function is most frequently employed in combination with linear gauge conditions like (98), where it almost certainly introduces errors globally (i.e., in non-perturbative analyses).

In the case of quantum gravity, where the gauge group has no canonical coordinates, it seems particularly inappropriate to confine our attention to Gaussian weight factors. We shall therefore introduce a more general weight factor, of the form $\exp(iU[\zeta])$, where we specify nothing about the functional $U[\zeta]$ except the following three conditions: (1) U becomes infinite on the boundary of the allowable domain of the ζ's (which domain we are assuming coincides with the range of the $P[\varphi]$'s). (2) U and all its first functional derivatives $U_{,\alpha}$ vanish at some chosen point (e.g., at $\zeta^\alpha = I^\alpha$ when the P's are group coordinates); its second functional derivatives $U_{,\alpha\beta}$, however, form a nonsingular continuous matrix at that point. (3) U vanishes nowhere else, and its first derivatives all vanish simultaneously nowhere else. The third condition is imposed mainly for convenience. Note that all three are satisfied by the Gaussian exponent $\frac{1}{2}\zeta^\alpha {}_\alpha M_\beta \zeta^\beta$ whenever it can be legitimately used.

Inserting $\exp(iU[\varphi])$ into the integrand of eq. (130) and integrating over the ζ's, one obtains

$$\langle \text{out} | \text{in} \rangle = N''[U] \int e^{i(S[\varphi] + U[P[\varphi]])} \det F[\varphi] \, d\varphi , \qquad (134)$$

$$N''[U] \stackrel{\text{def}}{=} N' / \int e^{iU[\zeta]} d\zeta \,, \tag{135}$$

the integration domain in (135) being understood to be the allowable domain of the ζ's. Equation (133) too may be replaced by a weighted average. Defining

$$T(A[\underset{\sim}{\varphi}]) \stackrel{\text{def}}{=} \left(\int e^{iU[\zeta]} d\zeta\right)^{-1} \int T(A[\underset{\sim}{\varphi}_\zeta]) e^{iU[\zeta]} d\zeta \,, \tag{136}$$

one may write

$$\langle \text{out} | T(A[\underset{\sim}{\varphi}]) | \text{in} \rangle = N''[U] \int A[\varphi] e^{i(S[\varphi] + U[P[\varphi]])} \det F[\varphi] \, d\varphi \,. \tag{137}$$

Equation (137), and generalizations of it, will be used frequently in the following sections. Definitions (132) and (136) reveal precisely what kind of averaged quantum operator is associated with each classical functional $A[\varphi]$ in this formalism. Note that if $f[\zeta]$ is any functional of the ζ's we have

$$T(f[P[\underset{\sim}{\varphi}_\zeta]]) = f[\zeta] \tag{138}$$

and

$$\langle \text{out} | T(f[P[\underset{\sim}{\varphi}]]) | \text{in} \rangle = \langle f \rangle \langle \text{out} | \text{in} \rangle \tag{139}$$

where

$$\langle f \rangle \stackrel{\text{def}}{=} \frac{\int f[\zeta] e^{iU[\zeta]} d\zeta}{\int e^{iU[\zeta]} d\zeta} \tag{140}$$

Having so freely manipulated formal expressions we should now check that no inconsistencies have crept into our results, by verifying directly that the right side of (134), for example, is truly independent of the choices we have made for the functionals $P^\alpha[\varphi]$ and $U[\zeta]$. Obviously the right side <u>will</u> be affected if we naively switch to P's for which the gauge condition (94) is no longer globally valid for all ζ's in the range of the P's. Therefore we must assume that the changes δP^α (which, without loss of generality, may be taken infinitesimal) maintain global validity.

We also confine our attention to changes δU that leave the location of the zero of U, as well as the three conditions that we imposed upon U, intact. It is not difficult to see that δU may then always be expressed in the form

$$\delta U[\zeta] = U_{,\alpha}[\zeta] \, \delta V^\alpha[\zeta] \,, \tag{141}$$

where the δV^α vanish at the zero of U. Note that under this change we have

$$\delta N''[U] = -iN''[U] \frac{\int e^{iU[\xi]} U_{,\alpha}[\xi]\,\delta V^{\alpha}[\xi]\,d\xi}{\int e^{iU[\xi]}\,d\xi} = N''[U]\langle(-1)^{\alpha}\delta V^{\alpha}{}_{,\alpha}\rangle, \tag{142}$$

the final form being obtained by an integration by parts in which the boundary of the integration domain contributes nothing because $\exp(iU[\xi])$ oscillates infinitely rapidly there.

Making use of eqs. (126), (139), (141) and (142) we now have

$$\begin{aligned}\delta\langle\text{out}|\text{in}\rangle = N''[U]\int e^{i(S[\varphi]+U[P[\varphi]])}\{&(-1)^{\alpha}\delta V^{\alpha}{}_{,\alpha}[P[\varphi]]\\ &+ iU_{,\alpha}[P[\varphi]](\delta V^{\alpha}[P[\varphi]] + \delta P^{\alpha}[\varphi])\\ &+(-1)^{\alpha}F^{-1\alpha}{}_{\beta}[\varphi]\,\delta P^{\beta}{}_{,i}[\varphi]Q^{i}{}_{\alpha}[\varphi]\}\det F[\varphi]\,d\varphi\,,\end{aligned} \tag{143}$$

where the inverse F^{-1}, if it is a Green's function (as it often will be), must satisfy the boundary conditions appropriate to the "in" and "out" states. The integral (143) does not obviously vanish. The way to show that it is nevertheless zero is as follows. Replace each φ^{i} in the integral (134) by $\bar{\varphi}^{i}$, where

$$\bar{\varphi}^{i} = \varphi^{i} + Q^{i}{}_{\alpha}[\varphi]\,\delta\xi^{\alpha}[\varphi]\,, \tag{144}$$

$$\delta\xi^{\alpha}[\varphi] = F^{-1\alpha}{}_{\beta}[\varphi](\delta V^{\beta}[P[\varphi]] + \delta P^{\beta}[\varphi])\,. \tag{145}$$

Since the φ's are just dummies this replacement has no effect. However, it is not difficult to show that the net apparent change in the integral is given precisely by (143) provided one is entitled to make the identifications

$$(-1)^{i(\alpha+1)}Q^{i}{}_{\alpha,i} = 0\,, \qquad (-1)^{\beta(\alpha+1)}c^{\beta}{}_{\alpha\beta} = 0\,. \tag{146}$$

We shall comment on these equations presently.

It is easy to see that the second term inside the curly brackets in (143) comes from the change that the replacement $\varphi \to \bar{\varphi}$ induces in the exponent of (134). That the first and third terms come from the change in the product $\det F[\varphi]\,d\varphi$ may be shown as follows. First compute

$$\begin{aligned}\bar{\varphi}^{i}{}_{,j} = \delta^{i}{}_{j} &+ (-1)^{j\alpha}Q^{i}{}_{\alpha,j}[\varphi]\,\delta\xi^{\alpha}[\varphi]\\ &- Q^{i}{}_{\alpha}[\varphi]F^{-1\alpha}{}_{\beta}[\varphi]\big((-1)^{jk}P^{\beta}{}_{,kj}[\varphi]Q^{k}{}_{\gamma}[\varphi] + (-1)^{j\gamma}P^{\beta}{}_{,k}[\varphi]Q^{k}{}_{\gamma,j}[\varphi]\big)\delta\xi^{\gamma}[\varphi]\\ &+ Q^{i}{}_{\alpha}[\varphi]F^{-1\alpha}{}_{\beta}[\varphi](\delta V^{\beta}{}_{,\gamma}[P[\varphi]]P^{\gamma}{}_{,j}[\varphi] + \delta P^{\beta}{}_{,j}[\varphi])\,,\end{aligned}$$

which, after rearrangement of some factors and use of eq. (126), yields the (super) Jacobian

$$\begin{aligned}\det(\bar{\varphi}^i{}_{,j}) = {} & 1 + (-1)^{i(\alpha+1)} Q^i{}_{\alpha,i}[\varphi]\,\delta\xi^\alpha[\varphi] \\ & - F^{-1\alpha}{}_\beta[\varphi]\Big((-1)^{\alpha(j+1)} P^\beta{}_{,ji}[\varphi]\,Q^i{}_\alpha[\varphi]\,Q^j{}_\gamma[\varphi] \\ & \qquad + (-1)^{\alpha(\gamma+1)} P^\beta{}_{,j}[\varphi]\,Q^j{}_{\gamma,i}[\varphi]\,Q^i{}_\alpha[\varphi]\Big)\delta\xi^\gamma[\varphi] \\ & + (-1)^\alpha \delta V^\alpha{}_{,\alpha}[P[\varphi]] + (-1)^\alpha F^{-1\alpha}{}_\beta[\varphi]\,\delta P^\beta{}_{,i}[\varphi]\,Q^i{}_\alpha[\varphi]. \end{aligned} \tag{147}$$

Combining $\delta d\varphi \overset{\text{def}}{=} d\bar{\varphi} - d\varphi = [\det(\bar{\varphi}^i{}_{,j}) - 1]$ with

$$\begin{aligned}\delta \det F[\varphi] \overset{\text{def}}{=} {} & \det F[\bar{\varphi}] - \det F[\varphi] \\ & = (-1)^\alpha \det F[\varphi]\,F^{-1\alpha}{}_\beta[\varphi]\,F^\beta{}_{\alpha,i}[\varphi]\,Q^i{}_\gamma[\varphi]\,\delta\xi^\gamma[\varphi] \\ & = (-1)^\alpha \det F[\varphi]\,F^{-1\alpha}{}_\beta[\varphi]\Big((-1)^{\alpha i} P^\beta{}_{,ij}[\varphi]\,Q^j{}_\alpha[\varphi]\,Q^i{}_\gamma[\varphi] \\ & \qquad + P^\beta{}_{,j}[\varphi]\,Q^j{}_{\alpha,i}[\varphi]\,Q^i{}_\gamma[\varphi]\Big)\delta\xi^\gamma[\varphi], \end{aligned}$$

and making use of eq. (29), one finds for the change in $\det F[\varphi]\,d\varphi$ under the replacement $\varphi \to \bar{\varphi}$,

$$\begin{aligned}\delta(\det F[\varphi]\,d\varphi) & \\ = \Big\{ & \big[(-1)^{i(\alpha+1)} Q^i{}_{\alpha,i} - (-1)^{\beta(\alpha+1)} c^\beta{}_{\alpha\beta}\big]\,\delta\xi^\alpha[\varphi] \\ & + (-1)^\alpha \delta V^\alpha{}_{,\alpha}[P[\varphi]] + (-1)^\alpha F^{-1\alpha}{}_\beta[\varphi]\,\delta P^\beta{}_{,i}[\varphi]\,Q^i{}_\alpha[\varphi]\Big\} \det F[\varphi]\,d\varphi. \end{aligned} \tag{148}$$

If eqs. (146) are assumed to hold one is left with precisely the first and third terms inside the curly brackets in (143).

Equations (146) were not needed in the derivation of eq. (134). Why are they needed now? When the gauge group has no anticommuting coordinates the answer is that eqs. (146) are forced on us by the procedure of factoring out the gauge group. Our interchanging the orders of integration in arriving at eq. (129), and our use of eq. (131), amount to adopting the rule that the gauge group is to be treated formally as if it were compact. For consistency the associated Lie algebra must likewise be treated as compact. The generators of real representations of compact Lie algebras all have vanishing trace. Hence eqs. (146). (Remember, we are assuming that the φ's transform linearly under the gauge group.)

In Yang-Mills theories eqs. (146) hold automatically because the generating group is always compact. In gravity theory the situation is more subtle. Both $Q^i{}_{\alpha,i}$ and $c^\beta{}_{\alpha\beta}$, if one tries to compute them from eqs. (28) and (35), are meaningless expressions involving derivatives of delta functions with coincident arguments. However, both are metric-independent covariant vector densities of unit weight. Any sensible regularization scheme must assign them the value zero, for otherwise spacetime would be endowed with a preferred direction even before a metric is imposed on it.

If G is a supergauge group, with anticommuting coordinates, the formal compactness argument fails. But the notion of simplicity, or semisimplicity, survives. The invariance groups of all known supergauge theories are semisimple, and the generators of real representations of such groups satisfy the supertrace laws (146). The semisimplicity argument can also be invoked in the case of the local frame group, which enters when the gravitational field is expressed in terms of local frame components rather than directly in terms of the metric tensor (e.g., when spinor fields are present).

If we choose ϕ's that do not transform linearly under the group then the functional $\mu[\phi]$ of eq. (118) has to be reintroduced into the theory. It is easy to verify that consistency of the above formalism is maintained under these circumstances provided the first of eqs. (146) is replaced by eq. (50), which is just the condition that the product $\mu[\phi]\,d\phi$ be gauge invariant. Equation (50) is, of course, consistent with the first of eqs. (146) when $\mu=1$.

At this point the student may object that, in the case of quantum gravity at least, there appears to be an inconsistency in what we have done. Consider the sets of variables defined by eqs. (40). All of these sets transform linearly under the diffeomorphism group. However, the Jacobian that arises in transforming from one set to another is not generally constant. How can one maintain μ = constant for all sets? The answer is that one must. If the Jacobian is replaced by the exponential of its logarithm, it contributes a formally divergent term of the form const. $\times\, \delta(0) \int \ln g\, d^n x$ to the exponents in the Feynman functional integrals. All terms of this kind must be suppressed by any viable regularization scheme. By this criterion the dimensional regularization method, for example, is a viable scheme.

GHOSTS. THE BRS TRANSFORMATION. THE GENERATING FUNCTIONAL

The perturbation rules to which eqs. (134) and (137) lead may be summarized as follows. The exponent in the integrands is, as usual, expanded about a stationary background ϕ_B, and the integrals are evaluated as series of Gaussian integrals. If exp (iU) is a Gaussian weight factor and the P^α are chosen (unwisely) to have the linear form (98), then the vertex functions are just the functional derivatives $S_{,i_1 \dots i_N}$, $N \geq 3$. Otherwise the vertex functions include contributions from $U[P[\phi]]$. In addition to the usual graphs that one can draw there is an infinite set of new graphs arising from the factor det $F[\phi]$, involving a new set of "formal particles" called ghosts. The inverse matrix $F^{-1\alpha}{}_{\beta}[\phi]$ is the bare ghost propagator in an arbitrary field ϕ, i.e., with an arbitrary number of ϕ-lines attached. The ghost propagators always enter in closed loops, never as external lines.

The conditions previously imposed on the functional U ensure that the ϕ-propagator exists. The presence of $U[P[\phi]]$ in the exponents of expressions (134) and (137) breaks the gauge symmetry and eliminates the redundancy that exists in the integration (118). An important symmetry nevertheless survives. It is most easily revealed by introducing two new fields, χ_α and ψ^α, that have the unusual property of being fermionic when the index α is bosonic and vice versa. Use of these fields together with the Berezin integration rules allows one to express det $F[\phi]$ in the form

$$\int e^{i\chi_\alpha F^\alpha{}_\beta[\phi]\psi^\beta}\, d\chi\, d\psi = C \det F[\phi], \tag{149}$$

where C is a (divergent) constant, and hence

$$\langle \text{out} \mid \text{in} \rangle = \bar{N}[U] \int e^{i(S[\phi] + U[P[\phi]] + \chi F[\phi]\psi)}\, d\phi\, d\chi\, d\psi \tag{150}$$

$$\bar{N}[U] \overset{\text{def}}{=} N''[U]/C. \tag{151}$$

Equation (150) shows that the fields χ_α, ψ^α are associated with the ghost particles, which are now placed on a common footing with the ϕ-particles.

It was discovered by Becchi, Rouet and Stora (BRS) (1975) that both the exponent and the volume element $d\phi\, d\chi\, d\psi$ in (150) are invariant under a set of transformations whose infinitesimal forms are given by

$$\left.\begin{aligned} \delta\phi^i &= Q^i{}_\alpha[\phi]\psi^\alpha \delta\lambda, \\ \delta\chi_\alpha &= \delta\lambda\, U_{,\alpha}[P[\phi]], \\ \delta\psi^\alpha &= -\tfrac{1}{2} c^\alpha{}_{\beta\gamma}\psi^\gamma \delta\lambda\, \psi^\beta, \end{aligned}\right\} \tag{152}$$

where $\delta\lambda$ is an arbitrary infinitesimal anticommuting constant. Using the special (anti)commutativity properties of the χ's and ψ's, together with the identity (29) and the definition , one readily verifies the invariance of the exponent. By computing the super-Jacobian of the BRS transformation one finds that the volume element $d\phi\, d\chi\, d\psi$ is likewise invariant, provided eqs. (146) are assumed to hold. It is also straightforward to verify that, if confined to the ϕ's and ψ's, the BRS transformations constitute an Abelian group. Inclusion of the χ's destroys the group property unless $F^{\alpha}{}_{\beta}[\phi]\psi^{\beta}=0$. Note that the BRS transformations do <u>not</u> constitute a local gauge group. The $\delta\lambda$'s are constants; they are not functions over spacetime. Thus the integral (150) contains no redundancy.

The BRS transformations play an important role in simplifying the derivation of the "Ward-Takahashi identity" satisfied by the so-called generating functional. What follows is a partial account, adapted to the case in which the P's are nonlinear and exp (iU) is non-Gaussian, of the theory of the generating functional given by B.W. Lee (1976) and originally due to Zinn-Justin.

One begins by replacing the exponent in eq. (150) by $\tilde{S}[\phi,\chi,\psi,K,L,M]+J_i\phi^i+\bar{J}^{\alpha}\chi_{\alpha}+\hat{J}_{\alpha}\psi^{\alpha}$, where

$$\tilde{S}[\phi,\chi,\psi,K,L,M] \overset{\text{def}}{=} S[\phi]+U[P[\phi]]+\chi_{\alpha}F^{\alpha}{}_{\beta}[\phi]\psi^{\beta} + \{K_i+M(U[P[\phi]])_{,i}\}Q^i{}_{\alpha}[\phi]\psi^{\alpha}-\tfrac{1}{2}(-1)^{\beta}L_{\alpha}c^{\alpha}{}_{\beta\gamma}\psi^{\gamma}\psi^{\beta} \tag{153}$$

and by generalizing eq. (137) to

$$\langle \text{out}|T(A[\underset{\sim}{\phi},\underset{\sim}{\chi},\underset{\sim}{\psi}])|\text{in}\rangle \overset{\text{def}}{=} \bar{N}[U]\int A[\phi,\chi,\psi]e^{i(\tilde{S}+J\phi+\bar{J}\chi+\hat{J}\psi)}d\phi\,d\chi\,d\psi. \tag{154}$$

J_i, $\bar{J}^{\alpha}$, $\hat{J}_{\alpha}$, K_i, L_{α} and M are <u>external sources</u>*, and the "matrix element" (154) is a functional of them. J_i and L_{α} are bosonic when their indices are bosonic and fermionic when their indices are fermionic. With $\bar{J}^{\alpha}$, $\hat{J}_{\alpha}$ and K_i the association is just the opposite. M is fermionic.

If the functional A in (154) is replaced by unity one gets a generalization of the "in-out" amplitude:

$$e^{iW[J,\bar{J},\hat{J},K,L,M]} \overset{\text{def}}{=} \langle \text{out}|\text{in}\rangle \overset{\text{def}}{=} \bar{N}[U]\int e^{i(\tilde{S}+J\phi+\bar{J}\chi+\hat{J}\psi)}d\phi\,d\chi\,d\psi. \tag{155}$$

This generalized amplitude is called the <u>generating functional</u>, because if it is expanded in a power series in the sources J_i, $\bar{J}^{\alpha}$ and $\hat{J}_{\alpha}$, the coefficients are the matrix elements of chronological products of field operators. The coefficient of zero order reduces to the original amplitude (150) when K_i, L_{α} and M vanish.

The functional $\tilde{S}$ may be viewed as a generalized action functional. With the aid of eqs. (29) and (36) one may readily

* M is a constant. The others depend, through their indices, on position in spacetime.

show that it is BRS invariant. Suppose the variables ϕ, χ, ψ in the integrand of (155), as well as in the volume element, are subjected to a BRS transformation. Since these variables are dummies the integral remains unaffected. Explicitly, however, the terms in J, $\bar{J}$, $\hat{J}$ change. Therefore

$$0 = i\bar{N}[U] \int \{ J_i Q^i_{\alpha}[\phi]\psi^{\alpha} + (-1)^{\alpha}\bar{J}^{\alpha} U_{,\alpha}[P[\phi]] + \tfrac{1}{2}(-1)^{\beta}\hat{J}_{\alpha} c^{\alpha}{}_{\beta\gamma}\psi^{\gamma}\psi^{\beta}\} \times e^{i(\tilde{S} + J\phi + \bar{J}\chi + \hat{J}\psi)} d\phi\, d\chi\, d\psi. \tag{156}$$

This result can be expressed in an alternative form through use of

$$0 = \int \frac{\delta}{\delta\chi_{\alpha}} \{ f[\phi] e^{i(\tilde{S} + J\phi + \bar{J}\chi + \hat{J}\psi)} \} d\phi\, d\chi\, d\psi$$
$$= i\int \{ F^{\alpha}{}_{\beta}[\phi]\psi^{\beta} - (-1)^{\alpha}\bar{J}^{\alpha}\} f[\phi] e^{i(\tilde{S} + J\phi + \bar{J}\chi + \hat{J}\psi)} d\phi\, d\chi\, d\psi, \tag{157}$$

where f is any functional of the ϕ's. One obtains

$$0 = i\bar{N}[U] \int \{ J_i Q^i_{\alpha}[\phi]\psi^{\alpha} + U_{,\alpha}[P[\phi]] F^{\alpha}{}_{\beta}[\phi]\psi^{\beta} + \tfrac{1}{2}(-1)^{\beta}\hat{J}_{\alpha} c^{\alpha}{}_{\beta\gamma}\psi^{\gamma}\psi^{\beta} \times e^{i(\tilde{S} + J\phi + \bar{J}\chi + \hat{J}\psi)} d\phi\, d\chi\, d\psi$$
$$= \bar{N}[U] \int \left(J_i \frac{\delta}{\delta K_i} + \frac{\partial}{\partial M} - \hat{J}_{\alpha}\frac{\delta}{\delta L_{\alpha}} \right) e^{i(\tilde{S} + J\phi + \bar{J}\chi + \hat{J}\psi)} d\phi\, d\chi\, d\psi, \tag{158}$$

in which use is made of the fact that the term containing M in $\tilde{S}$ may be written in the form $M U_{,\alpha}[P[\phi]] F^{\alpha}{}_{\beta}[\phi]\psi^{\beta}$. Multiplying eq. (158) by $-ie^{-iW}$, we finally get

$$0 = -ie^{-iW} \left(J_i \frac{\delta}{\delta K_i} + \frac{\partial}{M} - \hat{J}_{\alpha}\frac{\delta}{\delta L_{\alpha}} \right) e^{iW}$$
$$= J_i \frac{\delta W}{\delta K_i} + \frac{\partial W}{\partial M} - \hat{J}_{\alpha}\frac{\delta W}{\delta L_{\alpha}}. \tag{159}$$

This relation expresses an important symmetry property of the generating functional, which leads directly to the Ward-Takahashi identities to be discussed presently. First we must review some standard material on the so-called effective action.

MANY-PARTICLE GREEN'S FUNCTIONS. THE EFFECTIVE ACTION

In this section important use will be made of the Schwinger average:

$$\langle \underset{\sim}{A} \rangle \overset{\text{def}}{=} \frac{\langle \text{out} | T(\underset{\sim}{A}) | \text{in} \rangle}{\langle \text{out} | \text{in} \rangle}. \tag{160}$$

Here $\underset{\sim}{A}$ is an arbitrary functional of the operators $\underset{\sim}{\phi}^i, \underset{\sim}{\chi}_{\alpha}, \underset{\sim}{\psi}^{\alpha}$, and the numerator and denominator on the right are defined by eqs. (154) and (155) respectively. It will be convenient to define

$$\phi^i \overset{\text{def}}{=} \langle \underset{\sim}{\phi}^i \rangle, \quad \chi_{\alpha} \overset{\text{def}}{=} \langle \underset{\sim}{\chi}_{\alpha} \rangle, \quad \psi^{\alpha} \overset{\text{def}}{=} \langle \underset{\sim}{\psi}^{\alpha} \rangle. \tag{161}$$

When the sources $\bar{J}^\alpha$, $\hat{J}_\alpha$, K_i, L_α, M vanish, the averages χ_α and ψ^α vanish. Note that although the symbols ϕ^i, χ_α, ψ^α have previously been used for integration variables, no confusion about their meaning will arise in practice.

It will aslo be convenient to denote the operators $\underset{\sim}{\phi}^i$, $\underset{\sim}{\chi}_\alpha$, and $\underset{\sim}{\psi}^\alpha$ collectively by $\underset{\sim}{\phi}^A$, their averages by ϕ^A, and the sources J_i, $\bar{J}^\alpha$ and $\hat{J}_\alpha$ collectively by J_A. Let ΔJ_A be arbitrary finite increments in the sources. Then we may write

$$\sum_{n=0}^{\infty} \frac{i^n}{n!} \Delta J_{A_n} \cdots \Delta J_{A_1} \langle \text{out} | T(\underset{\sim}{\phi}^{A_1} \ldots \underset{\sim}{\phi}^{A_n}) | \text{in} \rangle$$

$$= \exp\left(\Delta J_A \frac{\delta}{\delta J_A}\right) \langle \text{out} | \text{in} \rangle = \left(e^{iW}\right)_{J \to J+\Delta J}$$

$$= \exp\left(iW + i\Delta J_A \phi^A + i \sum_{n=2}^{\infty} \frac{1}{n!} \Delta J_{A_n} \cdots \Delta J_{A_1} G^{A_1 \cdots A_n}\right), \tag{162}$$

where

$$\phi^A = \langle \underset{\sim}{\phi}^A \rangle = e^{-iW} \frac{\delta}{i\delta J_A} e^{iW} = \frac{\delta W}{\delta J_A}, \tag{163}$$

$$G^{A_1 \cdots A_n} \overset{\text{def}}{=} \frac{\delta}{\delta J_{A_1}} \cdots \frac{\delta}{\delta J_{A_n}} W. \tag{164}$$

Dividing both sides of eq. (162) by e^{iW} and comparing like powers of ΔJ_A, one obtains an infinite sequence of relations:

$$\left.\begin{aligned} \langle \underset{\sim}{\phi}^A \underset{\sim}{\phi}^B \rangle &= \phi^A \phi^B - iG^{AB}, \\ \langle \underset{\sim}{\phi}^A \underset{\sim}{\phi}^B \underset{\sim}{\phi}^C \rangle &= \phi^A \phi^B \phi^C - iP_3 \phi^A G^{BC} + (-i)^2 G^{ABC}, \text{ etc.,} \end{aligned}\right\} \tag{165}$$

where "P " means "sum over the N distinct permutations of indices, with a plus sign or a minus sign according to whether the permutation of the indices associated with fermionic fields is even or odd." G^{AB} is known as the <u>one-particle</u> propagator, and the $G^{A_1 \cdots A_n}$, $n \geq 3$ are known as <u>many-particle Green's functions</u>. They satisfy the boundary conditions specified by the "in" and "out" states.

Any functional of the sources J_A may be alternatively regarded as a functional of the averages ϕ^A. From equations (163) and (164) one sees that the one-particle propagator is the transformation matrix from one set of variables to the other:

$$G^{AB} = \frac{\delta \phi^B}{\delta J_A}. \tag{166}$$

This fact may be used to establish an important relation between the functional W and the Schwinger average of the operator field equations. The latter is obtained from the formal functional identity

$$0 = -ie^{-iW}\bar{N}[U]\int e^{i(\tilde{S}+J\phi+\bar{J}\chi+\hat{J}\psi)}\frac{\overleftarrow{\delta}}{\delta\phi^A}\,d\phi\,d\chi\,d\psi$$

$$= \langle\tilde{\underset{\sim}{S}}_{,A}\rangle + J_A \;, \tag{167}$$

where $\tilde{\underset{\sim}{S}}_{,A}$ is the operator corresponding to the functional $\tilde{S}_{,A}$.

If we differentiate eq. (167) on the left with respect to J_B and make use of (166) we obtain

$$G^{BC}\;{}_{C,}\langle\tilde{\underset{\sim}{S}}_{,A}\rangle = -\delta^B{}_A\,. \tag{168}$$

Here the functional derivative inside the brackets $\langle\;\rangle$ is with respect to the field operator $\underset{\sim}{\phi}^A$, and the functional derivative outside the brackets is with respect to the field average ϕ^C. ${}_{C,}\langle\tilde{\underset{\sim}{S}}_{,A}\rangle$ is seen to be the operator of which the one-particle propagator G^{BC} is the Green's function. Because of its boundary conditions G^{BC} may be shown to be both a left Green's function, as in eq. (168), and a right Green's function of ${}_{A,}\langle\tilde{\underset{\sim}{S}}_{,B}\rangle$ as well. It has the symmetry

$$G^{AB} = (-1)^{AB}G^{BA} \tag{169}$$

which implies

$${}_{A,}\langle\tilde{\underset{\sim}{S}}_{,B}\rangle = (-1)^{A+B+AB}\;{}_{B,}\langle\tilde{\underset{\sim}{S}}_{,A}\rangle\,. \tag{170}$$

But this is just the condition that there exist a functional $\tilde{\Gamma}[\phi,\chi,\psi,K,L,M]$ such that

$$\tilde{\Gamma}_{,A} = \langle\tilde{\underset{\sim}{S}}_{,A}\rangle\,. \tag{171}$$

$\tilde{\Gamma}$ is known as the <u>effective action</u>. It satisfies the equations

$$\tilde{\Gamma}_{,A} = -J_A\;, \tag{172}$$

$${}_{A,}\tilde{\Gamma}_{,C}\,G^{CB} = -\delta_A{}^B\,, \tag{173}$$

and is related to the functional W by a Legendre transformation:

$$W = \tilde{\Gamma} + J_A\phi^A. \tag{174}$$

This relation may be verified through differentiation with respect to J_B and use of eq. (172) in the form ${}_{A,}\tilde{\Gamma} = -(-1)^A J_A$. Thus

$$\frac{\delta W}{\delta J_B} = \frac{\delta\phi^A}{\delta J_B}\left[{}_{A,}\tilde{\Gamma} + (-1)^A J_A\right] + \phi^B = \phi^B\,,$$

which is just eq. (163). Since $\tilde{\Gamma}$ is determined only up to an arbitrary constant of integration, eq. (174) may be regarded as fixing it.

$\tilde{\Gamma}$ is also known as the generating functional for proper vertices. This stems from its relation to the may-particle Green's functions. By differentiating eq. (173) one can relate functional derivatives of the one-particle propagator to derivatives of . These relations yield, for example,

$$G^{ABC} = \frac{\delta}{\delta J_A} G^{BC} = G^{AD}{}_{D,}G^{BC}$$

$$= (-1)^{(B+C)D+(C+D)E+(D+E)F} G^{AD} G^{BE} G^{CF}{}_{DEF,}\tilde{\Gamma} . \qquad (175)$$

If the propagators are represented by lines and the third and higher derivatives of $\tilde{\Gamma}$ are represented by vertices, one easily sees that each new differentiation with respect to a source inserts a new line in all possible ways into the previous diagram. Each Green's function of given order is thus representable as a sum of all the possible tree diagrams of that order.

Suppose the spatial sections of spacetime are noncompact. Then an S-matrix can be introduced, connecting states defined "at infinity." The S-matrix is expressible in terms of the chronological products appearing in eq. (162). Because these products are expressible in terms of the Green's functions (eqs. (165)), it follows that when $\tilde{\Gamma}$ is used, only tree diagrams are needed in the construction of the S-matrix. No closed loops appear. The vertices generated by $\tilde{\Gamma}$ are the proper vertices, already containing all quantum corrections. By noting that identical tree diagrams occur in classical perturbation theory, but with $\tilde{\Gamma}$ replaced by $\tilde{S}$, one can show that $\tilde{\Gamma}$ describes the quantum-corrected dynamics of coherent large-amplitude fields. One must expect the same to be true also when the spatial sections are compact and there is no S-matrix.

THE WARD-TAKAHASHI IDENTITY

We now resume use of the symbols $\Phi^i, \chi_\alpha, \psi^\alpha, J_i, \bar{J}^\alpha, \hat{J}_\alpha$ and rewrite eq. (174) in the more explicit form

$$W[J,\bar{J},\hat{J},K,L,M] = \tilde{\Gamma}[\Phi,\chi,\psi,K,L,M] + J_i\Phi^i + \bar{J}^\alpha\chi_\alpha + \hat{J}_\alpha\psi^\alpha . \qquad (176)$$

The averages $\Phi^i, \chi_\alpha, \psi^\alpha$ depend on all six sources, but because K_i, L_α and M do not participate in the Legendre transformation one may show that

$$\frac{\delta W}{\delta K_i} = \frac{\delta \tilde{\Gamma}}{\delta K_i} , \qquad \frac{\delta W}{\delta L_\alpha} = \frac{\delta \tilde{\Gamma}}{\delta L_\alpha} , \qquad \frac{\partial W}{\partial M} = \frac{\partial \tilde{\Gamma}}{\partial M} , \qquad (177)$$

where the derivatives on the right refer only to the explicit dependence of $\tilde{\Gamma}$ on K_i, L_α and M. This result, combined with eq. (172) in the form $_{A,}\tilde{\Gamma} = -(-1)^A J_A$, allows eq. (159) to be rewritten as

$$-(-1)^i \frac{\delta\tilde{\Gamma}}{\delta\phi^i}\frac{\delta\tilde{\Gamma}}{\delta K_i} + \frac{\partial\tilde{\Gamma}}{\partial M} - (-1)^\alpha \frac{\delta\tilde{\Gamma}}{\delta\psi^\alpha}\frac{\delta\tilde{\Gamma}}{\delta L_\alpha} = 0, \tag{178}$$

all derivatives being left derivatives. This is the Ward-Takahashi identity.

The Ward-Takahashi identity has important implications for the structure of $\tilde{\Gamma}$. That it implies the existence of some sort of symmetry possessed by $\tilde{\Gamma}$ becomes obvious when one notes that, because of its BRS invariance, $\tilde{S}$ too satisfies the Ward-Takahashi identity. Unfortunately, to work from eq. (178) to the symmetry possessed by $\tilde{\Gamma}$ is a much harder task. In principle one might do the following. Assume that $\tilde{\Gamma}$ can be expanded as a power series in χ_α, ψ^α, K_i, L_α and M. Such an assumption has nothing a priori to do with perturbation theory since the expansion is to be carried out after the functional integration (155) has been performed. It is based on the reasonable belief that (155) varies smoothly (at least after appropriate renormalizations) as K_i, L_α, M, $\bar{J}^\alpha$ and $\hat{J}_\alpha$ (and hence χ_α and ψ^α) go to zero.

In determining the kinds of terms that can appear in the expansion it is useful to introduce the notion of "ghost number." If one assigns the ghost numbers 1 to ψ^α and $\bar{J}^\alpha$; 0 to ϕ^i and J_i; -1 to χ_α, $\hat{J}_\alpha$, K_i and M; and -2 to L_α; one easily sees that the integrand in (155) and the integral itself have total ghost number zero. Hence W and $\tilde{\Gamma}$ have total ghost number zero, and all the terms in the expansion of $\tilde{\Gamma}$ must have this property as well. Also the expansion can contain no terms in M of higher order than the first since M is an anticommuting constant.

If one inserts the expansion into (178) and groups together terms of like powers, one obtains an infinite sequence of subsidiary Ward-Takahashi identities relating the ϕ-dependent coefficients. Unfortunately there seems to be no easy way of drawing simple inferences from these identities en gros. So far (178) has been applied only to renormalizable models in perturbation theory. There it has proved to be of great service in the practical details of the renormalization program, as well as in the demonstration that the theory is indeed renormalizable to all orders and that unitarity is maintained. (The student is referred to the literature for details.) What role it is destined to play in quantum gravity remains to be seen.

REFERENCES

Becchi, C., A. Rouet and R. Stora (1975). Comm. Math. Phys. 42, 127.

Berezin, F.A. (1966), The Method of Second Quantization, Academic Press, New York.

DeWitt, B.S. (1965), Dynamical Theory of Groups and Fields, Gordon and Breach, New York.

DeWitt, B.S. (1967a), Phys. Rev. 160, 113.

DeWitt, B.S. (1967b), Phys. Rev. 162, 1195.

DeWitt, B.S. (1967c), Phys. Rev. 162, 1239.

Fadde'ev, L.D., and V.N. Popov (1967), Phys. Letters 25B, 29.

Fadde'ev, L.D. (1969), Theor. Math. Phys. 1, 3.

Fadde'ev, L.D. (1976), in Methods in Field Theory, R. Balian and J. Zinn-Justin, eds., 1975 Les Houches Lectures, North Holland, Amsterdam.

Feynman, R.P. (1948), Rev. Mod. Phys. 20, 267.

Feynman, R.P. (1950), Phys. Rev. 80, 440.

Feynman, R.P. (1963), Acta Phys. Polon. 24, 697. See also in Proceedings of the 1962 Warsaw Conference on the Theory of Gravitation, PWN-Editions Scientifiques de Pologne, Warszawa (1964).

Fischer, A.E. (1970), in Relativity: Proceedings of the Relativity Conference in the Midwest, Carmeli, Fickler and Witten, eds., Plenum, New York.

Freifeld, C. (1968), in Battelle Rencontres, 1967 Lectures in Mathematics and Physics, C.M. DeWitt and J.A. Wheeler eds., W.A. Benjamin, Inc., New York.

Gribov, V.N. (1977), Lecture at the Twelfth Winter School of the Leningrad Nuclear Physics Institute. (SLAC translation No. 176).

Isham, C.J., J. Strathdee and A. Salam (1971). Phys. Rev. D3, 1805.

Isham, C.J., J. Strathdee and A. Salam (1972). Phys. Rev. D5, 2584.

Kostant, B. (1977), in Differential Geometrical Methods in Mathematical Physics (Proceedings of the July 1-4, 1975 Symposium in Bonn), Bleuler and Reetz, eds. Lecture Notes in Mathematics No. 570, Springer, Berlin.

Lee, B.W. (1976), in Methods in Field Theory, R. Balian and J. Zinn-Justin, eds., 1975 Les Houches Lectures, North Holland, Amsterdam.

Misner, C.W., K.S. Thorne and J.A. Wheeler (1973). Gravitation, Freeman, San Francisco.

Nath, P. (1976), in Proceedings of the Conference on Gauge Theories and Modern Field Theory, Boston, 1975 (M.I.T. Press, Cambridge, Massachusetts).

Stelle, K.S. (1977), Phys. Rev. D16, 953.

Utiyama, R. (1956), Phys. Rev. 101, 1597.

Weinberg, S. (1965), Phys. Rev. 140, B516.

Weinberg, S. (1979), in Gravitational Theories Since Einstein, S.W. Hawking and W. Israel, eds., Cambridge University Press.

QUANTUM GRAVITY:

A FUNDAMENTAL PROBLEM AND SOME RADICAL IDEAS

G. 't Hooft

Institute for Theoretical Physics
University of Utrecht
The Netherlands

Introduction

A complete theory of the gravitational forces cannot possibly be constructed without taking into account the basic principles of quantum mechanics. Indeed, many attempts are made by general relativists to "quantize" their theory. Being a particle physicist with some experience in second quantization problems, I feel that I should give my point of view, which differs somewhat from most conventional approaches.

In the first part of my lectures I will show how one derives the Feynman rules of quantum gravity as a gauge theory. The method is the conventional one: that is through a perturbation expansion.

Even though, obviously, perturbation expansion cannot tell us everything about the system, it is crucial to understand its structure. The difficulties associated with this expansion will have to be taken into account in the real theory also. If you really want to go beyond perturbation expansion you are often tempted to make models and approximations that obscure these difficulties. However, on the basis of these I claim that if you really want to do quantum theory (quantum gravity) in a non-perturbative fashion then you might have to change radically the whole concept of the ideas you have about gravity itself.

For a start I will just show you very briefly what the perturbation theory looks like and also that the first phenomenon to arise then is that of divergencies. Well, particle physicists are quite used to divergencies of this nature and we have a good answer to these, which is renormalization of the theory. So we are going to investigate renormalization properties in quantum gravity and we find which counter terms would be necessary to obtain a finer theory. That is where the first real problems will arise, which will force me later on to abandon this scheme. But also there are some important results from this simple continuum theory with its ordinary perturbation expansion. There are some problems, which are occasionally discussed in the literature, but which I will show not be be really physical. Then I will discuss why pure gravity is called renormalizable up to the one loop level. In the last lecture, I will discuss something radically different.

A. CONVENTIONAL THEORY

1. Gauge Theory - Feynman Rules

The first procedure to be employed is the so-called Wick rotation into Euclidean space [1]. That implies that we obtain a flat background metric with signature (+,+,+,+). Physically this procedure corresponds to analytic continuation to imaginary time. The nice thing about it is that all field equations become elliptical in nature rather that hyperbolic which makes computations easier and more convergent. Oscillations about this flat metric are considered as a small perturbation :

$$g_{\mu\nu} = \delta_{\mu\nu} + h_{\mu\nu} \,. \tag{1}$$

As our Lagrangian we take the familiar one :

$$\mathcal{L} = -\sqrt{g}\, R \,. \tag{2}$$

(Note that now $g > 0$.) This is to be expanded with respect to $h_{\mu\nu}$. Let us write the terms quadratic in h explicitly :

$$\begin{aligned}\mathcal{L} = & -\tfrac{1}{4}(h_{\alpha\beta,\nu})^2 + \tfrac{1}{4}(h_{\alpha\alpha,\mu})^2 \\ & -\tfrac{1}{2} h_{\alpha\alpha,\beta}\, h_{\beta\mu,\mu} + \tfrac{1}{2} h_{\nu\beta,\alpha}\, h_{\alpha\beta,\nu} \\ & + \mathcal{O}(h^3) + \text{total derivative} \,.\end{aligned} \tag{3}$$

The commas denote ordinary derivatives, and the total derivative is irrelevant here.

The presence of higher order terms implies that the equations for gravitational waves are non-linear so that you can get scattering. The linear parts of the wave equations are trivial to quantize : the waves with given frequency must simply be quantized in energy units $h\nu$, called gravitons. The higher order terms in the Lagrangian now cause gravitons to interact. The scattering problem is really the problem we are going to address ourselves to, in particular the quantum corrections to the scattering amplitudes. The computational rules to obtain these amplitudes, and in principle also their quantum corrections have been obtained in particle physics.

The history of the derivation of the various steps in this procedure is rather remarkable [2]. The so-called non-Abelian gauge theories for elementary particles were investigated at first mainly because of their close resemblence to quantum gravity [3]. They served as a simplified model of gravity, and seemed to be of little relevance for the observed world of elementary particles. Then a breakthrough took place and they became extremely successful for the weak and electromagnetic interactions, a few years later the strong interactions. They were intensively studied. Now the details are well-understood and can be applied to quantum gravity.

The point is that there is an invariance in the system : invariance under infinitesimal coordinate transformations, corresponding to

$$h_{\mu\nu} \to h_{\mu\nu} + \xi_{\mu;\nu} + \xi_{\nu;\mu} + \mathcal{O}(\xi_\mu^2) , \tag{4}$$

where the semicolons are covariant derivatives. Any choice of the function $\xi_\mu(x)$ yields an equivalent set for $h_{\mu\nu}(x)$. This redundancy in the description of the solutions to the wave equations must first be removed. The prescription is as follows [4] : choose any four component function $C_\mu(x)$, composed of combinations of $h_{\alpha\beta}$, not invariant under eq.(4). It is preferred but not crucial to have C linear in h. We now require that the equation $C_\mu = o$ fixes the values of $\xi_\mu(x)$ (up to possible boundary effects). Or, if under (4)

$$C_\mu \to C_\mu + \hat{M}_{\mu\nu}\,\xi_\nu , \tag{5}$$

then the object $\hat{M}_{\mu\nu}$ which may be an operator in ξ-space containing derivatives, must have an inverse, $M^{-1}_{\mu\nu}(x,x')$.

Instead of Lagrangian (2) we now take

$$\mathcal{L} = -\sqrt{g}\,R - \tfrac{1}{2}C_\mu^2 \,, \tag{6}$$

which transforms under (4) as

$$\mathcal{L} \to \mathcal{L} - C_\mu \hat{M}_{\mu\nu}\,\xi_\nu \text{ + total derivative.} \tag{7}$$

Let us consider the Euler-Lagrange equations generated by Lagrangian (6). A function $h_{\mu\nu}(x)$ is a solution to those equations if any infinitesimal variation about $h_{\mu\nu}(x)$ does not change the total action $\int\mathcal{L}d^4x$ linearly. Let us in particular consider the infinitesimal variation (4) which happens to be also a coordinate transformation. According to eq.(7) $\int\mathcal{L}\,d^4x$ changes by an amount $\int C_\mu \hat{M}_{\mu\nu}\,\xi_\nu$. This must vanish for any choice of $\xi_\nu(x)$. Since $\hat{M}$ has an inverse it follows that $C_\mu = o$, if $h_{\mu\nu}$ satisfies these field equations. Thus the replacement (6) corresponds to imposing the "gauge condition" $C_\mu = o$ on our fields. A convenient choice for C_μ is

$$C_\mu = h_{\mu\alpha,\alpha} - \tfrac{1}{2}h_{\alpha\alpha,\mu} \,. \tag{8}$$

We then obtain

$$\mathcal{L} = -\tfrac{1}{4}(h_{\alpha\beta,\nu})^2 + \tfrac{1}{8}(h_{\alpha\alpha,\nu})^2 + \mathcal{O}(h^3) \text{+ total derivative.} \tag{9}$$

This Lagrangian is especially convenient because in the quadratic terms the differentiation index ν does not mix with the field indices α, β . Therefore the linear parts of the field equations become just Laplace's equation which is trivial to solve.

To solve the complete classical Euler Lagrange equations for a scattering process perturbatively is now straightforward. What is needed is the inverse of the Laplace operator, called propagator for the graviton, and the explicit form of the interaction terms $\mathcal{O}(h^3)$, called vertices. The combinatorial rules for these can also be interpreted as the history of a particular scattering process between gravitons. An example is given in Fig.1.

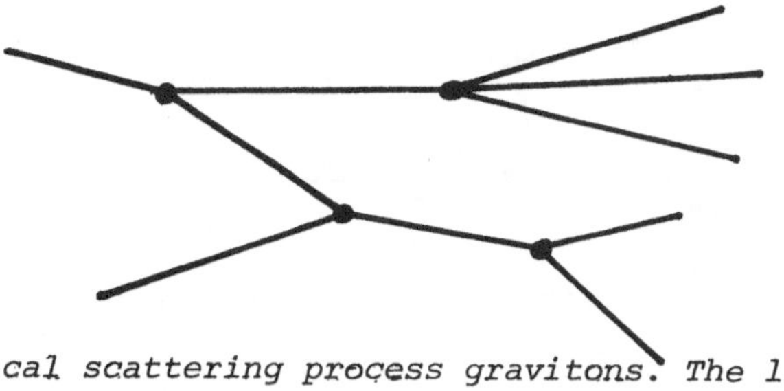

Fig.1 *Typical scattering process gravitons. The lines between dots are propagators (defined by the quadratic parts in (9).*

A quantum theory only differs from the classical theory by allowing in addition also diagrams with closed loops, see Fig.2.

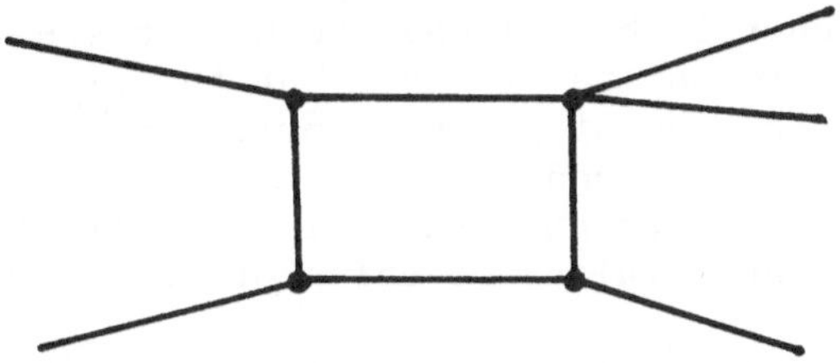

Fig.2 *Feynman Diagram with a loop.*

In simple field theories such as ϕ^4 the rules for the loops follow straightforwardly from the rules for trees such as Fig.1. Here however the rules suggested by the tree diagrams do not give the correct quantum theory. One must add a new so-called ghost field. This field does not correspond to any new physical particle but is needed to make the mathematics sound. It is obtained by adding to the Lagrangian

$$\mathcal{L}^{ghost} = \varphi^*_\mu \hat{M}_{\mu\nu} \varphi_\nu . \tag{10}$$

Here the object $\hat{M}$ is the same $\hat{M}$ as defined in eq.(5). It contains space-time derivatives acting on φ_ν on the right and terms very non-linear in h. Consequently the field equations for φ and φ^* have linear parts of the Laplace form and parts interacting with the h field. Because of the nature of the Feynman rules φ and φ^* can only occur in closed loops (see Fig.3). Finally, the weight factor for these diagrams is not the same as for ordinary theories without ghosts : one must insert an extra factor $(-1)^N$, where N is the number of ghost loops. Such a factor

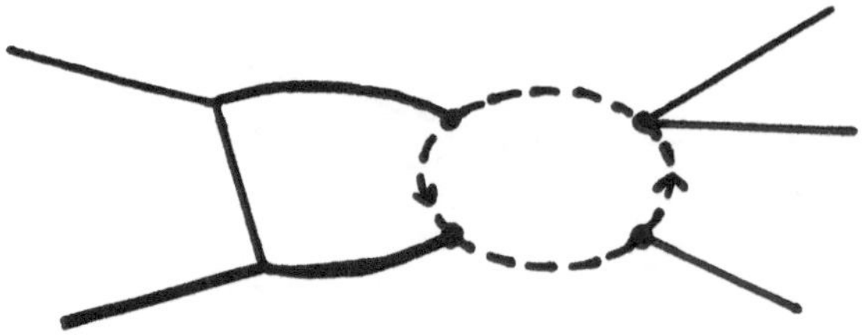

Fig.3 *Example of a diagram with ghost loop (and an ordinary loop).*

corresponds to attributing to this ghost wrong statistics : it is a fermion rather than a boson.

This completes the formal description of the Feynman rules. I don't know how much this has already been discussed in previous lectures, where I could not be present, but this is mainly the result of the investigation of gauge theories in particle physics. This part of it is all well understood. It is also well understood that this prescription is unique; there is very little room for playing around with these rules, or using other rules instead.

The reason why this prescription is so precise is that we want to describe scattering of gravitons by means of an S-matrix. This matrix must satisfy the requirements of causality and unitarity. If it is not unitary it would mean that probability is not conserved and that would make a totally senseless theory out of it. This unitarity condition is so restrictive that it prescribes in a unique fashion what the Feynman rules of this theory should be ; there is no way to play around with it. So that part of the conventional theory is nice and sound and understood. That does not mean that the theory will work all the way and the reason is of course something that you also know of in field theories - the divergencies, to which I now come.

2. Divergencies and Counter terms

Considering the natural unit of energy, $\sqrt{\hbar c^5/G} = 10^{22}$MeV, quantum effects in gravity are unlikely to ever observed experimentally. Nevertheless, the theory dictates that gravitons can split, form closed loops and rejoin while they scatter. Such processes, of which Figs. 2 and 3 are examples must necessarily occur in a good theory, otherwise the S-matrix will not be unitary and probabilities will not be conserved.

If one computes the contributions coming from such diagrams one finds that they are actually highly divergent. The momenta and energies of the gravitons in the intermediate states must be integrated over. One-loop diagrams typically correspond to integrals of the type

$$\int d^4k \, \frac{\mathrm{Pol}_1(k)}{\mathrm{Pol}_2(k)} \quad , \tag{11}$$

where the two polynomials can be of the same degree, so that the integral is quartically divergent where $k_\mu \to \infty$. But we have a scheme in gauge theories to deal with that, called dimensional renormalization [6] : in fact there are other renormalization schemes but this is the most convenient one. We replace d^4k by d^nk, saying that we are now going to do the same theory in n dimensions instead of 4. The crazy trick that we are going to do is that we choose n very close to 4 but just a non-integer number (maybe a complex number) near 4. Now complex dimension is of course a totally crazy concept if you are dealing with the full general relativistic theory. However, when you consider these Feynman rules of the system and the formulations of it you find that it makes perfectly good sense to define integrals over n-dimensional momenta instead of 4-dimensional momenta.

The point is that these integrals all have a very special form, and for this class of integrals a unique definition of the symbol $\int d^nk$ can be given. One basic ingredient is the extension of the expression for the surface of an n-dimensional sphere,

$$\int d^nk \quad \delta\,(k^2-r^2) = \frac{\pi^{n/2}}{\Gamma(n/2)}\, r^{n-2} \tag{12}$$

to a non-integer n. Another ingredient is a procedure to obtain the "finite part" of an integral, which works only if the degree of divergence is not integer, which is why we first choose n to be non-integer. But what happens if you let n tend to 4 ? You encounter poles and this is just another way of saying that at 4-dimensions you have problems. The prescription then turns out to be that you simply have to add a counter term to your Lagrangian:

$$\mathcal{L} \to \mathcal{L} + \frac{1}{8\pi^2(n-4)}\,\Delta\mathcal{L} \quad , \tag{13}$$

where $\Delta\mathcal{L}$ is now chosen in such a way that all poles in the physically relevant amplitudes cancel and a finite result is obtained after adding all contributions from all one-loop diagrams.

In gauge theories for elementary particles, the new Lagrangian (13) is of the same type as the previous one, so that only some coupling constants are renormalized this way. We can then preceed to include the effects of two-loop diagrams, which leads to $(n-4)^{-2}$ terms, and so on. We will always require the limit $n \to 4$ to exist for physically observable quantities.

Now this $\Delta\mathcal{L}$ in (13) has been computed for many gravity theories. The first theory for which this $\Delta\mathcal{L}$ has been computed is of course pure gravity [7] :

$$\Delta\mathcal{L} = \sqrt{g}\left[\frac{1}{120} R^2 + \frac{7}{20} R_{\mu\nu}^2\right] . \tag{14}$$

This is the theory with gravitons alone, no matter-fields in addition. Even though $\Delta\mathcal{L}$ is of a different form than $\mathcal{L}$ itself, it is rather harmless, because for pure gravitons $R = o$ and $R_{\mu\nu} = o$ classically. We come back to that in sect. 4.

If matter-fields are coupled to gravity in the form of one real, scalar field, then there is a harmful part in $\Delta\mathcal{L}$:

$$\Delta\mathcal{L} = \sqrt{g} \cdot \frac{203}{80} R^2 . \tag{15}$$

(see ref. 7). Note that here $R \neq o$, even classically.

3. Solution of the Measure Problem.

If one does not follow the diagrammatic approach to quantum gravity one might run into an apparent difficulty. I just want to mention it very briefly and that is the measure problem. This comes up every now and then in the literature. This is a problem which seems to occur if you write down the theory in terms of the functional integral. Should we write this as

$$\int \prod_{x,\mu,\nu} dg^{\mu\nu}(x)\; e^{-\int \mathcal{L} d^4x} ,$$

or as

$$\int \prod_{x,\mu,\nu} dg_{\mu\nu}(x)\; e^{-\int \mathcal{L} d^4x} , \tag{16}$$

or some other way ? My claim is that this question is senseless at any order in perturbation expansion because both expressions (16) are acceptable. In fact any local, 10 component function of $g^{\mu\nu}(x)$ may be chosen as integration variable at the point x in the functional integral. The reason is that, after renormalization, all choices give precisely the same amplitudes and are therefore equivalent. Let me illustrate this for the two expressions (16). The difference is a factor in the integrand which is the Jacobian of the transformation

$$g_{\mu\nu}(x) \rightarrow g^{\mu\nu}(x) :$$

$$\det_{\substack{x,x' \\ \mu\nu,\alpha\beta}}\left[\frac{\partial g_{\mu\nu}(x)}{\partial g^{\alpha\beta}(x')}\right] = \exp\left(\operatorname*{Tr}_{\substack{x,x' \\ \mu\nu,\alpha\beta}} \log \frac{\partial g_{\mu\nu}(x)}{\partial g^{\alpha\beta}(x')}\right) = \exp \sum_x 5 \log g(x). \tag{17}$$

The determinant in (17) is that of a huge matrix, whose rows and columns are labeled not only by the indices $\mu\nu$ and $\alpha\beta$ but also by the space-time points x and x'. The matrix is diagonal in x so that finally we obtained the exponent of a sum over all space time points x. One will now be tempted to replace $\sum_x$ by $\int dx$, but that is wrong ! The reason why that is wrong is that for the continuum limit of (17) to exist we need a quantity that produces the infinitesimal dx. Such a quantity is not there in (17). It is correct to say

$$\sum_x \rightarrow \frac{1}{d^4x} \int d^4x \ ,$$

in the continuum limit. This is quartically divergent ! Taking the limit more carefully we write

$$\frac{1}{d^4x} = \delta^4(0) = \int 1\, d^4k \rightarrow \infty . \tag{18}$$

Now this latter expression may remind you of something. We have already seen such divergences before in the theory (eq.11) and that is the message I want to convey to you in this little section - that there is a divergence in this determinant which is exactly comparable to the divergences of the one loop diagram. In fact it can be shown to be indeed, in a certain sense, a one loop divergence (which again I have no time to go into details about). But this is essentially a one loop divergence which is just as bad as that in Fig.2. But now remember for this one loop divergence we have a unique procedure to get the convergent theory. So the same unique procedure can be applied here to replace this infinity by a finite number. To explain what happens I have to go a little bit more into the details of this dimensional regularization procedure.

At n different from but close to 4 expression (18) still diverges. But the "finite part" of the integral is then uniquely defined. In gauge field theories integrals may be replaced by finite integrals (notation : $\int_F d^n k$) if they satisfy the following requirements :

(i) $\int_F f(k)\, d^n k = \int f(k)\, d^n k$,

if the right hand side converges.

(ii) $\int_F d^n k (A+B) = \int_F d^n k\, A + \int_F d^n k\, B$.

(iii) $\int_F d^n k\, (\lambda A) = \lambda \int_F d^n k\, A$. (19)

(iv) $\int_F d^n k A(k+p) = \int_F d^n k\, A(k)$.

(v) $\int_F d^n k\, A(\lambda k) = \lambda^{-n} \int_F d^n k\, A(k)$.

It follows that

$$\int_F d^n k\, (k^2)^\alpha = o \quad \text{if } \alpha \neq -n/2 . \qquad (20)$$

Note that eqs (19) and (20) allow us to make all divergent integrals finite except if the degree of divergence is integer. For instance $\int d^n k/(k^2+a)^{n/2}$ remains infinite. This is because equation (20) is divergent both at ∞ and at o if $2\alpha = -n$, so using (20) and (19,ii) merely enables us to replace the divergence at infinity by a divergence at the origin and (19,i) cannot be used to obtain a finite expression. This is how, in the dimensional renormalization procedure,poles develop at n = 4.

The procedure immediately tells us what to do with expression (18). The only finite number it can be replaced with is zero. Therefore the factor 17 is one, and the ambiguity disappears.

Less elegant renormalization procedures exist where cutoffs are introduced. The factor (17) then corresponds to an additional local term in $\mathcal{L}$. Whenever gauge-invariance is imposed (by requiring validity of identities between amplitudes called Slavnov-Taylor identities [4], [8]) this ambiguity is fixed up to a gauge invariant term without derivatives. Only one such term exists : the cosmological term $\sqrt{g}$, so also in such procedures, after renormalization, the metrix ambiguity can be absorbed in a redefinition of the cosmological term. Because of power-counting arguments however the cosmological constant is not renormalized in the dimensional renormalization scheme.

The conclusion of this section is that, provided proper gauge-invariant renormalization procedures are used such as dimensional renormalization, the metric of the functional integral may be chosen any way you like. Since the divergences of eq.(18) add to all other equally bad divergencies of the theory it is absolutely senseless to defend one choice of measure against one other.

4. Pure Gravity is One-Loop-Finite

The next point which I am sure many of you have heard before is the statement that pure gravity is one loop finite. Why is it so ? Well remember I had a counter term, and this counter term was of a very dangerous nature because it contains the square of the Riemann curvature (see eq. 14). That is different from the previous Lagrangian which we started off with, which only had a linear expression in the Riemann curvature. To have square terms will lead to having four derivatives in your Lagrangian or three derivatives in your field equations. In such systems it is impossible to write down a Hamiltonian that is bounded from below, and that will make the whole of Minkowski space unstable, and you will get very serious blowups of the system. Energy is no longer positive definite. The field theory, in fact, definitely does not work if you do this. A gauge theory for instance with such terms would be thrown out immediately.

However, we say nevertheless that pure gravity is one loop finite because if you have pure gravity we know that the classical equations will tell you that $R = o$ and $R_{\mu\nu}$ equals zero as well. So we say

$$\Delta\mathcal{L} = \sqrt{g}\left[\frac{1}{120}R^2 + \frac{7}{20}R_{\mu\nu}^2\right] = 0 . \qquad (21)$$

Why ? Usually it is not correct to insert an equation of motion back into a Lagrangian. The reason why it is allowed here is that $\Delta\mathcal{L}$ is only infinitesimal (in spite of the fact that $\frac{1}{n-4}$ is formally infinite, since only terms in a given order of perturbation expansion have been collected).

Theorem : $\mathcal{L}$ and $\mathcal{L} + \varepsilon\Delta\mathcal{L}$ are equivalent up to terms of order ε^2, if $\Delta\mathcal{L}$ vanishes for all those field configurations that satisfy the field equations generated by $\mathcal{L}$.

Proof : The equation of motion reads, in an abbreviated notation :

$$\frac{\delta \mathcal{L}}{\delta h_{\mu\nu}(x)} = 0 , \qquad (22)$$

where δ stands for derivative in the Euler sense. Therefore we may assume

$$\Delta\mathcal{L}(h) = \frac{\delta \mathcal{L}}{\delta h_{\mu\nu}(x)} Q_{\mu\nu}(h,x) \qquad (23)$$

where $Q_{\mu\nu}$ is any local function of the fields h at x. Otherwise the equations of motion for h would not imply that $\Delta\mathcal{L}$ vanishes. But then

$$\mathcal{L}(h) + \varepsilon\Delta\mathcal{L}(h) = \mathcal{L}(h_{\mu\nu}(x) + \varepsilon Q_{\mu\nu}(h,x)) + \mathcal{O}(\varepsilon^2) \qquad (24)$$

If we now take as our new field variables

$$h'_{\mu\nu}(x) = h_{\mu\nu}(x) + \varepsilon Q_{\mu\nu}(h,x) , \qquad (25)$$

then

$$\mathcal{L}(h) \Rightarrow \mathcal{L}(h') + \mathcal{O}(\varepsilon^2) , \qquad (26)$$

therefore, the two Lagrangians are equivalent up to terms of order ε^2. Physically, the fact that all terms of order ε can be absorbed by a field redefinition implies that they are not directly observable. In particular, when it comes to computing S matrix elements the results will be identical up to ε^2 terms. But the $\mathcal{O}(\varepsilon^2)$ terms would have to be added to all two-loop processes which are of the same order of magnitude and have been neglected so far.

5. Field Transformations and Gauge Dependence

To be more explicit in this given example of pure gravity, we have

$$\frac{\delta \mathcal{L}}{\delta g_{\mu\nu}} = R^{\mu\nu} - \tfrac{1}{2} R g^{\mu\nu} , \qquad (27)$$

so that the redefinition of the metric tensor has to be

$$g'_{\mu\nu} = g_{\mu\nu} + \delta g_{\mu\nu} \, ,$$
$$\delta g_{\mu\nu} \propto \tfrac{7}{20} R_{\mu\nu} - \tfrac{11}{60} R g_{\mu\nu} \tag{28}$$

On the other hand, when matter is present in the form of a scalar field, then $\Delta\mathcal{L}$ cannot be absorbed by a field renormalization and the first order infinity would be directly observable if it were not removed in any other way. This explains the usual statement that pure gravity is one-loop renormalizable and gravity with matter in general not.

A further remark is of order. Since the redefinition in (28) is not directly observable physically, one cannot exclude the possibility that $\delta g_{\mu\nu}$ is actually dependent on the way the procedure is being performed in quantizing gravity. It turns out that this whole procedure is very much dependent on the original choice of gauge. Remember I chose an object called C_μ to fix my gauge; had I chosen another C_μ I could have obtained something else. That's inherent in the procedure. Again, this is not in contradiction with general gauge invariance because such a redefinition is not physically observable - I cannot determine what my metric tensor is from scattering experiments. Anything that is not physically observable might be gauge dependent. That is a general law in gauge theory.

6. The Fate of the Metric Tensor

Ambiguities of the type of eq. (28) in the definition of the metric tensor are to my mind extremely important to observe. The coefficient in $\delta g_{\mu\nu}$ is not only formally infinite. Its finite part may be dependent on scale transformations (dilatations). That implies that after performing a scale transformation in the theory the metric tensor may have a different meaning than before. The conclusion from that is that there may not exist a unique definition of the metric tensor. There exists no formal procedure that distinguishes $g_{\mu\nu}$ as a basic field from say $g_{\mu\nu} + R_{\mu\nu}$ or $g_{\mu\nu} + Rg_{\mu\nu}$, or anything more complicated (at higher orders we encounter R^2 terms and so on).

Now what is the consequence of that? I think that perhaps here we are dealing with a basically new feature of quantum gravity. An illustration of that is given in Fig. 4.
In the conventional theory of general relativity the existence of a well-defined $g_{\mu\nu}$ was used as a starting point. The curvature $R_{\mu\nu}$ was a derived quantity. Now we see that at small distances quantum fluctuations of $R_{\mu\nu}$ (at higher orders, of objects such as

a b a b

(i) $R_{\mu\nu} = o$ (ii) $R_{\mu\nu}$ = large

Fig.4 *Measuring distances in curved space. The two units* a *and* b *are sensitive to tidal forces in a different way (one responds to* $g_{\mu\nu}$*, the other, say, to* $g_{\mu\nu} + R_{\mu\nu}$*). Whenever there is a curvature, the metric tensor is ambiguous, since there are no criterea to prefer* a *or* b *as fundamental units of distance.*

$R_{\mu\alpha\beta\gamma}\ R_{\nu\alpha\beta\gamma}$) become large and the definition of $g_{\mu\nu}$ becomes obscured by admixtures of these curvature tensors. Fundamental criterea to distinguish the pure $g_{\mu\nu}$ from the polluted $g_{\mu\nu}$ are absent. Suppose for instance that the unit of distance is defined by the wavelength of some spectral transition. Quantum effects then give effective couplings of the form $R^{\mu\nu}\ g^{\alpha\beta}\ F_{\mu\alpha}\ F_{\nu\beta}$ etc. (and with infinite coefficients !). This indeed implies that not the original metric tensor $g^{\mu\nu}$, but one of the polluted $g^{\mu\nu\prime}$ tensors is observed. My conclusion is that the metric tensor may be not at all as fundamental as is usually supposed. At distances smaller that the Planck length perhaps nothing even remotely resembling a metric tensor can be defined.

7. QGD versus QED and QCD

As we have seen the fact that the counter terms are not the same as the original Lagrangian implies the famous statement that "gravity is not renormalizable". (The arguments presented in sect.4 only work at the one-loop level.) Now nonrenormalizable theories are worse than one might think. Let me explain that by taking some examples. The gravitational coupling constant, G, in units where h and c are put equal to one, has the dimension of an inverse mass squared. That means that at smaller and smaller distance scales (corresponding to higher and higher mass scales) gravitational effects become stronger and stronger. Precisely the same is true for weak interactions in particle physics (Fig.5). The Fermi coupling constant (accidentally also called G) has the same dimensionality, so that that theory also is non-renormalizable. In particle physics the resolution of the problems that arise is that new physics must be going on at the distance scale where G

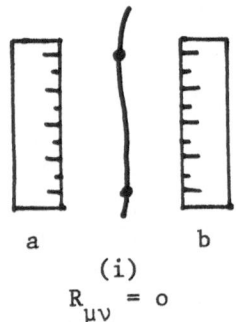

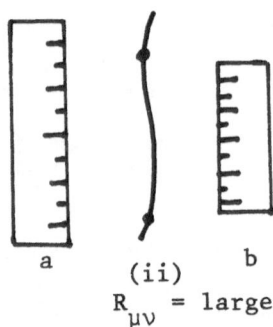

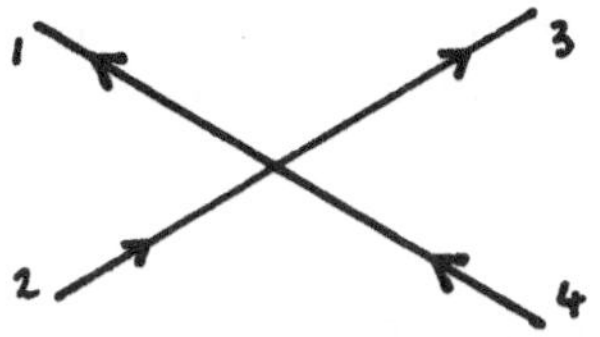

Fig. 5 *4-fermion interaction theory of weak interactions :*

$$\mathcal{L}^{int} = G\,\bar{\psi}_1\,\gamma_\mu(1+\gamma_5)\psi_2\,\bar{\psi}_3\,\psi_\mu(1+\gamma_5)\psi_4\ .$$

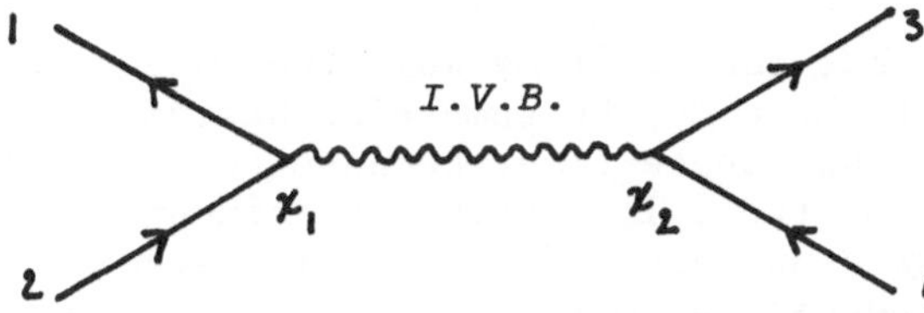

Fig. 6 *Intermediate Vector Boson theory : the interaction points x_1 and x_2 are now separated : the distance travelled by a virtual heavy particle (I.V.B.)*

becomes of order unity. In present-day theories it is generally expected that new heavy particles, called intermediate vector bosons, occur there, whose mass determines G, and they themselves only have dimensionless couplings (these couplings, by the way, have to obey the pattern of a gauge theory). The original 4-fermion interaction has now become a non-local one because it is described by an exchange process of a weak intermediate vector boson.

The well-known theory of quantum-electrodynamics (QED) is in much better shape, since it is described by a dimensionless small parameter α (the fine-structure constant). However, when scale transformations are performed more carefully one finds that α is not entirely dimensionless. At small distances it increases logarithmically, so that at exponentially large mass scales (10^{100} GeV, say) it reaches the value unity. Opinion on the interpretation of this differ, but my conclusion is that here also new physics is required.

The ideal theory in this respect is "quantum chromodynamics", a theory for strong interactions between quarks, where the coupling decreases with the mass-energy scale (though only logarithmically). This means that at small distances I have a less interactive theory and that may be a perfect theory, although mathematicians still put a question mark on that.

Clearly, basically new physics is needed to understand quantum gravitational dynamics at the Planck length-something like an intermediate boson being exchanged or something fundamentally different.

B. An alternative to the continuum theory ?

8. Discrete Space-time

So the big question is how could we possibly find an underlying theory that is even more basic that general relativity to explain quantum gravitational dynamics at the Planck length ? Now this is very difficult to conceive of and I just want to use the last part of these lectures to elucidate on the crazy idea about what possibly the physics could be. You see if my metric tensor no longer makes sense, if my momentum integrals diverge, this all means that there may be something basically wrong with working in a continuous space-time, because that is where all the difficulties came from. If I have a continuum, I can have infinitely large wave numbers and infinitely large momenta which give rise to divergences. Also my metric is no longer well-defined if I want to measure distances between two very close points because this distance is no longer a well-defined concept.

In particle theory we have learnt to be economical.If there is any possibility of reducing the number of physical degrees of freedom then we do. Thus, if a continuum makes no sense, then we replace it by a discrete space-time, called lattice. Lattices have been introduced before in physics. A famous lattice model for statistical mechanics is the exactly soluble Ising model (Fig. 7a) in two dimensions. Wilson [9] introduced a lattice version of gauge theories in four dimensions (fig. 7b). This lattice usually only serves to simplify the calculations and considerations, not because we really think that space-time in which quarks and gluons move is discrete.

In the case of gravity (Fig. 7c) we take the more ambitious point of view that it really does describe the physical situation accurately. The Planck length defines the distance scale on the lattice. However, there is a problem. In gravity the lattice which one would like to introduce would be totally different from the other lattices in field theory, and the reason is that we want to keep this beautiful notion of invariance under coordinate transformations. That would imply invariance under the interchange of two lattice points. That is, if there is interaction between two lattice points then there must be interaction between all other pairs of lattice points as well. (See Fig. 7c).

Is it, nevertheless, still possible to introduce some degree of order in this lattice ?

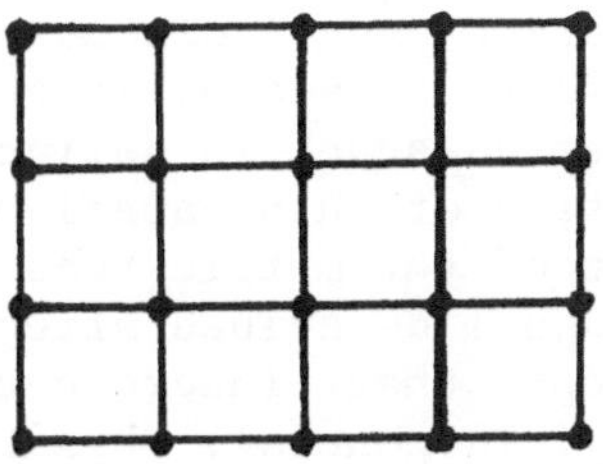

a) Ising model.

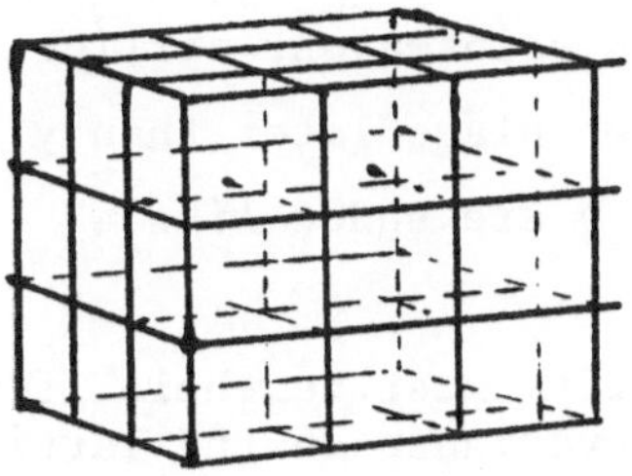

b) Wilson's gauge theory.

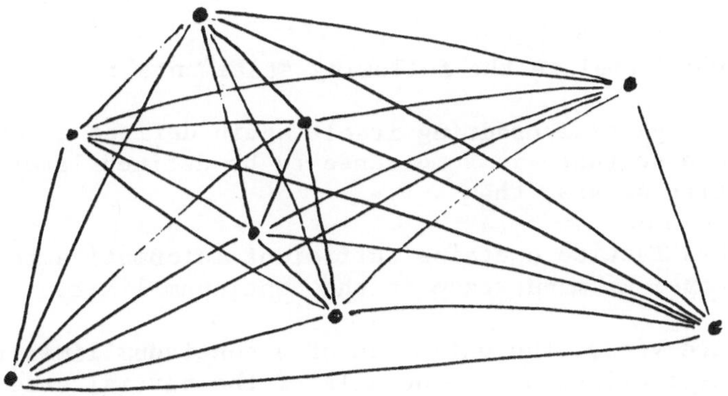

c) Gravity on a lattice.

Fig. 7

9. Causality on a lattice

I was asked a question when I discussed the ambiguities of the metric tensor in sects. 5 and 6. In general the causal ordering of points in Riemann space is defined by the metric. If one requires conservation of this causal ordering then only the conformal factor in front of the metric tensor is ambiguous. Evidently the redefinitions I described affect causal ordering also. This is a problem, I accept that. Indeed causality is a crucial principle in field theory. In general, physics will become pretty much impossible if one cannot tell whether one event is later or earlier than another. So let us impose a further structure on the lattice as follows. Even though a metric tensor does not make much sense on a lattice, we could at least define a causality relation for each pair of points (x,y). There must be three possibilities:

either	$x < y$	(x is a point-event earlier than y)	
or	$x > y$	(x takes place later than y)	
or	$x \mid y$	(x and y are space-like).	(29)

The latter possibility must be there if we wish to have some analogy to Lorentz invariance on the lattice. The existence of this relation between any pair of points must be dynamically determined. Of course one must require :

$$\text{if} \quad x < y \quad \text{and} \quad y < z \quad \text{then} \quad x < z \tag{30}$$

etc. We now formulate the following statements :

(i) This partial ordering itself again defines a lattice. Suppose that, in some sense to be defined later, a continuum limit exists, then

(ii) This lattice contains sufficient information to define a curved Riemann space in the continuum limit.

(iii) Vice versa, the existence of a continuum limit imposes restrictions on the details of the partial ordering relations (29) between all pairs of lattice points (not only partial ordering will enable us to make a continuum limit).

(iv) A curved Riemann space, if it has signature (+,+,+,-) contains all information on the physical history of the universe it describes.

10. The induced metric tensor

By statement (i) we mean the following. For any relatively time-like pair of points (x,y) with $x < y$ we can count the number of points z between them: $\rho\ (x,y)$ = number points z with $x < z < y$. If $\rho\ (x,y) = o$ then we join x and y with a line. In this way a lattice is obtained. We now assume that there is a continuum limit where this lattice is embedded in a four dimensional space. First we define the volume element in this space : $\sqrt{g}$ must be proportional to the number of lattice points inside a unit volume. To find the metric tensor induced in this continuum space we first define time-like distances. In continuous spaces the volume of all points z between two nearby timelike points x, y, with $x < z < y$, is easily found to be (see Fig. 8)

$$\rho(x,y) = \frac{\pi}{24}(dt)^4 .$$

We define now for the lattice

$$dt\ (x,y) = \sqrt[4]{\frac{24}{\pi}\rho(x,y) + 1} \tag{31}$$

where the +1 is needed for the case that $\rho = o$.

When $x|y$ the spacelike distance is harder to define. We simulate a formal measurement using light rays (see Fig. 9).

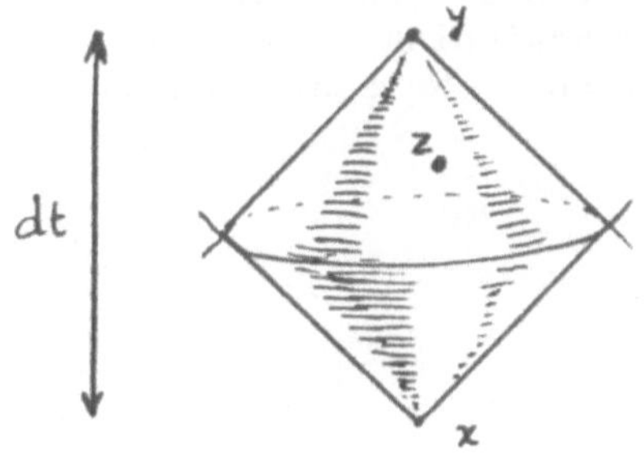

Fig.8 *Time-like distance between x and y (x<y) can be defined by counting the points z with x<z<y.*

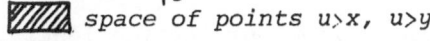

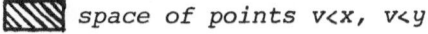

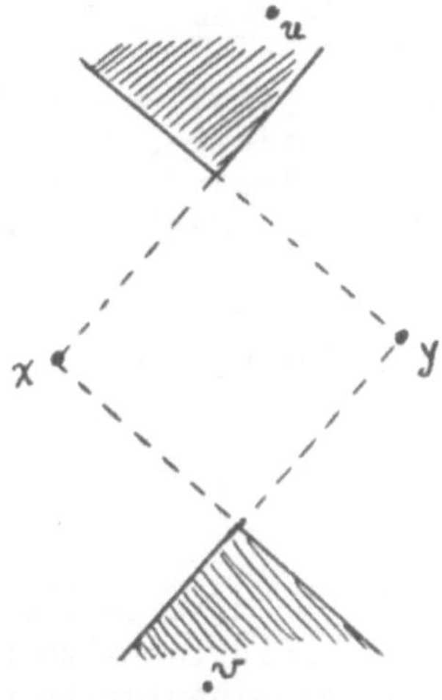

Fig.9 *Measuring spacelike distance for $x|y$.*
space of points $u > x$, $u > y$. (upper hatching)
space of points $v < x$, $v < y$. (lower hatching)

Consider all points u with $u > x$ and $u > y$.
Consider all points v with $v < x$ and $v < y$.

Then for all pairs (u,v) , $u > v$. An observer may move from v to u. He can emit a signal to both x and y and wait until both signals return. The minimal time he has to wait corresponds to the distance between x and y. Therefore,

$$ds(x,y) = \min_{u,v} dt(u,v)$$

$$\text{for all } u,v \text{ satisfying } u > x,\ u > y,\ v < x,\ v < y\ . \qquad (32)$$

Now that distances are well defined we can formulate more precisely the conditions for a continuum limit : the lattice must be embedded in a 4 dimensional continuous space such that a metric tensor $g_{\mu\nu}$ can be written down that describes to some degree of accuracy

$$ds^2 = g_{\mu\nu}\, dx^\mu\, dx^\nu \qquad (\text{if } x|y)$$

$$\text{or} \quad dt^2 = -g_{\mu\nu}\, dx^\mu\, dx^\nu \qquad (\text{if } x > y) \qquad (33)$$

11. The Cosmic Tree

At this point we could make connections with cosmology. Is the universe infinite, or does it have a beginning or an end ? I feel that at least a beginning is desirable. We could add as an axiom to the ordering relations :

There is a point o with

$$o < x \quad \text{for all} \quad x \neq o.$$

(Note added : in discussions the point was raised that this seems to be in contradiction with present-day big bang theories ; these give an initial singularity consisting of an infinity of points that are not causally connected.)

With this axiom our lattice looks like a tree (Fig. 10).

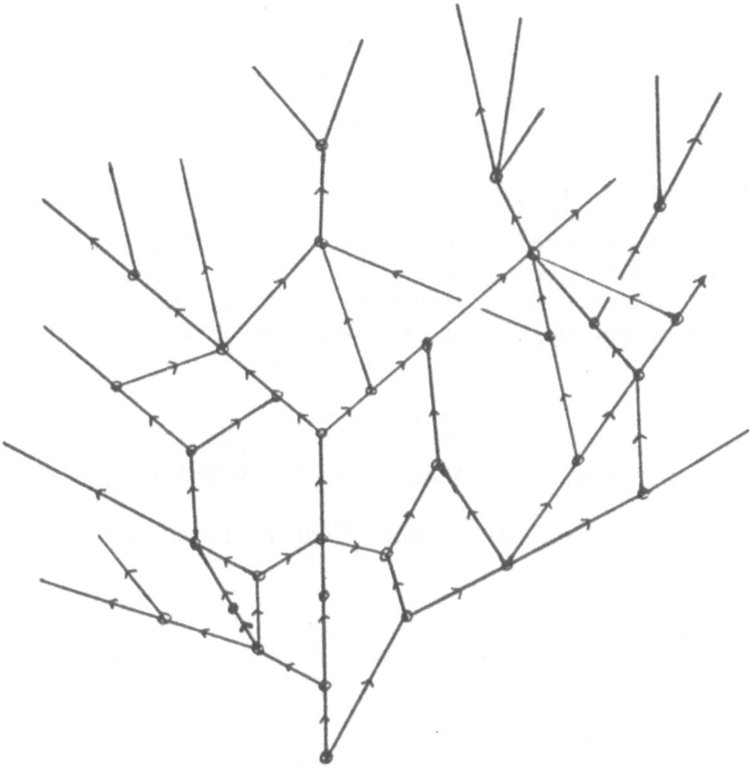

Fig. 10 *The Cosmic Tree.*

12. The Action ?

Perhaps we have here the seeds for a good theory. But we are not even half-way. Suppose that we do have the relevant variables here, in the sense that the details of the lattice are somehow determined by dynamics (the Einstein equations in the continuum limit). We still have to formulate this dynamics. That will be extremely hard. The functional integral formulation of field theories suggests that we have to sum the exponent of some action integral over all configurations. We have no idea how this action will look, but somehow it must approach the Einstein action in the continuum limit. There are two reasons why we expect such an action to be extremely non-local, however. One is that we insist on general coordinate invariance, implying that interactions between distant lattice points should be equal to that between close lattice points. The second is the absence of a cosmological term. We know that quantum fluctuations generate an energy-momentum tensor, even for the vacuum. The cancellation of this vacuum contribution by a "bare" cosmological term must be some mechanism that anticipates the (in itself non-local) quantum fluctuations. The question whether or not there is a vacuum at a point x can only be answered by looking at points surrounding x, even at fairly large distances. Thus the mechanism that renormalizes this cosmological constant must be non-local.

The above suggestions for a discrete gravity theory should not be taken for more than they are worth. The main message, and that is something I am certain of, is that it will not be sufficient to just improve our mathematical formalism of fields in a continuous Riemann space but that some more radical ideas are necessary and that totally new physics is to be expected in the region of the Planck length.

References

1) J. Schwinger, Proc. Nat. Acad. Sci. 44, 956 (1958).
K. Symanzik in Proc. Int. School "Enrico Fermi", R. Jost ed. Acad. Press (1969).
K. Osterwalder, R. Schrader, Comm. Math. Phys. 31, 83 (1973).

2) M. Veltman, Invited talk presented at the International Symposium on Electron and Photon Interactions at High Energies, Bonn 27-31 August 1973.

3) C.N. Yang and R. Mills, Phys. Rev. 96, 191 (1954).
R. Feynman, Acta Phys. Polonica 24, 697 (1963).

4) G. 't Hooft and M. Veltman, Nucl. Phys. B50, 318 (1972).

5) J.D. Bjorken and S.D. Drell, Relativistic Quantum Fields, McGraw-Hill (1965).
G. 't Hooft and M. Veltman, "Diagrammar", CERN report 73/9 (1973).

6) G. 't Hooft and M. Veltman, Nucl. Phys. B44, 189 (1972).
C.G. Bollini and J.J. Giambiagi, Phys. Lett. 40B, 566 (1972).
J.F. Ashmore, Lett. Nuovo Cim. 4, 289 (1972).

7) G. 't Hooft and M. Veltman, Ann. Inst. Henri Poincaré, 20, 69 (1974).

8) A.A. Slavnov, Teor i Mat. Fizika, 10, 153 (1972).

J.C. Taylor, Nucl. Phys. B33, 436 (1971).

9) K.G. Wilson, Phys. Rev. D10, 2245 (1974).

K.G. Wilson, Lectures given at the 1975 Intern. Summer School of Physics "Ettore Majorana".

QUANTUM FIELD THEORY AND FIBRE BUNDLES IN A GENERAL SPACE-TIME

S. J. Avis and C. J. Isham

Department of Physics
Blackett Laboratory, Imperial College
London SW7 2BZ

§§1. INTRODUCTION

There is as yet no complete and consistent quantum theory of gravity. Indeed some practitioners of this art still argue about what would even constitute such a theory. However, irrespective of the final format, it seems likely that something rather dramatic will happen if the system is probed at distances of the fundamental Planck length $L = (\frac{\hbar G}{c^3})^{1/2} \simeq 10^{-33}$ cms. It is often suggested that at this order of distance there will be violent quantum fluctuations in the space-time structure itself (rather than in just the metric tensor) and in particular the topological structure may be subject to quantum laws. Whether such an idea can be sensibly implemented is not yet clear and it might even be too conservative! However, in order to obtain any understanding of this problem it seems reasonable to first study the situation in which a quantum field is defined on a fixed (unquantized) manifold and to ask what rôle global space-time topology plays in the quantum field theory.

In these lectures we have chosen to study just one specific aspect of the problem. The starting point is the observation that an important property of topological structure is that it determines the types of inequivalent vector bundles that can be built over the space-time M. Recall that a vector bundle over M is essentially a 'twisted product' of M with some vector space V. More generally a fibre bundle (E, π, M) with <u>fibre</u> F consists of a <u>bundle space</u> E and a 'projection' map π from E onto M (the <u>base space</u>) with the property that locally E looks like M x F (the inverse image $\pi^{-1}(x)$ of $x \in M$ is called the <u>fibre</u> over x). Thus for any point $x \in M$ there exists an open set $U_1 \subset M$ containing x and a homeomorphism

$$h_{U_1} : U_1 \times F \rightarrow \pi^{-1}(U_1) \tag{1.1}$$

satisfying $\pi(h_{U_1}(x,v)) = x \ \forall x \in U_1, \forall v \in F$ (see Drechsler & Mayer 1977, Husemoller 1966, Steenrod 1951). If U_2 is another neighbourhood of x such that $U_1 \cap U_2 \neq \phi$ then there will also exist $h_{U_2} : U_2 \times F \rightarrow \pi^{-1}(U_2)$. Of course on $U_1 \cap U_2$ the two maps may well be different but there is some function $v'(v)$ such that

$$h_{U_1}(x,v) = h_{U_2}(x, v'(v)) \quad \forall x \in U_1 \cap U_2 \tag{1.2}$$

This function is sharply limited by requiring the existence of a lie group G of transformations of F such that $v'(v)$ is obtained by acting on v with some group element $g_{U_2 U_1}(x)$. Hence (1.2) becomes

$$h_{U_1}(x,v) = h_{U_2}(x, g_{U_2 U_1}(x) v) \tag{1.3}$$

and $x \rightsquigarrow g_{U_2 U_1}(x)$ is also required to be continuous. For vector bundles this structure group will be some subgroup of $GL(n,\mathbb{C})$ or $GL(n,\mathbb{R})$ such as $SU(n)$ or $SO(n)$.

From a physical point of view vector bundles have two important features. The first is the existence of cross sections. To motivate this idea consider a real valued scalar function ϕ defined on a circle viewed as the closed interval $[0,1]$ with 0 and 1 identified. The graph of ϕ is then drawn on a cylinder (Fig. 1). In effect we have placed a copy of the real line over each point of the circle S^1 and if π is the projection map, ϕ can be regarded as a function S from the circle into the cylinder satisfying

$$\pi(S(x)) = x \qquad \forall x \in S^1 \tag{1.4}$$

Note that the cylinder is actually a trivial product vector bundle $S^1 \times \mathbb{R}$ over S^1. However there is also a non-trivial vector bundle that can be built over the circle. This has an "open"

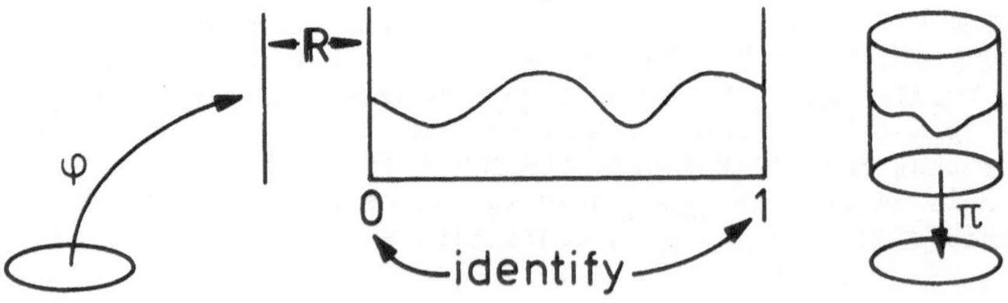

Figure 1

Möbius band as its bundle space, as shown in Fig. 2, and, as, in any bundle, a cross-section is defined to be a map S from the base space into the bundle space satisfying (1.4). Thus for a non-trivial bundle cross-sections are natural analogues of functions and we will refer to them as twisted fields. Now our general contention is the following. Suppose that we are interested in fields ϕ defined on a spacetime M, which transform under some finite dimensional representation of a 'gauge group' (based on a Lie group G) on a vector space V. Then rather than just looking at V-valued functions on M we should instead employ cross-sections of all possible vector bundles over M with fibre V and structure group G. The twisted fields may well behave differently from the untwisted fields when quantized. Furthermore the number and type of these vector bundles depends critically on the space-time topology. Thus by studying quantized twisted fields we are probing one aspect of the rôle played by space-time topology in quantum field theory.

Note that although cross-sections are not functions from M into V they do have local representations which are just V-valued functions. In the notation of eq. (1.1), the local representative of a cross-section S is a function $S_{U_1} : U_1 \to V$ such that

$$S(x) = h_{U_1}(x, S_{U_1}(x)) \qquad \forall\, x \in U_1 \tag{1.5}$$

If U_2 overlaps U_1 then we have another local representative $S_{U_2}(x)$ satisfying

$$S(x) = h_{U_2}(x, S_{U_2}(x)) \tag{1.6}$$

and eq. (1.3) implies

$$S_{U_2}(x) = g_{U_2 U_1}(x)\, S_{U_1}(x) \tag{1.7}$$

so that these two local representatives are gauge related by the

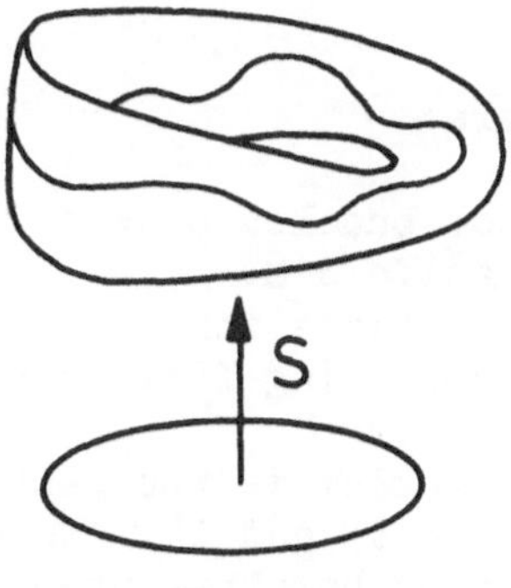

Figure 2

function $g_{U_2U_1}(x)$. Indeed it is widely recognised these days that the theory of fibre bundles is the appropriate framework within which to discuss the mathematical properties of gauge theories (e.g. Drechsler & Mayer 1977).

There is another physically significant aspect of vector bundles. Although a bundle is locally trivial and indeed all the fibres are homeomorphic, there is nevertheless no canonical way of associating one fibre with another. The way around this is to introduce a connection into the bundle. This is essentially a one-form in the bundle space which takes its values in the Lie algebra L(G) of the structure group G. It possesses local representatives which are L(G) values one-forms on M and are what are commonly called Yang-Mills fields. Different local representatives are gauge related via the usual Yang-Mills gauge transformation employing $g_{U_2U_1}(x)$ as the gauge function. The usual SU2 instanton structures are described mathematically as different bundles over the four sphere S^4. In this case the inequivalent bundles are labelled by an integer which in turn can be obtained by integrating over S^4 the (suitably normalised) quantity $\sum_{i=1} F^i_{\mu\nu} F^i_{\alpha\beta} \epsilon^{\mu\nu\alpha\beta}$ where $F^i_{\mu\nu}$ is the covariant curl of any Yang-Mills field associated with the bundle. One of the main aims of these lectures is to explore the construction of bundles in an arbitrary space-time M. Hence we wish to ask:

i) How are the vector bundles over M classified for an arbitrary gauge group G?
ii) Can the bundle be uniquely determined by an integral over functions of the Yang-Mills $F_{\mu\nu}$?

We will investigate these problems (especially (i)) for both compact and noncompact manifolds. Pseudo-riemannian space-times cannot be compact (unless closed timelike curves are admitted) but their 'Riemannian continuations' may be. Riemannian manifolds are currently employed in some functional integral approaches to quantum gravity (e.g. S. Hawking's lectures at this summer school) and since our methods apply to both cases we may as well discuss them simultaneously.

The main problems are:

1) Classify vector bundles over an arbitrary four-dimensional manifold M.
2) Construct lagrangians for the cross-sections.
3) Quantise the resulting field equations.

The classification problem is the central part of these lectures and is contained in §3 and §4. We have chosen to discuss this using various ideas from homotopy theory. There is no new mathematics in any of this but the techniques used are likely to be valuable in any investigation of 'quantum topology' and we hope that the rather heuristic treatment which we present may be useful.

In §2 we will discuss the quantisation problem for linear twisted scalar fields. It will be shown how such fields can be regarded as operators on a Hilbert space and that they yield significantly different results from an untwisted field. We will also discuss the 'solitonic' sector of an interacting $\lambda(\phi^2 - a^2)^2$ twisted field theory and show that the non trivial spatial topology has an interesting effect on the kink solutions. An alternative approach to quantization is via the use of functional integrals. This is mainly appropriate for a 'Riemmanianised' instanton structure and would be complementary to our pseudo-Riemannian solitonic results. Lack of time precludes an account of this but the lectures of Stephen Hawking may be consulted for a comprehensive account, albeit in a slightly different context.

Finally in §5 we will discuss the effects of space-time topology on spinor fields with the emphasis on the existence of inequivalent spin structures that can be present in a general space-time. This phenomenon would be especially interesting in a supergravity context where the presence of different spin multiplets and internal symmetry groups would lead to an interplay between the appropriate topological properties.

NOTATION.

It is convenient to gather together some of the definitions in homotopy theory that will be employed later. (Hu 1959, Spanier 1966, Switzer 1975).

Part of homotopy theory deals with pointed topological spaces. These are just topological spaces in which a base point has been chosen (which could in principle be any point of the space). The logically correct notation for a pointed space is for example (X,x_0); where X is a topological space and $x_0 \in X$ is its base point. Provided it is clear from context that pointed spaces are being considered it is usual to suppress the base point and simply write X instead of the correct and longer symbol (X,x_0).

A map between two pointed topological spaces (X,x_0) and (Y,y_0), written $f : (X,x_0) \to (Y,y_0)$ (or for brevity as $f : X \to Y$), is a continuous function from X into Y which satisfies $f(x_0) = y_0$. Two such maps $f,g : (X,x_0) \to (Y,y_0)$ are said to be homotopic (denoted by $f \sim g$) if there exists a continuous function $F : X \times I \to Y$ such that

$$F(x,o) = g(x) \qquad \forall x \in X$$

$$F(x,1) = f(x) \qquad \forall\, x \in X$$

and $$F(x_o,t) = y_o \qquad \forall\, t \in I.$$

Here $I = [o,1]$ and is the closed unit interval.

The relation of being homotopic is an equivalence relation on the set of all maps $h : (X,x_o) \to (Y,y_o)$. This relation partitions the set into disjoint equivalence classes; the collection of all equivalence classes is written as $[X,x_o;Y,y_o]$ (or for brevity simply as $[X,Y]$). A typical class $[f]$ consists of all maps $g : (X,x_o) \to (Y,y_o)$ which are homotopic to f.

If $f : (X,x_o) \to (Y,y_o)$ is a map and (M,m_o) is a space then f induces a function f_* from $[M,m_o;X,x_o]$ into $[M,m_o;Y,f(x_o)]$ defined by $(f_*\omega)(m) := f(\omega(m))$ where $\omega \in [M,m_o;X,x_o]$ and $m \in M$. This can be conveniently described by the commutative map diagram

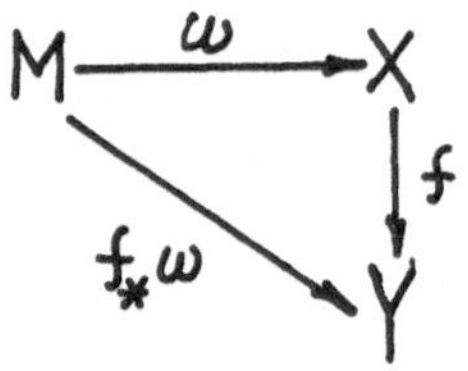

Thus $f_*\omega$ is defined in such a way as to make the diagram 'commute' (by which we mean $f_*\omega = f\cdot\omega$).

Especial significance is attached to the set (for each $n \geq 1$) $[S^n,s_o;M,m_o]$, where S^n is the n-sphere with base point s_o, known as the n'th homotopy group. This is precisely the object that is usually written as $\pi_n(M,m_o)$ or even $\pi_n(M)$. It is well known that the $\pi_n(M)$'s are in fact groups. In general the same is NOT true for the sets $[X,x_o;Y,y_o]$ unless X or Y are restricted to belong to a certain class of topological spaces.

We will assume throughout that the space-time M is connected. In general the set of path connected components of a topological space X will be denoted by $\pi_o(X,x_o)$. It is easy to see that $\pi_o(X,x_o) \simeq [S^o,s_o;X,x_o]$.

N.B.

As mentioned several times already, we will nearly always suppress from our notation the base points of the pointed system in the manner indicated by the foregoing section.

If maps $f : X \to Y$ and $h : Y \to X$ have the property that

$$f \cdot h \sim id_Y$$

$$h \cdot f \sim id_X$$

then X and Y are said to be homotopically equivalent and we denote this by $X \sim Y$. It should be noted that

(i) If $X \sim Y$ then $[M , X] = [M , Y]$ and $[X , M] = [Y , M]$ for any space M.

(ii) If $X \sim Y$ then the group homomorphisms $f_* : \pi_n(X) \to \pi_n(Y)$ and $h_* : \pi_n(Y) \to \pi_n(X)$ are isomorphisms for all n.

If X and Y are two topological spaces and $f : X \to Y$ induces isomorphisms $f_* : \pi_n(X) \to \pi_n(Y)$ for all n, then X and Y are said to be weakly homotopically equivalent. This does not in general mean that $X \sim Y$ although it is true if X and Y and CW complexes (§3).

When we refer to compact manifolds we mean compact without boundaries. Our general classification results, apply to any 4-manifolds including therefore compact with boundary. However the relevant cohomology groups are not the same as in the boundary-free case although they are readily computable using, for example, the Lefschetz duality theorem.

Note also that in many cases of physical interest a compact manifold arises because vanishing type boundary conditions are imposed on some fields defined originally on a noncompact space (as for example in the usual instanton case.) This vital physical ingredient should not be forgotten when employing our classification results.

§§2. TWISTED SCALAR FIELDS

§2.1 Classification

In this section we will discuss the Hilbert space quantization of a twisted scalar field. According to the philosophy espoused in §1 the first step is to classify vector bundles whose fibre is the real line. Let us first consider the general case where the fibre is any space and the structure group is G. There is one special and very important type of fibre bundle known as a principal fibre bundle in which the fibre is itself a lie group G with structure group some subgroup of G. Let $\zeta = (E,\pi,M)$ be such a principal G-bundle and suppose F is any space on which there is a left G-action. There exists a natural right action of G on E (which we will denote by $p \rightsquigarrow pg$) and hence a right-action of G on E x F can be defined by

$$(p, v) \rightsquigarrow (pg , g^{-1}v) \; ; \; p \in E , v \in F , g \in G \qquad (2.1)$$

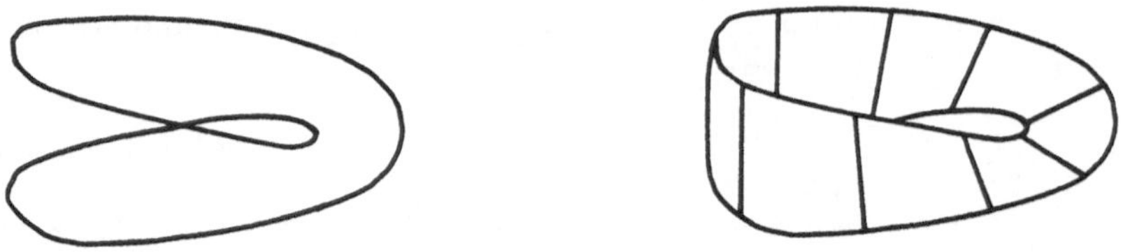

Figure 3

Now define (p_1, v_1) to be equivalent to (p_2, v_2) if there is some $g \in G$ such that $p_2 = p_1 g$, $v_2 = g^{-1} v_1$. The set of equivalence classes is denoted $E \times_G F$ and if $\pi_F : E \times_G F \to M$ is defined by $\pi_F(p,v) := \pi(p)$, then $(E \times_G F, \pi_F, M)$ is a fibre bundle over M with fibre F and structure group G (or some subgroup of G). Roughly speaking a copy of F has been glued onto each fibre of the principal bundle. This is illustrated below for the Möbius band which is obtained from the principal Z_2-bundle on the left hand side (Fig. 3).

An important result is that <u>every</u> fibre bundle can be obtained in this way from some principal bundle. Hence the problem of classifying fibre bundles reduces to that of classifying principal bundles. For vector bundles with fibres $\mathbb{R}$ or $\mathbb{C}$ this means looking for $GL(n,\mathbb{R})$ or $GL(n,\mathbb{C})$ bundles respectively. However, another important result in bundle theory is that if G is a noncompact group it is always possible to find a covering of M by open sets with the property that the structure functions $g_{U_2 U_1}$ defined in (1.3) lie in the maximal compact subgroup K of G. Thus without any loss of generality we can regard K as being <u>the</u> structure group of the bundle which in the case of vector bundles would be $O(N,\mathbb{R})$ or $U(N)$ (of course for specific bundles it might be smaller still).

In the case of twisted scalar fields the structure group is $GL(1,\mathbb{R})$ which has $O(1)$ as a maximal compact subgroup. However, $O(1)$ is simply the cyclic group of order two (Z_2) and so the problem reduces to classifying principal Z_2-bundles over M which in turn correspond to double coverings of M. The key tool is a crucial property possessed by fibre bundles - <u>the homotopy lifting property</u>:

<u>Prop 2.1</u> "Let (E, π, M) be a fibre bundle and let $F: X \times I \to M$ be a homotopy between maps $f, g : X \to M$. Suppose that f is 'covered' by a map $f^{\uparrow} : X \to E$ i.e. $\pi(f^{\uparrow}(x)) = f(x)\ \forall x \in M$.

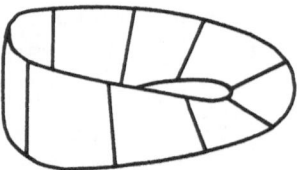

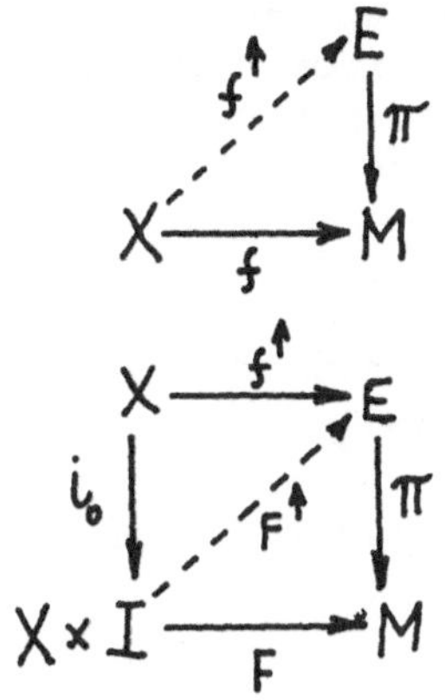

Then there is a homotopy $F^{\uparrow}: X \times I \to E$ such that

$$\text{(i)} \quad F^{\uparrow}(x,0) = f^{\uparrow}(x) \quad \forall x \in X$$

$$\text{(ii)} \quad \pi(F^{\uparrow}(x,t)) = F(x,t) \quad \forall x \in X \quad \forall t \in [0,1] \qquad (2.2)$$

$$i_0(x) := (x,0)$$

This feature is so important that it is used as a definition of the more general concept of a fibration in which the individual fibres are only homotopically equivalent to each other:

"A map π between any two topological spaces E and M is called a <u>fibration</u> if (E, π, M) has the homotopy lifting property for maps from an arbitrary topological space into M."

Consider now the situation where we have a principal bundle over M with some abelian discrete group D as fibre (e.g. Z_2). We will concentrate on the fibre over the base point $x_0 \in M$ and consider a map $w: I \to M$ with the property that $w(0) = w(1) = x_0$. This can be thought of as a homotopy of the trivial map which maps a set with a single point $\{*\}$ into $x_0 \in M$. This map is covered by the map from $\{*\}$ into the base point $p_0 \in \pi^{-1}(x_0)$ and by virtue of the homotopy lifting theorem there is a function $w^{\uparrow}: I \to E$ which covers the homotopy w. This map $w^{\uparrow}$ is a curve in E with $w^{\uparrow}(0) = p_0$ and $w^{\uparrow}(1) \in \pi^{-1}(x_0)$. (Figure 4).

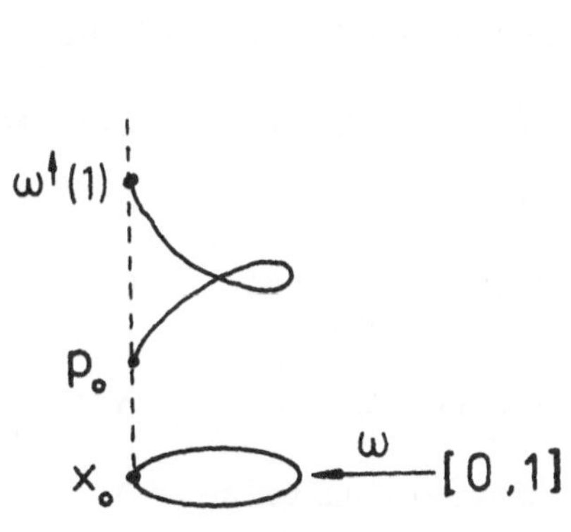

Figure 4

It can be shown (Steenrod 1951) that $w^{\dagger}(1)$ depends only on the homotopy class of w and that the resulting map from $\pi_1(M, x_0)$ into $\pi^{-1}(x_0) \approx D$ is a group homomorphism. Conversely any such homomorphism leads to a principal D-bundle and hence

$$\mathcal{B}_D(M) \approx \text{Hom.}(\pi_1(M), D)$$

where we have introduced the notation $\mathcal{B}_G(M)$ as the set of principal G-bundles over M (see §3 for more details).

In particular

$$\begin{aligned}\mathcal{B}_{Z_2}(M) &\approx \text{Hom.}(\pi_1(M), Z_2)\\ &\approx \text{Hom.}(H_1(M;Z), Z_2)\\ &\approx H^1(M;Z_2)\end{aligned} \tag{2.3}$$

where $H_1(M;Z)$ and $H^1(M;Z_2)$ are respectively the first integral homology group and first Z_2-valued cohomology group. (We will rederive (2.3) in §3 as part of the general scheme). The number of inequivalent twisted field structures (i.e. the number of different Z_2-bundles) is thus precisely the number of elements in $H^1(M;Z_2)$ (the 'Möbiusity' of M) with the trivial product bundle corresponding to the group identity. The cohomology element $W_1(\xi)$ corresponding to a specific bundle ξ is called the first Stieffel-Whitney class. The results for various model space-times are tabulated below.

M	Space-time	$H^1(M;Z_2)$	No. of bundles
$S^1 \times \mathbb{R}^1$	1 + 1 model	Z_2	2
$S^1 \times S^1 \times \mathbb{R}^1$	2 + 1 model	$Z_2 \oplus Z_2$	4
$S^1 \times S^1 \times S^1 \times \mathbb{R}^1$	Bianchi type I	$Z_2 \oplus Z_2 \oplus Z_2$	8
$S^2 \times \mathbb{R}^2$	Kruskal	0	1
$S^2 \times S^2$	'Riemannianised' Schwarzschild	0	1
S^4	'Instanton'	0	1
$S^3 \times \mathbb{R}^1$	Einstein Universe	0	1
$\mathbb{RP}^3 \times \mathbb{R}^1$		Z_2	2

§2.2 Quantization of linear twisted fields.

The quantization of linear (untwisted) fields propagating in a curved space-time is a subject that has attracted considerable attention over the last four years. (For recent reviews see DeWitt 1975, Parker 1977, Isham 1977b, Gibbons 1978). The cause of this activity was principally Hawking's remarkable discovery that a black hole of mass M quantum mechanically radiates particles with a thermal spectrum corresponding to a temperature of $kT = (8\pi M)^{-1}$ (Hawking 1975). For certain specific space-times there now exists well understood quantization algorithms but in general there does not appear to be any unambiguous Hilbert space based quantum field theory. There is certainly a natural C^*-algebraic canonical structure but the translation from the abstract algebra to a concrete algebra of field operators is not unique. (For a discussion of this see for example Hajicek 1977, Isham 1977a, Kay 1977).

Most Hilbert space oriented schemes have their roots in Segal's general framework for quantizing linear systems (e.g. Segal 1967). This in turn rests heavily on the structure of the space of classical solutions to the field equations. It is therefore natural to restrict attention to space-times which are globally hyperbolic and for which the field equation

$$(g^{\mu\nu}\nabla_\mu \partial_\nu + \mu^2)\phi = 0 \qquad (2.4)$$

possesses a decent Cauchy initial value problem. In special cases non globally hyperbolic manifolds can be handled but only by carefully specifying boundary data.

Globally hyperbolic space-times are topologically of the form $\mathbb{R} \times \Sigma$ (Geroch 1970) and hence any twists in the field must necessarily be associated only with the spatial hypersurface Σ. This is because any bundle over a contractible space like $\mathbb{R}$ is automatically trivial; (Husemoller 1966) or equivalently because

$$H^1(R\times\Sigma; Z_2) \approx H^1(\Sigma; Z_2) \qquad (2.5)$$

Thus the time evolution of a twisted field is similar to that for a normal field with the added feature of 'twist' conservation in time. All quantization schemes on a globally hyperbolic space-time relate in some way or an other to the canonical commutation relations (CCR) and this is one route for tackling the twisted case. The usu al classical Poisson Brackets are

$$\{\phi(f_1), \pi(f_2)\}_{P.B.} = \int_\Sigma f_1(\underline{x}) f_2(\underline{x}) \sqrt{{}^3g}\, d^3\underline{x} \qquad (2.6)$$

where f_1, f_2 are smooth, compact support test functions and

$$\phi(f) := \int_{\Sigma} \phi(\underline{x})\, f(\underline{x}) \sqrt{{}^3g}\, d^3\underline{x} \tag{2.7}$$

Now the right hand side of (2.7) would not make sense when $\phi(\underline{x})$ is twisted because it is not an $\mathbb{R}$-valued function. On the other hand if f were itself to be twisted the expression would be meaningful since $\phi(\underline{x})f(\underline{x})$ has an unambiguous real value. This is because for a Z_2 bundle the group acts on the fibre $\mathbb{R}$ by

$$\begin{aligned} er &= r \\ ar &= -r \end{aligned} \qquad Z_2 = \{e, a\}\ ,\ r \in \mathbb{R} \tag{2.8}$$

and hence the structure functions $g_{U_2U_1}(x)$ are all either +1 or −1. Thus eqn. (1.7) relating two local representatives of a section of the vector bundle becomes

$$S_{U_1}(x) = \pm S_{U_2}(x) \tag{2.9}$$

Hence if S and S' are two sections

$$S_{U_2}(x)\, S'_{U_2}(x) = S_{U_1}(x)\, S'_{U_1}(x) \tag{2.10}$$

and so $S(x)S'(x)$ can be unambiguously defined as the number associated with any local representative. The trick in our case is then clearly to smear the fields with smooth, compact support, cross-sections of the vector bundle (Drechsler and Mayer 1977, Isham 1978a). All of the usual CCR theory carries across to this new situation and we can talk about twisted Weyl C^*-algebras, twisted Fock spaces etc.

This technique completely solves the problem. However most 'practical' claculations of quantum field processes in a curved space-time have been performed using a heuristic, rather non rigorous version of covariant quantization and it would be helpful to know the twisted analogue. Typically (DeWitt 1975) the hermitian quantum field $\hat{\phi}$ is expanded as

$$\hat{\phi}(x) = \sum_i \{\hat{a}_i f_i(x) + \hat{a}_i^* f_i^*(x)\} \tag{2.11}$$

where $\{f_i\}$ are a 'complete' set of solutions to the field equations (2.4) and the operators $\hat{a}_i$, $\hat{a}_j^*$ obey the usual relations

$$[\hat{a}_i, \hat{a}_j^*] = \hbar\, \delta_{ij} \tag{2.12}$$

$$[\hat{a}_i, \hat{a}_j] = 0 \tag{2.13}$$

It is natural in the twisted case to adopt a similar scheme with the $\{f_i\}$ being a suitably complete set of twisted solutions to the field equation (2.4) (this concept makes sense since (2.4) is invariant under $\phi \to -\phi$). This procedure works and is equivalent to the more rigorous CCR method discussed previously.

As an example of this technique we can consider a model S^1 (space) x $\mathbb{R}$ (time) static space-time which admits two principal Z_2-bundles since $H^1(S^1 \times \mathbb{R}; Z_2) \approx Z_2$. These correspond to the cylinder and Möbius band constructed over the spatial circle. A suitable set of basis functions for the normal field is

$$f_n^N(\theta,t) \sim e^{in\theta} e^{-i(n^2+\mu^2)^{1/2}t} \quad ; \quad n \in Z, \quad 0 \leq \theta \leq 2\pi \tag{2.14}$$

On the other hand if a Möbius band is viewed as a rectangle with edges identified as shown below in Figure 5, it is clear that a cross-section of this non trivial bundle may be represented by a normal function which is anti-periodic. Hence a suitable set fo basis functions in this sense is

$$f_n^T(\theta,t) \sim e^{i(n+1/2)\theta} e^{-i[(n+1/2)^2+\mu^2]^{1/2}t} \tag{2.15}$$

These two systems have different Green's functions and lead to different physical results. For example there has been a considerable study over the last few years of the regularized vacuum expectation values of the energy-momentum tensor (see P.C.W. Davies 1977 for a review). In the present static case there is an unambiguous vacuum and hamiltonian operator H and (choosing μ=o for simplicity) one easily computes

$$\langle o|H|o\rangle^N = \frac{\hbar}{R}\sum_{n=0}^{\infty} n \tag{2.16}$$

$$\langle o|H|o\rangle^T = \frac{\hbar}{R}\sum_{n=0}^{\infty} (n+1/2) \tag{2.17}$$

where we have chosen a metric on the circle of $ds^2 = R^2 d\theta^2$. Both sums are divergent because of the infinite zero-point energy and must be regularized. The neatest way is to employ zeta-function methods (Dowker & Critchley 1976, Hawking 1977) and note that

$$\zeta(s) := \sum_{n=0}^{\infty} n^{-s} \qquad \zeta(s,q) := \sum_{n=0}^{\infty} (n+q)^{-s} \tag{2.18}$$

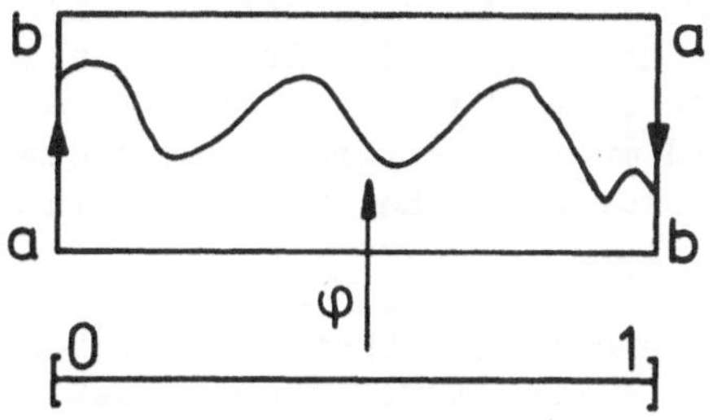

Figure 5

are complex analytic functions of s with a single pole at $s = +1$. Hence we define the regularized vacuum self energies:

$$\langle 0|H|0\rangle^{N}_{reg.} := \frac{\hbar}{R}\,\zeta(-1) = \frac{-\hbar}{12R} \tag{2.19}$$

$$\langle 0|H|0\rangle^{T}_{reg} := \frac{\hbar}{R}\,\zeta(-1,\tfrac{1}{2}) = \frac{\hbar}{24R} \tag{2.20}$$

The interesting feature of these results is not only are they different in magnitude but they have different signs. Thus quantum gravity schemes of the form

$$G_{\mu\nu}(g) = \langle\,|\,\hat{T}_{\mu\nu}\,|\,\rangle \tag{2.21}$$

would be radically affected by the presence of twisted fields. Further details of the above can be found in Isham 1978a. Banach and Dowker 1978 in a comprehensive and beautiful study have computed the regularised twisted field energy-momentum tensor in a number of different topological spaces and other examples can be found in DeWitt, Hart and Isham 1978. The sign change in the energy seems to be a fairly common feature of twisted fields.

§2.3 Non-linear twisted scalar fields

In this section we illustrate the differences between twisted and untwisted non-linear scalar fields by considering some specific models in both two and four dimensions.

The simplest two dimensional space-time which admits twisted scalar fields is the one just considered with the topology $\mathbb{R}^1$ (time) x S^1 (space) and the metric

$$ds^2 = dt^2 - R^2 d\theta^2$$

where R is the radius of the spatial circle.

For the sake of definiteness consider the lagrangian

$$\mathcal{L} = \frac{R}{2}\left[\left(\frac{\partial\phi}{\partial t}\right)^2 - \frac{1}{R^2}\left(\frac{\partial\phi}{\partial\theta}\right)^2 - \frac{\lambda^2}{2}(\phi^2 - a^2)^2\right] \tag{2.22}$$

where $\lambda > 0$ and $a > 0$. This is well defined for twisted fields since it is invariant under $\phi \to -\phi$. At the classical level the untwisted ground state is either $\phi = a$ or $\phi = -a$. On the other hand any twisted field must pass through zero somewhere, thus the classical ground states of the twisted fields cannot be $\phi^2 = a^2$.

It is plausible (and true!) that the twisted ground states are time-independent. Furthermore if we classify twisted fields according to the number of zeros on the spatial circle that they possess then it is found that, for a fixed R, there are only finitely many distinct classes of static solutions. They are

$$\phi \equiv 0$$

and

$$\phi_N(\theta;R) = (a^2 - w_N)^{1/2}\, \mathrm{sn}\left[\lambda R\left(\frac{a^2+w_N}{2}\right)^{1/2}(\theta+\theta_0);\ \gamma_N\right] \qquad (2.23)$$

where θ_0 is a constant, $\gamma_N = \frac{(a^2-w_N)}{(a^2+w_N)}$ and w_N satisfies the transcendental equation

$$\lambda R\left(\frac{a^2+w_N}{2}\right)^{1/2} = \frac{N}{\pi} K(\gamma_N) \qquad (2.24)$$

Here sn(u,γ) is a Jacobi elliptic function with parameter γ (this is a periodic function with period 4K(γ) and is rather like a sine curve in profile),

$$K(\gamma) = \int_0^{2\pi} \frac{du}{(1-\gamma \sin^2 u)^{1/2}} \qquad (2.25)$$

and N is an odd integer such that $\frac{N}{2a\lambda} < R$ and $N \geq 0$

The transcendental equation occurs because ϕ_N must be defined on the circle, and the restriction that N be odd arises from demanding that ϕ_N be a twisted field (and hence have an odd number of zeros). The main point to notice is that for $R \leqslant (2a\lambda)^{-1}$ there is only one static twisted field, namely $\phi \equiv 0$ (Avis and Isham 1978). Indeed it can be shown that for $R \leqslant (2a\lambda)^{-1}$ the $\phi \equiv 0$ solution is the absolutely stable twisted ground state, whilst for $R > (2a\lambda)^{-1}$ the ϕ_1 solution is 'the' absolutely twisted ground state. The energies of these solutions are shown in figure 6.

Thus there is a spontaneous breakdown of the translational invariance of the twisted ground state as the radius of the space-time cylinder increases across the critical value $R = (2a\lambda)^{-1}$.

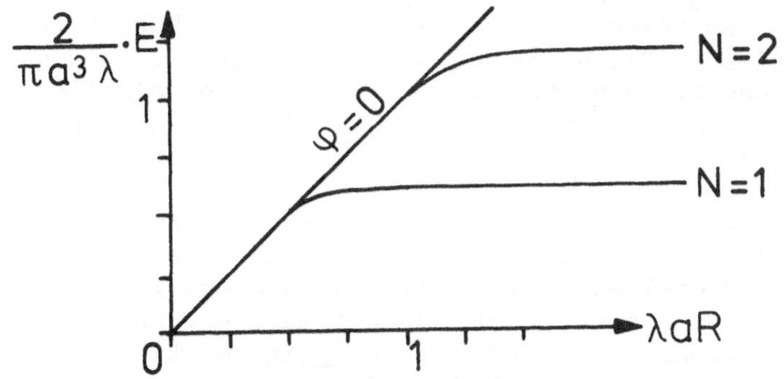

Figure 6

Of course, as $R \to \infty$ the twisted ground state goes into the usual two-dimensional kink, and so by one-point compactifying the real line into the circle we have separated the two-dimensional Minkowskian kink from the vacua $\phi = \pm a$ in a very rigid way.

In Minkowski space-time Derrick's theorem forbids the existence of position dependent, finite energy, static scalar fields in four dimensions. This theorem, however, does not hold on a general four-dimensional manifold. Thus there is nothing to prevent the two-dimensional model described above being generalised to four dimensions. Indeed with the space-time of topology $\mathbb{R}$ (time) x S^1 x S^1 x S^1 and metric

$$ds^2 = dt^2 - \sum_{i=1}^{3} R_i^2 \, d\theta_i^2 \tag{2.26}$$

there are now eight different types of scalar field, one untwisted and seven twisted. (They are characterised by their state of "twistedness" over each of the three circles).

With the Lagrangian

$$\mathcal{L} = \frac{R_1 R_2 R_3}{2}\left[\left(\frac{\partial\phi}{\partial t}\right)^2 - \sum_{i=1}^{3}\frac{1}{R_i^2}\left(\frac{\partial\phi}{\partial\theta_i}\right)^2 - \frac{\lambda^2}{2}(\phi^2 - a^2)^2\right] \tag{2.27}$$

the absolutely stable twisted ground states are found to be

$$\phi \equiv 0 \qquad \text{for} \qquad R \leqslant (2a\lambda)^{-1}$$

and

$$\phi = \phi_1(n_1\theta_1 + n_2\theta_2 + n_3\theta_3 + \theta_0; R) \qquad \text{for } R > (2a\lambda)^{-1}$$

where $$\frac{1}{R^2} = \frac{n_1^2}{R_1^2} + \frac{n_2^2}{R_2^2} + \frac{n_3^2}{R_3^2}$$

Here the integers n_1, n_2, n_3 are either 0 or 1 and satisfy $0 < n_1^2 + n_2^2 + n_3^2 \leq 3$. The different solutions of this inequality correspond to the seven types of twisted field. Thus with this four dimensional model, as in two dimensions, there is a spontaneous breakdown of the translational invariance of the (classical) twisted ground states as a characteristic length increases across a critical value.

Similar phenomena will occur for other Z_2-invariant lagrangians for example the sine-Gordon model. However there are also four dimensional space-times and Lagrangians which although they admit twisted scalar fileds, do not have any static twisted field configurations whose energy is a finite multiple of the untwisted ground state. For further details we refer the reader to our paper Avis and Isham 1978.

§§3. UNIVERSAL BUNDLES

The problem of usefully classifying principal G-bundles (E, π, M) is non trivial and involves a delicate interplay between the topologies of the space-time M and the Lie group G. There is indeed always _a_ classification in terms of homotopy classes of maps from M into a certain topological space BG (the base space of a universal G-bundle, see below). However in general, sets of the form [X, Y] are not easily analysed and the knowledge that G-bundles are in bijective correspondence with elements of [M, BG] is not necessarily of much practical use. Fortunately when M is four dimensional (or less) there is an intimate connection between [M, BG] and certain cohomology groups of M (that _are_ in principle computable). Most of the next two sections will be devoted to developing the mathematical tools necessary to derive these results.

Let us start by defining a _bundle map_ between two bundles (E, π, M) and (E', π', M'). This is a pair of maps u: E → E', f : M → M' which map fibres into fibres so that

$$u(\pi^{-1}(x)) \subset \pi'^{-1}(f(x)) \quad \forall x \in M \qquad (3.1)$$

or equivalently $\pi' \cdot u = f \cdot \pi$

which can be neatly summarised in the commutative map diagram

$$\begin{array}{ccc} E & \xrightarrow{u} & E' \\ \downarrow \pi & & \downarrow \pi' \\ M & \xrightarrow{f} & M' \end{array}$$

A bundle map is called an _isomorphism_ if there is another bundle map u' : E' → E, f' : M' → M such that

$$u' \cdot u = id_E \qquad u \cdot u' = id_{E'}$$

$$f' \cdot f = id_M \qquad f \cdot f' = id_{M'} \qquad (3.3)$$

We will usually be concerned in situations where M = M' and both f and f' are equal to the identity map. In this case a bundle map is a map u : E → E' such that

$$\begin{array}{ccc} E & \xrightarrow{u} & E' \\ & {\searrow}\pi \quad \pi'{\swarrow} & \\ & M & \end{array}$$

is commutative. Isomorphisms of this type are called _M-isomorphisms_ and it is a classification of the set $\mathcal{B}_G(M)$ of M-isomorphic

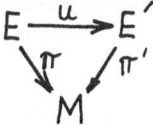

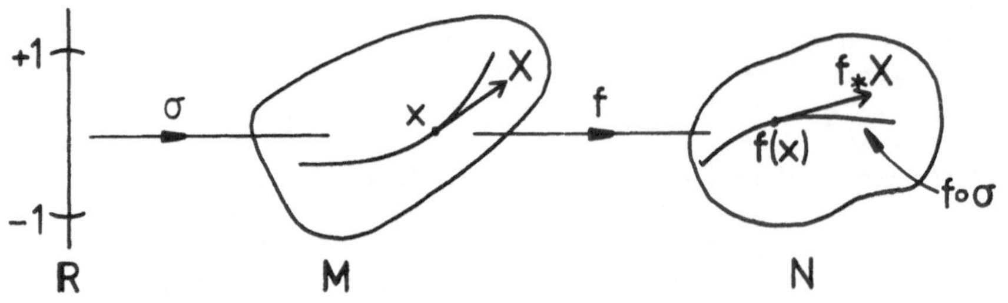

Figure 7

principal G-bundles that we wish to obtain. If (E, π, M) is a principal G-bundle then a bundle map is required to be compatible with the right G actions in the sense that u(pg) = u(p).g for all p ε E and g ε G. Note that all maps are required to be continuous (and even smooth if E and E' are manifolds).

In elementary differential geometry considerable importance is attached to structures which 'push forward' or 'pull back' under the action of a differentiable map f between two manifolds M and N. Thus if X is a tangent vector at x ε M, tangent to the curve $\sigma : (-1,1) \to M (\sigma(o) = x)$ then $f_* X$ is defined to be the tangent vector at f(x) ε N which is tangent to the curve $f \cdot \sigma$ On the other hand if $\omega \in T^*_y(N)$ is a cotangent vector (regarded as an element of the dual space of $T_y(N)$) then the pull back $f^*\omega$ is the cotangent vector at xεM (where x is any point such that f(x) = y) defined by

$$(f^*\omega)(Y) := \omega(f_*(Y)) \quad \forall Y \in T_x(M) \tag{3.4}$$

It is noteworthy that vector <u>fields</u> (unlike tangent <u>vectors</u>) do not necessarily push forward in this way (for example if f is a many-to-one map) whereas differential forms always pull back.

There is an analogous situation in bundle theory. Let $\xi = (E, \pi, B)$ be any fibre bundle and let f be a map of M into B. Then an <u>induced bundle</u> $f^*\xi$ can be constructed over M by 'gluing' a copy of the fibre over any point y ε B onto all points x ε M satisfying f(x) = y. (Figure 8).

More precisely the bundlespace of $f^*\xi$ is defined as

$$E(f^*\xi) = \{(x,p) \in M \times E \mid f(x) = \pi(p)\} \quad ; \pi_f(x,p) := x \tag{3.5}$$

and there is a natural bundle map

$$\begin{array}{ccc} E(f^*\xi) & \xrightarrow{\bar{f}} & E \\ \downarrow \pi_f & & \downarrow \pi \\ M & \xrightarrow{f} & B \end{array} \qquad \bar{f}(x,p) := p \qquad (3.6)$$

Note that if ξ is a principal G-bundle then so is $f^*(\xi)$ with the obvious group action $(x,p)g := (x, pg)$ which is well defined since $\pi(pg) = \pi(p)$. A <u>universal G-bundle</u> $\mathcal{U}_G = (EG, \pi, BG)$ is essentially (see below), a principal G-bundle with the property that <u>every</u> principal G-bundle over an arbitrary manifold M can be obtained as a 'pull back' $f^*(\mathcal{U}_G)$ induced from some map $f : M \to BG$. To investigate the existence of such potentially important objects it is necessary to study the homotopy properties of fibre bundles.

One natural question to ask is, supposing a $\mathcal{U}_G$ does exist, when will two maps $f_1, f_2 : M \to BG$ lead to induced bundles that are M-isomorphic? A partial answer is provided by the following result (Steenrod 1951, Husemoller 1966).

<u>Prop 3.1</u> "Let $f_1, f_2 : M \to N$ and let $\xi = (E, \pi, N)$ be a principal G-bundle. Then if f_1 and f_2 are homotopic, the induced bundles $f_1^*(\xi)$ and $f_2^*(\xi)$ are isomorphic". This is a direct result of the homotopy lifting property of fibre bundles. Thus there is a many-to-one map from $[M, N]$ into the set of G-bundles over M defined by $[f] \rightsquigarrow f^*(\xi)$. This motivates the definition:

<u>Defn 3.2</u> "A <u>n-universal G-bundle</u> (EG, π, BG) is a principal G-bundle with the property that the map $[M, BG] \to$ {principal G-bundles over M} is one-to-one and onto for any n-dimensional manifold M".

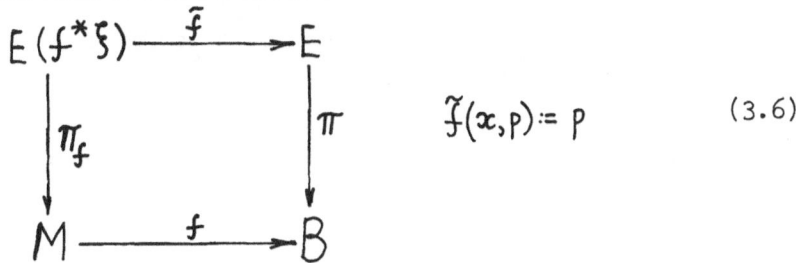

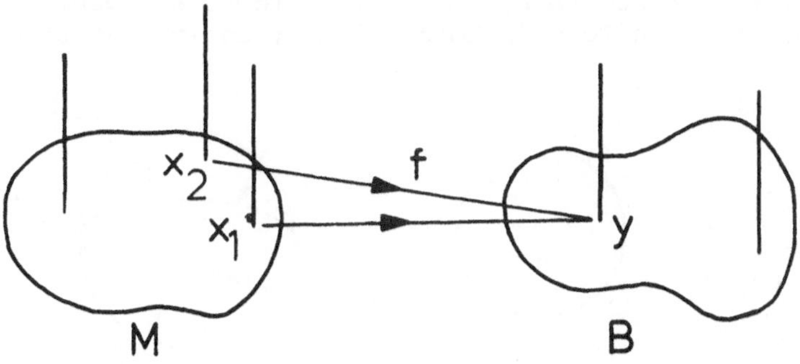

Figure 8

To study this problem further it is useful to dissect M into a collection of 'building blocks'. This concept finds its formal realisation in the definition of a cell complex K on a topological space M. This is a collection of subsets (or m-cells) e_i^m of M, $i \in J_m$ (an index set) with the following properties:

1) $M = \bigcup_{m,i} e_i^m \qquad (m = 0,1,2....)$ (3.7)

2) Let i) $K^n = \{e_i^m \mid m \leq n,\ i \in J_m\}$ denote the n-skeleton of K.

ii) $|K^n| := \bigcup_{\substack{m \leq n \\ i \in J_m}} e_i^m$

iii) $\dot{e}_i^n := e_i^n \cap |K^{n-1}|$ the boundary of e_i^n

iv) $\mathring{e}_i^n := e_i^n - \dot{e}_i^n$ the interior of e_i^n

Then we require

$$\mathring{e}_i^n \cap \mathring{e}_j^m \neq \phi \Rightarrow n = m \ \&\ i = j \qquad (3.8)$$

3) For each cell e_i^n there is a characteristic map $f_i^n : (B^n, \partial B^n \approx S^{n-1}) \to (e_i^n, \dot{e}_i^n)$ which is surjective and which maps $B^n - S^{n-1}$ homeomorphically onto $\mathring{e}_i^n$

Thus M is built up of balls of various dimensions with their boundaries glued into M in a specific way. The (n+1)-skeleton is obtained from the n-skeleton by fixing in (n+1)-cells. A few diagrammatic illustrations of this are given below in Figure 9.

To avoid pathological topological properties when there are infinitely many cells it is usual to require K to be a CW complex which is a cell complex with some additional topological restrictions. It is unnecessary to delve into the details of this and we refer the reader to the literature (Spanier 1966, Switzer 1975). Since any manifold is homotopically equivalent to a CW

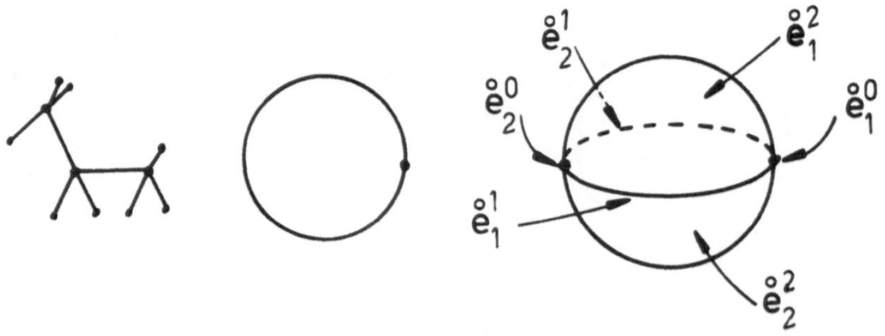

Figure 9

complex we lose nothing by considering such spaces and indeed their many desirable properties make them a natural class of objects to consider in a genuine 'quantum topology' scheme when we have dropped the differentiable manifold requirement.

Cell complexes have the advantage over generic topological spaces that their homology and cohomology spaces can (in principle) be computed by purely combinatorial means. Their great virtue for us however is the following. Suppose M is a CW complex (by abuse of language we will call by this name the topological space that carries the complex K) and we are interested in maps from M into some space X. Then one can contemplate the possibility of constructing such functions by starting with a map defined on K^0 and then extending it to K^1, K^2 etc. Suppose we have a map h from K^n into X. Then in order to extend it to K^{n+1} we have to extend it from the boundaries of the various n+1 cells into the cells themselves. But each such map $h\cdot f_i^{n+1}/S^n$ determines an element of the homotopy group $\pi_n(X)$ and this element must vanish if the extension is to be possible. This recasting of the problem into homotopy group language (and its cohomological ramification §4) is most profitable. For example if M is a n-dim complex (i.e. it contains no cells of dimension > n) and if X is <u>(n - 1) connected</u> (i.e. $\pi_0(X) = \pi_1(X) = \ldots = \pi_{n-2}(X) = \pi_{n-1}(X) = 0$) then maps from M into X can always be constructed in this way. This is an especially useful tool for constructing cross-sections of bundles and is a major ingredient in proving the theorem (Steenrod 1951, Husemoller 1966):

<u>Prop 3.4</u> "A principal G-bundle (E, π, M) is n-universal if and only if $\pi_0(E) = \pi_1(E) = \ldots \pi_n(E) = 0$".

[Caution - Steenrod calls this (n - 1)-universal but most recent texts use the convention above.]

If the bundle is ∞-universal (i.e. n-universal for all n) it is denoted (EG, π, BG). There are a number of simple examples of universal bundles:

1) Let the projective space $\mathbb{R}P_5$ be obtained from S^5 by identifying antipodal points. Then (S^5, π, $\mathbb{R}P_5$) is a principal Z_2-bundle with $\pi(v) := [v]$. Since $\pi_0(S^5) = \pi_1(S^5) = \ldots \pi_4(S^5) = 0$, this is 4-universal and hence any of the twisted scalar field bundles in §2 can be induced by a map from M into $\mathbb{R}P_5$. Clearly (S^{n+1}, π, $\mathbb{R}P_{n+1}$) is n-universal and (S^∞, π, $\mathbb{R}P_\infty$) is an example of an <u>∞-universal</u> bundle.

2) Let $S^3 = \{(z_1, z_2) \in \mathbb{C}^2 \mid |z_1|^2 + |z_2|^2 = 1\}$ and let S^2 be regarded as the complex projective plane $\mathbb{C}P_1$ with homogeneous co-ordinates $(z_1 : z_2)$. Then the Hopf map $p: S^3_{(z_1,z_2) \mapsto (z_1:z_2)} S^2$ turns S^3 into a principal U1-bundle over

S^2. This is clearly 2-universal since $\pi_0(S^3) = \pi_1(S^3) = \pi_2(S^3) = 0$. Thus if Σ is any two-dimensional manifold

$$\mathcal{B}_{U1}(\Sigma) \approx [\Sigma, S^2] \tag{3.9}$$

and in particular if $\Sigma \approx S^2$ we find $\mathcal{B}_{U1}(S^2) \approx [S^2, S^2] = \pi_2(S^2) \approx Z$. This also works for CW complexes of the form $R^2 \times \Sigma$. In particular there is a countable family of inequivalent U1-bundles over the full Kruskal space-time. The cross-sections of the associated vector bundles would be twisted complex scalar fields. Note that if dim $\Sigma>2$ the Hopf bundle is of no use. We will show using more powerful methods in §4 that for any CW complex M:

$$\mathcal{B}_{U1}(M) \approx H^2(M; Z) \tag{3.10}$$

3) There is an analogous Hopf map $p: S^7 \to S^4$ obtained using quaternions rather than complex numbers. The preimage of any point is an S^3 and in fact (S^7, p, S^4) is a principal SU2-bundle which is clearly 6-universal. In particular if M is any four dimensional space-time (compact, noncompact, Riemannian or pseudo-Riemannian) we have

$$\mathcal{B}_{SU2}(M) \approx [M, S^4] \tag{3.11}$$

For the usual 'instanton' model $M \approx S^4$ and hence

$$\mathcal{B}_{SU2}(S^4) \approx [S^4, S^4] \approx \pi_4(S^4) \approx Z \tag{3.12}$$

thus recovering the well known existence of a countable family of inequivalent SU2-structures. The Hopf bundle (S^7, p, S^4) is of course a particular example and corresponds to 'winding number one'. Thus the well known Belavin et al n = 1 instanton structure is actually universal for <u>all</u> space-time manifolds! For general four-dimensional manifolds M the structure of $[M, S^4]$ is not immediately obvious but in §§4 we will see that

$$[M, S^4] \approx H^4(M; Z) \tag{3.13}$$

In order to exhibit examples of universal bundles for other groups it is evidently necessary to gain information on the homotopy groups of the total space of a principal bundle. This is contained in the well known long exact sequence connecting the homotopy groups of base space, bundle space and fibre. We will indicate the derivation of this invaluable result using a technique (Switzer 1975) which has the virtue of introducing ideas that will be useful later.

<u>Defn 3.5</u> 1) "The (reduced) <u>suspension</u> SX of a pointed space (X, x_0) is obtained by taking $X \times [0,1]$ and identifying $X \times \{0\}$, $X \times \{1\}$ and $\{x_0\} \times [0,1]$ to a point"

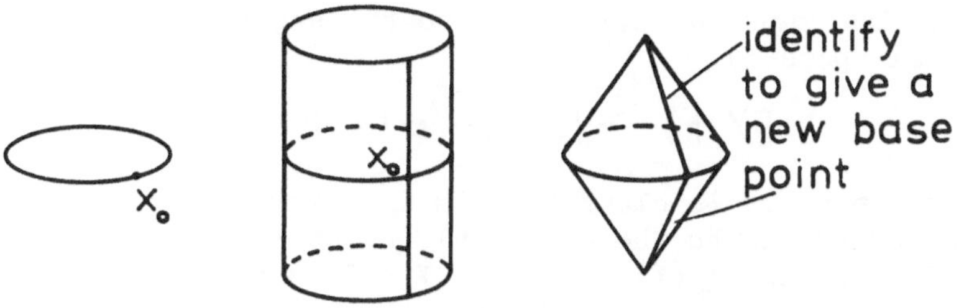

Figure 10

2) The loop space ΩX of (X, x_0) is the space of functions $f : (S^1, s_0) \to (X, x_0)$ with base point the constant loop $\omega_0 : S^1 \to \{x_0\} \subset X$ (This function space is equipped with the compact-open topology).

Different dimensional spheres are related by suspension:

$$S(S^k) = S^{k+1} \qquad k \geqslant 0$$

so that

$$S^n(S^0) = S^n \tag{3.14}$$

There is a natural duality between looping and suspension epitomised in the relation (Spanier 1966, Switzer 1975)

$$[X, \Omega Y] = [SX, Y] \tag{3.15}$$

for any pair of (pointed) topological spaces X and Y. Thus if $f : X \to \Omega Y$ one can define $\tilde{f} : SX \to Y$ by

$$\tilde{f}(x, t) := (f(x))(t) \qquad x \in X,\ t \in [0,1] \tag{3.16}$$

In particular eq. (3.15) gives

$$\pi_n(Y) = [S^n, Y] = [S^n(S^0), Y] = [S^0, \Omega^n Y] \tag{3.17}$$

Now let (E, π, B) be any bundle with fibre F. We will assume that E and B have base points e_0 and b_0 respectively and that F is identified with $\pi^{-1}(b_0) \subset E$ via a map $i : F \to E$ giving a diagram

$$\begin{array}{ccc} F & \xrightarrow{i} & E \\ & & \downarrow \pi \\ & & B \end{array} \tag{3.18}$$

with $\pi \cdot i = b_0$ (the constant map which takes everything into b_0). The desired long exact sequence will be derived in a series of steps.

1) The maps in (3.18) lead to the sequence (for any topological space X)

$$[X, F] \xrightarrow{i_*} [X, E] \xrightarrow{\pi_*} [X, B] \qquad (3.19)$$

Since $\pi \cdot i = b_0$ it is clear that Im $i_* \subset$ ker π_* where the kernel of π_* is defined to be the set of all homotopy classes of maps $f : X \to E$ such that $\pi_*(f) = \pi \cdot f : X \to B$ is homotopic to the constant map $b_0 : X \to \{b_0\}$. The homotopy lifting property implies that ker $\pi_* \subset$ Im i_* and so

$$\ker \pi_* = \mathrm{Im}\, i_* \qquad (3.20)$$

(a sequence with this property is called <u>exact</u>).

2) Let $\overset{\circ}{E} = \{(e, \omega) \in E \times B^I \mid \pi(e) = \omega(0)\}$ where B^I denotes the space of continuous functions from $I = [0,1]$ into B (this function space is equipped with the compact opentopology). Since π is a fibration a path lifting function exists. This is a map $\lambda : \overset{\circ}{E} \to E$ with $(\lambda(e, \omega))(0) = e$ and $\pi(\lambda(e, \omega)) = \omega$ (Figure 11)

3) Define

$$\rho : \Omega B \longrightarrow F \qquad \omega \longmapsto (\lambda(e_0, \omega^{-1}))(1) \qquad (3.21)$$

where ω^{-1} denotes the inverse path of ω, $\omega^{-1}(t) := \omega(1 - t)$. There are natural definitions

$$\Omega\pi : \Omega E \longrightarrow \Omega B \qquad \omega \rightsquigarrow \pi \cdot \omega \qquad\qquad \Omega i : \Omega F \longrightarrow \Omega E \qquad \omega \rightsquigarrow i \cdot \omega \qquad (3.22)$$

giving the series of maps

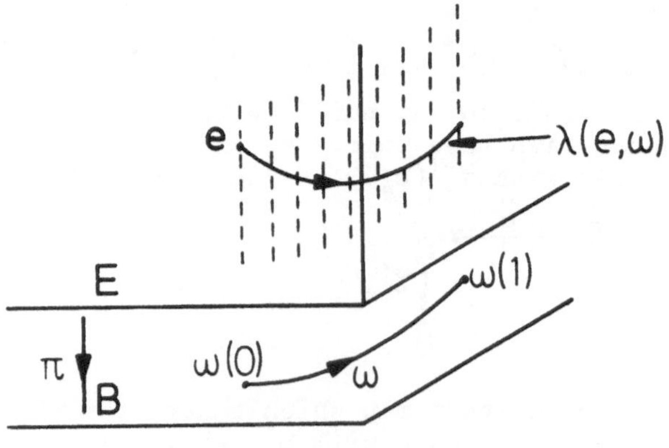

Figure 11

$$\Omega F \xrightarrow{\Omega i} \Omega E \xrightarrow{\Omega \pi} \Omega B \xrightarrow{\rho} F \xrightarrow{i} E \xrightarrow{\pi} B \qquad (3.23)$$

This sequence may be extended indefinitely to the left to give

$$\to \Omega^3 B \xrightarrow{\Omega^2 \rho} \Omega^2 F \xrightarrow{\Omega^2 i} \Omega^2 E \xrightarrow{\Omega^2 \pi} \Omega^2 B \xrightarrow{\Omega \rho} \Omega F \xrightarrow{\Omega i} \Omega E \xrightarrow{\Omega \pi} \Omega B \xrightarrow{\rho} F \xrightarrow{i} E \xrightarrow{\pi} B \qquad (3.24)$$

and the analogous extension of (3.19) is the Puppe sequence:

$$\to [X, \Omega^2 B] \xrightarrow{\Omega \rho_*} [X, \Omega F] \xrightarrow{\Omega i_*} [X, \Omega E] \xrightarrow{\Omega \pi_*} [X, \Omega B] \xrightarrow{\rho_*} [X, F] \xrightarrow{i_*} [X, E] \xrightarrow{\pi_*} [X, B] \qquad (3.25)$$

For fibrations over a CW complex B (Which is the case we are interested in) λ can be chosen to satisfy $\lambda(e_0, \omega_0^{-1})(t) = e_0$ $\forall t \in I$: the sequence (3.25) is then exact.

It is worth remarking that spaces of the form $[X, \Omega^p Y]$ possess a group structure when $p \geq 1$ and in this case the connecting maps in (3.25) of the form $\Omega\alpha$ are group homomorphisms and exactness holds in the algebraic sense.

4) The long exact homotopy sequence of a fibre bundle is obtained from (3.25) by choosing $X = S^0$ and using (3.14) and (3.17) to give

$$\to \pi_3(B) \to \pi_2(F) \to \pi_2(E) \to \pi_2(B) \to \pi_1(F) \to \pi_1(E) \to \pi_1(B) \to$$
$$\to \pi_0(F) \to \pi_0(E) \to \pi_0(B) \qquad (3.26)$$

Notice that if (EG, π, BG) is a ∞-universal bundle then since $\pi_n(EG) = 0$ $\forall n$, the sequence in (3.26) breaks up into an infinite set of short exact sequences of the form

$$0 \to \pi_n(BG) \to \pi_{n-1}(G) \to 0 \qquad n \geq 1 \qquad (3.27)$$

which, by exactness, implies

$$\pi_n(BG) \simeq \pi_{n-1}(G) \qquad \forall n \geq 1 \qquad (3.28)$$

As an application of this consider the problem of classifying G-bundles over S^n. These are in bijective correspondence with elements of $[S^n, BG] \simeq \pi_n(BG)$ and hence using (3.28)

$$\mathcal{B}_G(S^n) \simeq \pi_{n-1}(G) \qquad (3.29)$$

In particular any connected, simple, non abelian Lie group has $\pi_3(G) \simeq Z$ as does $U(n)$, $n \geq 2$. Hence in all these cases

$$\mathcal{B}_G(S^4) \simeq Z \qquad (3.30)$$

From this point of view conventional S^4-based instanton physics is not very interesting because the topological effects are similar for all the usual matrix groups. As we shall see in §4 this is definitely not the case for a general space-time M where, for example, there can be a big difference between a SU2 and a SO3 gauge theory. As another (not very serious!) application of (3.29) we might look at SO8 'instanton' physics in a 11-dimensional space-time (this arises in a supergravity context discussed by Cremmer et al 1978). A partial list of homotopy groups of the classical Lie groups may be found in Borel 1955. In particular

$$\mathcal{B}_{SO8}(S^{11}) \approx \pi_{10}(SO8) \approx Z_{24} \oplus Z_{24} \tag{3.31}$$

so in this case there are precisely 576 inequivalent bundles/ instanton sectors!

Eq. (3.26) may be employed in the construction of n-universal bundles. For example using the relations

$$SO(p+1)/SO(p) \approx S^p \quad ; \quad \pi_i(S^p) = 0, \; 0 \leq i < p \tag{3.32}$$

it is easy to show that $\pi_i(SO(p))/SO(q)) = 0$ for $0 \lesssim i < q$. Hence the SO(m)-bundle

$$\begin{array}{ccc} SO(m) & \longrightarrow & SO(n+m+1)/SO(n+1) \\ & & \downarrow \\ & & SO(n+m+1)/SO(n+1) \times SO(m) \end{array}$$

is n-universal.

Similarly

$$\begin{array}{ccc} SU(m) & \longrightarrow & SU(n+m)/SU(n) \\ & & \downarrow \\ & & SU(n+m)/SU(n) \times SU(m) \end{array}$$

is a 2n-universal SU (m)-bundle. By taking the limit $n \to \infty$, ∞-universal bundles are obtained.

It is of course comforting to know that universal bundles do actually exist but the structure of the base spaces BG is usually sufficiently complicated to preclude any elementary way of analysing [M, BG] and hence solving the classification problem in a practical sense. The three examples given earlier are exceptions in this respect as is the result

$$\mathcal{B}_G(S^n) \approx \pi_{n-1}(G) \tag{3.33}$$

or its generalisation

$$\mathcal{B}_G(S\Sigma) \approx [\Sigma, G] \tag{3.34}$$

If space-times other than spheres are to be studied it is necessary to find more powerful ways of analysing [M, BG]. Fortunately for low dimensional space-times considerable progress can be made by approximating BG by a series of spaces with relatively simple homotopy properties as we will now explain.

§§4. CLASSIFICATION OF G-BUNDLES

§4.1 Eilenberg-Maclane spaces

In investigating [M, BG], approximations to either M or BG might be employed. However, for a given group G, BG is fixed whereas M is 'variable' and hence it is probably going to be easier to apply approximation techniques to BG rather than to M. This suggests looking at sets of the form [M, K] where K is a space with some especially simple homotopic properties. A profitable choice for K are the various Eilenberg-Maclane spaces $K(\pi,n)$ which satisfy

i) $K(\pi,n)$ is path connected

ii) $\pi_i(K(\pi,n)) \simeq \pi$ if $i = n$
$\simeq 0$ if $i \neq n$

where π is any abelian group and n is an integer $\geqslant 1$.

CW complexes of this type are unique (up to homotopy equivalence) and may be explicitly constructed (Mosher and Tangora 1968, Bott and Mather 1968). In general these spaces are infinite dimensional. An exception to this is K(Z,1) which may be chosen to be the circle since,

$$\pi_i(S^1) \simeq Z \quad , i=1$$
$$\simeq 0 \quad i \neq 1$$

We will show later that

$$K(\pi, n) \sim \Omega K(\pi, n+1) \tag{4.1}$$

and hence $[M, K(\pi,n)]$ carries a natural abelian group structure (Switzer 1975). This group is called the n'th cohomology group of M with values in π and is isomorphic to the usual cohomology groups defined using triangulations of M (or more generally using the CW structure). It is denoted $H^n(M;\pi)$.

Clearly $[M, K(\pi,n)]$ stands in some sort of dual relation to the homotopy groups $\pi_n(M) = [S^n, M]$. In particular if N is a contractible subspace of M, the series

$$N \longrightarrow M \longrightarrow M/N$$

yields a cohomology exact sequence analagous to the homotopy exact sequence of eq. (3.26) obtained from the fibration in eq. (3.18). Note that M/N is defined as the space M with N collapsed to a point x_o and that the pair $(M/N, x_o)$ is homotopically equivalent to (M,N). Hence the exact sequence referred to above is precisely the usual one in cohomology theory.

Cohomology groups are often easier to compute than homotopy groups. For example if M is a p-sphere we see at once that

$$H^n(S^p;\pi) = [S^p, K(n,\pi)] = \pi_p(K(n,\pi)) = \pi \quad \text{if } p=n$$
$$= 0 \quad \text{if } p\neq n \qquad (4.2)$$

As another example let M be a ($n<\infty$) n-dimensional CW complex. Then

$$H^q(M;\pi) = [M, K(\pi,q)] = 0 \quad \text{if } q>n \qquad (4.3)$$

i.e. any function f from M into $K(\pi,q)$ is homotopic to the constant map. This is essentially because a function from (B^p,S^{p-1}) into a pointed space (X,x_o), can be deformed into the constant map if $\pi_{p-1}(X,x_o) = 0$. However, M is made up of cells of dimension $\leq n$ and hence a homotopy function from M x I into $K(\pi,q)$ can be extended to M x I cell by cell without meeting any obstruction (if $q > n$) since the only non vanishing homotopy group of $K(\pi,q)$ is the q'th one. In particular if M is a four-dimensional manifold there can be at most four non vanishing cohomology groups (plus $H^o(M;\pi) \simeq \pi$) in contrast to the potentially infinite number of non vanishing homotopy groups. (In this sense the cohomology groups carry less topological information than the homotopy groups, which is one reason why they are easier to compute). It will thus be a distinct advance if [M, BG] can be expressed purely in terms of the cohomological properties of M.

For the special cases when G is Z_2 or U1 we can do this at once. Applying the exact homotopy sequence to the fibre bundle $(S^n,\pi,\mathbb{R}P_n)$ (cf §3), gives

$$\pi_1(\mathbb{R}P_n) \simeq Z_2$$
$$\pi_i(\mathbb{R}P_n) \simeq 0 \quad 1<i<n \qquad (4.4)$$

and hence $\pi_i(\mathbb{R}P_\infty) = Z_2$ if $i = 1$
$= 0$ if $i \neq 1$ (4.5)

Thus $\mathbb{R}P_\infty$ is a $K(Z_2, 1)$ space. On the other hand we know (§3) that $(S^\infty, \pi, \mathbb{R}P_\infty)$ is a ∞-universal Z_2-bundle and so

$$\mathcal{B}_{Z_2}(M) \simeq [M, BZ_2] = [M, \mathbb{R}P_\infty]$$
$$\simeq [M, K(Z_2,1)] = H^1(M;Z_2) \qquad (4.6)$$

This justifies the remark in §2 that Z_2 bundles over an arbitrary space-time manifold (with any finite dimension) are classified by the cohomology group $H^1(M; Z_2)$.

For the U1 case recall that the Hopf map $p : S^3 \to S^2 \simeq \mathbb{C}P_1$ defined a 2-universal U1 bundle. This can be generalised by letting

$$S^{2n-1} = \{(z_1, z_2 \ldots z_n) \in \mathbb{C}^n \mid \sum_{i=1}^{n} |z_i|^2 = 1\}$$ and defining

$$p : S^{2n-1} \longrightarrow \mathbb{C}P_{n-1} \quad ; \quad (z_1, z_2 \ldots z_n) \rightsquigarrow (z_1 : z_2 : \ldots z_n)$$

which exhibits S^{2n-1} as a bundle over $\mathbb{C}P_n$ with fibre S^1. This is a principal U1-bundle which is (2n - 2)-universal and so $(S^\infty, p, \mathbb{C}P_\infty)$ is ∞-universal. Hence eq. (3.28) implies

$$\pi_2(\mathbb{C}P_\infty) \approx \pi_1(U1) \approx Z$$
$$\pi_i(\mathbb{C}P_\infty) \approx 0 \quad \text{if } i \neq 2 \tag{4.7}$$

and thus $\mathbb{C}P_\infty$ is a K(Z,2) space. This means that U(1) bundles are classified as elements of $H^2(M;Z)$ since

$$\mathcal{B}_{U1}(M) = [M, BU1] = [M, \mathbb{C}P_\infty]$$
$$= [M, K(Z,2)] = H^2(M; Z) \tag{4.8}$$

The element in $H^2(M;Z)$ corresponding to a particular principal U1 bundle ξ is known as the <u>first (integral) Chern class</u> of ξ and is denoted $C_1(\xi)$.

§4.2 Postnikov systems

We must now consider the problem of analysing [M, BG] in a tractable, cohomological way for the (usual) case when BG is not simply a $K(\pi,n)$ space. Although there are concrete examples of universal bundles for any Lie group G the base spaces are topologically complicated and, at the moment, the only general information available to us is that, for ∞-universal spaces,

$$\pi_p(BG) \approx \pi_{p-1}(G) \tag{4.9}$$

(see eq. (3.28)).

The crucial observation is that if we are interested in homotopy classes of maps from a n-dimensional manifold M into BG, the homotopy groups of BG beyond the n'th one should not play any role in classifying these classes. This follows from the same type of argument employed previously when the manifold M is viewed as a CW complex whose cells have maximum dimension n. More formally a continuous map f between two path connected spaces X and Y is said to be a <u>p-equivalence</u> if for all $x \in X$, f_* is an isomorphism of $\pi_q(X,x)$ onto $\pi_q(Y,f(x))$ for $0 < q < p$ and an epimorphism for q=p. Then if $f : BG \to BG_n$ is an (n + 1) equivalence for some space BG_n it can be shown that (if dim M $\leqslant$ n)

$$\mathcal{B}_G(M) = [M, BG] \approx [M, BG_n] \tag{4.10}$$

and hopefully if BG_n is topologically simpler than BG the computation of $[M, BG_n]$ will be simpler than that of $[M, BG]$. This seems especially likely if the homotopy groups of BG_n all vanish beyond the $(n + 1)$'th which in turn suggests using Eilenberg-Maclane spaces in some way.

One obvious possibility is the following. Let π_i denote the i'th homotopy group of BG. Then since in general

$$\pi_p(X \times Y) \approx \pi_p(X) \oplus \pi_p(Y) \tag{4.11}$$

it follows that the CW complex

$$K := K(\pi_1, 1) \times K(\pi_2, 2) \times \ldots \times K(\pi_n, n) \tag{4.12}$$

has the homotopy groups

$$\begin{aligned} \pi_i(K) &= \pi_i & 1 \leq i \leq n \\ &= 0 & i > n \end{aligned} \tag{4.13}$$

Thus the first n homotopy groups of BG and K are identical. Unfortunately this does not mean that BG and K are $(n + 1)$ equivalent as there may be no map f from BG into K which induces these isomorphisms (and the existence of such a map is necessary for the validity of (4.10)). On the other hand there is a 2-equivalence between BG and $K(\pi_1, 1)$ arising from the relation between $K(\pi_1, 1)$ and $H^1(BG; \pi_1)$. If $\pi_1(BG) = \pi_2(BG) = \ldots \pi_p(BG) = 0$, then one starts instead with a $(p + 2)$ equivalence between BG and $K(\pi_{p+1}, p + 1)$. This suggests that it may be possible to find the desired BG_n by building inductively on $K(\pi_1, 1)$ to obtain first a BG_2, then a BG_3 etc. Now as remarked already we cannot in general simply chose $BG_2 = K(\pi_1, 1) \times K(\pi_2, 2)$. However there is another way of assembling these two Eilenberg-Maclane spaces and that is to form a twisted product. In other words consider building a fibre bundle over $K(\pi_1, 1)$ with a fibre $K(\pi_2, 2)$. The exact homotopy sequence shows at once that the bundle space E has the desired homotopy groups

$$\pi_1(E) = \pi_1 \quad , \quad \pi_2(E) = \pi_2 \quad , \quad \pi_p(E) = 0 \quad p > 2 \tag{4.14}$$

and since there are possible many such bundles the chances are improved of finding one which is a 3-equivalence with BG.

The first question is how does one build such bundles? There is one natural bundle with fibre $K(\pi_2, 2)$ obtained by constructing the path space $PK(\pi_2, 3)$. This is the space of paths (maps from $[0,1]$ into $K(\pi_2, 3)$) whose initial points are the base point y_0 in

$K(\pi_2,3)$. There is a map $p : PK(\pi_2,3) \to K(\pi_2,3)$ defined by $p\omega := \omega(1)$, $\omega \in PK(\pi_2,3)$ and $(PK(\pi_2,3), p, K(\pi_2,3))$ is a fibration with fibre the loop space $\Omega K(\pi_2,3)$ - the space of paths in $K(\pi_2,3)$ satisfying $\omega(0) = \omega(1) = y_0$. Now $PK(\pi_2,3)$ is a contractible space (just run all the paths back to their common base point) and hence all its homotopy groups vanish. The exact homotopy sequence (or alternatively (3.15)) shows that

$$\pi_p(\Omega K(\pi_2,3)) \simeq \pi_{p+1}(K(\pi_2,3)) \quad \forall p$$

and hence

$$\Omega K(\pi_2,3) \sim K(\pi_2,2)$$

These results extend of course to arbitrary Eilenberg-Maclane spaces and in general $\xi_{\pi,q} := (PK(\pi,q), p, K(\pi,q))$ is a fibration whose fibre is

$$\Omega K(\pi,q) \sim K(\pi,q-1) \tag{4.15}$$

Thus one might attempt to build the BG_2 space by constructing a map $\theta_1 : BG_1 = K(\pi_1,1) \to K(\pi_2,3)$ and pulling back the natural $K(\pi_2,2)$ bundle $\xi_{\pi_2,3}$. The map θ_1 must be chosen so that the bundle space of this induced bundle $\theta_1^*(\xi_{\pi_2,3})$ is a 3-equivalence with BG. If this is possible then the next step is to find a map $\theta_2 : BG_2 \to K(\pi_3,4)$ to construct BG_3 and so on. Such a collection of maps and spaces is called a <u>Postnikov system over BG</u>. The following result is crucial:

<u>Prop 4.1</u> "If X is any simply connected space (or more generally a space in which $\pi_1(X)$ induces trivial automorphisms of the higher homotopy groups) there exists a Postnikov system".

For simplicity we will restrict our attention to connected groups. Hence by (4.9) $\pi_1(BG) = 0$ and so Prop. 4.1 is applicable to the base space of universal bundles. The 'approximating' spaces BG_n are determined up to homotopy type but the maps $\theta_n : BG_n \to K(\pi_{n+1}, n+2)$ are not necessarily unique. We are implicitly chosing a set of maps θ_n which is consistent with the usual concept of characteristic class and will briefly indicate at the end of §4 how the maps are constructed. The collection of approximating maps and spaces will be indicated:

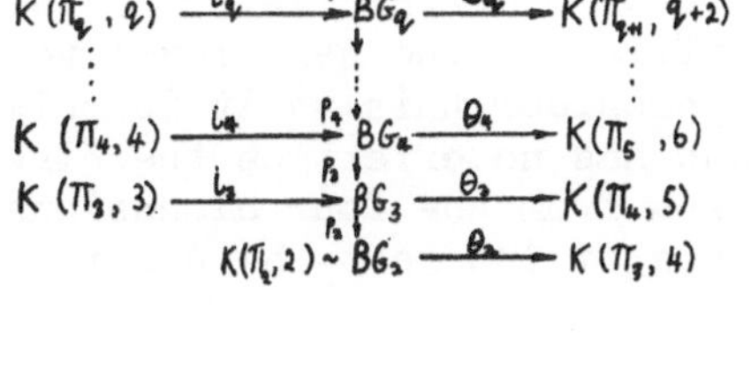

Fig.12

There are (q + 1) homotopy equivalence maps $f_q: BG \to BG_q$ q=1,2... compatible with the fibre bundle projection maps p_q in the sense that

$$p_q \cdot f_{q+1} \sim f_q \tag{4.16}$$

The maps $i_q: K(\pi_q, q) \to BG_q$ are the inclusion maps taking the fibre $K(\pi_q, q)$ into the bundle space of the bundle $(BG_q, p_{q-1}, BG_{q-1})$.

Note that since we are assuming

$$\pi_1(BG) = 0 \tag{4.17}$$

the space BG_2 is simply $K(\pi_2, 2)$ in accord with the remark made earlier about starting with the first non vanishing homotopy group of BG.

Let us now consider the implications of this Postnikov factorization of BG in the special case of interest when M is a four-dimensional manifold. Using (4.9) gives

$$\pi_2(BG) \approx \pi_1(G) \tag{4.18}$$

which may or may not vanish. If the group G is simply connected (so that $\pi_1(G) = 0$) then the diagram simplifies further. However, it does no harm to assume that $\pi_1(G)$ is not zero as the correct answer in the other case is simply obtained by setting $\pi_1(G) = 0$ in the final results.

The next layer of the diagram is determined by

$$\pi_3(BG) \approx \pi_2(G) \tag{4.19}$$

However one of the crucial topological properties of semisimple Lie groups is that $\pi_2(G)$ always vanishes. This means that BG_3 can be simply taken as a copy of BG_2 with $p_2 : BG_3 \to BG_2$ being the identity map. A study of the next level uses

$$\pi_4(BG) \approx \pi_3(G) \tag{4.20}$$

and since in general $\pi_3(G) \neq 0$, this layer of the Postnikov construction is non trivial. The next level would involve $\pi_4(G)$ but we do not need to consider this as it is only used in the construction of BG_5 and has no effect on the classification of homotopy classes of maps from the four dimensional space-time into BG. Hence in the cases of interest the diagram collapses to

$$\begin{array}{ccc} K(\Pi_4(BG),4) & \xrightarrow{i} & BG_4 \\ & & \downarrow p \\ & & K(\Pi_2(BG),2) \xrightarrow{\theta} K(\Pi_4(BG,5) \end{array}$$

Fig.13

The groups π_2 and π_4 are the second and fourth homotopy groups of BG and hence the first and third homotopy groups of G. Bearing this in mind it is convenient to relabel Fig.13 as:

$$\begin{array}{ccc} K(\Pi_3,4) & \xrightarrow{i} & BG_4 \\ & & \downarrow p \\ & & K(\Pi_1,2) \xrightarrow{\theta} K(\Pi_3,5) \end{array}$$

Fig.14

with the understanding that in future the symbol π_q will always refer to the q'th homotopy group of G.

We are now very close to the desired result. Essentially we consider homotopy classes of maps from M into the various parts of the bundle $(BG_4, p, K(\pi_1,2))$ and use the exact sequence in eq.(3.25). This sequence usually terminates in

$$\longrightarrow [M,F] \longrightarrow [M,E] \longrightarrow [M,B]$$

but when (as in the present case) the bundle is induced from a map θ into a 'universal' $K(\pi,q)$ bundle $(PK(\pi,q+1),\ p,\ K(\pi,q+1))$ this can be extended to

$$\longrightarrow [M,F] \longrightarrow [M,E] \longrightarrow [M,B] \longrightarrow [M,K(\pi,q+1)]$$

Thus Fig.(14) and (3.25) together give the long exact sequence:

$$\begin{array}{l} \xrightarrow{\Omega^3\theta_*} [M,\Omega^2K(\Pi_3,4)] \xrightarrow{\Omega^2 i_*} [M,\Omega^2BG_4] \xrightarrow{\Omega^2 p_*} [M,\Omega^2K(\Pi_1,2)] \xrightarrow{\Omega^2\theta_*} \\ \xrightarrow{\Omega^2\theta_*} [M,\Omega K(\Pi_3,4)] \xrightarrow{\Omega i_*} [M,\Omega BG_4] \xrightarrow{\Omega p_*} [M,\Omega K(\Pi_1,2)] \xrightarrow{\Omega\theta_*} \\ \xrightarrow{\Omega\theta_*} [M,K(\Pi_3,4)] \xrightarrow{i_*} [M,BG_4] \xrightarrow{p_*} [M,K(\Pi_1,2)] \xrightarrow{\theta_*} [M,K(\Pi_3,5)] \end{array} \tag{4.21}$$

The various homotopy classes appearing in this sequence may be interpreted as follows:

1) $[M, K(\pi_3,5)] \simeq H^5(M;\pi_3) \simeq 0$ in a four-dimensional manifold M.

2) $[M, K(\pi_1,2)] \simeq H^2(M;\pi_1)$

3) $[M, BG_4] \simeq [M, BG]$ (for a four-dimensional space-time) $\simeq \mathcal{B}_G(M)$

- this is the space of inequivalent principal G-bundles and is what we are trying to determine.

4) $[M, K(\pi_3, 4)] \simeq H^4(M; \pi_3)$

5) $[M, \Omega K(\pi_1, 2)] \simeq [M, K(\pi_1, 1)]$ (from 4.15)
$\simeq H^1(M; \pi_1)$

and similar reductions can be made for the other sets involving $\Omega^p K(\pi,q)$. In addition $\Omega K(\pi,1)$ is homotopically equivalent to the (discrete) group π. Hence the classifying exact sequence (CES) becomes

$$\begin{aligned}&\to H^2(M;\pi_3) \to [M, \Omega^2(BG_4)] \to [M,\pi_1] \to \\ &\to H^3(M;\pi_3) \to [M, \Omega(BG_4)] \to H^1(M;\pi_1) \to \\ &\to H^4(M;\pi_3) \xrightarrow{i_*} \mathcal{B}_G(M) \xrightarrow{p_*} H^2(M;\pi_1) \xrightarrow{\theta_*} 0\end{aligned} \tag{4.22}$$

Note that $\mathcal{B}_G(M)$ appears in this expression flanked by space-time cohomology groups. These are in principle computable and in many cases (see below) the exact sequence then allows the desired space $\mathcal{B}_G(M)$ to be found. In five or higher dimensional space-times, there would be additional layers in the Postnikov factorization leading to a much harder classification problem.

Let us now study the implications of (4.22) for various compact simple gauge groups G (the noncompact or semisimple cases present no further problems).

1. <u>G simply connected</u> (SU(n), Spin(n) ($n \geq 5$), Sp(n), G_2, F_4, E_8 and universal covering spaces of E_6 and E_7).

   By definition $\pi_1(G) = 0$ and any compact, connected, simple and non abelian Lie group has $\pi_3(G) \simeq Z$. Thus the CES becomes

$$0 \to H^4(M;Z) \to \mathcal{B}_G(M) \to 0 \tag{4.23}$$

The factorisation of Fig.12 degenerates in this case to $BG_4 \sim K(Z,4)$ and the result in (4.23) can be sharpened to

$$\mathcal{B}_G(M) \simeq H^4(M;Z) \tag{4.24}$$

If M is a noncompact manifold then $H^4(M;Z) \simeq 0$ and so in this case there are no trivial G-bundles. If M is a compact and oriented manifold then $H^4(M;Z) \simeq Z$ (Dold 1972) and there are a countable number of inequivalent G-bundles. This result of course includes as a very special case the SU2 instanton results on S^4.

If ξ is a SU(n) (resp. Sp(n) bundle then the corresponding cohomology class in $H^4(M;Z)$ is called the <u>second Chern class</u> $C_2(\xi)$ (resp. <u>first symplectic Pontryagin class</u> $P_1(\xi)$). These are special examples of characteristic classes. In general let G be any Lie group and let $f : M \to BG$ represent a principal G-bundle over M. Suppose that $\chi \in H^p(BG;\pi)$ for some integer p and some abelian group π. Then $f^* \chi \in H^p(M;\pi)$ is defined by $f^*\chi : M \to K(\pi,p)$; $BG \xrightarrow{\chi} K(\pi,p)$

$$x \rightsquigarrow \chi(f(x)) \qquad \begin{array}{ccc} BG & \xrightarrow{\chi} & K(\pi,p) \\ {\scriptstyle f}\uparrow & \nearrow {\scriptstyle f^*\chi} & \\ M & & \end{array}$$

and is known as a <u>characteristic class</u> of the bundle. The utility of this concept lies in the observation that a necessary condition for two G-bundles ξ_1 and ξ_2 to be isomorphic is that all their characteristic classes must agree. Unfortunately bundles cannot be easily classified by these classes because the map

$$\mathcal{B}_G(M) \longrightarrow H^p(M;\pi) \qquad (4.25)$$

$$\xi \rightsquigarrow (p,\pi) \text{ characteristic class of } \xi$$

is in general many-to-one and into. The Postnikov factorization method can be made to tell us what classes are sufficient for a classification and relations between them.

Note that by using De Rham's theorem, real cohomology classes may be represented by closed differential forms. The Chern-Weil theory (see for example Milnor and Stasheff 1974, Kobayashi and Nomizu 1969) shows that the forms corresponding to real Chern and Pontryagin classes may be constructed from arbitrary connections (i.e. 'Yang-Mills' fields) in the principal bundle. This is well documented in both the mathematical and physics literature (e.g. Drechsler and Mayer 1977) and will not be discussed here. It is worth noting however that although the characteristic classes corresponding to the $H^2(M;\pi_1)$ group in (4.22) which appear when $\pi_1 \neq 0$ (see next section) do have a differential form representation (Allendoerfer and Eells 1958, Vaisman 1973) they are not apparently related to the Yang-Mills fields. This is one reason why fibre bundle theory over a general space-time is structurally more complicated than in the simple S^4 instanton case (where $H^2(S^4;\pi_1) \simeq 0$).

2. <u>U(n)</u>

If $n = 1$, $\pi_1(U(n)) \simeq Z$, $\pi_3(U(n)) \simeq 0$ and the CES collapses to

$$0 \longrightarrow \mathcal{B}_{U1}(M) \longrightarrow H^2(M;Z) \longrightarrow 0$$

which implies (since Fig.12 becomes just $(BU1)_4 \sim K(Z,2)$)

$$\mathcal{B}_{U1}(M) \simeq H^2(M;Z) \qquad (4.26)$$

thus reproducing the earlier result in eq. (4.8) (N.B. this works for any dimension space-time).

If $n \geqslant 2$ $\pi_1(U(n)) \simeq Z$, $\pi_3(U(n)) \simeq Z$ and superficially there is no simplification in (4.22). However we know that the crucial map θ in Fig.14 maps $K(Z,2)$ into $K(Z,5)$ and hence determines an element of $H^5(K(Z,2);Z)$. Further more $K(Z,2) \simeq CP_\infty$ and the cohomology of CP_n is known to possess only even degree elements (Spanier 1966). This makes it plausible that $H^5(K(Z,2);Z) \simeq 0$ and indeed a detailed argument confirms this (Mosher and Tangora 1968). Therefore θ must be homotopic to the constant map as must $\Omega\theta$. Hence the image of $\Omega\theta_*$ in $H^4(M;Z)$ is 0 and, by exactness, so is the kernel of i_*. This leads to the collapsed CES

$$0 \longrightarrow H^4(M;Z) \xrightarrow{i_*} \mathcal{B}_{U1}(M) \xrightarrow{p_*} H^2(M;Z) \longrightarrow 0 \tag{4.27}$$

In fact the triviality of θ implies that $BU(n)_4 \sim K(Z,2) \times K(Z,4)$ and hence (4.27) can be sharpened to

$$\mathcal{B}_{U(n)}(M) \simeq H^2(M;Z) \oplus H^4(M;Z) \tag{4.28}$$

The cohomology elements in $H^2(M;Z)$ and $H^4(M;Z)$ corresponding to a specific $U(n)$ bundle are called the <u>first</u> and <u>second</u> Chern class respectively. In a noncompact space-time $H^4(M;Z) \simeq 0$ and so the $U(n)$ bundles are determined entirely by the first Chern class.

Notice that on the usual 'instanton 4-sphere' only the second Chern class is involved since $H^2(S^4;Z) \simeq 0$. However, on for example an $S^2 \times S^2$ compactified Schwaschild 'space-time' (Duff and Madore 1977) both characteristic classes are involved since

$$\begin{aligned} H^2(S^2\times S^2;Z) &\simeq Z\oplus Z \\ H^4(S^2\times S^2;Z) &\simeq Z \end{aligned} \tag{4.29}$$

and so $U(n)$ bundles are classified by <u>three</u> integers

3. <u>$SO(n)$ $n = 3$, $n \geqslant 5$.</u>

We now have $\pi_1(SO(n)) \simeq Z_2$, $\pi_3(SO(n)) \simeq Z$ and again there is no obvious simplification of (4.22). The map θ now determines an element of $H^5(K(Z_2,2);Z)$ and there is no reason to expect it to be trivial. The portion of the CES immediately surrounding $\mathcal{B}_{SO(n)}(M)$ is

$$\longrightarrow H^1(M;Z_2) \xrightarrow{\Omega\theta_*} H^4(M;Z) \xrightarrow{i_*} \mathcal{B}_{SO(n)}(M) \xrightarrow{p_*} H^2(M;Z_2) \xrightarrow{\theta_*} 0 \tag{4.30}$$

If M is noncompact, $H^4(M;Z) = 0$ and using a lemma on page 8 of Thomas 1966, the CES result can be sharpened to

$$\mathcal{B}_{SO(n)}(M) \simeq H^2(M;Z_2) \tag{4.31}$$

where the characteristic cohomology element representing a bundle ξ is known as the second <u>Stieffel-Whitney</u> class $W_2(\xi)$. Even if $H^4(M;Z)$

does not vanish eq. (4.30) still shows that p_* is surjective and one defines $W_2(\xi) := p_*(\xi)$. For example on a Kruskal $S^2 \times R^2$ space-time there are precisely two classes of SO(n) bundle (one of which is trivial of course). This clearly demonstrates that in a general space-time SU2 and SO3 gauge theories may be topologically inequivalent.

If M is a compact oriented four-manifold then $H^4(M;Z) \simeq Z$. Now let α be a homorphism from $H^1(M;Z_2)$ into Z, i.e.

$$\alpha \in Hom(H^1(M;Z_2), Z) = Hom(Hom(H_1(M;Z), Z_2), Z) \qquad (4.32)$$

and let $f \in Hom(H_1(M;Z), Z_2)$. Then $f + f = 0$ and hence $\alpha(f + f) = 2\alpha(f) = 0$ which implies $\alpha(f) = 0$ and since this is true for arbitrary f, we get $\alpha = 0$. In particular even though θ may be homotopically nontrivial map, we see that $\Omega\theta_* \in Hom(H^1(M;Z_2), Z)$ is trivial. Thus the image of $\Omega\theta_*$ is just 0 and hence by exactness, so is the kernel of i_*.

Thus the exact sequence collapses to

$$0 \longrightarrow Z \longrightarrow \mathcal{B}_{SO(n)}(M) \longrightarrow H^2(M;Z_2) \longrightarrow 0 \qquad (4.33)$$

Hence roughly speaking

$$\mathcal{B}_{SO(n)}(M)/Z \approx H^2(M;Z_2) \qquad (4.34)$$

although it must be remembered that $\mathcal{B}_{SO(n)}(M)$ is, a priori, a set and not a group! The sequence (4.33) is an exact sequence of pointed sets and care must be exercised in interpreting it. A more detailed examination of the fibration in Fig.14 which led to (4.33) shows (Spanier 1966) that this result can be extended:- the various bundles mapped into any given element in $H^2(M;Z_2)$ are distinguished by the integers.

Note that the arguments above are readily extended to other non simply connected groups such as $SU3/Z_3$. For example in a non compact space-time

$$\mathcal{B}_{SU3/Z_3}(M) \approx H^2(M;Z_3)$$

The corresponding characteristic class doesn't seem to have any special name in the mathematical literature so we will call it the second <u>triality class</u> $t_2(\xi)$.

§4.3 Further characteristic classes

The classification of SO(n) bundles provided by eq.(4.33) is not very satisfactory as it is unclear how to specify the integer which distinguishes different bundles with the same second Stieffel-Whitney class. Ideally this specification would be in terms of other characteristic classes. The most important of these is the (orthogonal) Pontryagin class which arises in the following way. Any SO(n) -bundle ξ can clearly be regarded as a SU(n)-bundle (denoted $J(\xi)$) and the i'th Pontagin class $P_i(\xi) \in H^{4i}(M;Z)$ is defined as

$$P_i(\xi) := (-1)^i \, C_{2i}(J(\xi)) \tag{4.35}$$

where $C_{2i}(J(\xi))$ is the 2 i'th Chern class. As we are only concerned with four dimensional manifolds M, only $P_1(\xi)$ is relevant with $C_2(J(\xi))$ being defined as in §4.2.

In order to incorporate this new characteristic class it is useful to extend the Postnikov factorization in §4.2 to an arbitrary fibration

$$\begin{array}{ccc} F & \xrightarrow{i} & E \\ & & \downarrow p \\ & & B \end{array} \tag{4.36}$$

where B is simply connected. It may be shown (e.g. Thomas 1966) that this fibration may be decomposed in the form (cf. Fig.12)

$$\begin{array}{ccccc}
K(\pi_4(F),4) & \xrightarrow{i_4} & E_4 & & \\
 & & \downarrow p_3 & & \\
K(\pi_3(F),3) & \xrightarrow{i_3} & E_3 & \xrightarrow{\theta_3} & K(\pi_4(F),5) \\
 & & \downarrow p_2 & & \\
K(\pi_2(F),2) & \xrightarrow{i_2} & E_2 & \xrightarrow{\theta_2} & K(\pi_3(F),4) \\
 & & \downarrow p_1 & & \\
K(\pi_1(F),1) & \xrightarrow{i_1} & E_1 & \xrightarrow{\theta_1} & K(\pi_2(F),3) \\
 & & \downarrow p_0 & & \\
 & & E_0 := B & \xrightarrow{\theta_0} & K(\pi_1(F),2)
\end{array}$$

Fig.15

in which every triple (E_q, p_q, E_{q-1}) is a fibration with fibre $K(\pi_q(F), q)$ induced, via the map θ_{q-1} from the bundle $PK(\pi_q(F), q+1)$ over $K(\pi_q(F), q+1)$. Furthermore there are maps $f_q : E \to E_q$ which are (q + 1)-homotopy equivalences and which satisfy (cf. eq.4.16)

$$p_q \cdot f_{q+1} \sim f_q \tag{4.37}$$

Clearly the Postnikov factorization of BG in Fig.12 is a special case of Fig.15 with F = E = BG, $i = id_F$, B = (*) and with p the constant map from BG onto (*).

As a first application of this factorization, consider an ∞-universal SO6-bundle

$$\begin{array}{ccc} SO6 & \longrightarrow & ESO6 \\ & & \downarrow p \\ & & BSO6 \end{array} \tag{4.38}$$

The group SO6 acts in the usual way on the coset space SO6/SO5 $\simeq S^5$

leading to an associated bundle with total space $ES0 6 \times_{S06} (S06/S05) \approx (ES06)/S05$

$$\begin{array}{ccc} S^5 & \longrightarrow & (ES06)/S05 \\ & & \downarrow \rho \\ & & BS06 \end{array} \qquad (4.39)$$

Now ES06 is clearly a bundle over (ES06)/S05 with fibre S05 and since for all i, $\pi_i(ES06) \approx 0$ it follows that this bundle is an ∞-universal S05 bundle and hence $(ES06)/S05 \sim BS05$. Thus (4.39) becomes

$$\begin{array}{ccc} S^5 & \longrightarrow & BS05 \\ & & \downarrow \rho \\ & & BS06 \end{array} \qquad (4.40)$$

in which ρ also corresponds to the obvious way in which any S05 bundle may be viewed as an S06 bundle. Now $\pi_i(S^5) = 0 \quad i < 5$, $\pi_5(S^5) \approx Z$ and hence the lowest layer in Fig.15 is effectively

$$\begin{array}{ccc} K(Z,5) & \longrightarrow & (BS05)_5 \\ & & \downarrow \\ & & BS06 \longrightarrow K(Z,6) \end{array} \qquad (4.41)$$

giving rise to the Puppe sequence (for a 4-dimensional manifold M),

$$0 \longrightarrow [M, (BS05)_5] \longrightarrow [M, BS06] \longrightarrow 0 \qquad (4.42)$$

which implies

$$\mathcal{B}_{S05}(M) \approx \mathcal{B}_{S06}(M) \qquad (4.43)$$

This method may be trivially extended by induction to prove

$$\forall n \geqslant 6 \qquad \mathcal{B}_{S05}(M) \approx \mathcal{B}_{SO(n)}(M) \qquad (4.44)$$

and hence in order to classify all SO(n) bundles over a 4-dimensional manifold it suffices to classify the S05 bundles and of course the two special cases, S03 and S04, not included in (4.44).

We will now apply these techniques to the Pontryagin class. The embedding $j : SO(n) \to SU(n)$ leads to a fibration (by the same route that (4.40) was obtained)

$$\begin{array}{ccc} SU(n)/SO(n) & \longrightarrow & BSO(n) \\ & & \downarrow J \\ & & BSU(n) \end{array} \qquad (4.45)$$

and a Puppe sequence,

$$\to [M, SO(n)] \to [M, SU(n)] \to [M, SU(n)/SO(n)] \to \mathcal{B}_{SO(n)}(M) \xrightarrow{J_*} \mathcal{B}_{SU(n)}(M) \qquad (4.46)$$

in which, after making the identification $B_{SU(n)}(M) \approx H^4(M;Z)$

(eq. 4.24) we have

$$J_*(\xi) = -P_1(\xi)$$

In order to probe (4.45) more closely with a Postnikov factorization we need to know the first four homotopy groups of SU(n)/SO(n) (the higher ones are irrelevant if dim. M = 4 since $[M,(BSO(n))_4] \underset{\sim}{\sim} [M,BSO(n)]$). Chasing around the homotopy exact sequences of the fiberings

$$\begin{array}{ccc} SO(n) & \text{---} & SU(n) \\ & & \downarrow \\ & & SU(n)/SO(n) \end{array} \qquad \begin{array}{ccc} S^n & \text{---} & SU(n+1)/SO(n) \\ & & \downarrow \\ & & SU(n+1)/SO(n+1) \end{array} \tag{4.47}$$

one can show $(Q_n := SU(n)/SO(n))$

$$\begin{array}{lll} \pi_1(Q_n) \approx 0 & & \forall n \\ \pi_2(Q_n) \approx Z_2 & & \forall n \\ \pi_3(Q_n) \approx \pi_3(Q_4) & & n \geqslant 4 \\ \pi_4(Q_n) \approx \pi_4(Q_5) & & n \geqslant 5 \end{array} \tag{4.48}$$

Using the homeomorphism SU4/SO4 $\approx$ SO6/(SO3 x SO3) and the exact sequence for the fibering

$$\begin{array}{ccc} SO3 \times SO3 & \longrightarrow & SO6 \\ & & \downarrow \\ & & SO6/SO3 \times SO3 \end{array} \tag{4.49}$$

we obtain

$$\begin{array}{l} \pi_3(Q_4) \approx Z_2 \\ \pi_4(Q_4) \approx Z \\ \pi_4(Q_5) \approx 0 \end{array} \tag{4.50}$$

Finally using the exact sequences for the fiberings in the commutative diagram

$$\begin{array}{ccccc} SO3 & \longrightarrow & SU3 & \longrightarrow & SU3/SO3 \\ \downarrow & & \downarrow & & \\ SO6 & \longrightarrow & SU6 & \longrightarrow & SU6/SO6 \\ \downarrow & & \downarrow & & \\ SO6/SO3 & \longrightarrow & SU6/SU3 & & \end{array} \tag{4.51}$$

we find

$$\begin{array}{l} \pi_3(Q_3) \approx Z_4 \\ \pi_4(Q_3) \approx 0 \end{array} \tag{4.52}$$

For convenience we have tabulated our results below (Q_n : = SU(n)/SO(n))

	Q_2	Q_3	Q_4	$Q_5 \ldots\ldots$	Q_n $n \geq 5$
π_0	0	0	0	0	0
π_1	0	0	0	0	0
π_2	Z	Z_2	Z_2	Z_2	Z_2
π_3	Z	Z_4	Z_2	Z_2	Z_2
π_4	Z_2	0	Z	0	0

If we consider first SO(n) bundles with $n \geq 5$, the Postnikov system for the fibration in (4.45) is

$$\begin{array}{ccccc} K(Z_2,3) & \xrightarrow{i_3} & (BSO(n))_3 \sim (BSO(n))_4 & & \\ & & \downarrow & & \\ K(Z_2,2) & \xrightarrow{i_2} & (BSO(n))_2 & \xrightarrow{\theta_2} & K(Z_2,4) \\ & & \downarrow & & \\ & & BSU(n) & \xrightarrow{\theta_1} & K(Z_2,3) \end{array} \qquad (4.53)$$

Now θ_1 determines an element of $H^3(BSU(n)\,;\,Z_2) \simeq 0$ since BSU(n) is 3-connected and hence, as a map, θ_1 is homotopic to a constant. Thus $(BSO(n))_2 \sim BSU(n) \times K(Z_2,2)$ and (4.53) collapses to

$$\begin{array}{ccccc} K(Z_2,3) & \xrightarrow{i_3} & (BSO(n))_4 & & \\ & & \downarrow J_3 & & \\ & & BSU(n) \times K(Z_2,2) & \xrightarrow{\theta_2} & K(Z_2,4) \end{array} \qquad (4.54)$$

with an associated Puppe sequence

$$\longrightarrow H^3(M;\,Z_2) \longrightarrow \mathcal{B}_{SO(n)}(M) \longrightarrow [M, BSU(n) \times K(Z_2,2)] \longrightarrow H^4(M;Z_2) \qquad (4.56)$$

Since BSU(n) is 3-connected with $\pi_4(BSU(n)) \simeq Z$ it follows that

$$(BSU(n))_4 \sim K(Z,4) \qquad (4.57)$$

$$(BSU(n) \times K(Z_2,2))_4 \sim K(Z,4) \qquad (4.58)$$

and hence $[M, BSU(n) \times K(Z_2,2)] \simeq H^4(M;Z) \oplus H^2(M;Z_2)$ (4.59)

Thus (4.56) becomes

$$\longrightarrow H^3(M;\,Z_2) \longrightarrow \mathcal{B}_{SO(n)}(M) \xrightarrow{J_{3*}} H^4(M;Z) \oplus H^2(M;Z_2) \xrightarrow{\theta_{2*}} H^4(M;Z_2) \qquad (4.60)$$

and it is quite easy to see that if ξ is an SO(n) bundle then

$$J_{3*}(\xi) = (-P_1(\xi), W_2(\xi)) \tag{4.61}$$

It is clearly highly desirable to know the precise form of the map θ_{2*} since its kernel corresponds (by the exactness of eq. 4.60) to the subset of $H^4(M;Z) \oplus H^2(M;Z_2)$ of pairs (α,β) of cohomology classes that can be respectively the Pontryagin class (α) and second Stieffel-Whitney class (β) of an SO(n)-bundle. The structure of θ_2 can be discovered by employing the Serre short cohomology sequence associated with the fibration in (4.54) (Spanier 1966, Mosher and Tangora 1968). Such methods lie outside the scope of these notes so we will content ourselves with stating the result:

$$\theta_{2*} : \underset{(\alpha,\beta)}{H^4(M;Z) \oplus H^2(M;Z_2)} \longrightarrow \underset{\alpha \bmod 2 + \beta^2}{H^4(M;Z_2)} \tag{4.62}$$

where $\beta^2 = \beta \cup \beta$ (the cup product - an analogue of exterior derivative of differential forms) and α mod 2 is the image of α by the map mod 2 : $H^4(M;Z) \to H^4(M;Z_2)$ induced by the map mod 2 :

$$\bmod 2 : Z \longrightarrow Z_2 = \{e, a\} \qquad \begin{matrix} 2m \rightsquigarrow e \\ 2n+1 \rightsquigarrow a \end{matrix} \tag{4.63}$$

(see Spanier 1966).

The exactness of (4.60) plus (4.62) shows that the Pontryagin and Stieffel-Whitney classes can not be specified independently but must satisfy the well-known constraint

$$P_1(\xi) \bmod 2 = (W_2(\xi))^2 \tag{4.64}$$

In general J_{3*} (eqn. (4.60)) is a many-to-one map of $\mathcal{B}_{SO(n)}(M)$ onto the set of cohomology classes $\langle\alpha,\beta\rangle \in H^4(M;Z) \oplus H^2(M;Z_2)$ satisfying $\alpha \bmod 2 + \beta^2 = 0$. However if M is simply connected

$$H^3(M;Z_2) \approx H_1(M;Z_2) \approx 0 \tag{4.65}$$

and J_{3*} is injective. Thus in this case we have a complete classification of the SO(n) bundles $(n \geqslant 5)$ in terms of characteristic classes.

As is clear from Fig. 16, the classification of SO3 bundles proceeds in an identical way but with the Z_2 group replaced by Z_4. This makes the structure of the θ_{2*} map a little harder to describe but it leads again to (4.64) as a necessary constraint on the characteristic classes.

The SO4 case is somewhat more complicated. Since $\pi_3(SO4) \simeq Z\oplus Z$ the exact sequence of eqn. (4.33) becomes

$$0 \longrightarrow Z\oplus Z \longrightarrow \mathcal{B}_{SO4}(M) \longrightarrow H^2(M;Z_2) \longrightarrow 0 \tag{4.66}$$

Furthermore since $\pi_4(Q_4) \simeq Z$ (rather than 0) we have to use the three level Postnikov factorization of the fibration (4.45);

$$\begin{array}{lll} K(Z,4) \xrightarrow{i_4} (BSO4)_4 & & \\ \qquad\qquad\qquad \downarrow J_3 & & \\ K(Z_2,3) \xrightarrow{i_3} (BSO4)_3 & \xrightarrow{\theta_3} & K(Z,5) \\ \qquad\qquad\qquad \downarrow J_2 & & \\ \qquad BSU4\times K(Z_2,2) & \xrightarrow{\theta_2} & K(Z_2,4) \end{array} \tag{4.67}$$

with the associated Puppe sequences

$$\longrightarrow H^4(M;Z) \xrightarrow{i_{4*}} \mathcal{B}_{SO4}(M) \xrightarrow{J_{3*}} [M,\ (BSO4)_3] \xrightarrow{\theta_{3*}} 0 \tag{4.68}$$

$$\longrightarrow H^3(M;Z_2) \xrightarrow{i_{3*}} [M,(BSO4)_3] \xrightarrow{J_{2*}} H^4(M;Z)\oplus H^2(M;Z_2) \xrightarrow{\theta_{2*}} H^4(M;Z_2) \tag{4.69}$$

Since, as is clear from (4.66), SO4-bundles with vanishing second Stieffel-Whitney class are labelled by a <u>pair</u> of integers it seems likely that there should be another $H^4(M;Z) \simeq Z$ characteristic class of relevance. This new class comes from the fibration

$$\begin{array}{ccc} S^3 & \longrightarrow & BSO3 \\ & & \downarrow \\ & & BSO4 \end{array} \tag{4.70}$$

with the Postnikov decomposition ($\pi_3(S^3) \simeq Z$, $\pi_4(S^3) \simeq Z_2$)

$$\begin{array}{lll} K(Z_2,4) \longrightarrow (BSO3)_4 & & \\ \qquad\qquad\qquad \downarrow & & \\ K(Z,3) \longrightarrow (BSO3)_3 & \xrightarrow{\chi_3} & K(Z_2,5) \\ \qquad\qquad\qquad \downarrow & & \\ \qquad\qquad BSO4 & \xrightarrow{\chi_2} & K(Z,4) \end{array} \tag{4.71}$$

and associated Puppe sequences (dim. M = 4)

$$\longrightarrow H^4(M;Z_2) \longrightarrow \mathcal{B}_{SO3}(M) \longrightarrow [M,\ (BSO3)_3] \xrightarrow{\chi_{3*}} 0$$

$$\cdots\cdots \longrightarrow H^3(M;Z) \to [M,\ (BSO3)_3] \to \mathcal{B}_{SO4}(M) \xrightarrow{\chi_{2*}} H^4(M;Z) \tag{4.72}$$

If ξ is a SO4-bundle the cohomology class $\chi_{2*}(\xi) \in H^4(M;Z)$ is the <u>Euler class</u> of the bundle.

These methods may be applied in the investigation of other characteristic classes. For example in the fibration ($n \geqslant 5$)

$$\begin{array}{ccc} SO(n)/SO3 & \longrightarrow & BSO3 \\ & & \downarrow \\ & & BSO(n) \end{array} \tag{4.73}$$

we have (Husemoller 1966)

$$\begin{aligned} &\pi_3(SO(n)/SO3) \approx Z \qquad (n \geqslant 5) \\ &\pi_4(SO(n)/SO3) \approx 0 \qquad (n \geqslant 6) \end{aligned} \tag{4.74}$$

leading to the Puppe sequence ($n \geqslant 6$)

$$\rightarrow H^3(M;Z_2) \rightarrow \mathcal{B}_{SO3}(M) \rightarrow \mathcal{B}_{SO(n)}(M) \rightarrow H^4(M;Z_2) \tag{4.75}$$

The image of a SO(n)-bundle ξ in $H^4(M;Z_2)$ is the 4th <u>Stieffel-Whitney</u> class $W_4(\xi)$. For n = 4 or 5 the corresponding class may be defined by embedding the group in SO6. The embedding SO5 → SO6 was shown by eq.(4.43) to lead to a bijective correspondence between SO5 and SO6 bundles. On the other hand the embedding SO4 ⊂ SO5 leads to the fibration

$$\begin{array}{ccc} S^4 & \xrightarrow{i} & BSO4 \\ & & \downarrow k \\ & & BSO5 \end{array} \tag{4.76}$$

and the Puppe sequence

$$\longrightarrow H^4(M;Z) \xrightarrow{i_*} \mathcal{B}_{SO4}(M) \xrightarrow{k_*} \mathcal{B}_{SO5}(M) \longrightarrow 0 \tag{4.77}$$

Thus there is a many-to-one map k_* from the set of SO4 bundles onto the set of SO5 bundles, whose kernel is measured by the image of i_*. Put in a slightly different way, the structure group of an SO5-bundle over a four-dimensional manifold can always be reduced to SO4 but in general there will be many such reductions which, viewed as SO4 bundles, are inequivalent to each other. The 4th Stieffel-Whitney class of an SO4-bundle ξ may be unambiguously defined as $W_4(k_* \xi)$. Note that this technique applied to a SO3 bundle ξ implies, because of the exactness of eq. (4.75), that $W_4(\xi) = 0$.

We should conclude this section by discussing the method by which the crucial Postnikov maps θ_n are constructed. If $F \xrightarrow{i} E \xrightarrow{p} B$ is a fibration over a simply connected space B and if B and

F are respectively (n-1) and (m-1) connected then the theory of spectral sequences leads to the Serre short exact sequence (for any coefficient group)

$$0 \longrightarrow H^1(B) \xrightarrow{p^*} H^1(E) \xrightarrow{i^*} H^1(F) \xrightarrow{\tau} H^2(B) \xrightarrow{p^*} H^2(E) \xrightarrow{i^*} H^2(F) \xrightarrow{\tau} \cdots\cdots \tag{4.78}$$
$$\cdots \longrightarrow H^{n+m-2}(F) \xrightarrow{\tau} H^{n+m-1}(B) \xrightarrow{p^*} H^{n+m-1}(E) \xrightarrow{i^*} H^{n+m-1}(F)$$

This is a very non-trivial result and a proof can be found in Spanier 1966. For our purposes the crucial maps are the <u>transgressions</u> $\tau : H^i(F) \to H^{i+1}(B)$. As an example consider the fibration of eq. (4.70)

$$\begin{array}{ccc} S^3 & \longrightarrow & BSO3 \\ & & \downarrow \\ & & BSO4 \end{array} \tag{4.79}$$

Since S^3 and BSO4 are respectively 2 and 1-connected, there is a transgression map $\tau : H^3(S^3; Z) \to H^4(BSO4; Z)$. Now $H^3(S^3; Z) \simeq Z$ and choosing a generator α of this group we obtain an element $\tau(\alpha) \in H^4(BSO4; Z)$. This in turn determines a map χ_2 : BSO4 $K(Z; 4)$ and it is precisely this map which is chosen in the Postnikov construction in (4.71). Of course there are two generators of Z and so ultimately one needs to make some choice of sign convention (this applies to many characteristic classes). In general if π_ℓ is the lowest nonvanishing homotopy group of the fibre f in a fibration, there is a canonical element in $H^\ell(F;\pi_\ell)$ whose image by transgression determines the lowest Postnikov map $\theta : B \to K(\pi_\ell, \ell + 1)$. The higher levels are built up in a similar way as described for example in Mosher and Tangora 1968 or Thomas 1966.

This is all related to Borel's fundamental work on the construction of the cohomology groups of the base space BG of a universal bundle and, as remarked earlier on, a characteristic class is just a pull back of an element of one of these groups. The virtue of employing the Postnikov method to analyse maps $f:M \to BG$ is that, as we have seen, it does lead to classification schemes. However it should be emphasised that the actual computation of a characteristic class for a <u>specific</u> bundle will depend very much on the way in which the bundle is specified.

Finally let us list a few problems that could be usefully tackled by the methods in these notes augmented with more sophisticated machinery:

1) The SO(n) ($n \geq 5$) classification needs improving in the situation when $H^3(M;Z_2) \neq 0$. A careful analysis of the Postnikov maps shows that, even in this case, eqn.(4.60) degenerates into

$$0 \longrightarrow \mathcal{B}_{SO(n)}(M) \xrightarrow{J_{3*}} H^4(M;Z) \oplus H^2(M;Z_2) \xrightarrow{\theta_{2*}} H^4(M;Z_2) \qquad (4.80)$$

Unfortunately since $\mathcal{B}_{SO(n)}$ is a set, not a group, this does not necessarily mean that J_{3*} is injective (and hence give a complete specification).

2) The SO(4) results need collecting together into a proper classification scheme.

3) It is a mathematically interesting and physically potentially very important problem to obtain a complete classification of $SU(n)/Z_n$ bundles over a compact manifold. For example the classification for $SU3/Z_3$ in terms of triality classes is only complete for a noncompact manifold. What is wanted is clearly the analogue of the Pontryagin class of a SO(n) bundle. The analogue of the embedding $SO(n) \subset SU(n)$ would seem to be $SU3/Z_3 \subset SO8$, i.e. we might attempt to achieve a $SU3/Z_3$ classification in terms of the second Stieffel-Whitney and first Pontryagin classes of the associated SO8 bundles. The Postnikov factorization of the relevant fibration.

$$\begin{array}{ccc} SO8/(SU3/Z_3) & \longrightarrow & B(SU3/Z_3) \\ & & \downarrow \\ & & BSO8 \end{array} \qquad (4.81)$$

requires knowledge of the first four homotopy groups of $SO8/(SU3/Z_3)$ which regrettfully, at the time of writing, we do not possess.

§§5. SPINOR FIELDS IN TOPOLOGICALLY NON TRIVIAL SPACE-TIMES

§5.1 Existence of Spinor Structures.

It has been known for some time that spinor fields can only be constructed if the space-time satisfies certain topological restrictions (Borel and Hirzebruch 1959). We wish to demonstrate this using the techniques developed in §4 which will also be deployed to show the existence of inequivalent spin structures (Milnor 1963).

One possible starting point for either a rigorous or a heuristic treatment of spinor fields is the bundle of oriented orthonormal frames. An orthonormal frame (or tetrad or vierbein) at a point $x \in M$ is a set of four basis vectors $\{L^{(a)}; a = 0, 1, 2, 3\}$ for the tangent space $T_x(M)$ satisfying:

$$L^{(a)}{}_{\mu}(x)\, L^{(b)}{}_{\nu}(x)\, \eta_{ab} = g_{\mu\nu}(x) \qquad (5.1)$$

$$L^{(a)}{}_{\mu}(x)\, L^{(b)}{}_{\nu}(x)\, g^{\mu\nu}(x) = \eta^{ab} \qquad (5.2)$$

where η_{ab} is the Minkowski metric diag. (1, -1, -1, -1) and $g_{\mu\nu}$ is the pseudo-riemannian metric on the space-time M. The manifold M is assumed to be both time and space oriented and correspondingly $L^{(0)}$ is chosen to point along the positive time direction and the three vectors $(L^{(1)}, L^{(2)}, L^{(3)})$ form a spatially oriented set. The set of all oriented vierbeins forms a principal $SO(3,1)_0$ $(\equiv L_+^\uparrow)$ bundle ξ over M with the right $SO(3,1)_0$ action:

$$\begin{aligned} E(\xi) \times SO(3,1)_0 &\to E(\xi) \\ (\{L^{(a)}\}, O) &\rightsquigarrow L^{(b)} O_b{}^a \end{aligned} \tag{5.3}$$

A <u>spinor structure</u> (η, f) is defined to be a principal $SL(2,\mathbb{C})$ bundle η over M together with a map $f : E(\eta) \to E(\xi)$ which is a two-to-one map from each fibre in $E(\eta)$ onto a fibre in $E(\xi)$:

$$\begin{array}{ccc} & & E(\eta) \\ & \swarrow^{\pi_\eta} & \downarrow f \\ M & \xleftarrow{\ \pi_\xi\ } & E(\xi) \end{array} \qquad \pi_\xi \cdot f = \pi_\eta$$

Furthermore if $\Lambda : SL(2,\mathbb{C}) \to SO(3,1)$ is the usual double covering, the map f is required to be compatible with the group actions in the sense that

$$f(pA) = f(p)\Lambda(A) \qquad \forall p \in E(\eta), \forall A \in SL(2,\mathbb{C}) \tag{5.5}$$

i.e.

$$\begin{array}{ccc} E(\eta) \times SL(2,\mathbb{C}) & \longrightarrow & E(\eta) \\ \Big\downarrow f \times \Lambda & & \Big\downarrow f \\ E(\xi) \times SO(3,1) & \longrightarrow & E(\xi) \end{array}$$

Now let V be a vector space carrying some spinorial representation of $SL(2,\mathbb{C})$ and form the vector bundle $(E(\eta) \times_{SL(2,\mathbb{C})} V, \pi, M)$ using the technique discussed at the beginning of §2.1. Cross sections of this bundle are called <u>spinor fields</u>.

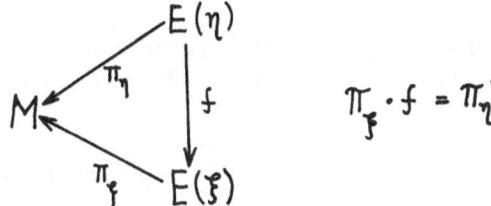

Note that the spinor structure consists of both the bundle η and the map f and both may be expected to have some physical significance in the theory of spinor fields. Roughly speaking non trivial bundles η will give rise to twisted spinor fields and non trivial maps f will lead to 'twisted spin connections' and 'twisted covariant derivatives'. The role played by connections will be briefly discussed in §5.2 but for the moment let us concentrate on investigating the various bundles η which can cover the given frame bundle ξ. A priori there may be none at all. We may as well consider the more general situation where G is any simple, simply connected Lie group, N is a normal subgroup (with p elements say) and ξ is some given G/N bundle. The aim is to find 'generalised' spin structures (η, f) where η is a principal G bundle and $f : E(\eta) \to E(\xi)$ is a p-to-1 map satisfying (5.4) and (5.5) with $\Lambda : G \to G/N$. Let $\mathcal{U}_G = (EG, \pi_G, BG)$ be a ∞-universal G-bundle. There is a natural principal G/N bundle over BG with total space (EG)/N and group action

$$\begin{array}{ccc} (EG)/N \times G/N & \longrightarrow & (EG)/N \\ ([p]_N, [g]_N) & \leadsto & [pg]_N \end{array} \quad ; \; p \in EG, \; g \in G \tag{5.6}$$

If $\mathcal{U}_{G/N} = (E(G/N), \pi_{G/N}, B(G/N))$ is an ∞-universal G/N bundle then there must exist some map $J : BG \to B(G/N)$ such that $J^* (\mathcal{U}_{G/N}) \simeq ((EG)/N, \pi, BG)$:

$$\begin{array}{ccccc} EG & \xrightarrow{\rho} & (EG)/N & \xrightarrow{\tilde{J}} & E(G/N) \\ & \searrow^{\pi_G} & \downarrow{\scriptstyle \pi} & & \downarrow{\scriptstyle \pi_{G/N}} \\ & & BG & \xrightarrow{J} & B(G/N) \end{array} \tag{5.7}$$

where

$$\rho(p) := [p]_N \tag{5.8}$$

$$\pi([p]_N) := \pi_G(p) \tag{5.9}$$

Define $j := \tilde{J} \cdot \rho$ and note that

$$\begin{aligned} j(pg) &= \tilde{J}(\rho(pg)) = \tilde{J}([pg]_N) = \tilde{J}([p]_N [g]_N) = (\tilde{J}([p]_N))[g]_N \\ &= j(p)\Lambda(g) \qquad \forall g \in G, \; \forall p \in EG \end{aligned} \tag{5.10}$$

and hence $(j, J) : (EG, BG) \to (E(G/N), B(G/N))$ may be viewed as a 'universal' generalised spin structure. Indeed if $\alpha : M \to B(G/N)$ represents a G/N bundle ξ then generalised spin structures **on M** correspond to maps $\beta : M \to BG$ such that

$$J \cdot \beta = \alpha \qquad \text{i.e.} \qquad \begin{array}{ccc} M & \xrightarrow{\beta} & BG \\ & \searrow^{\alpha} & \downarrow{\scriptstyle J} \\ & & B(G/N) \end{array} \tag{5.11}$$

This can be summarized in the homotopy commutative map diagram

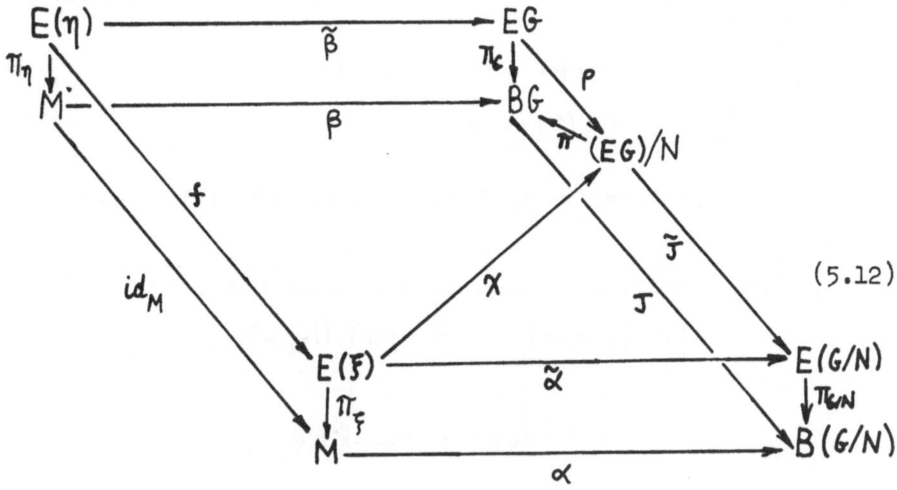

(5.12)

Note that $\lambda := (EG, \rho, (EG/N))$ is a principal N-bundle which is ∞-universal since $\pi_i(EG) = 0$ for all i. The generalised spin structure (η, f) gives rise to a principal N-bundle $(E(\eta), f, E(\xi))$ which is induced from λ by the map $\chi := \rho \cdot \tilde{\beta} \cdot f^{-1}$ (this is independent of the choice of inverse image in f^{-1} and hence well defined).

It is clear that to study generalised spin structures it suffices to investigate the map $J: BG \to B(G/N)$. We do this in a series of steps sketched below.

1) Any map $h : X \to Y$ between topological spaces is homotopically equivalent to a fibration $h' : X' \to Y$ (Mosher & Tangora 1968)

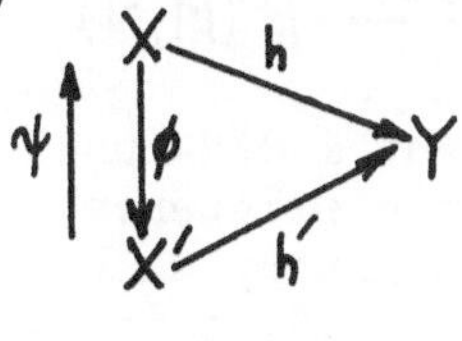

with $\psi \cdot \phi = id_X$ (5.13)

$\phi \cdot \psi \sim id_{X'}$

$$h \cdot \psi \sim h'$$
$$h' \cdot \phi \sim h \qquad (5.14)$$

In particular J leads to a map $J' : BG' \to B(G/N)$. Furthermore J'_* induces isomorphisms

$$J'_* : \Pi_n(BG') \xrightarrow{\sim} \Pi_n(B(G/N)) , \ n \neq 2 \qquad (5.15)$$

2) The fibres $\pi^{-1}(y)$ in a fibration $\pi: X \to Y$ are all homotopically equivalent and lead to a homotopy exact sequence analogous to (3.26). In the present case this shows that the homotopy groups of a typical fibre F are

$$\begin{aligned} \pi_n(F) &\approx 0 \qquad n \neq 1 \\ \pi_1(F) &\approx N \end{aligned} \tag{5.16}$$

and since F may be taken to a CW complex (5.16) implies that $F \sim K(N,1)$.

3) This fibration is induced from some map $\varepsilon : B(G/N) \to K(N,2)$:

$$\begin{array}{ccccc} K(N,1) & \xrightarrow{i} & BG' & \xrightarrow{\tilde{\epsilon}} & PK(N,2) \\ & & \downarrow J' & & \downarrow \\ & & B(G/N) & \xrightarrow{\epsilon} & K(N,2) \end{array} \tag{5.17}$$

and leads to the Puppe sequence (3.25)

$$\begin{aligned} &\to [M, \Omega^2 K(N,1)] \to [M, \Omega^2 BG'] \to [M, \Omega^2 B(G/N)] \to \\ &\to [M, \Omega K(N,1)] \to [M, \Omega BG'] \to [M, \Omega B(G/N)] \to \\ &\to [M, K(N,1)] \xrightarrow{i_*} [M, BG'] \xrightarrow{J_*} [M, B(G/N)] \xrightarrow{\epsilon_*} [M, K(N,2)] \end{aligned} \tag{5.18}$$

This can be rewritten using eqn. (4.15), the homotopy equivalence of ΩBG with G (cf. (4.9)) and $[M,BG'] = [M,BG]$, as :

$$\begin{aligned} &0 \longrightarrow [M, \Omega G] \longrightarrow [M, \Omega(G/N)] \longrightarrow \\ &\to [M, N] \to [M, G] \longrightarrow [M, G/N] \longrightarrow \\ &\to H^1(M;N) \xrightarrow{i_*} \mathcal{B}_G(M) \xrightarrow{J_*} \mathcal{B}_{G/N}(M) \xrightarrow{\epsilon_*} H^2(M;N) \end{aligned} \tag{5.19}$$

4) Now let $\beta : M \to B(G/N)$ represent a G/N-bundle ξ. A necessary and sufficient condition for the existence of a generalised spin structure is that the homotopy class of β lies in the image of J_* and by the exactness of (5.19) this is equivalent to $\varepsilon_*(\xi) = 0$.

In the special case when ξ is a SO(n) bundle it is not difficult to show that

$$\epsilon_*(\xi) = W_2(\xi) \tag{5.20}$$

hence recovering the well known result (Borel and Hirzebruck 1959, Milnor 1963) normally derived using spectral sequences:

"A SO(n) bundle ξ admits a spin structure if and only if its second Stieffel-Whitney class vanishes".

5.2 Inequivalent Spin Structures

We will now assume that the given G/N-bundle ξ does admit a generalized spin structure and examine the question of how many inequivalent spin structures (η, f) exist. The spectral sequence argument referred to earlier shows that this number is equal to the number of elements in the cohomology group $H^1(M;N)$ (Milnor 1963). Spectral sequences are complicated things to describe and we are deliberately avoiding using them. We will prove the result using simpler methods in the special case when M is a four dimensional non compact space-time:

1) Two spin structures (η_1, f_1) and (η_2, f_2) are regarded as being equivalent if there is an isomorphism α between the G-bundles η_1 and η_2 such that

$$f_2 = f_1 \cdot \alpha \qquad \text{i.e.} \qquad \begin{array}{ccc} E(\eta_1) & \xrightarrow{\ \alpha\ } & E(\eta_2) \\ {\scriptstyle f_2}\downarrow & \swarrow {\scriptstyle f_1} & \\ E(\xi) & & \end{array} \tag{5.21}$$

Since M is a non compact 4-manifold $H^4(M;Z) \approx 0$ and hence eqn. (4.24) implies $\mathcal{B}_G(M) \simeq 0$. Thus η is necessarily equivalent to the product bundle and the inequivalence of different spin structures (η, f) must lie in the maps

$$f : E(\eta) \approx M \times G \longrightarrow E(\xi) \tag{5.22}$$

rather than in bundle η.

The existence of such a map implies that $E(\xi)$ is also trivial and spin structure maps f are then in bijective correspondence with maps $\tilde{f} : M \to G/N$ satisfying

$$f(x, g) = (x, \tilde{f}(x)\Lambda(g)) \tag{5.23}$$

2) Bundle isomorphisms of a trivial bundle are necessarily of the form

$$\begin{array}{rcl} \alpha : M \times G & \longrightarrow & M \times G \\ (x, g) & \rightsquigarrow & (x, \tilde{\alpha}(x) g) \end{array} \tag{5.24}$$

for some map $\tilde{\alpha} : M \to G$ and the condition in (5.21) becomes

$$\tilde{f}_2(x) = \tilde{f}_1(x)\Lambda(\tilde{\Omega}(x)) \tag{5.25}$$

Defining
$$\Omega_{12}(x) = (\tilde{f}_1(x))^{-1}\tilde{f}_2(x) \tag{5.26}$$

eqn. (5.25) states that the spin structures corresponding to the maps $\tilde{f}_1$ and $\tilde{f}_2$ are equivalent if and only if Ω_{12} : M → G/N can be lifted to a function $\tilde{\Omega}$: M → G

$$\begin{array}{ccc} & & G \\ & \nearrow^{\tilde{\Omega}} & \downarrow \Lambda \\ M & \xrightarrow[\Omega_{12}]{} & G/N \end{array} \tag{5.27}$$

(In particular if $\tilde{f}_1$ and $\tilde{f}_2$ are homotopic they lead to equivalent structures). Hence the set of all spin structures is in bijective correspondence with [M,G/N] / Im Λ_* where Λ_* : [M,G] → [M,G/N] is induced by Λ : G → G/N.

3) In (5.19) it can be readily shown that $\Omega J_* = \Lambda_*$. Furthermore $\mathcal{B}_G(M) = 0$ and so (5.19) gives the short exact sequence

$$[M,G] \xrightarrow{\Lambda_*} [M,G/N] \longrightarrow H^1(M;N) \longrightarrow 0 \tag{5.28}$$

which implies

$$[M,G/N]/\mathrm{Im}\,\Lambda_* \approx H^1(M;N) \tag{5.29}$$

Hence the number of inequivalent generalised spin structures is equal to the number of elements in the cohomology group $H^1(M;N)$.

Note that there is a 'trivial' spin structure (corresponding to the identity element in $H^1(M;N)$ and with respect to the given trivialisation of E(ξ)) :

$$T(x,g) = (x,\Lambda(g)) \tag{5.30}$$

Any other one f_Ω is obtained from a map Ω : M → G/N with

$$f_\Omega(x,g) = (x,\Omega(x)\Lambda(g)) \tag{5.31}$$

and f_Ω is equivalent to T if and only if Ω lifts to a map from M into G.

Generalised spinor fields are defined as cross sections of any vector bundle associated with η via a 'spin' representation of G on some vector space V. Since η is trivial there are no twisted spinors and the presence of inequivalent spinor structures manifests itself only in the form of the spin connection in the following sense. Let B be a connection one-form on $E(\xi)$ (i.e. a 'Yang-Mills' field) taking its values in the lie algebra $\mathcal{L}(G/N)$ of G/N. This one-form is pulled back by f to give a $\mathcal{L}(G/N) \simeq \mathcal{L}(G)$ valued one-form on $E(\eta)$ and it is this new form that is for example, used in the construction of covariant derivatives of generalised spinor fields. If A_μ denotes a local representative (an $\mathcal{L}(G)$ valued one-form on some open subset of M) of the connection corresponding to the trivial spin structure and A_μ^Ω corresponds to the map $\Omega : M \to G/N$ then

$$A_\mu^\Omega(x) = \Omega(x) A_\mu \Omega^{-1}(x) + \partial_\mu \Omega(x) . \Omega^{-1}(x) \tag{5.32}$$

Normally two 'Yang-Mills' fields which are gauge related in this way would be regarded as equivalent. However if Ω represents a spin structure inequivalent to the trivial one then it cannot be lifted to a function $\tilde{\Omega} : M \to G$ and hence (5.32) can <u>not</u> be compensated by a gauge transformation on the spinors

$$\psi(x) \rightsquigarrow \tilde{\Omega}(\Omega(x)) \psi(x) \tag{5.33}$$

Hence inequivalent spin structures really do lead to inequivalent interactions with spinor fields. Further details of this in the special case $G \simeq SL(2,\mathbb{C})$ and $G/N \simeq SO(3,1)_0$ can be found in Isham 1978b. The results above are also relevant for the interesting use of generalized spin structures suggested by Hawking and Pope 1978. (Caution: they use the word 'generalised' in a different way.)

REFERENCES

Allendoerfer C.B. and Eells J. Comm. Math. Helv. 32 165 (1958)
Avis S.J. and Isham C.J. Imperial College preprint ICTP/77-78/9 To appear Proc. Roy. Soc. (1978)
Banach and Dowker S. Manchester University preprint (1978)
Borel A. and Hirzebruch F. Amer. Jour. Math. 81 315 (1959)
Borel A. Bull. Am. Math. Soc. 61 397 (1955)
Bott R. and Mather J. In 'Batelle Rencontres 1967' Eds. C.M. DeWitt and Wheeler J; Benjamin, New York (1968)
Cremner E. Julia B. Scherk J. Preprint Ecole Normale Superieure 1978
Davies P.C.W. In "Proceedings of the eighth Texas symposium on relativistic astrophysics". New York Academy of Sciences (1977)
DeWitt B.S. Phys. Rep. 196 297 (1975)
DeWitt B.S. Hart C. and Isham C.J. Austin University preprint (1978)
Dold A. "Lectures in algebraic topology", Springer-Verlag, New York (1972)
Dowker J.S. and Critchley R. Phys. Rev. D13 3224 (1976)
Drechsler V. and Mayer M.E. "Fibre bundle techniques in gauge theories", Springer-Verlag, New York (1977)
Duff M.J. and Madore J. Brandeis University preprint (1977) to appear Phys. Rev. D.
Geroch R. Jour. Math. Phys. 11 343 (1970)
Hajicek P. In "Proceedings of the eight Texas symposium on relativistic astrophysics". New York Academy of Sciences (1977)
Hawking S.W. Comm. Math. Phys. 43 199 (1975)
Hawking S.W. Comm. Math. Phys. 55 133 (1977)
Hawking S.W. and Pope C.N. Phys. Letts. 73B 42 (1978)
Hirzebruch F. "Topological methods in algebraic geometry" Springer-Verlag New York (1966)
Hu. S-T. "Homotopy theory". Academic Press, New York (1959)
Husemoller D. "Fibre bundles". McGraw Hill, New York (1966)
Isham C.J. In "Proceedings of the 1977 Bonn conference on applications of differential geometry to theoretical physics". In press Springer-Veralg (1977a)
Isham C.J. In "Proceedings of the eighth Texas symposium on relativistic astrophysics". New York Academy of Sciences (1977b)
Isham C.J. Imperial College preprint ICTP/77-78/3. To appear Proc. Roy. Soc. (1978a)
Isham C.J. Imperial College preprint ICTP/77-78/13. To appear Proc. Roy. Soc. (1978b)
Kay B. Trieste preprints. To appear Comm. Math. Phys. (1977)
Kobayashi S. and Nomizu K. "Foundations of differential geometry". Vol. II Interscience, London (1969)
Milnor J.W. L'Enseignement Math. 9 198 (1963)
Milnor J.W. and Stasheff J. "Characteristic classes". Princeton University Press (1974)
Mosher R.E. and Tangora M.C. "Cohomology operations and applications in homotopy theory". Harper and Row, London (1968)

Parker L. In "Proceedings of the symposium on asymptotic properties of space-time". Plenum, New York (1977)
Segal I.E. In "Applications of mathematics to problems in theoretical physics". Ed. F. Luciat. Gordon and Breach, New York (1967)
Spanier E.H. "Algebraic Topology". McGraw Hill, New York (1966)
Steenrod N. "The topology of fibre bundles". Princeton University Press (1951)
Switzer R.M. "Algebraic topology - Homotopy and Homology". Springer-Verlag, New York (1975)
Thomas E. "Seminar on fibre spaces". Lecture notes in mathematics, Vol. 13 (1966) Springer Verlag.
Vaisman I. "Cohomology and differential forms". Marcel Dekker, New York (1973)

Part III
Supergravity

SUPERGRAVITY AND SUPERSPACE

Bruno Zumino

C.E.R.N.
Theory Division
CH-1211, Geneva 23

INTRODUCTION

In this first lecture I shall try to review briefly what seem to me the more important aspects of supersymmetry and supergravity. Many of the points I shall barely touch upon are treated in more detail by other lecturers at this school. In the following lectures I shall then concentrate mainly on the superspace techniques in supergravity.

Supersymmetry [1] is a (graded) extension of the Poincaré algebra. In simple supersymmetry [2-4] the generators consist of the usual P_a, M_{ab}, of the Poincaré algebra plus the fermionic generators Q_α (the index α is a four-component spinor index and the Q_α form a Majorana spinor ; it is convenient to use a Majorana representation, in which the four Dirac gamma matrices are real). In addition to the usual commutation relations of the Poincaré algebra

$$
\begin{aligned}
&[P_a, P_b] = 0 \\
&i[M_{ab}, P_c] = \eta_{ac} P_b - \eta_{bc} P_a \qquad (1.1) \\
&i[M_{ab}, M_{cd}] = \eta_{ac} M_{bd} - \eta_{bc} M_{ad} - \eta_{ad} M_{bc} + \eta_{bd} M_{ac}
\end{aligned}
$$

and the relations stating that Q_α is a translationally invariant spinor

$$[P_a, Q_\alpha] = 0$$
$$i[M_{ab}, Q_\alpha] = \left(\Sigma_{ab} Q\right)_\alpha \tag{1.2}$$

one has the basic anticommutation relation

$$\{Q_\alpha, \bar{Q}_\beta\} = -2(\gamma^a)_{\alpha\beta} P_a \tag{1.3}$$

where $\bar{Q} = Q\gamma^0$. One knows many examples of local, renormalizable supersymmetric quantum field theories. In local field theory the fermionic generators can be represented as integrals over three-space of the time component of a conserved vector-spinor current.

The representation of the supersymmetry algebra in terms of one-particle states (supermultiplets) contain both integral and half-integral spins. For finite mass one can easily find their content by going to the rest frame, where (1.3) becomes essentially a Clifford algebra. One finds that the supermultiplet consists of four states of equal masses and of spins $s-\frac{1}{2}$, s, s, $s+\frac{1}{2}$ (s is a positive integer or half integer ; for s = 0, there are three states of spins 0, 0, $\frac{1}{2}$). For zero mass it is convenient to work with helicity eigenstates and one finds that the supermultiplet consists of two states of spins s, $s+\frac{1}{2}$ (this again does not apply to s = 0, in which case one has three states of spins 0, 0, $\frac{1}{2}$). If one applies a spinor generator Q_α to a one-particle state, one obtains another state of the supermultiplet having the same mass but spin (helicity) differing by one half. There are two especially interesting supermultiplets. The first is that with spins 1/2 and 1 : it occurs in the widely studied supersymmetric Yang-Mills theories [5, 6] (it is certainly very tempting to imagine that the photon and the electron neutrino form a supermultiplet). The second, with spins 3/2 and 2 is the supermultiplet of supergravity, where the graviton is taken to be the partner of a spin 3/2 Rarita-Schwinger [7] field.

Supersymmetry can be broken either explicitly (but softly)[8] or spontaneously [9]. In field theories with soft explicit or spontaneous breaking the particles of the supermultiplet no longer have the same mass but satisfy instead certain mass sum rules. The various relations among masses and coupling constants valid in the supersymmetric limit are then modified by finite calculable corrections. The most appealing alternative is that of spontaneous breaking, however when supersymmetry is spontaneously broken, a "Goldstone" massless particle emerges [8,10] which is a spin 1/2 fermion, corresponding to the spinorial nature of the generator Q_α which no longer leaves the vacuum invariant. It has been suggested [3] that the (electron ?) neutrino may be this Goldstone fermion, but the neutrino spectrum does not satisfy the low energy theorems which must be valid for a Goldstone fermion [11] (see below for a way out of this difficulty, the supersymmetric Higgs effect).

Supersymmetry can be combined with internal symmetry by attributing each supermultiplet to some representation of the internal symmetry. The generators Q_α are taken as singlets under the internal symmetry, just as the generators of the Poincaré subalgebra. Most attempts at constructing realistic supersymmetric field theories for particle physics are based on this idea. A more interesting way of combining supersymmetry with internal symmetry [12] attributes the spinorial generators to a representation of the internal symmetry group, while naturally the Poincaré generators are still singlets. Although in principle one could imagine other possibilites, for various reasons one is led to an internal symmetry O(N). One has N spinor charges Q^i_α (i=1,... N) belonging to the vectorial representation of O(N). Instead of(1.3) one writes

$$\{ Q^i_\alpha , \bar{Q}^j_\beta \} = -2 \delta^{ij} (\gamma^a)_{\alpha\beta} P_a \qquad (1.4)$$

The supermultiplets are now richer. Let us consider the case of zero mass. Starting with a state of maximum helicity λ (singlet under O(N) and applying to it the operators Q^i_α one obtains N states of helicity $\lambda - \frac{1}{2}$ (in the vector representation of O(N). Applying Q^i_α twice one goes to helicity λ - 1 and multiplicity N(N-1). Applying Q^i_α k times one obtains N!/k!(N-k!) states of helicity λ -k/2. Finally, applying Q^i_α N times one obtains one state (again a singlet) of helicity λ -N/2. If the helicity equals $-\lambda$ (which means that N=4λ) the states form a set which is complete under TCP and one can expect that a corresponding field theory exists. Otherwise, one must adjoin the TCP conjugate states with opposite helicities.

It is clear from the above that, if one fixes the maximum spin of the supermultiplet one cannot have an arbitrary number N of spinorial generators. In particular, if the maximum spin is 1, N cannot exceed 4 and, if the maximum spin is 2, N cannot exceed 8. The first case is realized in the O(4) supersymmetric Yang-Mills theory [13, 14] . In this theory the group O(4) is not gauged, but there is a gauge group which is to a large extent arbitrary and under which the spinor charges are singlets. All fields are in the adjoint representation of the gauge group and the number of adjoint representations is 1 for the vector, 4 for the spin $\frac{1}{2}$, 3 for the scalars and 3 for the pseudoscalars. This theory has the interesting property that the renormalization group function $\beta(g)$ vanishes in the one-and-two-loop approximation [15,16] (it is not known what happens in higher orders).

If one assumes that there is only one graviton (a singlet under the one O(N) group), then the largest supergravity theory is that for N=8. In any case, the existence of consistent interacting theories with fields of spin higher than 2 is very problematic. The particle content of supergravity theories (i.e. maximum spin 2) for various values of N is given in the table. In these theories one achieves the unification of different spin fields with the graviton field in a single supermultiplet. In the theories with N from 4 to 8 even fields with spin zero are part of the supermultiplet. Observe that the theories with N = 7 and N = 8 have the same particle content.

Table

Particle multiplicity of supergravity theories for various values of N.

Spin	N=1	2	3	4	5	6	7	8
2	1	1	1	1	1	1	1	1
3/2	1	2	3	4	5	6	7	8
1		1	3	6	10	16	28	28
1/2			1	4	11	26	56	56
0				2	10	30	70	70

The Lagrangians for the theories with $N \leq 4$ are explicitly known [17-24]. For the N = 8 theory (which will contain all those for lower N) one knows [25] the lower order vertices in an expansion in the gravitational constant, but it will not take long before the full Lagrangian is known.

It is known [26] that the supergravity theories for various N are one loop finite (on the mass shell) and that at least the leading two-loop divergences cancel. This result should be compared with the situation in ordinary gravity, which is one-loop finite, but probably two-loop divergent, and with the situation with gravity coupled to spin 1, $\frac{1}{2}$ or 0 (or any combination of them) which is already one-loop divergent. Observe that these theories are contained as sub-theories in the supergravity theories ; gravity coupled to spin 1 (Maxwell-Einstein) in the theories with $N \geq 2$, gravity coupled to spin 1/2 (Dirac Einstein) in the theories with $N \geq 3$, and gravity coupled with spin 0 in the theories with $N \geq 4$. Unfortunately, at the three-loop level there appear to arise divergences in supergravity [27] and the same is true for the higher loops. However disappointing this may be, the cancellation of divergences due to diagrams involving the spin 3/2 fields improves the convergence of the theories dramatically. It is essential for this that all fields belong to the same irreducible supermultiplet. For instance, it is known that simple supergravity coupled to spin 1/2 - spin 1 supermatter is not even one-loop finite.

I shall now describe briefly simple supergravity [17,18], the gauge theory of simple supersymmetry. The Lagrangian can be formulated by introducing the gauge fields associated with the generators P_a, M_{ab}, Q_α of the supersymmetry algebra. We shall call them respectively $e_m{}^a$ (the vierbein field), $\omega_m{}^{ab}$ (the connection field), and ψ_m^α (the Rarita-Schwinger spin 3/2 field). These fields must be varied independently. One takes

$$L = L_E + L_{RS} \tag{1.5}$$

where the Einstein Lagrangian is

$$L_E = -1/2 \; Re \tag{1.6}$$

and the Rarita-Schwinger Lagrangian is

$$L_{RS} = -\frac{i}{2} \, \varepsilon^{\ell m n r} \, \bar{\psi}_\ell \gamma_5 \gamma_m \mathcal{D}_n \psi_r \tag{1.7}$$

For simplicity we have chosen units in which the gravitational constant equals one. In (1.6) appear the determinant of the vierbein field

$$e = \det e_m{}^a \tag{1.8}$$

and the contracted curvature tensor

$$R = R_{mn}{}^{ab} \cdot e_a{}^m \cdot e_b{}^n$$

$$R_{mn}{}^{ab} = \partial_m \omega_n{}^{ab} - \partial_n \omega_m{}^{ab} - \omega_m{}^{a}{}_{c}\,\omega_n{}^{cb} + \omega_n{}^{a}{}_{c}\,\omega_m{}^{cb} \tag{1.9}$$

The spinorial covariant derivative in (7) is defined as

$$\mathcal{D}_m = \frac{\partial}{\partial x^m} - \frac{1}{2}\,\omega_{mab}\,\Sigma^{ab} \tag{1.10}$$

The expression is covariant in spite of the world index r carried by the spin 3/2 field because of the antisymmetrization in n and r due to the epsilon symbol. By construction (1.5) is invariant under general coordinate transformations and under local Lorentz transformations (see Section 2 for more details). The remarkable fact is that it is also invariant under the local supersymmetry transformation

$$\delta e_m{}^a = i\,\bar{\zeta}\,\gamma^a\,\psi_m$$

$$\delta \psi_m = 2\,\mathcal{D}_m \zeta \tag{1.11}$$

$$\delta \omega_m{}^{ab} = B_m{}^{ab} - \frac{1}{2}\,e_m{}^a\,B_c{}^{bc} + \frac{1}{2}\,e_m{}^b\,B_c{}^{ac}$$

where $\zeta(x)$ is a spinorial (Majorana) anticommuting infinitesimal parameter which can be an arbitrary function of x and

$$B_a{}^{mn} = i\,\bar{\zeta}\,\gamma_5\,\gamma_a\,\mathcal{D}_\ell\,\psi_r\;\varepsilon^{mn\ell r} \tag{1.12}$$

The above is a first order formulation. Eliminating $\omega_m{}^{ab}$ in the Lagrangian by solving for it from its equation of motion, one obtains an equivalent second order form [17] which exhibits explicitly a quartic contact interaction of the spin 3/2 field with itself.

The transformation (1.11) on the Rarita-Schwinger field ψ_m is the generalization to curved space of the spinorial gauge invariance

$$\delta \psi_m = 2\,\partial_m \zeta \tag{1.13}$$

observed by Rarita and Schwinger [7] for the free massless spin 3/2 Lagrangian, which would be given by (1.7) if one replaces the covariant derivative by an ordinary derivative.

In checking the invariance of (1.5) under the local supersymmetry transformation (1.11) one makes essential use of the fermionic nature of the spin 3/2 field. Both the Rarita-Schwinger field ψ_m and the parameter $\zeta(x)$ are treated as totally anticommuting (Grassmannian) quantities in performing the necessary Fierz rearrangements.

If one calculates the commutator of two local supersymmetry transformation (1.11) of parameters ζ_1 and ζ_2, one finds a general coordinate transformation of parameter $\zeta^m = 2i\,\bar{\zeta}_1\gamma^a\zeta_2 e_a{}^m$ accompanied by a Lorentz transformation and by a supersymmetry transformation (both field dependent) plus terms which vanish by use of the spin 3/2 equations of motion : the algebra closes on the mass shell. Actually, if one introduces suitable auxiliary fields (see P. Van Nieuwenhuizen's lectures) the terms containing the spin 3/2 equations of motion do not even appear. In this well defined sense the general coordinate invariance is a consequence of local supersymmetry, which appears as the primary gauge principle.

The equations of motion derived from (1.5) were the first known example of consistent equations for coupled higher spin fields and, together with those of extended supergravity, probably the only consistent equations of this kind. The standard consistency checks proceeds by deriving constraints from the field equations by differentiation. Here the constraint turn out to vanish identically, provided one makes use of the anticommuting nature of the spinor fields. The vanishing of the constraints has also the consequence that no acausal wave propagation occurs [18]. This can be shown by computing the characteristic surfaces by the standard methods of the theory of partial differential equations, as applied to theories with gauge groups, such as gravitation. All these facts are direct consequences of the new gauge principle, local supersymmetry.

There is an interesting generalization of (1.5) which also admits local supersymmetry [28-30]. One adds to (1.5) a cosmological term and a spin 3/2"mass" term in the particular combination

$$3\mu^2 e - \frac{i}{2}\mu\,\varepsilon^{\ell m n r}\,\bar{\psi}_\ell\gamma_5\Sigma_{mn}\psi_r \qquad (1.14)$$

The sum is invariant under a modified local supersymmetry

$$\delta e_m{}^a = i\,\bar{\zeta}\gamma^a\psi_m$$

$$\delta\psi_m = 2\mathcal{D}_m\zeta + \mu\gamma_m\zeta \tag{1.15}$$

(with a corresponding modification for $\delta\omega_m{}^{ab}$). The physical meaning of this theory is not easy to extract, because the cosmological term requires quantization in a background de Sitter space. However, it is interesting that a cosmological term with the opposite sign from that in (1.14) is generated when supergravity is coupled to supersymmetric matter with spontaneous breaking of supersymmetry. These two cosmological terms can actually be cancelled against each other, by adjusting suitably the constants in the theory. The resulting theory can be quantized in a background Minkowski space.

When supergravity is coupled to supersymmetric matter with spontaneous breaking of supersymmetry, supersymmetric version of the Higgs effect occurs [30]. The originally massless spin 3/2 field (which is the gauge field of supersymmetry) absorbs the degrees of freedom of the spin 1/2 Goldstone fermion associated with the spontaneous breaking of supersymmetry and becomes massive. The absence in nature of the Goldstone fermion can thus be understood. This is satisfactory because it leaves one the freedom to use models of matter with spontaneous breaking of supersymmetry.

For extended supergravity theories a generalization analogous to (1.14) allows the gauging of the group O(N). Observe that the number of spin 1 fields dictated by supersymmetry is exactly right for them to be the gauge fields of O(N). It is indeed possible [28] to introduce, besides the gravitational couplings, a gauge coupling with constant g and preserve local supersymmetry by adding also a spin 3/2 mass term and a cosmological term. This has been done explicitly for $N \leq 4$ and presumably can also be done for the higher N's. Unfortunately, the resulting cosmological term is fo the order of $(g/\kappa^2)^2$ (where we have introduced explicitely the gravitational constant κ). If one takes $g \sim 1$, it is enormous and in obvious drastic disagreement with the known limits on the cosmological constant. Irrespective of this difficulty, even the gauged O(8) theory does not seem to be sufficiently ample to describe the presently known elementary forces. According to the commonly accepted picture, one requires a gauge group containing SU(2) x U(1) x SU(3) (for flavour and color) ; O(8) is too small. Another aspect of the same difficulty is that the supermultiplet of the O(8) theory, however large it may appear, does not contain enough fields to accommodate the known vectors, leptons and quarks [31] .

Supergravity as it is now is more a general framework than a concrete physical theory. The extended supergravity theories represent an obvious qualitative success. The gauge principle of extended local supersymmetry provides a way to unify the graviton with the lower spin fields and Einstein gauge invariance with internal symmetry. The improvement in convergence with respect to ordinary gravity coupled with matter is also remarkable. In spite of these positive results, some new input seems necessary for supergravity to become a physically relevant theory. Perhaps one should give up the strong requirement that all fields belong to the same irreducible supermultiplet. Indeed this requirement loses much of its motivation if, as seems to be the case, supergravity theories are not finite after all. Without this very strong restriction one has much more freedom in model building. For instance, within the framework of simple supergravity coupled to simply supersymmetric matter one can construct models which are almost realistic. Another case which deserves more study is that of N=4 supergravity coupled to N=4 Yang-Mills theory.

2. DIFFERENTIAL GEOMETRY

In this lecture we develop some basic notions of differential geometry. We take a local point of view, i.e. we work in a finite region in which the points of our manifold can be identified by their coordinates x^m. The study of the properties of the manifold in the large (topology) transcends our present interest ; there is enough structure in supergravity in the small. Our main purpose is to establish a certain number of notions and formulas which we shall later generalize to the case of superspace (section3).

In a differentiable manifold one can define scalars, vectors and tensors as well as densities. They are characterized by their transformation properties under an infinitesimal coordinate transformation of parameters $\xi^m(x)$ (clearly one can use finite transformations but we shall not consider them explicitly here, for the sake of brevity). For instance, a scalar f(x) transforms as

$$\delta f = \xi^m \partial_m f \qquad \partial_m = \frac{\partial}{\partial x^m} \tag{2.1}$$

a covariant vector u_m and a contravariant vector v^m as

$$\begin{aligned} \delta u_m &= \xi^\ell \partial_\ell u_m + \partial_m \xi^\ell u_\ell \\ \delta v^m &= \xi^\ell \partial_\ell v^m - \partial_\ell \xi^m v^\ell \end{aligned} \tag{2.2}$$

and so on, for tensor with several indices. A density $\mathcal{h}$ of weight p transforms as

$$\delta \mathcal{h} = \xi^m \partial_m \mathcal{h} + p\, \partial_m \xi^m \mathcal{h} \tag{2.3}$$

a vector density k^m of weight p as

$$\delta k^m = \xi^\ell \partial_\ell k^m - \partial_\ell \xi^m k^\ell + p\, \partial_\ell \xi^\ell k^m \tag{2.4}$$

Certain operations of differentiation have a geometric meaning in a differentiable manifold. One can take the gradient of a scalar, the curl of a covariant vector etc.. One can also take the divergence of a contravariant vector density of weight p = 1 and more generally of an antisymmetric tensor density of weight p = 1. If one wishes to give general meaning to differentiation one must introduce more structure and work with an affine space.

In an affine space one has a connection, given by its coefficients $\Gamma_{mn}{}^\ell$, in terms of which one can construct covariant derivatives, for instance

$$\begin{aligned} \mathcal{D}_m u_n &= \partial_m u_n - \Gamma_{mn}{}^\ell u_\ell \\ \mathcal{D}_m v^n &= \partial_m v^n + \Gamma_{m\ell}{}^n v^\ell \end{aligned} \tag{2.5}$$

The transformation property of the connection coefficients is chosen so that the covariant derivatives transforms as tensors. One finds

$$\delta \Gamma_{mn}{}^\ell = \xi^s \partial_s \Gamma_{mn}{}^\ell + \partial_m \xi^s \Gamma_{sn}{}^\ell + \partial_n \xi^s \Gamma_{ms}{}^\ell - \partial_s \xi^\ell \Gamma_{mn}{}^s + \partial_m \partial_n \xi^\ell \tag{2.6}$$

Except for the last term (which is symmetric in m and n) this is the transformation law of a tensor. Therefore $\Gamma_{mn}{}^\ell$ is not itself a tensor, but the antisymmetric combination

$$T_{mn}{}^\ell = \Gamma_{mn}{}^\ell - \Gamma_{nm}{}^\ell \tag{2.7}$$

is a tensor, called the torsion. Covariant derivatives do not commute. It is easy to see that

$$[\mathcal{D}_m, \mathcal{D}_n] v^\ell = R_{mns}{}^\ell v^s - T_{mn}{}^s \mathcal{D}_s v^\ell \tag{2.8}$$

where the curvature tensor, antisymmetric in the first two indices, is given by

$$R_{mns}{}^\ell = \partial_m \Gamma_{ns}{}^\ell - \partial_n \Gamma_{ms}{}^\ell - \Gamma_{ms}{}^r \cdot \Gamma_{nr}{}^\ell + \Gamma_{ns}{}^r \cdot \Gamma_{mr}{}^\ell \tag{2.9}$$

If both torsion and curvature vanish, one can find a coordinate frame in which the connection coefficients vanish : the affine space is flat.

The cyclic summ over r, m, n

$$\oint_{rmn} [\mathcal{D}_r, [\mathcal{D}_m, \mathcal{D}_n]] v^\ell = 0 \tag{2.10}$$

vanishes by the Jacobi identity.

Using (2.8) and remembering that the vector $v^\ell(x)$ is arbitrary, this gives the two Bianchi identities

$$\oint_{rmn} (\mathcal{D}_r T_{mn}{}^\ell + T_{rm}{}^s T_{sn}{}^\ell - R_{rmn}{}^\ell) = 0 \tag{2.11}$$

and

$$\oint_{rmn} (\mathcal{D}_r R_{mns}{}^\ell + T_{rm}{}^t R_{tns}{}^\ell) = 0 \tag{2.12}$$

Instead of referring vectors and tensor to the local curvilinear coordinates, one can consider the linear tangent space at each point of the manifold and specify vectors and tensor by their components with respect to a linear frame in the tangent space. The basis vectors $e_a(x)$ of this local frame have curvilinear components which we denot by $e_a{}^m(x)$ and their duals $e^a(x)$ curvilinear components $e_m{}^a(x)$. The duality relation is expressed by

$$e_a{}^m \cdot e_m{}^b = \delta_a{}^b \tag{2.13}$$

The matrices $e_m{}^a$ and $e_a{}^m$ are inverses of each other. Since in four dimensions there are four basis vectors it is customary to call $e_m{}^a(x)$ the "vierbein" field (from the German ; in more than four dimensions the name vielbein, started as a joke, is becoming popular among physicists). If one knows the tangent space components of a contravariant vector, one can obtain its curvilinear components.

$$v^m = v^a e_a{}^m \tag{2.14}$$

and viceversa

$$v^a = v^m e_m{}^a \tag{2.15}$$

Similarly for a covariant vector

$$u_m = e_m{}^a u_a \quad , \quad u_a = e_a{}^m u_m \tag{2.16}$$

An infinitesimal linear change of the local tangent space frame changes the components of vectors as

$$\begin{aligned} \delta v^a &= v^b X_b{}^a \\ \delta u_a &= -X_a{}^b u_b \end{aligned} \tag{2.17}$$

so that $v^a u_a$ is invariant. The infinitesimal matrix $X_a{}^b(x)$ can be an arbitrary function of x. If we wish to define covariant derivatives for tangent space vectors we must introduce quantities $\omega_{ma}{}^b(x)$ (rotation coefficients) which transform as

$$\delta\omega_{ma}{}^b = \omega_{ma}{}^c X_c{}^b - X_a{}^c \omega_{mc}{}^b - \partial_m X_a{}^b \tag{2.18}$$

so that the covariant derivatives

$$\begin{aligned} \mathcal{D}_m v^a &= \partial_m v^a + v^b \omega_{mb}{}^a \\ \mathcal{D}_m u_a &= \partial_m u_a - \omega_{ma}{}^b u_b \end{aligned} \tag{2.19}$$

transform in the same way as the vectors which are being differentiated. Under general coordinate transformations, tangent space vectors behave as scalars and the $\omega_{ma}{}^{b}$ as covariant vectors with index m. Then the covariant derivatives (2.19) are covariant vectors. One can also consider quantities with both tangent space indices a, b, etc.. and curvilinear or "world" indices m, n, etc. In this case the covariant derivative, constructed in analogy with (2.5) and (2.19) contains both $\Gamma_{mn}{}^{\ell}$ and $\omega_{ma}{}^{b}$. For instance the covariant derivative of the vielbein field is

$$\mathcal{D}_m e_n{}^a \equiv \partial_m e_n{}^a - \Gamma_{mn}{}^{\ell} e_\ell{}^a + e_n{}^b \omega_{mb}{}^a \tag{2.20}$$

If one wants the definitions (2.5) and (2.19) to be consistent with (2.14), (2.15) and (2.16) one must require that the covariant derivative (2.20) of the vielbein vanish ; that of the inverse vielbein will then also vanish. Setting

$$\mathcal{D}_m e_n{}^a = 0 \tag{2.21}$$

one obtains a relation which permits us to express $\Gamma_{mn}{}^{\ell}$ in terms of $\omega_{ma}{}^{b}$ and the vielbein and its derivatives (or viceversa)

$$\Gamma_{mn}{}^{\ell} e_\ell{}^a = \tilde{\mathcal{D}}_m e_n{}^a \tag{2.22}$$

where we indicate in general with a tilde a derivative which covariantizes only with respect to the tangent space index, in this case

$$\tilde{\mathcal{D}}_m e_n{}^a \equiv \partial_m e_n{}^a + e_n{}^b \omega_{mb}{}^a \tag{2.23}$$

Naturally, a tilde derivative is not covariant if the object on which it operates has world indices, as in (2.22). Antisymmetrizing (2.22) we obtain, form (2.7),

$$T_{mn}{}^a = T_{mn}{}^{\ell} e_\ell{}^a = \tilde{\mathcal{D}}_m e_n{}^a - \tilde{\mathcal{D}}_n e_m{}^a \tag{2.24}$$

the curl-like combination on the right hand side is covariant.

It is easy to check that the commutator of two tilde derivatives gives

$$[\tilde{\mathcal{D}}_m, \tilde{\mathcal{D}}_n]\, v^a = R_{mnb}{}^a\, v^b \tag{2.25}$$

where

$$R_{mna}{}^b = \partial_m \omega_{na}{}^b - \partial_n \omega_{ma}{}^b - \omega_{ma}{}^c \cdot \omega_{nc}{}^b + \omega_{na}{}^c \cdot \omega_{mc}{}^b \tag{2.26}$$

is a world tensor with respect to the indices m, n and a tangent space tensor with respect to the indices a, b. Indeed, the commutator in (2.25) is a curl-like construction and thus covariant. On the other hand the commutator of two fully covariant derivatives has an extra term

$$[\mathcal{D}_m, \mathcal{D}_n]\, v^a = R_{mnb}{}^a\, v^b - T_{mn}{}^\ell \cdot \mathcal{D}_\ell\, v^a \tag{2.27}$$

Comparing this with (2.8) and using (2.21) we see that

$$R_{mna}{}^b = R_{mn\ell}{}^s \cdot e_a{}^\ell \cdot e_s{}^b \tag{2.28}$$

If we define the covariant derivatives referred to the tangent space frame

$$\mathcal{D}_a = e_a{}^m\, \mathcal{D}_m \tag{2.29}$$

(2.27) together with (2.21) implies

$$[\mathcal{D}_c, \mathcal{D}_d]\, v^a = R_{cdb}{}^a \cdot v^b - T_{cd}{}^b \cdot \mathcal{D}_b\, v^a \tag{2.30}$$

Because of (2.21), the Bianchi identities (2.11) and (2.12) can be written with tangent space indices

$$\oint_{abc} \left(\mathcal{D}_a T_{bc}{}^d + T_{ab}{}^f \cdot T_{fc}{}^d - R_{abc}{}^d \right) = 0 \tag{2.31}$$

and

$$\oint_{abc} \left(\mathcal{D}_a R_{bcd}{}^{f} + T_{ab}{}^{g} R_{gcd}{}^{f} \right) = 0 \tag{2.32}$$

In these two equations one could use equally well tilde derivatives, since no world index occurs. Slightly simpler forms corresponding to (2.24) and (2.25), are given by

$$\oint_{rmn} \left(\tilde{\mathcal{D}}_r T_{mn}{}^{a} - R_{rmn}{}^{a} \right) = 0 \tag{2.33}$$

and

$$\oint_{rmn} \tilde{\mathcal{D}}_r R_{mna}{}^{b} = 0 \tag{2.34}$$

where only the indices over which one sums cyclically are world indices.

In an affine space the tangent space group is the general linear groups described in infinitesimal form by (2.17). The connection $\omega_{ma}{}^{b}$ and the curvature $R_{mna}{}^{b}$, considered as matrices in a, b belong to the algebra of this group. One can introduce more structure into the space by restricting the tangent space group to be the (pseudo-) orthogonal group. This amounts to postulating the existence of a numerically invariant symmetric tensor η_{ab} which, together with the inverse η^{ab} can be used to raise and lower tangent space indices. Then

$$X_{ab} \equiv X_a{}^{c} \eta_{cb} = -X_{ba} \tag{2.35}$$

By choosing suitably the tangent space frame, η_{ab} can be taken to be the Euclidean (or respectively Minkowski) metric. Now connection and curvature belong to the algebra of the orthogonal (or Lorentz) group

$$\omega_{mab} = \omega_{ma}{}^{c} \eta_{cb} = -\omega_{mba} \tag{2.36}$$

$$R_{mnab} = -R_{mnba} \tag{2.37}$$

The metric tensor

$$g_{mn} = e_m{}^a \cdot e_n{}^b \cdot \eta_{ab} \tag{2.38}$$

is covariantly conserved (metric postulate), as a consequence of (2.21)

$$\mathcal{D}_\ell g_{mn} = \partial_\ell g_{mn} - \Gamma_{\ell m}{}^s g_{sn} - \Gamma_{\ell n}{}^s g_{ms} = 0 \tag{2.39}$$

If in addition the torsion vanishes we have a Riemann space. Eq. (2.39) can be solved for $\Gamma_{mn}{}^\ell$ which is symmetric in m and n

$$\Gamma_{mn}{}^\ell = \frac{1}{2} g^{\ell s} \left(\partial_m g_{ns} + \partial_n g_{ms} - \partial_s g_{mn} \right) \tag{2.40}$$

Equivalently one can set the expression (2.24) for the torsion equal to zero

$$\partial_m e_n{}^a - \partial_n e_m{}^a + \omega_{mb}{}^a \cdot e_n{}^b - \omega_{nb}{}^a \cdot e_m{}^b = 0 \tag{2.41}$$

Making use of (2.36), this equation can be solved for

$$\omega_{mab} = e_m{}^c \cdot \omega_{cab} \quad \text{with}$$

$$\omega_{cab} = -\frac{1}{2} \left(C_{cab} - C_{abc} + C_{bca} \right) \tag{2.42}$$

and

$$C_{bca} = \left(\partial_m e_{na} - \partial_n e_{ma} \right) e_b{}^m e_c{}^n \tag{2.43}$$

In physical applications the torsion needn't vanish; instead it is sometimes determined dynamically, for instance it is proportional to the spin density. Then the solution of (2.24) is given by a formula similar to (2.42) but with C_{abc} replaced by $C_{abc} - T_{abc}$.

When the tangent space group is (pseudo-) orthogonal one can introduce spinors by referring then to the local orthonormal tangent frame. For instance one can use the numerical Dirac matrices satisfying

$$\gamma_a \gamma_b + \gamma_b \gamma_a = 2 \eta_{ab} \tag{2.44}$$

and the matrices

$$\Sigma_{ab} = \frac{1}{4} [\gamma_a, \gamma_b] \tag{2.45}$$

The spinorial covariant derivative of a spinor ψ is

$$\mathcal{D}_m \psi = \left(\partial_m - \frac{1}{2} \omega_{m\,ab} \Sigma^{ab} \right) \psi \tag{2.46}$$

For a Rarita-Schwinger vector spinor ψ_m we have

$$\tilde{\mathcal{D}}_m \psi_n = \left(\partial_m - \frac{1}{2} \omega_{m\,ab} \Sigma^{ab} \right) \psi_n \tag{2.47}$$

while the fully covariant derivative is

$$\mathcal{D}_m \psi_n = \tilde{\mathcal{D}}_m \psi_n - \Gamma_{mn}{}^{\ell} \psi_{\ell} \tag{2.48}$$

Observe that again the curl-like expression

$$\psi_{mn} = \tilde{\mathcal{D}}_m \psi_n - \tilde{\mathcal{D}}_n \psi_m \tag{2.49}$$

is a covariant tensor-spinor.

With the tools developed above one can easily write Lagrangians which are densities (of weight one) under general coordinate transformations and are invariant under local (x dependent) Lorentz transformations on the tangent space indices.
To obtain a density one multiplies a scalar by the basic density

$$e = \det e_m{}^a \tag{2.50}$$

The totally antisymmetric tensor density $\varepsilon^{mn\ell r}$, with $\varepsilon^{0123} = 1$, is also useful. The construction of the Einstein Lagrangian in the form (1.6) and of the Rarita-Schwinger Lagrangian in the form (1.7) is now clear.

It is worth observing that a complete formulation can be obtained by using only the variables $e_m{}^a$ $e_a{}^m$ and $\omega_{ma}{}^b$, and the tilde derivatives. The torsion can be defined by (2.24) and the curvature by (2.25) or both by (2.30) where the derivatives can be taken to be tilde derivatives, since no world indices occur. The Bianchi identities are then given by (2.31) and (2.32). (This is the formulation we shall take over to superspace, see section 3. Since we shall use only tilde derivatives, we shall drop the tilde). This formulation is extremely natural if one uses the language of differential forms. Define the vielbein forms

$$e^a \equiv dx^m e_m{}^a \tag{2.51}$$

and the connection forms

$$\omega_a{}^b \equiv dx^m \omega_{ma}{}^b \tag{2.52}$$

corresponding to (2.19), the covariant differential is given by*

$$\begin{aligned} \mathcal{D}v^a &= dv^a + v^b . \omega_b{}^a \\ \mathcal{D}u_a &= du_a - \omega_a{}^b . u_b \end{aligned} \tag{2.53}$$

The torsion form is defined by

$$T^a = \mathcal{D}e^a = de^a + e^b \omega_b{}^a = \frac{1}{2} dx^n dx^m T_{mn}{}^a \tag{2.54}$$

and the curvature form by

$$R_a{}^b = d\omega_a{}^b + \omega_a{}^c \omega_c{}^b = \frac{1}{2} dx^n . dx^m R_{mna}{}^b \tag{2.55}$$

---

* Since we write the differentials dx^m on the left the d operation always starts from the factor at the extreme right. Thus, for two forms α and β

$$d(\alpha\beta) = \alpha\, d\beta + (-1)^s d\alpha . \beta$$

where s is the degree of β .

The Bianchi identities follow immediately from these definitions. They are, in the present notation,

$$dT^a + T^b \omega_b{}^a - e^b R_b{}^a = 0 \tag{2.56}$$

and

$$dR_a{}^b + R_a{}^c \cdot \omega_c{}^b - \omega_a{}^c \cdot R_c{}^b = 0 \tag{2.57}$$

In the formalism of differential forms a Lagrangian density becomes a 4-form, to be integrated over 4-space. For instance the Einstein Lagrangian takes the very simple form

$$L_E d^4x = \frac{1}{8} R^{ab} e^c e^d \varepsilon_{abcd} \tag{2.58}$$

where ε_{abcd} is the numerical totally antisymmetric tensor referred to the local tangent space. To write the Rarita-Schwinger Lagrangian we introduce the matrix 1-form

$$\gamma = dx^m \cdot e_m{}^a \gamma_a \tag{2.59}$$

and the spinor 1-form

$$\psi = dx^m \cdot \psi_m \tag{2.60}$$

whose covariant differential is

$$\mathcal{D}\psi = d\psi - \frac{1}{2} \Sigma^{ab} \cdot \psi \cdot \omega_{ab} = \frac{1}{2} dx^n dx^m \psi_{mn} \tag{2.61}$$

The Rarita-Schwinger Lagrangian is simply

$$L_{RS} d^4x = \frac{i}{2} \bar{\psi} \gamma_5 \gamma \mathcal{D} \psi \tag{2.62}$$

3. GEOMETRY OF SUPERSPACE

In global supersymmetry, superspace [32-34] provides a technique for constructing representations of the supersymmetry algebra by fields. In one introduces anticommuting infinitesimal Majorana spinor parameters ζ_1 and ζ_2 (they anticommute also with the generators Q), the basic anticommutation relation (1.3) can be written as a commutation relation

$$[\bar{\zeta}_1 Q, \bar{\zeta}_2 Q] = -2\bar{\zeta}_1 \gamma^a \zeta_2 P_a \tag{3.1}$$

The algebra can now be exponentiated to a group. This kind of (super-)groups are of course not ordinary Lie groups. The group parameters are c-numbers for the even generators P_a, M_{ab} and anticommuting numbers for the odd generators Q_α (more precisely, they are respectively even and odd elements of a Grassmann algebra).

Rigid superspace is the quotient, in the group-theoretic sense, of the full supersymmetry group divided by the Lorentz group. Its points are parametrized by coordinates x^a (for the translations) and θ^α (for the supersymmetry transformations). It is easy to see that the effect of a supersymmetry transformation on this quotient space is (infinitesimally)

$$\delta x^a = -i\bar{\zeta}\gamma^a\theta \qquad \delta\theta^\alpha = \zeta^\alpha . \tag{3.2}$$

Indeed, the commutator of two transformations (3.2)

$$\begin{aligned} [\delta_2, \delta_1] x^a &= -2i\bar{\zeta}_1 \gamma^a \zeta_2 \\ [\delta_2, \delta_1]\theta^\alpha &= 0 \end{aligned} \tag{3.3}$$

is a translation of parameter $-2i\bar{\zeta}_1\gamma^a\zeta_2$, as required by the algebra (3.1).

A superfield $V(x,\theta)$ is a function of superspace which transforms like a scalar under (3.2)

$$\delta V = \left(\zeta^\alpha \frac{\partial}{\partial\theta^\alpha} - i\bar{\zeta}\gamma^a\theta\partial_a\right)V \tag{3.4}$$

Since the four components θ^α anticommute, in particular the square of each one is zero, a power series in θ^α is at most a polynomial of fourth degree. Therefore a superfield is equivalent to a finite collection of ordinary field (a supermultiplet), carrying various spinorial indices and satisfying corresponding statistics. When the superfield transforms as in (3.4) the coefficients of its expansion in θ^{α} transform into each other (with derivatives). Such a representation by fields is not irreducible. The "covariant derivatives"

$$D_{\alpha} = \frac{\partial}{\partial\theta^{\alpha}} + i\,(\gamma^{a}\theta)_{\alpha}\,\partial_{a} \qquad\qquad D_{a} = \partial_{a} \tag{3.5}$$

commute with the infinitesimal operator in (3.4) and can be used to impose covariant constraints on superfields. They satisfy

$$\{D_{\alpha}, \bar{D}_{\beta}\} = -\,2i\,(\gamma^{a})_{\alpha\beta}\,D_{a} \tag{3.6}$$

showing the superspace has torsion, even though its curvature vanishes. This "rigid" superspace has been extensively used in the study of supersymmetric field theories in Minkowski space. We refer for this to the original literature (references can be found in the review article by Fayet and Ferrara [1]). We shall instead describe the geometric theory of curved superspace. Just as Einstein's theory is obtained when one generalizes from flat Minkowski space to curved space-time, supergravity can be obtained by considering a suitable curved superspace.

The first work on curved superspace was that of Arnowitt and Nath [35] , who developed the differential geometry of a super-Riemannian superspace. However, the flat superspace of rigid supersymmetry does not fit naturally as a special case of a super-Riemannian geometry. For this reason a different geometry of superspace was introduced by Wess and the author [36-42] and by Akulov, Volkov and Soroka [43] . In this geometry the tangent space group at each point of superspace is not taken to be the graded Lorentz group (orthosimpletic (4,4)), but just the ordinary Lorentz group. Furthermore the superspace is allowed to have torsion. More recently this geometry has been adopted by other authors [44] . In these lectures we shall follow our own work, as developed in ref. [36-42] . Ours is a second order formulation in what we impose on the supertorsion certain conditions ("constraints"), which are sufficient to express the superconnection in terms of the supervielbein. These conditions also restrict to a certain extent the supervielbein itself. The dynamics is given in terms of a superspace action which must be varied keeping the constraints on the torsion satisfied. Our formulation is exactly equivalent to the recently developed formalism for supergravity with a minimal set of auxiliary fields (See P. Van Nieuwenhuizen's lectures at this school). Indeed this auxiliary field formalism can be derived from our geometrical formulation [41] .

A non-geometric description of supergravity in superspace has been developed by Ogievestsky and Sokatchev [45] and by Warren Siegel [46] (see also ref. [47]. The supergravity fields are contained in a superfield $U^m(x,\theta)$ endowed with a vector index m. The theory, based directly on this superfield, has a certain appeal because it is free of constraints. However, the lack of a complete geometrical structure in such a theory makes the construction of invariants a difficult task. On the other hand, from our point of view, the superfield U^m emerges in solving the constraints on the supertorsion. The supervielbein and the superconnection are expressed in terms of it and we can use the powerful techniques of differential geometry for the construction of invariants.

The geometric superspace approach gives a complete and satisfactory method for the study of simple supergravity and its couplings to supersymmetric matter (both for the vector-spinor and for the spinor-scalar-pseudoscalar multiplets). All possible invariants of given dimension are easily constructed and their knowledge can be used [47] to discuss the divergences which arise in perturbation theory. The situation is quite different in the case of extended supergravity. Clearly one must use here a larger superspace in which the fermionic coordinates carry, in addition to the spinorial index, and internal symmetry index, and the tangent space group should be the product of the Lorentz group with the internal symmetry O(N). The difficulty consists in finding the correct separation of the equations into kinematical constaints (which are part of the specification of the geometry) and equations of motion (which should follow from the variation of an action, subject to the constraints). The results obtained so far [48] for $N = 3$ extended supergravity in superspace do not make this separation. These difficulties in the superspace method correspond to our present ignorance as to the correct auxiliary fields for extended supergravity. In the following we limit ourselves to the case of simple supergravity, which is fully understood.

After these introductory remarks we now develop the geometry of superspace. Let us first define an affine superspace, in complete analogy with sec. 2. Its points are parametrized by coordinates $z^M \equiv (x^m, \theta^\mu)$. Small latin letters denote bosonic (vectorial), small greek letters fermionic (spinorial) indices. Capital letters run over all right values. We use late alphabet letters (L, M, N, etc..) for (super-) world indices, early alphabet letters (A, B, C, etc.) for tangent space indices. The coordinates x^m

commute with each other and with the θ^μ ; the (real) coordinates θ^μ anticommute with each other. In an affine superspace there is a supervielbein matrix $E_M{}^A(z)$, where $A \equiv (a, \alpha)$, and its inverse $E_A{}^M(z)$ as well as a superconnection $\phi_{MA}{}^B(z)$ (which is the analogue of the $\omega_{ma}{}^b$ of Sect. 2). The supervielbein can be used to transform world tensors into tangent space tensors and viceversa. The matrix elements $E_m{}^a$ and $E_\mu{}^\alpha$ are bosonic, E_m^α and $E_\mu{}^a$ fermionic quantities. In analogy with (2.19), the covariant derivative of a tangent space vector u_A is defined as

$$\mathcal{D}_M u_A = \partial_M u_A - \phi_{MA}{}^B u_B \tag{3.7}$$

where $\partial_M \equiv \partial/\partial z^M$. For a vector v^A with an upper index it is

$$\mathcal{D}_M v^A = \partial_M v^A + (-1)^{mb} v^B \phi_{MB}{}^A \tag{3.8}$$

where the sign factor $(-1)^{mb}$ is specified by the convention that $m = 0$ if M is vectorial and $m = 1$ if M is spinorial, and similarly for B. The need for the sign factor comes from the fermionic nature of the ordinary derivative with respect to a fermionic coordinate. For the ordinary derivatives one has

$$\partial_M (v^A u_A) = \partial_M v^A \; u_A + (-1)^{ma} v^A \partial_M u_A \tag{3.9}$$

The sign factor in (3.8) assures that the covariant derivatives also satisfy this (graded) Leibnitz rule and that the covariant derivative of the scalar $v^A u_A$ agrees with its ordinary derivative. The generalization of (3.7) and (3.8) to tangent space tensors with more indices is obvious.

The supercurvature $R_{ABC}{}^D$ and the supertorsion $T_{AB}{}^C$ can be defined generalizing directly to superspace the relations (2.24) and (2.25) or (2.29) and (2.30). So, defining

$$\mathcal{D}_A \equiv E_A{}^M \mathcal{D}_M \tag{3.10}$$

one has

$$[\mathcal{D}_C, \mathcal{D}_D]_\pm = -R_{CDA}{}^B u_B - T_{CD}{}^B \mathcal{D}_B u_A \tag{3.11}$$

where the bracket is a commutator if the indices C and D are both bosonic or one bosonic and one fermionic, an anticommutator otherwise. We shall not write out here the explicit expressions for the supercurvature and the supertorsion, which differ from those of sec. 2 only by sign factors. The generalized Jacobi identity (involving commutators and anticommutators) applied to (3.11) gives the Bianchi identities, analoguous to (2.31) and (2.32)

$$\oint_{ABC} \left(\mathcal{D}_A T_{BC}{}^{D} + T_{AB}{}^{F} \cdot T_{FC}{}^{D} - R_{ABC}{}^{D} \right) = 0 \tag{3.12}$$

and

$$\oint_{ABC} \left(\mathcal{D}_A R_{BCD}{}^{F} + T_{AB}{}^{G} \cdot R_{GCD}{}^{F} \right) = 0 \tag{3.13}$$

The cyclic sums are taken with appropriate signs : a permutation of two bosonic or one bosonic and one fermionic index gives rise to a change in sign, while for two fermionic indices there is no change in sign.

In an affine superspace the geometric formulation requires covariance under general coordinate transformations of x^m and θ^μ (with the restriction that the new x^m should remain bosonic and the new θ^μ fermionic) and under tangent space transformations belonging to the general graded linear group. To specify the geometry further we restrict the tangent space group to be a local (x and θ dependent) Lorentz group. Considered as a matrix in the last two indices, the curvature belongs to the algebra of the tangent space group. Our restriction of this group implies that it should be a Lorentz transformation, the same for vectors and spinors

$$R_{CD,ab} = - R_{CD,ba}$$

$$R_{CD,\alpha\beta} = \frac{1}{2} R_{CD,ab} \left(\Sigma^{ab}\right)_{\alpha\beta} \tag{3.14}$$

$$R_{CD,a\beta} = R_{CD,\alpha b} = 0$$

The superconnection also belongs to the algebra of the tangent space group and satisfies analogous restrictions

$$\phi_{M\,ab} = -\,\phi_{M\,ba}$$

$$\phi_{M\alpha\beta} = \frac{1}{2}\,\phi_{Mab}\left(\Sigma^{ab}\right)_{\alpha\beta} \tag{3.15}$$

$$\phi_{M\,a\beta} = \phi_{M\alpha b} = 0$$

The final restriction on the geometry consists in **imposing** constraints on the supertorsion. One cannot prescribe all its components. This, together with the tangent group restrictions (3.14), would show too restrictive and, at best, one would be left only with the rigid superspace described at the beginning of this section. Instead, we impose the constraints

$$T_{\alpha\beta}{}^{c} = 2i\left(\gamma^{c}\right)_{\alpha\beta}\,, \qquad T_{\alpha\beta}{}^{\gamma} = 0$$

$$T_{ab}{}^{c} = 0 \qquad , \qquad T_{\alpha b}{}^{c} = 0 \tag{3.16}$$

but leave $T_{\alpha b}{}^{\gamma}$ and $T_{ab}{}^{\gamma}$ undertermined (this last components is related to the Rarita-Schwinger field strength $\psi_{ab}{}^{\gamma}$). The constraints (3.16) together with the restrictions (3.15) are sufficient to express all components of the superconnection in terms of the supervielbein and its derivatives. Therefore, they serve an analogous purpose in superspace as the constraints of vanishing torsion in a Riemann space.

Using the Bianchi identities (3.12) and (3.13) one can show [38, 40, 42] that (3.14) and (3.16) imply that all components of the supercurvature and supertorsion are expressible in terms of the three superfields $G_{\alpha\dot\beta}$, $W_{\alpha\beta\gamma}$, R and their conjugates. Here we have used two-component spinor notation (see Appendix A). $G_{\alpha\dot\beta}$ is hermitean, $W_{\alpha\beta\gamma}$ totally symmetric. For instance

$$R_{\alpha\beta,\gamma\delta} = -4\left(\varepsilon_{\alpha\gamma}\,\varepsilon_{\beta\delta} + \varepsilon_{\alpha\delta}\,\varepsilon_{\beta\gamma}\right) R^{*}$$

$$T_{\alpha,b,\gamma} \equiv T_{\alpha,\beta\dot\beta,\gamma} = \frac{i}{4}\left(\varepsilon_{\beta\gamma}\,G_{\alpha\dot\beta} - 3\,\varepsilon_{\alpha\gamma}\,G_{\beta\dot\beta} - 3\,\varepsilon_{\alpha\beta}\,G_{\gamma\dot\beta}\right)$$

$$T_{\alpha,b,\dot\gamma} \equiv T_{\alpha,\beta\dot\beta,\dot\gamma} = -2i\,\varepsilon_{\alpha\beta}\,\varepsilon_{\dot\beta\dot\gamma}\,R^{*} \tag{3.17}$$

$$T_{ab\gamma} = T_{\alpha\dot\alpha,\beta\dot\beta,\gamma} = -2\varepsilon_{\dot\alpha\dot\beta} W_{\alpha\beta\gamma} + \frac{1}{2}\varepsilon_{\alpha\beta}\left(\mathscr{D}_{\dot\alpha} G_{\gamma\dot\beta} + \mathscr{D}_{\dot\beta} G_{\gamma\dot\alpha}\right) + \frac{1}{2}\varepsilon_{\dot\alpha\dot\beta}\left(\varepsilon_{\beta\gamma}\mathscr{D}_{\alpha} R + \varepsilon_{\alpha\gamma}\mathscr{D}_{\beta} R\right)$$

The remaining content of the Bianchi identities is expressed by the differential relations

$$\mathscr{D}^{\alpha} W_{\alpha\beta\gamma} = \mathscr{D}_{\beta}{}^{\dot\varepsilon} G_{\gamma\dot\varepsilon} + \mathscr{D}_{\gamma}{}^{\dot\varepsilon} G_{\delta\dot\varepsilon} \;, \qquad \mathscr{D}^{\alpha} G_{\alpha\dot\beta} = \mathscr{D}_{\dot\beta} R^{*}$$

$$\mathscr{D}_{\dot\varepsilon} W_{\alpha\beta\gamma} = 0 \qquad\qquad \mathscr{D}_{\dot\varepsilon} R = 0 \tag{3.18}$$

$$\mathscr{D}_{\beta\dot\varepsilon} \equiv (\sigma^{a})_{\beta\dot\varepsilon}\, \mathscr{D}_{a}$$

The superfield $G_{\alpha\dot\alpha}$ has a simple physical meaning, which can be exhibited [47] by considering the coefficients of its expansion in θ . For instance, at the $\theta\bar\theta$ level one finds a tensor which contains the Einstein tensor, at the $\theta\theta\bar\theta$ level a spinor which is the Rarita - Schwinger operator (left-hand side of the Rarita-Schwinger equation). Similarly, R contains the contracted (scalar) curvature tensor, at the $\theta\theta$ level. $W_{\alpha\beta\gamma}$ contains the Weyl conformal spinor, at the θ level ; for $\theta = 0$, it contains the Rarita Schwinger field strength. Observe that some components of R are obtained from those of $G_{\alpha\dot\alpha}$ by tracing over certain indices : in superspace this is expressed by the differential identities (3.18) in θ .

In a general affine superspace one has the identity

$$\partial_{M}\left(E v^{A} E_{A}{}^{M}\right)(-1)^{a} = E\left(\mathscr{D}_{A} v^{A} + v^{B} T_{BA}{}^{A}\right)(-1)^{a} \tag{3.19}$$

where v^A is an arbitrary vector and

$$E = \det E_{M}{}^{A} \tag{3.20}$$

is the graded determinant of the supervielbein matrix, which is a density in superspace (see Appendix B). With our constraints (3.16) on the torsion, the last term $T_{BA}{}^{A}(-1)^{a}$ vanishes (some torsion components vanish and others cancel in the trace) and one has simply

$$\partial_{M}\left(E v^{A} E_{A}{}^{M}\right)(-1)^{a} = E\, \mathscr{D}_{A} v^{A}(-1)^{a} \tag{3.21}$$

Integrating over all superspace, and assuming that there are no boundary terms, one obtains

$$\int dx\, d\theta\; E\; \mathcal{D}_A v^A (-1)^a = 0 \tag{3.22}$$

a very useful formula which permits systematic use of integration by parts (for the definition of superspace integration see Appendix C).

The above constraints are naturally invariant under general coordinate transformations in superspace and local Lorentz transformations. They are also invariant under certain conformal transformations. In infinitesimal form these transformations involve an infinitesimal parameter function Σ which is covariantly chiral, i.e. satisfies

$$\mathcal{D}_{\dot\alpha}\Sigma = 0 \tag{3.23}$$

They are (Howe and Tucker [49])

$$\delta E_M{}^A = (\Sigma + \Sigma^*)\, E_M{}^A$$

$$\delta E_M{}^\alpha = (2\Sigma^* - \Sigma) E_M{}^\alpha - \frac{i}{2} E_M{}^a (\sigma_a)^{\alpha\dot\alpha} \mathcal{D}_{\dot\alpha} \Sigma^* \tag{3.24}$$

$$\delta E_M{}^{\dot\alpha} = (2\Sigma - \Sigma^*) E_M{}^{\dot\alpha} + \frac{i}{2} E_M{}^a (\sigma_a)^{\alpha\dot\alpha} \mathcal{D}_\alpha \Sigma$$

and

$$\delta\Omega_{M,\alpha\beta} = E_{M\alpha}\mathcal{D}_\beta\Sigma + E_{M\beta}\mathcal{D}_\alpha\Sigma + (\sigma^{ab})_{\alpha\beta} E_{Ma}\mathcal{D}_b(\Sigma + \Sigma^*) \tag{3.25}$$

For the determinant of the supervielbein, (3.24) imply

$$\delta E = 2(\Sigma + \Sigma^*) E \tag{3.26}$$

while curvature and torsion behave as given by

$$\delta R = (2\Sigma^* - 4\Sigma) R - \frac{1}{4} \mathcal{D}_{\dot\alpha} \mathcal{D}^{\dot\alpha} \Sigma^*$$

$$\delta G_{\alpha\dot\alpha} = -(\Sigma + \Sigma^*) G_{\alpha\dot\alpha} + i \mathcal{D}_{\alpha\dot\alpha} (\Sigma^* - \Sigma)$$

$$\delta W_{\alpha\beta\gamma} = -3\Sigma \cdot W_{\alpha\beta\gamma} \qquad (3.27)$$

Since the constraints are invariant under these transformations, we can say that the geometry (kinematics) is invariant under conformal transformations. The dynamics may of may not be, depending on the choice of Lagrangian. For instance, the action of supergravity, given in the next section, is only invariant under general coordinate transformations in superspace and under local Lorentz transformations, but not under conformal transformations. The action of conformal supergravity, also given below, is instead also conformally invariant.

Observe that the commutation relations (3.11) together with the solution (3.17) of the Bianchi identities (plus the analogous expressions [42] for the other components of the curvature and torsion), gives the commutator or anticommutator of any two covariant derivatives in terms of the superfields $G_{\alpha\dot\beta}$, $W_{\alpha\beta\gamma}$, R and their conjugates. Using these commutation relations it is easy to verify that the expression

$$\phi = (\mathcal{D}_{\dot\alpha} \mathcal{D}^{\dot\alpha} - 8R)\, U \qquad (3.28)$$

satisfies identically, for any U, the chirality constraint

$$\mathcal{D}_{\dot\alpha} \phi = 0 \qquad (3.29)$$

Viceversa, any chiral superfield can be written in the form (3.28). More generally, if $U_{\alpha\beta\gamma\ldots}$ has only undotted indices,

$$\phi_{\alpha\beta\gamma\ldots} = (\mathcal{D}_{\dot\alpha} \mathcal{D}^{\dot\alpha} - 8R)\, U_{\alpha\beta\gamma\ldots} \qquad (3.30)$$

implies

$$\mathcal{D}_{\dot\alpha} \phi_{\alpha\beta\gamma\ldots} = 0 \qquad (3.31)$$

and viceversa. One also has the identity

$$\left(\mathcal{D}_{\dot{\alpha}}\mathcal{D}^{\dot{\alpha}} - 8\mathcal{R}\right)\mathcal{D}_{\dot{\beta}}\Lambda^{\dot{\beta}} = 0 \tag{3.32}$$

These relations are generalizations to curved superspace of well known identities which, in flat superspace, follow from (3.6)

4. DYNAMICS IN SUPERSPACE

In the previous section we have completely specified the geometry of superspace by the tangent group restrictions (3.14) and (3.15) and by the torsion constraints (3.16). We refer to this specification of the geometry as kinematics : it is the analogue of the specification of the geometry in Einstein's theory as being Riemannian. Just as in Einstein's theory the dynamics is given by giving an action (or the field equations), we shall now specify the dynamics by giving the action for supergravity in superspace. It is simply

$$I = \int dx\, d\theta \;\; \det E_M{}^A \tag{4.1}$$

the superspace integral of the simplest density, the determinant of the supervielbein. The action (4.1) must be varied keeping the kinematical constraints satisfied. The variation of the action is

$$\delta I = \int dx\, d\theta\, E\; E_A{}^M\; \delta E_M{}^A\, (-1)^a \tag{4.2}$$

but the variations $\delta E_M{}^A$ are not arbitrary, because the constraints on the supertorsion restrict them. The simplest way to find these restrictions [40] is to perform an infinitesimal variation of the supertorsion constraints and to solve explicitly the linear equations for $\delta E_M{}^A$ which one obtains in this way. The result can be expressed in terms of the quantities

$$H_A{}^B = E_A{}^M\, \delta E_M{}^B \tag{4.3}$$

The trace entering (4.2) turns out to be [40]

$$H_A^A(-1)^a = -\frac{1}{4}[\mathcal{D}_\alpha, \mathcal{D}_{\dot\alpha}]v^{\alpha\dot\alpha} + v^b\left(T_{b\alpha}{}^{\alpha} + T_{b\dot\alpha}{}^{\dot\alpha}\right) + X + X^* \tag{4.4}$$

where

$$v^{\alpha\dot\alpha} = (\sigma^b)^{\alpha\dot\alpha}\, v_b \tag{4.5}$$

is an arbitrary superfield with a vector index and

$$\mathcal{D}_{\dot\alpha} X = 0 \tag{4.6}$$

The last two terms in (4.4) correspond to the ambiguity in the solution of the constraints due to their invariance under a conformal transformation (see 3.26). According to (3.28), (3.29), they can be written as

$$X + X^* = \left(\mathcal{D}_{\dot\alpha}\mathcal{D}^{\dot\alpha} - 8R\right)U + \left(\mathcal{D}^{\alpha}\mathcal{D}_{\alpha} - 8R^*\right)U^* \tag{4.7}$$

where U is arbitrary. If we now insert (4.4) with (4.7) into (4.2), observe that all terms with derivatives integrate to zero by (3.22) and remember that V_a and U are arbitrary, we find the equations of motion

$$R = R^* = 0 \tag{4.8}$$

and

$$T_{a\alpha}{}^{\alpha} + T_{a\dot\alpha}{}^{\dot\alpha} = 0 \tag{4.9}$$

This last equation becomes, using the second of (3.17),

$$G_{\alpha\dot\alpha} = 0 \tag{4.10}$$

Eqs. (4.8) and (4.10) are the full non-linear field equations of supergravity in superspace (for a detailed study of their component form in the linearized approximation see Ref. [47]).

If we use the identities (3.18), we see that they imply

$$\mathcal{D}^{\alpha} W_{\alpha\beta\gamma} = 0 \qquad , \qquad \mathcal{D}_{\dot{\alpha}} W_{\alpha\beta\gamma} = 0 \tag{4.11}$$

Applying $\mathcal{D}^{\dot{\alpha}}$ to the first of these equations and using the second and the commutation relations among covariant derivatives, it follows that

$$\mathcal{D}^{\alpha\dot{\alpha}} W_{\alpha\beta\gamma} = 0 \tag{4.12}$$

This equation looks formally like the conformal invariant (but inconsistent) equation for spin 3/2 in curved space expressed in terms of the Rarita-Schwinger field strength. In superspace, however, there is no consistency problem [38] , because the commutator of two covariant derivatives contains additional terms.

In view of (4.8) and (4.10), on the mass shell of supergravity all components of the supertorsion and of the supercurvature are expressible in terms of the symmetric spinor superfield $W_{\alpha\beta\gamma}$ satisfying (4.11) and of its derivatives (and, of cours, of the complex conjugate quantities). It is therefore easy to list all possible non vanishing scalars of given dimension and to discuss the question of possible perturbation theory divergences (see Ref. [47] : the discussion given there in the linearized approximation carries over now verbatim to the full non-linear case).

A general scalar in superspace, upon multiplication by the basic density (3.20), the determinant of the supervielbein, gives rise to a density which can be used as a Lagrangian. If the scalar is chiral, however, one has to proceed differently. In flat superspace a chiral superfield ϕ can be written in the form $\phi = D_{\dot{\alpha}} D^{\dot{\alpha}} U$. One can use ϕ as a Lagrangian by integrating only over $\theta^{\dot{\alpha}}$ (and x) but not over θ^{α}. Alternatively, one can integrate U over all four θ components (and x) with the same result. The first procedure has no analogue in curved superspace in a general gauge, the second does. Any chiral superfield ϕ , (3.29), can be written in the form (3.28), where U has a certain arbitrarity, given by (3.30). The action obtained from U can be transformed as follows

$$\int E\,\mathcal{U} \equiv -\frac{1}{8}\int \frac{E}{R}\left(\phi - \mathcal{D}_{\dot{\alpha}}\mathcal{D}^{\dot{\alpha}}\mathcal{U}\right) = -\frac{1}{8}\int \frac{E}{R}\,\phi \tag{4.13}$$

where the last equality follows by partial integration because R is chiral. We see that, in order to construct a Lagrangian density from a chiral superfield, we must multiply it by -E/8R, not by E. As an example, consider the chiral superfield $W_{\alpha\beta\gamma} W^{\alpha\beta\gamma}$. The corresponding action is proportional to

$$\int \frac{E}{R} W_{\alpha\beta\gamma} W^{\alpha\beta\gamma} \tag{4.14}$$

It is easy to verify that it is invariant under the conformal transformations (3.26) and (3.27). Indeed, (4.14) is the superspace form of the action for conformal supergravity. Observe that the action of non conformal supergravity (4.1) can be interpreted in two equivalent ways. One can think of it as being obtained by multiplying the identity superfield by E of by multiplying the chiral superfield R by E/R.

5. COUPLINGS TO SUPERSYMMETRIC MATTER

We first describe the coupling to the vector-spinor multiplet. In analogy with flat superspace, this matter supermultiplet is described by a real scalar superfield V which can undergo gauge transformations of the form

$$V \to V + i\Lambda - i\Lambda^*, \qquad \mathcal{D}_{\dot\alpha}\Lambda = 0 \tag{5.1}$$

The superfield

$$W_\alpha = \left(\mathcal{D}_{\dot\sigma}\mathcal{D}^{\dot\sigma} - 8R\right)\mathcal{D}_\alpha V$$

is gauge invariant under (5.1). This can be verified by using the commutation relations among the covariant derivatives. From (3.30) and (3.31) we see that it is chiral

$$\mathcal{D}_{\dot\alpha} W_\alpha = 0 \tag{5.2}$$

It also satisfies identically

$$\mathcal{D}^\alpha W_\alpha = \mathcal{D}_{\dot\alpha} W^{\dot\alpha} \tag{5.3}$$

which is the supersymmetric form of the electromagnetic Bianchi identities. The gauge invariant superfield W_α gives the spin 1/2 field of the supermultiplet for $\theta = 0$ and contains the electromagnetic field strength at the θ level. We can take as action, up to a proportionality constant,

$$\int E \left(\frac{1}{R} W^\alpha W_\alpha + \frac{1}{R^*} W_{\dot\alpha} W^{\dot\alpha} \right) \tag{5.4}$$

(observe that we use the density E/R appropriate to a chiral Lagrangian). The corresponding equation of motion is

$$\mathscr{D}^\alpha W_\alpha + \mathscr{D}_{\dot\alpha} W^{\dot\alpha} = 0 \tag{5.5}$$

which, combined with (5.3), gives

$$\mathscr{D}^\alpha W_\alpha = \mathscr{D}_{\dot\alpha} W^{\dot\alpha} = 0 \tag{5.6}$$

The action (5.4) is invariant under the gauge transformation (5.1). It is also invariant under conformal transformations provided one assumes the conformal property

$$\delta V = 0 \tag{5.7}$$

which implies

$$\delta W_\alpha = -3 \Sigma W_\alpha \tag{5.8}$$

This conformal invariance can be easily verified by combining (5.8) with (3.26) and (3.27) and integrating by parts.

If one adds the actions (4.1) and (5.4), one has the superspace formulation of the theory given first in component form in Ref. [50, 51]. Conformal invariance is spoiled by the supergravity term. An interesting alternative is obtained by adding to (5.4) the action

$$\int E\, e^V \;. \tag{5.9}$$

Expanding the exponential, the first term gives the supergravity action and the second term (linear in V) is a generalization of the so called Fayet-Illiopoulos terms, which in rigid supersymmetry gives rise to spontaneous breaking of supersymmetry [9] . The action (5.9) is neither gauge invariant (under (5.1)) nor conformal invariant (under (3.26), (5.7)) but is invariant under the combination of a gauge transformation and a conformal transformation provided their parameters (which are both chiral) are related

$$\Sigma + i\Lambda = 0 \tag{5.10}$$

Since (5.4) is both gauge and conformal invariant one can add it to (5.9) and one obtains the curved superspace version of the Fayet-Illiopoulos term. It is gauge invariant, if we define as gauge transformation the combination specified by (5.10). Using this gauge invariance one can partially choose the gauge (so called Wess-Zumino gauge) so that the gauge dependent fields of the vector spinor supermultiplet (except for the vector field) vanish. One is left with an ordinary gauge invariance for the vector field and the theory takes the polynomial form originally given by Freedman [52] . The superspace description given here corresponds to the component formulation of Stelle and West [53] .

We now consider the coupling of supergravity to spinor-scalar-pseudoscalar matter. This supermultiplet is described by a chiral superfield ϕ

$$\mathcal{D}_{\dot\alpha}\,\phi = 0 \qquad , \qquad \mathcal{D}_{\alpha}\,\phi^* = 0 \tag{5.11}$$

As action, to add to that of supergravity, one can take

$$\int E\,\phi\,\phi^* + E\left(\frac{1}{R}F(\phi) + \frac{1}{R^*}F(\phi^*)\right) \tag{5.12}$$

where $F(\phi)$, the potential, is a (real) function of ϕ . In order to derive the equations of motion it is simplest to satisfy the conditions (5.11) by setting, according to (3.28),

$$\begin{aligned} \phi &= \left(\mathcal{D}_{\dot\alpha}\mathcal{D}^{\dot\alpha} - 8R\right)\mathcal{U} \\ \phi^* &= \left(\mathcal{D}^{\alpha}\mathcal{D}_{\alpha} - 8R^*\right)\mathcal{U}^* \end{aligned} \tag{5.13}$$

The superfields U and U^* can be subjected to arbitrary variations. One finds the equation of motion

$$\left(\mathcal{D}^\alpha \mathcal{D}_\alpha - 8R^*\right)\phi - 8F'(\phi^*) = 0 \tag{5.14}$$

and its complex conjugate. It is not difficult to verify that (5.11) and (5.14) imply that the supercurrent

$$J_{\alpha\dot\alpha} = i\,\phi\left(\overrightarrow{\mathcal{D}}_{\alpha\dot\alpha} - \overleftarrow{\mathcal{D}}_{\alpha\dot\alpha}\right)\phi^* + \frac{1}{2}\mathcal{D}_\alpha\phi\cdot\mathcal{D}_{\dot\alpha}\phi^* + 2G_{\alpha\dot\alpha}\,\phi\phi^* \tag{5.15}$$

satisfies

$$\mathcal{D}^\alpha J_{\alpha\dot\alpha} = \mathcal{D}_{\dot\alpha}\, S^* \tag{5.16}$$

where

$$S = 6\,F(\phi) - 2\,\phi\,F'(\phi)\,, \qquad \mathcal{D}_{\dot\alpha} S = 0 \tag{5.17}$$

The matter superfields $J_{\alpha\dot\alpha}$ and S, S^* (which contain the energy momentum tensor and the spinor current [54] satisfy an identity analogous to that (see (3.18)) satisfied by the supergravity superfields $G_{\alpha\dot\alpha}$ and R, R^* (which contain the Einstein tensor and the Rarita-Schwinger operator). Indeed the supergravity equations state the proportionality of these two sets of fields

$$G_{\alpha\dot\alpha} = c\cdot J_{\alpha\dot\alpha} \quad , \quad R = c\, S \tag{5.18}$$

If one attibutes to the superfield ϕ the conformal property

$$\delta\phi = -2\Sigma\ \phi \tag{5.19}$$

one finds that the first (kinetic) term in the action (5.12) is conformally invariant. For the potential term, one obtains

$$\delta\int\frac{E}{R}F(\phi) = 2\int\frac{E}{R}\cdot S\cdot\Sigma \tag{5.20}$$

The superfield S, given by (5.17), appears as a measure of the violation of conformal invariance in the matter action. For $F(\phi) \propto \phi^3$, one has S = 0 and the entire matter action is conformally invariant. In general, however, for instance if the matter supermultiplet has a mass ($F \propto \phi^2$)R will not vanish. The coupling described by (5.12) is only one of many possible forms which have been studied in detail in terms of component fields (see the lectures by S. Ferrara and P. Van Nieuwenhuizen, where many references can be found). It is an amusing exercise to translate the other forms into superspace language.

6. AUXILIARY FIELDS FROM SUPERSPACE

Recently, a minimal set of auxiliary fields for simple supergravity has been found, and a tensor calculus developed [57, 58] (see also P. Van Nieuwenhuizen's lectures at this school). This work generalizes to the full non linear theory some earlier results obtained in the linearized approximation [45, 47]. The auxiliary field formalism is contained in the superspace formulation as given in the previous sections. There are two ways of extracting it. The first is to solve the superspace constraints on the torsion [59, 60] and will be discussed at the end of this section. The second consists in using the constraints, without actually solving them, by means of the Bianchi identities. We now describe briefly this second method.

Let us first identify, in superspace, the transformations of local supersymmetry. For any superfield V, with any number of tangent space indices, which we do not indicate explicitly, consider the transformation

$$\delta V = \xi^A \mathcal{D}_A V \tag{6.1}$$

with the covariant derivative (3.10). From (3.7), (3.8) we see that it can be interpreted as the combination of a general coordinate transformation and of a Lorentz transformation of respective parameters

$$\xi^M = \xi^A E_A{}^M \qquad L_{ab} = -\xi^M \phi_{M,ab} \tag{6.2}$$

These parameters are field dependent, since we take ξ^A to be field independent. The commutator of two transformations can be obtained using

$$\delta_2 \delta_1 V = \xi_1^A \xi_2^B \mathcal{D}_B \mathcal{D}_A V \tag{6.3}$$

and the commutator of two covariant derivatives (3.11). One finds

$$[\delta_2, \delta_1] V = -\xi_1^A \xi_2^B \left(\frac{1}{2} R_{BA,cd} \mathcal{M}^{cd} + T_{BA}{}^C \mathcal{D}_C \right) V \tag{6.4}$$

This is the combination of a field dependent transformation like (6.1) and of a local Lorentz transformation ($\mathcal{M}^{cd}$ is the infinitesimal Lorentz operator). For the supervielbein itself (6.2) give

$$\delta E_M{}^A = \mathcal{D}_M \xi^A + \xi^B T_{BM,}{}^A \tag{6.5}$$

It is convenient to work in the gauge in which, for $\theta = 0$, the components of the supervielbein take the values

$$E_m{}^a = e_m{}^a \qquad E_m{}^\alpha = \frac{1}{2} \psi_m{}^\alpha \tag{6.6}$$

$$E_\mu{}^a = 0 \qquad E_\mu{}^\alpha = \delta_\mu{}^\alpha \tag{6.7}$$

and $\phi_{\mu,ab}$ vanishes, while $\phi_{m,ab}$ equals the co-ordinate space connection*. This still leaves a considerable gauge arbitrariness which can be used to give a convenient form to the higher θ coefficients of the supervielbein, but we shall not need these coefficients here. For a supersymmetry transformation we take $\xi^a = 0$ and $\xi^\alpha = \zeta^\alpha(x)$ for $\theta = 0$. Using the constraints (3.16) on the torsion and setting $\theta = 0$, it is easy to see that (6.5) will preserve the gauge conditions (6.7) provided

$$\xi^a = -2i \bar{\zeta} \gamma^a \theta \tag{6.8}$$

On the other hand, one finds

$$\delta e_m{}^a = \frac{1}{2} \psi_m{}^\alpha \zeta^\beta 2i (\gamma^a)_{\alpha\beta} = i \bar{\zeta} \gamma^a \psi_m \tag{6.9}$$

* By the constraint $T_{ab}{}^c = 0$, $\phi_{m,ab}$ contains the contribution from the spin 3/2 field.

and

$$\frac{1}{2}\delta\psi_m{}^\alpha = \mathcal{D}_m\zeta^\alpha + e_m{}^c\,\zeta^\beta\,T_{\beta c}{}^\alpha \tag{6.10}$$

where $\mathcal{D}_m$ is just the coordinate space covariant derivative (1.10). Using (3.17), the torsion components $T_{\beta c}{}^\alpha$, are expressible in terms of the superfields G_a and R. We identify the values of these superfields at $\theta = 0$ with the auxiliary fields

$$G_a = -\frac{1}{3}b_a \ , \qquad R = -\frac{1}{6}M \ , \qquad \theta = 0 \tag{6.11}$$

The complex field M equals (1/2)(M+iN) of Ref. [58]. We shall now use two-component spinor notation. One finds

$$\delta\psi_{m\alpha} = 2\mathcal{D}_m\zeta_\alpha - i b_m\zeta_\alpha - \frac{i}{3}(\sigma_m\bar{\sigma}_n\zeta)_\alpha b^n - \frac{i}{3}(\sigma_m\bar{\zeta})_\alpha M \tag{6.12}$$

Equations (6.9) and (6.12) agree with the transformation laws postulated in [57, 58].

The transformation laws of the auxiliary fields can also be derived [41], but we shall not do it here. Instead we consider a chiral superfield ϕ

$$\mathcal{D}_{\dot\alpha}\,\phi = 0 \tag{6.13}$$

Its components are defined as the values for $\theta = 0$ of certain spinorial tangent space derivatives

$$A = \phi \ , \quad \chi_\alpha = \mathcal{D}_\alpha\phi \ , \quad F = -\frac{1}{4}\mathcal{D}^\alpha\mathcal{D}_\alpha\phi \ , \ (\theta = 0) \tag{6.14}$$

In curved superspace this definition is more convenient than that in terms of the θ expansion and gives component fields with well defined Lorentz character. These components are complete in the sense that their knowledge determines the chiral superfield. By means of the known commutation relations, higher spinorial derivatives can be reduced to the components above and to space derivatives of them. Applying (6.1) and setting $\theta = 0$ one obtains, from (6.13),

$$\delta A = \xi^{\alpha} \mathcal{D}_{\alpha} \phi = \zeta^{\alpha} \chi_{\alpha} \tag{6.15}$$

Similarly

$$\begin{aligned} \delta\chi_{\alpha} &= \left(\xi^{\beta} \mathcal{D}_{\beta} - \xi^{\dot\beta} \mathcal{D}_{\dot\beta}\right) \mathcal{D}_{\alpha} \phi \\ &= -\frac{1}{2} \xi_{\alpha} \mathcal{D}^{\gamma} \mathcal{D}_{\gamma} \phi + 2i\, \xi^{\dot\beta} \mathcal{D}_{\alpha\dot\beta} \phi \end{aligned} \tag{6.16}$$

where we have used (6.13) and the commutation relations (3.11) among the covariant derivatives. Now we observe that, for any component, the vectorial tangent space covariant derivative in superspace becomes, for θ = 0, the supercovariant derivative of the component formalism *. Therefore, (6.16) gives

$$\delta\chi_{\alpha} = 2 \zeta_{\alpha} F + 2i\, \zeta^{\dot\beta} \hat{\mathcal{D}}_{\alpha\dot\beta} A \tag{6.17}$$

$$\hat{\mathcal{D}}_{a} A = \mathcal{D}_{a} A - \frac{1}{2} \psi_{a} \chi \tag{6.18}$$

Similarly,

$$-4\delta F = \left(\xi^{\alpha} \mathcal{D}_{\alpha} - \xi^{\dot\alpha} \mathcal{D}_{\dot\alpha}\right) \mathcal{D}^{\beta} \mathcal{D}_{\beta} \phi \tag{6.19}$$

---

* For instance

$$\mathcal{D}_{a} \phi = E_{a}{}^{m} \mathcal{D}_{m} \phi + E_{a}{}^{\mu} \mathcal{D}_{\mu} \phi$$

$$\mathcal{D}_{\alpha} \phi = E_{\alpha}{}^{m} \mathcal{D}_{m} \phi + E_{\alpha}{}^{\mu} \mathcal{D}_{\mu} \phi$$

For θ = 0, one finds $\mathcal{D}_{\alpha} \phi = \delta_{\alpha}{}^{\mu} \mathcal{D}_{\mu} \phi$ and

$$\mathcal{D}_{a} \phi = e_{a}{}^{m} \partial_{m} \phi - \frac{1}{2} \psi_{a}{}^{\alpha} \mathcal{D}_{\alpha} \phi = \mathcal{D}_{a} A - \frac{1}{2} \psi_{a}{}^{\alpha} \chi_{\alpha}$$

Here the commutation relation among the covariant derivatives introduce curvative and torsion terms expressible in G_a and R^*

$$\delta F = -2\xi^{\alpha} R^* \mathscr{D}_{\alpha}\phi + \xi^{\dot{\alpha}}\left(i\mathscr{D}_{\alpha\dot{\alpha}}\mathscr{D}^{\alpha}\phi + \frac{1}{2}G_{\alpha\dot{\alpha}}\mathscr{D}^{\alpha}\phi\right) \tag{6.20}$$

For $\theta = 0$ this gives

$$\delta F = \frac{1}{3}M^*\zeta^{\alpha}\chi_{\alpha} - \frac{1}{2}\zeta^{\dot{\alpha}} b_{\alpha\dot{\alpha}}\chi^{\alpha} + i\zeta^{\dot{\alpha}}\hat{\mathscr{D}}_{\alpha\dot{\alpha}}\chi^{\alpha} \tag{6.21}$$

$$\hat{\mathscr{D}}_a\chi_{\alpha} = \mathscr{D}_a\mathscr{D}_{\alpha}\phi = \mathscr{D}_a\chi_{\alpha} - \psi_{a\alpha}F - i\psi_a{}^{\dot{\beta}}\hat{\mathscr{D}}_{\alpha\dot{\beta}}A \tag{6.22}$$

Equations (6.15), (6.17), and (6.21) agree with Ref.[57] . Observe that the terms with the auxiliary fields appear naturally and are not part of the supercovariant derivative. The components of the chiral multiplet

$$\Xi = \left(\mathscr{D}_{\dot{\alpha}}\mathscr{D}^{\dot{\alpha}} - 8R\right)\phi^* , \quad \mathscr{D}_{\dot{\alpha}}\phi = 0, \quad \mathscr{D}_{\dot{\alpha}}\Xi = 0 \tag{6.23}$$

can be calculated by the same technique in terms of those of ϕ . One uses the definitions (6.14) on Ξ and commutes through the covariant derivatives. This is, up to a factor, the "kinetic multiplet" of Ref. [57] . For other applications of this technique, in particular for the derivation of the transformation properties of a vector-spinor multiplet we refer to [41] .

We now indicate briefly the procedure by which the constraints can actually be solved. We give only the starting point of this procedure and refer to [60] for further details. It is convenient to choose a gauge in which certain components of the inverse vielbein have a simple form

$$E_{\dot{\alpha}}{}^{M} = \rho\,\delta_{\dot{\alpha}}{}^{M} \tag{6.24}$$

where ρ is a function of x, θ and $\bar{\theta}$. Then, for a scalar superfield U

$$\mathcal{D}^{\dot\alpha}\,\mathcal{U} = \rho\,\partial^{\dot\alpha}\mathcal{U} \tag{6.25}$$

where $\partial^{\dot\alpha}$ is an ordinary anticommuting derivative. We see that in this gauge a chiral superfield is actually independent of $\bar\theta$: we call it the chiral gauge.

Let us now take the torsion constraint

$$T_{\dot\alpha\dot\beta}{}^{N} = 0 \tag{6.26}$$

or, more explicitely

$$E_{\dot\alpha}{}^{M}\partial_M E_{\dot\beta}{}^{N} + E_{\dot\beta}{}^{M}\partial_M E_{\dot\alpha}{}^{N} + \left(\phi_{\dot\alpha\dot\beta}{}^{\dot\gamma} + \phi_{\dot\beta\dot\alpha}{}^{\dot\gamma}\right)E_{\dot\gamma}{}^{N} = 0 \tag{6.27}$$

It can be solved for the connection, with the result

$$\phi_{\dot\alpha\dot\beta\dot\gamma} = \partial_{\dot\beta}\rho\ \varepsilon_{\dot\alpha\dot\gamma} + \partial_{\dot\gamma}\rho\ \varepsilon_{\dot\alpha\dot\beta} \tag{6.28}$$

This component of the connection enters in the expression for $R_{\dot\varepsilon\dot\alpha\dot\beta\dot\gamma}$. Making use of the complex conjugate of the first of (3.17), this is expressible in terms of the chiral superfield R . In this way one finds

$$\partial_{\dot\alpha}\partial^{\dot\alpha}\rho^2 = -\,8R \tag{6.29}$$

which is obviously chiral in the gauge (6.25). This equation shows that it would not have been possible, in general, to choose the gauge so that in (6.24) ρ was equal to unity. We have now all the ingredients to calculate an expression of the form (3.28). Using (6.24), (6.25), (6.28) and (6.29), the various terms quadratic in ρ and its derivatives combine in the very simple expression

$$(\mathcal{D}_{\dot\gamma}\mathcal{D}^{\dot\gamma} - 8R)\,\mathcal{U} = \partial_{\dot\gamma}\partial^{\dot\gamma}(\rho^2\mathcal{U}) \tag{6.30}$$

Again, the result is obviously chiral, as we know it must be. In the chiral gauge an integral of the form (4.13) takes a particularly simple form. Remembering (see Appendix C) that a θ integration can be written as a θ derivative, we have

$$-\frac{1}{8}\int dx\, d\theta\, d\bar{\theta}\, \frac{E}{R}\phi = -\frac{1}{8}\int dx\, d\theta\, \partial_{\dot{\alpha}}\partial^{\dot{\alpha}}\left(\frac{E}{R}\phi\right)$$
$$= -\frac{1}{8}\int dx\, d\theta \left(\partial_{\dot{\alpha}}\partial^{\dot{\alpha}} E\right)\frac{1}{R}\phi \equiv \int dx\, d\theta\, \mathcal{E}\phi \tag{6.31}$$

where we have used the chirality conditions

$$\partial_{\dot{\alpha}}\phi = \partial_{\dot{\alpha}} R = 0 \tag{6.32}$$

We see that the density $\mathcal{E}$ appropriate to a chiral superfield satisfies

$$\partial_{\dot{\alpha}}\partial^{\dot{\alpha}} E = -8 R \mathcal{E} \tag{6.33}$$

or, from (6.29)

$$\mathcal{E} = \partial_{\dot{\alpha}}\partial^{\dot{\alpha}}\left(E/\rho^2\right) \tag{6.34}$$

These relations can indeed be verified explicitly in terms of components.

The chiral gauge can be obtained from a general gauge by making a suitable Lorentz transformation accompanied by a general coordinate transformation. One could have chosen a different chiral gauge obtained from the above by complex conjugation, in which the vielbein with an undotted index had a simple form analogous to (6.24). These two chiral gauges are the analogues in curved superspace, of the two representations in flat superspace (called basis 1 and 2 in Ref. [34]), in which the two types of chiral fields are respectively independent of θ and of $\bar{\theta}$. In flat superspace these two representations are related by a pure imaginary translation. In curved superspace the two chiral gauges are related by a pure imaginary general coordinate transformation. Two quantities, like $E_\alpha{}^M$ and $E_{\dot{\alpha}}{}^M$, which ordinarily are related by complex conjugation, in the gauge (6.24) are obtained from each other by performing a complex conjugation followed by a pure imaginary general coordinate transformation. Thus, when (6.25) is valid, one has

$$E_\alpha{}^M \partial_M = e^{-i U^M \partial_M} \rho^* \partial_\alpha \, e^{i U^M \partial_M} \qquad (6.35)$$

where $-iU^M(x, \theta)$ is the pure imaginary parameter of the general coordinate transformation relating the two gauges. The two equations (6.24) and (6.35) express all components of the inverse supervielbein with spinorial tangent space indices in terms of the two superfields ρ and U^M. Using these expressions in the supertorsion constraints one can express the other components of the supervielbein (and therefore also all components of the super connection) in terms of the same two superfields. Furthermore, it is easy to see that the conditions (6.24) determining the chiral gauge does not fix the gauge completely. Using the remaining gauge arbitrariness one can impose conditions such that everything is expressed in terms of the components U^m alone (vectorial values of M only). At this point the constraints are completely solved in terms of this superfield U^m and the action of supergravity can be expressed in terms of it. This superfield U^m is the generalisation of that used in the linearized formulation of supergravity [45, 47]. To establish the connection with the component formalism with auxiliary fields one must fix the gauge even further (so called Wess-Zumino gauge) so that the first few coefficients in the θ, $\bar{\theta}$ expansion of the superfield U^m vanish. One is finally reduced to the fields M (complex), $e_a{}^m$, $\psi_\alpha{}^m$ and b^m which correspond to those defined in the first part of this section, and in the component fields formalism. One must realize, however, that the fields defined in terms of the expansion of U^m and those defined in the first part of this section are not actually identical, rather they are functions of each other and thus provide equivalent descriptions of supergravity.

ADDED NOTE

Since these lectures were delivered some further progress has been made. Ref.[42], containing the full solution of the superspace Bianchi identities, has appeared as a preprint from Karlsruhe University, submitted to Nuclear Physics.

E. Cremmer, B. Julia and J. Scherk, Phys. Lett. 76 B (1978)409, have constructed the Lagrangian of simple supergravity in 11 space-time dimensions. Using this result, E. Cremmer and B. Julia, LPTENS 78/23, (Ecole Normale, Paris, Preprint), have given a Lagrangian for O(8) supergravity in 4 space-time dimensions.

A. Two Component Spinors.

The two component spinor notation is very convenient when dealing with higher spin fields in four dimensions. Its usefulness in gravitation, even when there are no half-integral spin fields, has been recognized for some time (see e.g. Penrose's paper [56] which contains many references). It is even more natural in supergravity where such fields occur.

To establish contact with the usual four component spinor notation, one uses a representation of the gamma matrices where γ_5 is diagonal (Weyl representation)

$$\gamma_5 = \begin{pmatrix} i & 0 \\ 0 & -i \end{pmatrix} \tag{A.1}$$

and

$$\gamma_a = \begin{pmatrix} 0 & i\sigma_a \\ i\bar{\sigma}_a & 0 \end{pmatrix} \tag{A.2}$$

Here

$$\begin{aligned} (\sigma_a)_{\alpha\dot{\beta}} &= (1, \vec{\sigma}) \\ (\bar{\sigma}_a)^{\dot{\alpha}\beta} &= (1, -\vec{\sigma}) \end{aligned} \tag{A.3}$$

where $\vec{\sigma}$ denotes the usual Pauli matrices. The undotted indices take the values 1 and 2 and refer to the spinorial representation SL(2, C) of the Lorentz group, the dotted indices to its complex conjugate. Dotted and undotted indices can be raised of lowered (always fro the left) by means of the antisymmetric constant bispinor.

$$\begin{aligned} &\psi^{\alpha} = \varepsilon^{\alpha\beta}\psi_{\beta} \quad , \quad \bar{\psi}^{\dot{\alpha}} = \varepsilon^{\dot{\alpha}\dot{\beta}}\bar{\psi}_{\dot{\beta}} \\ &\varepsilon^{12} = -\varepsilon^{21} = \varepsilon^{\dot{1}\dot{2}} = -\varepsilon^{\dot{2}\dot{1}} = 1 \\ &\varepsilon^{\alpha\beta} = -\varepsilon_{\alpha\beta} \quad , \quad \varepsilon^{\dot{\alpha}\dot{\beta}} = -\varepsilon_{\dot{\alpha}\dot{\beta}} \end{aligned} \tag{A.4}$$

For instance, for anticommuting two component spinors,

$$\theta^{\alpha}\psi_{\alpha} = \varepsilon^{\alpha\beta}\theta_{\beta}\psi_{\alpha} = -\theta_{\beta}\varepsilon^{\beta\alpha}\psi_{\alpha} = -\theta_{\beta}\psi^{\beta} = \psi^{\beta}\theta_{\beta} \tag{A.5}$$

In the Weyl representation a four component Majorana spinor has the form

$$\psi = \begin{pmatrix} \psi_\alpha \\ \bar{\psi}^{\dot\alpha} \end{pmatrix} \qquad \bar{\psi}_{\dot\alpha} = (\psi_\alpha)^* \tag{A.6}$$

where the star means complex conjugation. The usual four dimensional Pauli invariant of two Majorana spinors is

$$\bar{\psi}\chi = \psi\gamma^0\chi = -i\left(\psi^\alpha\chi_\alpha + \bar{\psi}_{\dot\alpha}\bar{\chi}^{\dot\alpha}\right) \tag{A.7}$$

The sigma matrices satisfy

$$\left[(\sigma_a)_{\alpha\dot\beta}\right]^* = (\bar{\sigma}_a)_{\dot\alpha\beta} \tag{A.8}$$

since they are hermitean, this implies that the transposed of σ_a equals $\bar{\sigma}_a$. Clearly

$$\bar{\sigma}_a\sigma_b + \bar{\sigma}_b\sigma_a = \sigma_a\bar{\sigma}_b + \sigma_b\bar{\sigma}_a = -2\eta_{ab} \tag{A.9}$$

A vector can be transformed into a mixed spinor with one undotted and one dotted index

$$V_{\alpha\dot\alpha} = V^a(\sigma_a)_{\alpha\dot\alpha} \quad , \quad V^a = -\frac{1}{2}(\bar{\sigma})^{\dot\alpha\alpha}V_{\alpha\dot\alpha} \tag{A.10}$$

If V^a is real, then $V_{\alpha\dot\alpha}$ is a hermitean matrix. In general, irreducible spinors have a certain number of dotted and undotted indices with particular symmetry and reality properties. For instance, an antisymmetric tensor $F_{ab} = -F_{ba}$, can be decomposed as follows

$$(\sigma^a)_{\alpha\dot\alpha}(\sigma^b)_{\beta\dot\beta}F_{ab} = \varepsilon_{\alpha\beta}F_{\dot\alpha\dot\beta} + \varepsilon_{\dot\alpha\dot\beta}F_{\alpha\beta} \tag{A.11}$$

where the spinors $F_{\alpha\beta}$ and $F_{\dot\alpha\dot\beta}$ are symmetric and, if F_{ab} is real, they are complex conjugates of each other. Similarly, the Riemann tensor can be decomposed into irreducible spinors [56] .

B. Superdeterminants.

Let M be a matrix of the form

$$M = \begin{pmatrix} A & \Gamma \\ \Delta & B \end{pmatrix} \tag{B.1}$$

where the submatrices A and B have bosonic matrix elements while Γ and Δ have fermionic elements (respectively even and odd elements of a Grassmann algebra). Matrices like M can be combined linearly and multiplied with each other and the generic matrix of this kind has an inverse. The supervielbein matrix is of this type. We define the trace by

$$\mathrm{Tr}\, M = \mathrm{Tr}\, A - \mathrm{Tr}\, B \tag{B.2}$$

The reason for the minus sign in this definition is that we wish the basic property of the trace

$$\mathrm{Tr}\, M_1 M_2 = \mathrm{Tr}\, M_2 M_1 \tag{B.3}$$

to hold. Indeed we have

$$\begin{aligned} \mathrm{Tr}\, M_1 M_2 &= \mathrm{Tr}\left(A_1 A_2 + \Gamma_1 \Delta_2\right) - \mathrm{Tr}\left(\Delta_1 \Gamma_2 + B_1 B_2\right) \\ &= \mathrm{Tr}\left(A_2 A_1 - \Delta_2 \Gamma_1\right) - \mathrm{Tr}\left(-\Gamma_2 \Delta_1 + B_2 B_1\right) \\ &= \mathrm{Tr}\, M_2 M_1 \end{aligned} \tag{B.4}$$

The determinant can be defined from

$$\det M = \exp \mathrm{Tr} \ln M \tag{B.5}$$

Because of (B.3), it satisfies the product property

$$\det\left(M_1 M_2\right) = \left(\det M_1\right)\left(\det M_2\right) \tag{B.6}$$

Indeed, writing

$$\ln M = N\ , \tag{B.7}$$

one has

$$\ln(M_1 M_2) = N_1 + N_2 + \frac{1}{2}[N_1, N_2] + \cdots \tag{B.8}$$

where the dots denote multiple commutators. Taking the trace, all commutator terms vanish by (B.3), and one has

$$\mathrm{Tr} \ln (M_1 M_2) = \mathrm{Tr} \ln M_1 + \mathrm{Tr} \ln M_2 \tag{B.9}$$

which establishes (B.6). It follows from (B.5) that, if

$$M = 1 + X \tag{B.10}$$

where 1 is the unit matrix and X is infinitesimal then

$$\det M = 1 + \mathrm{Tr}\, X \tag{B.11}$$

More generally, one finds in the standard way, combining (B.6) and (B.11), that

$$\delta \det M = (\det M)\, \mathrm{Tr}\, M^{-1} \delta M \tag{B.12}$$

for any infinitesimal variation δM . An explicit form for the determinant can be obtained by writing M in the standard form

$$M = \begin{pmatrix} C & \Sigma \\ 0 & 1 \end{pmatrix} \begin{pmatrix} 1 & 0 \\ \phi & D \end{pmatrix} \tag{B.13}$$

Comparing with (B.1) one finds

$$\begin{aligned} C &= A - \Gamma B^{-1} \Delta , & \Sigma &= \Gamma . B^{-1} \\ \phi &= \Delta , & D &= B \end{aligned} \tag{B.14}$$

On the other hand (B.13) gives, using (B.2), (B.5) and (B.6),

$$\det M = \det C / \det D \tag{B.15}$$

Therefore

$$\det M = \frac{\det(A - \Gamma B^{-1}\Delta)}{\det B} \tag{B.16}$$

An equivalent form is

$$\det M = \frac{\det A}{\det(B - \Delta A^{-1}\Gamma)} \tag{B.17}$$

In (B.16) and (B.17) the determinants occurring in the right hand side are ordinary determinants of matrices with bosonic elements.

It is now easy to see that the determinant of the supervielbein matrix

$$E = \det E_M{}^A \tag{B.10}$$

transforms like a density under general coordinate transformations. Infinitesimally the supervielbein transforms as

$$\delta E_M{}^A = \xi^L \partial_L E_M{}^A + \partial_M \xi^L E_L{}^A \tag{B.19}$$

Using (B.12) one finds

$$\begin{aligned} \delta E &= \xi^L \partial_L E + E \cdot E_A{}^M \partial_M \xi^L E_L{}^A (-1)^a \\ &= \xi^L \partial_L E + E \cdot \partial_L \xi^L (-1)^\ell \\ &= \partial_L (\xi^L E)(-1)^\ell \end{aligned} \tag{B.20}$$

which is the transformation property of a superspace density. Under local Lorentz transformations E is invariant.

In this Appendix we have followed the first of Ref. [35]. The concept of superdeterminant was developped independently (and earlier) by F. A. Berezin.

C. Integration in Superspace.

Integration over anticommuting variables has been used in quantum field theory for a long time. Its properties have been codified by Berezin [55] . We have in mind here the analogue of the definite integral (from $-\infty$ to $+\infty$) for ordinary functions, not the analogue of the primitive of a function. Consider a function of one variable. When it exists, the definite integral is translationally invariant

$$\int_{-\infty}^{+\infty} dx\, f(x+a) = \int_{-\infty}^{+\infty} dx\, f(x) \tag{C.1}$$

For functions of one anticommuting variable θ , which means that

$$\theta^2 = 0 \tag{C.2}$$

we shall postulate the analogous property

$$\int d\theta\, f(\theta+\alpha) = \int d\theta\, f(\theta) \tag{C.3}$$

Now, because of (C.2), the most general function of one anticommuting variable θ is linear in θ

$$f(\theta) = a + \theta\beta \tag{C.4}$$

Therefore (C.3) means

$$\int d\theta \left(a + \theta\beta + \alpha\beta\right) = \int d\theta \left(a + \theta\beta\right) \tag{C.5}$$

or, assuming linearity,

$$\int d\theta\, \alpha\beta = 0 \tag{C.6}$$

This equation is generalized by postulating that the integral of any constant vanishes. In order to obtain a non trivial result in (C.5) one must then assume that the integral of θ itself does not vanish, and it is usual to normalize it to 1. In conclusion, one assumes the basic relations

$$\int d\theta = 1 \qquad\qquad \int d\theta \cdot \theta = 1 \tag{C.7}$$

together with linearity. Observe that this corresponds to taking

$$\int d\theta \, f(\theta) \equiv \frac{\partial}{\partial\theta} f(\theta) \tag{C.8}$$

In spite of this identity, it is often convenient to use the integral notation, expecially when there are accompanying bosonic integrations. In many cases the integral notation is more inspiring for instance there exists a theory of Fourier transforms (with completeness relations etc.) for functions of anticommuting variables. The delta function for anticommuting variables is

$$\delta(\theta-\theta') = \theta - \theta' \tag{C.9}$$

Indeed, one sees from (C.7) that

$$\int d\theta \, (\theta-\theta') \, (a+\theta\beta) = a + \theta'\beta \tag{C.10}$$

The integral of a derivative vanishes

$$\int d\theta \, \frac{\partial}{\partial\theta} f(\theta) = 0 \tag{C.11}$$

This follows e.g. from (C.8) since the square of $\frac{\partial}{\partial\theta}$ vanishes. Therefore, integration by parts can be used

$$\int d\theta \, f(\theta) \frac{\partial}{\partial\theta} g(\theta) = \mp \int d\theta \left[\frac{\partial}{\partial\theta} f(\theta) \right] g(\theta) \tag{C.12}$$

where the plus sign has to be used if $f(\theta)$ is itself fermionic.

For a function of several variables one just applies the definition (C.7) successively for each variable. Observe, however, that the order counts. From (C.8) one sees that

$$\int d\theta_1 \, d\theta_2 \, f = - \int d\theta_2 \, d\theta_1 \, f \quad . \tag{C.13}$$

when the integrand depends also on ordinary commuting variables, the integrations over them are defined in the usual way. Changes of variables can be performed in an integral. The integrand must be multiplied by the Jacobian of the transformation, which is defined in the usual way but by using the concept of superdeterminant discussed in Appendix A. Finally we observe that, while integration over a commuting variable changes the dimensions of a quantity in the same way as multiplication by x, on the contrary integration over an anticommuting variable θ changes it as division by θ. This is obvious from (C.8).

REFERENCES

1. For reviews of supersymmetry with extensive references see for instance
   B. Zumino, Proc.17th International Conf. on High Energy Physics, London, 1974 (Science Research Council, Didcot 1974) p.I-254;
   P. Fayet and S. Ferrara, Physics Reports 32 (1977) 251.

2. Y.A. Golfand and E.P. Lichtman, JETP Letters 13 (1971) 323.

3. D.V. Volkov and V.P. Akulov, Phys. Letters 46B (1973) 109.

4. J. Wess and B. Zumino, Nuclear Phys. B70 (1974) 39.

5. S. Ferrara and B. Zumino, Nuclear Phys. B79 (1974) 413.

6. A. Salam and J. Strathdee, Phys. Letters 51B (1974) 353.

7. W. Rarita and J. Schwinger, Phys. Rev. 60 (1941) 61.

8. J. Iliopoulos and B. Zumino, Nuclear Phys. B76 (1974) 310

9. P. Fayet and J. Iliopoulos, Phys. Letters 51B (1974) 461 ; L. O'Raifeartaigh, Nuclear Phys. B96 (1975) 331.

10. A. Salam and J. Strathdee, Phys. Letters 49B (1974) 465.

11. W. Bardeen, unpublished ; B. de Wit and D.Z. Freedman, Phys. Rev. D12 (1975) 2286.

12. See R. Haag, J.T. Łopuszanski and M. Sohnius, Nucl. Phys. B88 (1975) 257 and references therein.

13. L. Brink, J.H. Schwarz and J. Scherk, Nucl. Phys. B211 (1977) 77.

14. F. Gliozzi, J. Scherk and D. Olive, Nucl. Phys. B122(1977)253.

15. D.R.T. Jones, Phys. Letters 72B (1977) 199.

16. E.C. Poggio and H.N. Pendleton, Phys. Letters 72B (1977) 200.

17. D.Z. Freedman, P. Van Nieuwenhuizen and S. Ferrara, Phys. Rev. D13 (1976) 3214.

18. S. Deser and B. Zumino, Phys. Letters B.62 (1976) 335.

19. S. Ferrara and P. Van Nieuwenhuizen, Phys. Rev. Letters 37 (1976) 1669.

20. S. Ferrara, J. Scherk and B. Zumino, Phys.Letters 66B (1977) 35.

21. D.Z. Freedman, Phys.Rev. Letters 38 (1977) 105.

22. A. Das, Phys. Rev. D15 (1977) 2805.

23. E. Cremmer, J. Scherk and S. Ferrara, Phys. Letters 68B(1977)234.

24. E. Cremmer and J. Scherk, Nucl. Phys. B127 (1977) 259.

25. B. de Wit and D.Z. Freedman, Nucl. Phys. B130 (1977) 105.

26. See the review by M. Grisaru and P. Van Nieuwenhuizen in Deeper Pathways in High-energy Physics, Proc. Orbis Scientiae Coral Gables 1977 (Plenum Press, New York, 1977).

27. S. Deser, J.H. Kay and K.S. Stelle, Phys. Rev. Letters 38 (1977) 527.

28. D.Z. Freedman and A. Das, Nucl. Phys. B120 (1977) 221.

29. P.K. Townsend, Phys. Rev. D15 (1977) 2802.

30. S. Deser and B. Zumino, Phys. Rev. Letters 38 (1977) 1433.

31. M. Gell-Mann, unpublished.

32. D.V. Volkov and V.P. Akulov, Phys. Lett. 46B (1973) 109.

33. A. Salam and S. Strathdee, Nucl. Phys. 76B (1974) 477.

34. S. Ferrara, B. Zumino and J. Wess, Phys. Lett. 51B (1974)239.

35. R. Arnowitt, P. Nath and B. Zumino, Phys. Lett. 56B (1975)81 ; R. Arnowitt and P. Nath, Phys. Lett. 56B(1975)117 ; Phys. Lett. 65B (1976) 73. For a more recent account of this approach see the review by R. Arnowitt in Deeper Pathways in High Energy Physics, Proc. Orbis Scientiae, Coral Gables 1977 (Plenum Press, New York, 1977).

36. B. Zumino, Proc. of the Conf. on Gauge Theories and Modern Field Theory, Northeastern University, September 1975 eds. R. Arnowitt and P. Nath (MIT Press).

37. J. Wess and B. Zumino, Phys. Lett. 66B (1977) 361.

38. R. Grimm, J. Wess and B. Zumino, Phys. Lett. 73B (1978) 415.

39. J. Wess, Supersymmetry-Supergravity, lectures given at the VIII G.I.F.T. Seminar (Salamanca, 1977) Karlsruhe Univ.

preprint, to be published in Springer Tracts.

40. J. Wess and B. Zumino, Phys. Lett. 74B (1978) 51.

41. J. Wess and B. Zumino, CERN preprint TH.2553 (1978), to appear in Phys. Letters.

42. J. Wess and B. Zumino, in preparation.

43. V.P. Akulov, D.V. Volkov and V.A. Soroka, JETP Letters 22(1975) 396 (English, 187).

44. L. Brink, M. Gell Mann, P. Ramond and J.H. Scharz, Phys. Lett. 74B(1978)336.

45. V. Ogievetsky and E. Sokatchev, Nucl. Phys. B124 (1977) 309 ; Dubna preprint E2-11702 (1978).

46. W. Siegel, various Harward University preprints (1977-78).

47. S. Ferrara and B. Zumino, Nucl. Phys. B134(1978)301.

48. L. Brink, M. Gell-Mann, P. Ramond and J.H. Schwarz, Phys. Lett. 76B(1978) 417.

49. P.S. Howe and R.W. Tucker, CERN preprint TH 2524.

50. D.Z. Freedman and J.H. Schwarz, Phys. Rev. D15(1977)1007

51. S. Ferrara, J. Scherk and P. van Nieuwenhuizen, Phys. Rev. Lett. 37(1976)1035 ; S. Ferrara, F. Gliozzi, J. Scherk and P. van Nieuwenhuizen, Nucl.Phys. B117 (1976)333.

52. D.Z. Freedman, Phys. Rev. D15 (1977)1173.

53. K.S. Stelle and P.C. West, Imperial College Preprint ICTP/77-78/24.

54. S. Ferrara and B. Zumino, Nucl.Phys. B87(1975)207.

55. F.A. Berezin, The Method of Second Quantization, New York, Academic Press (1966).

56. R. Penrose, Annals of Phys. 10(1960)171.

57. S. Ferrara and P. van Nieuwenhuizen, Phys. Lett. 74B (1978)333 ; 76B(1978) 404 ; Ecole Normale preprint LPTENS 78/14.

58. K.S. Stelle and P.C. West, Phys.Lett. 74B (1978)330 ; Imperial College preprints ICTP/77-78/15 ; ICTP/77-78/24.

59. W. Siegel, Harvard University preprint.

60. J. Wess and B. Zumino, in preparation.

THE DYNAMICS OF SUPERGRAVITY

S. Deser

Department of Physics
Brandeis University
Waltham, Massachusetts 02154 USA

I. INTRODUCTION

In these lectures I will deal with supergravity as a dynamical system, necessarily self-coupled, of spin 2 plus $3/_2$ fields. I will discuss its Hamiltonian formulation and those of its properties which can best be understood in this way. The Hamiltonian approach, while often neglected in quantum field theory because it is not immediately related to diagrams and calculations, has in fact been essential in understanding the structure of other gauge theories, notably general relativity itself. In addition to the many other insights this approach offers into gauge theories (as we shall see) is the matter of principle that canonical quantization is really the only sure starting-point from which to get the correct path integral form (and Feynman rules) complete with ghosts and correct integration measures - see for example [1].

Since we are dealing with higher spin systems, invariances under fermionic transformations, graded algebras and other less than familiar concepts, study of the initial value system with its constraints and degrees of freedom will provide guidance to these notions from a different point of view. I will also stress the "square root of gravity" ideas of Teitelboim and collaborators, which constitute a beautiful application of Dirac's original idea to a quantum field system. The Hamiltonian approach has also yielded new results, particularly the demonstration that supergravity has positive energy, a requirement essential to any physical system. This in turn has been exploited to show that classical gravity therefore also has positive energy, a conjecture which despite decades of intensive work had never been completely demonstrated there. The very fact that the square root of gravity may be taken thus tells us a great deal about the properties of general relativity itself.

The basic references I will use freely without detailed mention are as follows: For the original supergravity papers, [2]; for Hamiltonian formulation, [3] and [4]; we will use notation and conventions of [3]; for positive energy in supergravity, [5], and its application to classical gravity, [6]; the square root of gravity approach is developed in [7]; background on the Hamiltonian formulation of gravity may be found in [8] and in [9], and on general higher spin actions in [10] and [13]; Lagrangian analysis of spin $3/2$ is treated in [14].

II. THE FREE SPIN 3/2 FIELD

The action of a free pure spin $3/2$ field was formulated by Rarita and Schwinger in 1941. To describe spin $3/2$, the field must be a vector-spinor, $\psi_\mu^{(\alpha)}$ where (α) denotes the spinor and μ the vector index (the former is usually suppressed). In the massless case, one expects a (spinorial!) gauge invariance associated with the vector index of the form

$$\delta\psi_\mu = \partial_\mu \alpha(x) \tag{2.1}$$

and since ψ_μ is a fermion, it must be described by a first order action. Clearly, the quantities

$$f_{\mu\nu} = \partial_\mu\psi_\nu - \partial_\nu\psi_\mu \quad , \quad {}^*f^{\mu\nu} \equiv \tfrac{1}{2}\,\varepsilon^{\mu\nu\alpha\beta} f_{\alpha\beta} \tag{2.2}$$

are gauge invariant, but since the action is bilinear in ψ, how do we simultaneously "protect" the invariance of the other $\delta\psi$? The answer is of course through antisymmetry between the derivative and both ψ indices, namely by writing the Lagrangian as

$$\mathcal{L}_{3/2} = -i/2\;\varepsilon^{\mu\nu\alpha\beta}\,\bar\psi_\mu\gamma_5\gamma_\nu\partial_\alpha\psi_\beta \equiv -\tfrac{i}{2}\,\bar\psi_\mu\gamma_5\gamma_\nu\,{}^*f^{\mu\nu} \tag{2.3}$$

and recalling the identity $\partial_\mu {}^*f^{\mu\nu} = 0$. Here γ_ν is introduced to saturate the remaining index and γ_5 is needed for parity. [Also, if we had omitted γ_5, the action would be a total divergence since the ψ are anticommuting Majorana (real) spinors.] Incidentally it is possible to write the spin 1/2 action in this form as well,

$$\mathcal{L}_{1/2} = -i/12\;\varepsilon^{\mu\nu\alpha\beta}\,\bar\lambda\gamma_5\gamma_\mu\gamma_\nu\gamma_\alpha\partial_\beta\lambda \equiv -i/2\;\bar\lambda\,\not\partial\,\lambda \tag{2.4}$$

with a mass term

$$im\,\varepsilon^{\mu\nu\alpha\beta}\,\bar\lambda\gamma_5\gamma_\mu\gamma_\nu\gamma_\alpha\gamma_\beta\lambda \tag{2.5}$$

Indeed we will have to check that there is no helicity 1/2, but only $3/2$, in (2.3) since it seems to contain a part like (2.4) from the piece of ψ_μ which is proportional to $\gamma_\mu\gamma\cdot\psi$.

Now consider the space-time decomposition of (2.3) in order to separate out time derivatives and constraints. It is clear that the latter will be present, first because a local gauge invariance like (2.1) always implies them, second because the $\mathcal{E}$ symbol keeps the two ψ and the derivative indices disjoint, so that ψ_0 can only appear linearly and with a spatial derivative only, i.e., as a Lagrange multiplier, (this has significance to full supergravity, as we shall see). We have

$$\mathcal{L}_{3/2} = -\tfrac{i}{2}\, \varepsilon^{0ijk}\left[\bar{\psi}_i \gamma_5 (\gamma_j \partial_0 - \gamma_0 \partial_j)\psi_k + 2\bar{\psi}_0 \gamma_5 \gamma_i \partial_j \psi_k\right]$$

$$\equiv \mathcal{L}_0 + \mathcal{L}_K + \mathcal{L}_C\,. \tag{2.6}$$

The promised constraint is the coefficient of ψ_0 in $\mathcal{L}_C$, namely

$$(\varepsilon^{0ijk}\gamma_5\gamma_i)\partial_j\psi_k \equiv (2\gamma^0\sigma^{jk})\partial_j\psi_k = 0\,. \tag{2.7}$$

It involves only the curl of ψ_k , and so only its transverse part in the usual decomposition of any vector:

$$\psi_i = \psi_i^T + \partial_i\phi\,, \quad \psi_i^T \equiv P_{ij}\psi_j = \tfrac{1}{2}(\delta_{ij} - \nabla^{-2}\partial^2_{ij})\psi_j\,. \tag{2.8}$$

This is also manifestly the case for the kinetic energy part of the action, $\int \mathcal{L}_K\, d^4x$. To see that only ψ_i^T enters in the time derivative part $\int \mathcal{L}_0$, use the γ -identity in (2.7) to rewrite

$$\int \mathcal{L}_0\, d^4x = i\int \bar{\psi}_i \gamma^0 \sigma^{ij}\partial_0 \psi_j\, d^4x\,. \tag{2.9}$$

The integral is easily seen to depend only on ψ^T by use of the constraint to eliminate the $\partial_i\phi$ parts of ψ_i . But ψ^T had better not be the final variable, since it only satisfies one condition ($\partial_i\psi_i^T \equiv 0$) on its three vector indices, whereas we want a single final variable to describe the two $\pm 3/2$ helicity components just as one λ describes two $\pm 1/2$ helicity neutrino states. The second requirement is given by the constraint itself which we must in fact still solve. It may be written as

$$\sigma^{ij}\partial_i\psi_j^T \equiv \tfrac{1}{2}[-\delta_{ij} + \gamma_i\gamma_j]\partial_i\psi_j^T \equiv \underline{\gamma}\cdot\underline{\nabla}\,(\underline{\gamma}\cdot\underline{\psi}^T) = 0, \tag{2.10}$$

whose only solution is $\underline{\gamma}\cdot\underline{\psi}^T = 0$. We therefore need a quantity which is both transverse and γ-traceless,

$$\underline{\nabla}\cdot\underline{\psi}^{TT} = 0 = \underline{\gamma}\cdot\underline{\psi}^{TT}. \tag{2.11}$$

A ψ_i^{TT} is obtained from an arbitrary ψ_i by the projection

$$\psi_i^{TT} = \tfrac{1}{2} P_{ik}\gamma_\ell\gamma_k P_{\ell j}\psi_j\,. \tag{2.12}$$

Note that both $\underline{\psi}^T$ and $\underline{\psi}^{TT}$ are gauge invariant under (2.1); alternately one may choose the gauge $\underline{\psi} = \underline{\psi}^T$ or $\underline{\gamma}\cdot\underline{\psi} = 0$ by appropriate choice of $\alpha(x)$, namely $\nabla^2\alpha = -\underline{\nabla}\cdot\underline{\psi}$ or $\not\nabla\alpha = -\underline{\gamma}\cdot\underline{\psi}$; this route also leads to pure $\underline{\psi}^{TT}$ dependence. The action of course depends only on the ψ^{TT} in this abelian theory, as may easily be checked. It reads

$$\int \mathcal{L}_{3/2} = -\tfrac{i}{2}\int \varepsilon^{0ijk}\bar{\psi}_i^{TT}\gamma_5(\gamma_j\partial_0 - \gamma_0\partial_j)\psi_k^{TT} = -\tfrac{i}{2}\int \underline{\bar{\psi}}^{TT}\not\partial\,\underline{\psi}^{TT} \tag{2.13}$$

where the final Dirac form for the single field $\underline{\psi}^{TT}$ follows by γ-identities. There is a close analogy here to electrodynamics whose action may be written as $1/2\int \underset{\sim}{A}^T \,\Box\, \cdot \underset{\sim}{A}^T$ in terms of the gauge invariant purely transverse components $\underset{\sim}{A}^T$. Incidentally, Hamiltonian decomposition of free higher spin fermions is very similar [19].

The fact that the physical quantity here is the field strength of (2.2) means that we can write the field equations in terms of it. Its components are just

$$E_i = -\dot{\psi}_i^{TT} \,, \quad B^i = \tfrac{1}{2}\,\varepsilon^{ijk}(\partial_j \psi_k^{TT} - \partial_k \psi_j^{TT}) \equiv (\nabla \times \psi^{TT})^i \tag{2.14}$$

and obey the simple field equation

$$\underset{\sim}{E} + \gamma_5 \underset{\sim}{B} = 0 \tag{2.15}$$

which also holds in supergravity with appropriate covariant derivatives.

A final important element of the theory is of course the canonical anti-commutation relation. Because ψ is real and so there is no ψ^*, one must be slightly careful (as for Majorana neutrinos) and define, for example, the complex chiral (Weyl) components $\underset{\sim}{\psi}^{TT}_{L,R} \equiv 2^{-1/2}(1 \mp i\gamma_5)\underset{\sim}{\psi}^{TT}$ in terms of which one has

$$\{\underset{\sim}{\psi}_L^{TT*}(\underset{\sim}{r}),\, \underset{\sim}{\psi}_L^{TT}(\underset{\sim}{r}')\} = \overleftrightarrow{\delta}_L^{TT}(\underset{\sim}{r}-\underset{\sim}{r}') \tag{2.16a}$$

where $\overleftrightarrow{\delta}_L^{TT}$ is the unit matrix in the TT space, namely the TT projection of the spin plus space unit operator; ψ_R^{TT} behaves identically and $\{\psi_R, \psi_L'\} = 0$. It is important to note that for the TT components, the Hilbert space is positive, since δ^{TT} is just the unit operator. This would not be the case for the anti-commutator obtained for the ψ_i in the usual way from the action before use of constraints; that would read*

$$\{\psi_i^{\alpha}(\underset{\sim}{r}),\, \psi_j^{\beta}(\underset{\sim}{r}')\} \sim (\gamma_j\cdot\gamma_i\cdot)^{\alpha\beta}\, \delta^3(\underset{\sim}{r}-\underset{\sim}{r}') \tag{2.16b}$$

which is clearly not positive since $\gamma_j\cdot\gamma_i\cdot$ has a part antisymmetric in ($\alpha\beta$) for example. As in all gauge theories, positivity properties hold only after gauge conditions and constraints have been used to eliminate redundant variables. This point will be particularly relevant to the energy discussion.

Before leaving the free massless field, we mention briefly its duality and chirality transformation properties. In covariant form the field equations have many equivalent expressions, including the sets

---

*This form arises because the coefficient of $\dot{\psi}_k$ is just $\bar{\psi}_i\cdot\gamma^0\sigma^{ik} \sim \varepsilon^{ijk}\bar{\psi}_i\cdot\gamma_5\gamma_j \sim \bar{\psi}_i\cdot\sigma_{ik}$. We then take this to be the conjugate momentum, and Eq (2.16b) is just equivalent to $\{\psi_i\cdot\sigma_{ij}\,,\,\psi_k'\} = \delta_{jk}\,\delta^3(\underset{\sim}{r}-\underset{\sim}{r}')$.

$$\gamma_\mu {}^*f^{\mu\nu} = 0\,,\quad \gamma_\mu f^{\mu\nu} = 0\,,\quad f_{\mu\nu} + \gamma_5 {}^*f_{\mu\nu} = 0 = \sigma^{\mu\nu} f_{\mu\nu} \tag{2.17}$$

in terms of the field strengths of (2.2). [The last set includes the $\sigma\cdot f = 0$ condition separately, as the σ-trace of the left side vanishes identically and this left side consists only of the three independent equations (2.15)]. There are therefore two apparently independent invariances of the equations: under the fermion-like chirality invariance, $\delta\psi_\mu = \gamma_5\psi_\mu$ and under the vector-like duality invariance, $\delta f_{\mu\nu} = {}^*f_{\mu\nu}$, $\delta {}^*f_{\mu\nu} = -f_{\mu\nu}$. It turns out, however, that these two alternate ways of rotating helicity are in fact equivalent. Indeed, the free-field duality rotation states that (2.14) transform as

$$\delta(\nabla\times\underline{\psi}) = \dot{\underline{\psi}}\,,\quad \delta(-\dot{\underline{\psi}}) = \nabla\times\underline{\psi} \tag{2.18}$$

where we have dropped the "TT". But $\dot{\psi}$ is determined by the field equation to be $\gamma_5\nabla\times\underline{\psi}$, and the duality transformation simply reduces to a chiral one, $\delta\underline{\psi} = \gamma_5\underline{\psi}$, with corresponding generator

$$G = i/2\int d^3r\, \bar{\underline{\psi}}^{TT}\gamma_5\gamma_0\cdot\underline{\psi}^{TT} \equiv \int d^3r\, \mathcal{A}^0 \equiv \int d\sigma_\mu\, \tfrac{1}{2}\epsilon^{\mu\nu\alpha\beta}\bar{\psi}_\nu\gamma_\alpha\psi_\beta \tag{2.19}$$

The axial current $\mathcal{A}^\mu$ is just the dual of the spin density, $S^{\nu\alpha\beta}$, and is of course conserved by virtue of the field equations, all as for spin 1/2 where $\mathcal{A}^\mu = i\bar{\lambda}\gamma_5\gamma^\mu\lambda$ and $S^{\nu\alpha\beta} \equiv \epsilon^{\nu\alpha\beta\mu}\mathcal{A}_\mu$. The significance of the above transformations and their properties become more complicated when interaction with gravity is included; we do not discuss them further here.

We conclude this section with a brief excursion into the massive theory, because the latter appears formally in some variants of supergravity [11], although it is known that it is actually "massless" [12] in that context, and that flat space ideas are not relevant there. Nevertheless it is worth noting the way in which the two helicity 1/2 states reappear with a mass term, which takes the unusual form (compare (2.5)),

$$\mathcal{L}_m = im\,\bar{\psi}_\mu\sigma^{\mu\nu}\psi_\nu = -i\frac{m}{2}\epsilon^{\mu\nu\alpha\beta}\bar{\psi}_\mu\gamma_5\sigma_{\alpha\beta}\psi_\nu \tag{2.20}$$

That this is correct will emerge when we obtain the appropriate Dirac equations for the right (2s+1 = 4) number of components. In the Hamiltonian form, we still have, surprisingly, a constraint since ψ_0 clearly appears only linearly in (2.20). However there is also a new variable as well, namely $\underline{\gamma}\cdot\underline{\psi}^L$, the longitudinal part of ψ (in the $m\bar{\psi}_i\sigma_{ij}\psi_j$ term). The constraint now reads

$$(\not\partial - m)\chi - m\underline{\gamma}\cdot\underline{\psi}^L = 0\,,\quad \chi \equiv \underline{\gamma}\cdot\underline{\psi}^T \tag{2.21}$$

and must be used to eliminate either variable in favor of the other, rather than just stating that χ vanishes. When this is done, and the result inserted into the rest of the action, one finds that

$$I = -\tfrac{1}{2}\int\bar{\underline{\psi}}^{TT}(\not\partial + m)\cdot\underline{\psi}^{TT} - \tfrac{1}{2}\int\bar{\chi}'(\not\partial + m)\chi'\,,\quad \chi' \equiv \sqrt{\tfrac{3}{2}}\,\gamma_5\chi\,. \tag{2.22}$$

The algebra is straightforward except perhaps for use of the fact that $\bar{\chi}\gamma^0\gamma^i\partial_{0i}\chi$ is a total derivative because of the anti-commutation of the χ's. So this is indeed a helicity ($3/2 + 1/2$) system, just as massive electrodynamics acquires a longitudinal helicity zero part. Once introduced, of course, these lower helicity pieces do not vanish as $m \rightarrow 0$ in either system. However, for spin 1 their coupling to the electromagnetic current vanishes as $m \rightarrow 0$, so they become free and hence unobservable (except gravitationally). The spin 3/2 particles are more like their spin 2 counterparts in that their lower helicity states do not decouple from the sources as $m \rightarrow 0$. We mention here only that in both cases there is a similar finite difference (even as $m \rightarrow 0$) in the source-source interactions by one-quantum exchange when compared to pure $m = 0$ exchange, which is not unexpected since spin 3/2 and 2 do form a (super) multiplet. That is, if linearized gravity couples to a necessarily conserved prescribed source $T_{\mu\nu}$ then the massless one-graviton exchange leads to an effective source-source interaction,

$$I_2(m=0) \sim \int d^4x\, d^4x' [T_{\mu\nu} - \tfrac{1}{2}\eta_{\mu\nu} T^\alpha_\alpha] D(x-x') T'_{\mu\nu} \tag{2.23a}$$

while the massless limit of massive spin 2 gives [15]

$$I_2(m\rightarrow 0) \sim \int d^4x\, d^4x' [T_{\mu\nu} - \tfrac{1}{3}\eta_{\mu\nu} T^\alpha_\alpha] D(x-x') T'_{\mu\nu} \tag{2.23b}$$

a finite difference due to the non-decoupling of the helicity zero mode here. Correspondingly, if the spin 3/2 field couples to a necessarily conserved prescribed vector-spinor source $\mathcal{J}^{(\alpha)}_\mu$, the corresponding interactions read

$$I_{3/2}(m=0) \sim i\int d^4x\, d^4x' [\bar{\mathcal{J}}^\mu S(x-x') \mathcal{J}'_\mu - \tfrac{1}{2} \overline{\gamma\cdot\mathcal{J}}\, S(x-x')\, \gamma\cdot\mathcal{J}'] \tag{2.24a}$$

and

$$I_{3/2}(m\rightarrow 0) \sim i\int d^4x\, d^4x' [\bar{\mathcal{J}}^\mu S(x-x') \mathcal{J}'_\mu - \tfrac{1}{3} \overline{\gamma\cdot\mathcal{J}}\, S(x-x')\, \gamma\cdot\mathcal{J}'] \tag{2.24b}$$

where $S(x)$ is the massless spin 1/2 propagator. There is even a supersymmetric reason for the numerical agreement between the two systems. Of course, since we do not at present have an acceptable massive gravity (let alone supergravity) model, the flat space $m \neq 0$ cases are purely academic.

III. SUPERGRAVITY

The action for supergravity [2] is, in first order form,

$$\int \mathcal{L}_{SG}(e,\omega,\psi) = -\tfrac{1}{2}\int eR(e,\omega) - \tfrac{i}{2}\epsilon^{\lambda\mu\nu\rho}\bar{\psi}_\lambda \gamma_5 e_{\mu a}\gamma^a D_\nu(\omega)\psi_\rho \,. \tag{3.1}$$

Here the independent variables are: the gauge fields of local translations, the vierbeins $e_{\mu a}(x)$; the gauge fields of local Lorentz transformations, the Ricci rotation coefficients $\omega_\mu{}^{ab}(x)$, and the gauge fields of supersymmetry transformations, $\psi_\mu(x)$. The curvature scalar density eR is the contraction of the full curvature with the inverses $e^{\mu a}$ of $e_{\mu a}$ ($e^{\mu a} e_{\mu b} = \delta^a_b$) times the determinant (e) of the $e_{\mu a}$:

$$R \equiv e^{\mu a} e^{\nu b} R_{\mu\nu ab} \,, \quad R_{\mu\nu ab} \equiv [\partial_\mu \omega_{\nu ab} + \omega_{\mu ac} \omega_{\nu cb}] - (\mu \leftrightarrow \nu) \tag{3.2}$$

The covariant curl appearing in (3.1),

$$D_\mu \equiv \partial_\mu - \tfrac{1}{2} \omega_{\mu ab} \sigma^{ab} \tag{3.3}$$

is related to the curvature through the Ricci identity,

$$[D_\mu, D_\nu] \alpha(x) \equiv -\tfrac{1}{2} R_{\mu\nu ab} \sigma^{ab} \alpha(x) \,. \tag{3.4}$$

Although it is not our purpose here to derive (3.1) or to discuss its consistency and invariances, we note that it can be understood as the minimally self-coupled [16] ("covariantized") version of the sum of the linearized spin 2 plus spin $3/2$ actions. The former is just the quadratic part of eR when $e_{\mu a}(x)$ is expanded about its flat space value $\eta_{\mu a}$, the latter being of course the free action $I_{3/2}$ discussed in the previous section. When we couple the Noether current

$$J^\mu \equiv + \tfrac{i}{4} \varepsilon^{\mu\nu\alpha\beta} \gamma_5 \eta_{\nu c} \gamma^c \omega_{\alpha ab} \sigma^{ab} \psi_\beta \tag{3.5}$$

(which is conserved by virtue of the combined linearized equations and gravitational Bianchi identities) to its gauge field $\bar{\psi}_\mu$, and couple in the stress tensor as well by setting $\eta_{\nu c} \rightarrow e_{\nu c}$ everywhere (including the spin 2 action), we just obtain (3.1). Because everything is in first order form, no extra contact terms are needed or permitted. Indeed, we can "predict" this, because any quartic contact terms in ψ are necessarily at least bilinear in ψ_0 , which would destroy the Lagrange multiplier nature of ψ_0 that we know goes with the gauge invariance of the spin $3/2$ part. There is one quadratic term, as we have seen, namely $\bar{\psi}_\mu \sigma^{\mu\nu} \psi_\nu$ which is only linear in ψ_0 and that is why "massive" supergravity [11] can exist.

Knowing that (3.1) represents a doubly gauge invariant system, one gauge being general coordinate invariance, the other the local symmetry fermionic gauge $\alpha(x)$, tells us immediately what the general Hamiltonian form of (3.1) is without any calculation. First, we recall that for any bosonic matter coupled to ordinary gravity, the action becomes [8]

$$I_{2+B} = \int \sum_1^6 \pi^{ij} \dot{g}_{ij} + \sum_{A=1}^{\tilde{N}} \pi^A \dot{\phi}_A - N_\mu R^\mu(\pi^{ij}, g_{ij}, \pi^A, \phi_A) - \lambda_A R^A \tag{3.6}$$

in terms of the initial value data of the gravitational field - the spatial metric and its conjugate -, and of the matter field (π^A, ϕ_A). The "Hamiltonian" necessarily vanishes by general coordinate invariance and must be the sum of four constraints R^μ times four Lagrange multipliers N_μ , where R^μ depends only on the dynamical variables and their spatial derivatives. If the matter field is itself gauge invariant, as in Yang-Mills theory, then we have additional constraint terms such as $A^a_0 \partial_i \mathcal{E}^{ia}$ where the electric field $\mathcal{E}^{ia}$ is anisovector and a contravariant vector density; they are represented by $\lambda_A R^A$ above. We may now proceed backwards and count the degrees of freedom which are left when the constraints are eliminated: the four $R^\mu = 0$, together with the four corresponding gauge freedoms remove four pairs of variables, while the Lagrange multipliers disappear once the constraints are solved and reinserted. This leaves 6 - 4 = 2 pairs of conjugate graviton variables (2 helicity degrees of freedom as m = 0), together with a similar count for the other gauge fields, which proceeds exactly as in flat space for them.

In the present case some generalization is needed because we must use vierbeins rather than metrics to couple with fermions. But there are more vierbeins than metric components (16 $e_{\mu a}$ versus 10 $g_{\mu\nu}$) and the constraints which reduce the redundancy must be the six generators of local Lorentz rotations [9] under which the theory is also manifestly invariant (indeed, $\omega_{\mu ab}$ are the gauge fields corresponding to these). We can therefore predict that there will be a "$p\dot{q}$" term for gravity which now looks like $\Sigma_i^n p^{i\alpha}\dot{e}_{i\alpha}$ in terms of the spatial components of the vierbeins. [Henceforth we use the range i,j,... for spatial world indices, a,b... for spatial local indices, with $\mu,\nu,\ldots$; $\alpha,\beta\ldots$ for the corresponding space time indices and we always place local indices last.] There must be an appropriate fermionic term $\sim\bar{\phi}\Gamma\dot{\phi}$ where Γ are some γ-matrices and ϕ are essentially the spatial components of ψ_μ. Indeed, we expect to see $\epsilon^{ijk}\bar{\phi}_i\gamma_5\gamma_j\dot{\phi}_k$ here, after appropriate absorption of vierbein powers in the ψ_i to define the ϕ_i. We must still have the $N_\mu R^\mu(p,e,\phi)$ term by general covariance and since spin 3/2 is a gauge field, some corresponding form of the $\bar{\psi}_0\sigma^{ij}\partial_i\psi_j$ constraint which may now depend on $p^{i\alpha}$ and $e_{i\alpha}$, just as the R^μ depend on ϕ_i (it is only for spin 1 that the matter constraints can be kept free of metric dependence; for higher spin this is impossible because covariant derivatives become unavoidable). Let us write this constraint as $\bar{\eta}\,S(p,e,\phi)$ where η is essentially ψ_0. Finally we must include the generators of local Lorentz rotations - i.e. the angular momenta $J^{\alpha\beta}(p,e,\phi)$ multiplied by appropriate multipliers, $\lambda^{\alpha\beta}$ say. Thus, we finally predict that (3.1) will have the form

$$I_{SG} = \int \Sigma_i^n p^{i\alpha}\dot{e}_{i\alpha} - \tfrac{i}{2}\epsilon^{ijk}\bar{\phi}_i\gamma_5\gamma_j\dot{\phi}_k - N_\mu R^\mu(p,e,\phi) - \lambda^{\alpha\beta}J_{\alpha\beta}(p,e,\phi) - \bar{\eta}S(p,e,\phi). \qquad (3.7)$$

The count is also in order, for each constraint removes its multiplier and two of the "dynamical" variables. Thus, the 12 graviton pairs reduce to two because there are 4 R^μ and 6 $J_{\alpha\beta}$ constraints, while we have already seen in the free case how S = 0 removes two of the three ϕ_i variables and the multiplier η. [This is a general property of higher spin fermionic gauge invariance: each invariance removes 3 components of the field.]

The actual work involved in reaching (3.7) from (3.1) is rather heavy even in the absence of spin 3/2, primarily because of the extra vierbein components and their conjugates. Since this aspect is not very interesting anyway, one may immediately fix this gauge in the most natural way, which consists of setting $e^0{}_a = 0 = e_i{}^0$, (this is called time gauge because it aligns the local and world time directions) and making the spatial components symmetric, $e_{ia} - e_{ai} = 0$. Fixing the local vierbein gauge at the outset then eliminates the $\lambda^{\alpha\beta} J_{\alpha\beta}$ constraints as well. [Also, the transformations generated by the remaining constraints are not quite the general set; for example the $\delta e_{\mu a}$ rules will have to respect the gauge conditions, but the local vierbein alignment freedom is not very difficult to understand.] A further simplification consists in referring the ψ_i to the local frame in a way which makes the $\emptyset\emptyset$ term vierbein - independent; a priori, we had $\epsilon^{ijk}\bar{\psi}_i \gamma_5 \gamma^a e_{ja} \psi_k$. This is accomplished by defining

$$\phi_a \equiv \sqrt{^3e}\,\psi_a \equiv \sqrt{e}\, e^\mu{}_a \psi_\mu , \quad \eta \equiv e_0{}^0 \sqrt{e}\, e^\mu{}_0 \psi_\mu \tag{3.8}$$

where (^{3}e) is the determinant of e_{ia}. A great deal of algebra goes into removing those ω components which do not involve time derivatives of the vierbeins, and in appropriate grouping of terms into gravitational and spin 3/2 sectors. The presence of torsion requires redefinition of the gravitational canonical momentum (as in any derivative coupling theory) and this is also rather tedious.

The final result is given in detail (though in different terms) in [3,4] and there is no point in quoting it here, except that the predicted structure of (3.7) is indeed observed.* The coefficient of N_a, which is the momentum constraint is always the simplest since the momentum density for lower spin field always looks like $\sim \pi^A \partial_i \phi_A$. In our case, this is almost true;

$$R^a \sim D_i(e)p^{ia} + \tfrac{i}{2}\epsilon^{bcd}\bar{\phi}_b \gamma_5 \gamma_c e^i{}_a D_i(e)\phi_d \tag{3.9}$$

where D_i(e) is the usual covariant derivative with respect to the (spatial) vierbeins. The first term is the purely gravitational contribution already familiar from the metric formulation, while the second is the (covariantized) momentum density of the free spin 3/2 field; its metric dependence is unavoidable because the spin exceeds 1, where ∂_i suffices. The energy constraint R^0, (apart from various quartic terms in $\emptyset$) is the sum of the usual gravitational energy constraint, the kinetic part of the spin 3/2 action which is just the covariantized form of $\mathcal{L}_K$ in (2.6), and a cross term between the gravitational momentum and a part quadratic

*In [4] the vierbein gauge freedom is reinstated at the end and the ψ_i rather than ψ_a are used. Some apparent differences between the N_i constraints are due to effective use of one constraint to simplify another, which is always permitted. In particular those corresponding to (3.9) differ by multiples of the η constraint.

in $\emptyset$. This term has its counterpart in the $3/_2$ constraint; it is due to the covariant derivative coupling and will be better understood in the "square root of gravity" analysis. The total $3/_2$ constraint term $\bar{\eta}\mathcal{S}$ is again simply the covariantized form of the flat space constraint, together with the above-mentioned essential $\bar{\eta}\gamma_a\psi_j p^{ia}$ cross-term. Thus the final Hamiltonian reduction yields precisely what the invariances of the theory required and this provides a clear count of the degrees of freedom and of the physics of the constraints. Incidentally, the detailed form of the latter also agrees with general properties characteristic of higher spin coupling to gravitation [13]. We shall examine the significance of the constraints and of their algebra from a different point of view in the next section, where we look at the "square root" idea. As a final note, it was through the Hamiltonian analysis (second paper in [7]) that it was first realized that the complications in that algebra would lead to quartic ghost terms in the action, something which was far from obvious in the covariant form.

IV. THE SQUARE ROOT OF GRAVITY

Having obtained the detailed Hamiltonian form of supergravity, we could try to form the commutators and anti-commutators of the constraints with each other in terms of the initial (non-reduced) canonical commutation relations, and see how these extend the algebra satisfied by the R^{μ} in general relativity. It is known there that these relations really contain all the structure of the Einstein theory, and the same should be the case here. The "square root approach" is essentially concerned with this problem, and it can be tackled from opposite ends. The first way asks: can one take some sort of Dirac square root of the constraints of pure gravity, thereby obtaining supergravity directly from Einstein theory? The second concentrates on the algebra which would have to be satisfied by the enlarged constraint system (with square root) in order that they generate the appropriate geometrical transformations.

In view of the quadratic dependence of the Einstein energy constraint on the canonical momenta, the attempt is a natural one, though a priori unlikely to succeed in view of the terrible nonlinearity of the constraints. But given the new physics, uncovered by Dirac when he found his square root, and the unification of spin and geometry to be sought here, it is worth trying. Let us recall that Dirac faced (in a particle context) the on-shell constraint $H\underline{\Psi} \equiv P^2\underline{\Psi} = 0$ for all physical states, and that while "taking the square root" $Q = \not{P}$ in the free case by means of the new anti-commuting objects γ^{μ} was literally just that* since $(\not{P})^2$ reproduces

*Note incidentally the foreshadowing of global supersymmetry in $[H,H] = 0 = [H,Q]$, $\{Q,Q\} = H$ and of the field transformations in $i[Q,x^{\mu}] = \gamma^{\mu}$, $\{Q,\gamma^{\mu}\} = p^{\mu}$. The Dirac particle can in fact be treated as an example of matter-supergravity coupling in one space-time dimension [17].

P^2, as soon as coupling was present and one took $\gamma p \to \gamma(P-eA) \equiv \not{\pi}$, squaring gave an extra term, namely the correct gyromagnetic factor of the electron as an automatic new coupling, $(\not{\pi})^2 = \pi^2 - ie\, \sigma \cdot F$. Following Teitelboim, let us try the same trick, this time for the four constraints of general relativity,

$$g_{ij}\, g_{kl}\,(\pi^{ik}\pi^{jl} - \tfrac{1}{2}\pi^{ij}\pi^{kl}) + ({}^3g)^3 R = 0 = D_i \pi^{ij} \tag{4.1}$$

and particularly for the energy constraint which is precisely quadratic in the field momenta. Obviously since we are dealing here with fields, the generalization of γ_μ will be some anti-commuting field $\psi_i(x)$ obeying the sort of "Clifford algebra" relation $\{\psi_i^\alpha, \psi_j^{\beta\prime}\} = M_{ij}^{\alpha\beta}\, \delta^3(r-r')$ we have discussed earlier. Now it is not so difficult to imagine taking the square root of the first term (although it is really not all that quadratic due to the g_{ij} coefficients), through some spinorial combination $\sim \gamma_a \psi_i\, p^{ia}$, but how can one expect to take the square root of a curvature? To be sure, there is a hint of it in the Ricci identity which states that the commutator of two covariant derivatives is a curvature, but here we need an anti-commutator since the square root will be spinorial. But once armed with various properties of anti-commutators of products, and use of γ-matrix identities it is at least conceivable that some spinor $\chi \sim p^{ij} D_i \psi_j$ exists that will give the required result. Nevertheless it is still an amazing fact that the following is the leading part of an identity:

$$\{\sigma^{ij} D_i \psi_j(r), \sigma^{lm} D_l' \psi_m(r')\} \sim {}^3R\, \delta^3(r-r') + \ldots \tag{4.2}$$

whose verification is a relatively simple exercise in juggling γ-algebra, the basic anti-commutation relation (2.16b) and the Ricci identity (3.4). Note that $\gamma_a \phi_j \pi^{aj} + \sigma^{ij} D_i \phi_j$ are the leading parts of the spin $3/2$ constraint arrived at in the end of the last section; we are indeed building up to an algebra involving both S and the R^α.

The complementary approach concentrates on the algebra which the hypothetical square root constraint system must satisfy in order to generate the correct geometrical and spinorial transformations. It must of course be a vanishing "Hamiltonian" as a function of the Cauchy data (because still generally covariant) and involve a new spinorial part. This is precisely the form of (3.7). The local algebra is then found, and it has rather complicated local operator "structure constants," which are proportional to the tangent space projections of the curvatures $R_{\alpha\beta\gamma\delta}$ and $f_{\alpha\beta}$ (they generalize their gravitational counterparts which involve only $R_{\alpha\beta\gamma\delta}$). The most interesting aspect is the fundamental square root relation

$$\{\bar{S}(\vec{r}), S(\vec{r}')\} = \gamma^\alpha R_\alpha(\underline{r})\, \delta^3(\underline{r}-\underline{r}') \tag{4.3}$$

which means that $S\psi = 0$ implies

$$R_\alpha \underline{\bar{\Psi}} = 0 \tag{4.4}$$

just as for spinning particle(and similarly the R_α now include the $T_{0\alpha}(\psi)$).

The above square root motivations have yielded certain formal demands. Can these be implemented, and if so, is this uniquely accomplished by supergravity? The answer is yes, by virtue of the identities we sketched and of course by the known Hamiltonian decomposition of the previous section. We shall only look at the leading terms, which are easy to check; one can fall back on the detailed results of the Hamiltonian analysis to take care of the various polynomials in ψ etc., which would be very hard to verify straightforwardly. So let us be a little more systematic and first ask for the square root of the four constraints,

$$S^\alpha(x): \{S^\alpha, S^{\beta\prime}\} \sim T_\mu^{\alpha\beta} \tilde{R}^\mu_{(\underline{r})} \delta^3(\underline{r}-\underline{r}'),\ \tilde{R}^\mu = R^\mu + \ldots \tag{4.5}$$

with the expectation that $T_\mu \to \gamma_\mu$, $\tilde{R}_\mu \to R_{\mu} - T_{0\mu}(\psi)$ in the end.

The spinor operator square root must be a linear function of the π^{ij}, of the form

$$S^\alpha = \Gamma^\alpha_{ij}\, \pi^{ij} + V^\alpha \tag{4.6}$$

and yield some combination of the constraints plus possible terms depending on the "Γ" field (this corresponds to the "square root" of $(p-eA)^2$ squaring back to $(P-eA)^2 - ie\,\sigma\cdot F$). The choice $\Gamma^\alpha_{ij} \sim (\gamma_i \gamma_j \lambda)^\alpha$ corresponding to a spin 1/2 field λ cannot work since $\Gamma^\alpha_{ij} \pi^{ij} \sim \delta_{ij}\pi^{ij}\lambda^\alpha \sim \pi\lambda^\alpha$. Nor can the spin $5/2$ choice $\Gamma^\alpha_{ij} \sim \phi^\alpha_{ij}$ with ϕ^α_{ij} obeying irreducible tensor anti-commutator relations. This leaves the spin $3/2$ alternative, $\Gamma^\alpha_{ij} \sim (\gamma_i \psi_j + \gamma_j \psi_i)^\alpha$. One indeed gets the correct kinetic $(\pi_{ij}^2 - 1/2\, \pi^2)$ term iff ψ_i obeys the relation (2.16b). There will of course be additional terms from the $[\pi,e] \sim \delta^3$ relation which will give terms $\sim \bar{\psi}\gamma\pi$ and these did appear in the Hamiltonian analysis. So this is a good start. To get the leading part of V, we have the identity (4.2). Furthermore $(\Gamma\pi + V)$ is indeed the leading part of the S constraint obtained in the previous section. The remainder consists of bilinears in the ψ_i. Finally, the cross term $\{\gamma\psi\pi,\ \sigma\cdot\nabla \times \psi\}$ clearly gives term such as $D_j \pi^{ij}$ which make up the spatial Einstein constraint, as well as terms like $\psi\nabla\psi$ which represent the ψ stress tensor.

The above dynamical treatment shows very clearly how spin is unified with geometry in supergravity, and in particular that the spin $3/2$ field is the spin counterpart of gravitation exactly as spin (through the γ^μ) is the fermionic counterpart of the originally structureless particle after the square root is taken.

V. POSITIVE ENERGY

Before being able to pose the problem of positive energy in either supergravity or classical relativity, we must first review what that concept means in the framework of a generally covariant theory, and why it is difficult to establish its sign there. Energy is an interesting quantity in flat space because it is conserved by time translation invariance. In curved space, there is in general no such symmetry (except for stationary spacetimes) and furthermore, as we have seen, the whole "Hamiltonian" in fact vanishes. Clearly the concept of mass or energy can only make sense for an isolated system outside which one may stand and measure its $\underline{P}^{\mu}$ or $\underline{P}^{\mu}\underline{P}_{\mu} = -m^2$ in essentially cartesian coordinates, that is to say, if the space is asymptotically flat. Then, and only then, is there an asymptotic Poincaré group available (this is just the remaining global symmetry from the local Poincaré group in the interior; the latter alone is of no relevance to the concept of total energy).

With respect to this asymptotic group, the notion of the ten conserved quantities (P_{μ}, $J_{\alpha\beta}$) is then legitimately the usual one. It should be added that the delicate question of rate of approach to flatness, $g_{\mu\nu} \sim \eta_{\mu\nu} + O(1/r)$, $\partial g \sim O(1/r^2)$, etc. have all been worked out in detail. They are derivable from the constraint equations themsleves, as might be expected since energy is defined for a system purely in terms of Cauchy data on a given hypersurface. Furthermore, because of the underlying local gauge group, energy is a flux integral at spatial infinity just as is charge in electrodynamics. The primary definition of the Hamiltonian in every given coordinate frame is indeed as the sum of local contributions from each spatial point, just as Q is defined to be $\int \rho \, d^3r$, but its value is also obtainable from $\oint \underline{E}^L \cdot d\underline{S}$ by Gauss's law. The total value of the energy (but not the density) is furthermore an invariant, independent of the chosen frame. It is defined in a very natural way from the energy constraint equation by writing it as

$$G^{L}_{oo} = - G^{NL}_{oo} + T^{Matter}_{oo} \tag{5.1}$$

Where G^{L}_{oo} is the linearized part of the Einstein tensor (and G^{NL} the rest), and behaves like $-\nabla^2 g^T$ where g^T is an appropriate component of the spatial metric. Thus the value of energy is read off as

$$E = -\oint \nabla g^T \cdot dS \tag{5.2}$$

or alternatively, $g^T \sim (1 + E/r)$, just as in electrodynamics, $Q = \oint \nabla \phi \cdot dS$ or $\phi \sim Q/r$ To infer that this value is positive is quite another matter, however. One has (in principle) to solve (5.1) for g^T as an explicit functional of the remaining dynamical variables and show that its generic form is positive. In linearized approximation that is easy to do explicitly, and one finds that $$-\nabla^2 g^T \sim \tfrac{1}{2}\left[(\nabla g^{TT})^2 + (\pi^{TT})^2\right] \tag{5.3}$$

where (g^{TT}, π^{TT}) are the transverse-traceless conjugate pairs of graviton variables. But, although a great deal of progress has been made on the general case, there was no complete proof available within classical gravity because (5.1) is too hard to solve explicitly. How then can supergravity, which is necessarily a quantum field theory (remember that ψ obeys anti-commutation rules with a δ-function right-hand side, i.e., is second quantized and therefore so must the gravitational field be), provide any proof as to its far more complicated constraint for energy? The answer lies in the fact that for isolated systems, both gauge aspects now lead to flux integrals. That is, we can also write the fermionic constraint $S = j^0$(if there are sources) as

$$S^L = - S^N + j^0 \tag{5.4}$$

and solve it for the leading part, $S^L = \partial_i(\sigma^{ij}\psi_j)$; the value of the total supercharge is a surface integral,

$$Q = \oint \sigma^{ij}\psi_j \, dS_i \, . \tag{5.5}$$

As in gravity this is fixed for a given system once a ψ-gauge is chosen, but again the value of Q is independent of this gauge. "Asymptotic ψ -flatness" here agrees with the metric concept: ψ_i must fall off as $1/r^2$ in order for the energy and Q to be well-defined. Furthermore, the set of quantities (P^μ, $J_{\alpha\beta}$, Q) must satisfy the <u>global</u> supersymmetry remainder of local supersymmetry. This is not the place to explain how flux integrals at infinity can have commutation relations usually associated with volume definitions of these quantities in global supersymmetry, but we remark that once all gauges are fixed then in fact the dynamical fields become non-local, and the asymptotic algebra does make sense. This is in fact an essential point; for we will be deriving positive energy with no work at all from the simple global relation

$$\{Q^\alpha, \bar{Q}^\beta\} = \gamma_\mu^{\alpha\beta} P^\mu \tag{5.6}$$

from which it follows, by tracing with $\gamma_0^{\alpha\beta}$, that

$$P^0 = \tfrac{1}{2} \sum_\alpha Q^\alpha Q^\alpha \geq 0 \, . \tag{5.7}$$

The positivity statement is not true for any old Majorana spinors, but only if the Hilbert space of the fields from which they are constructed is positive. This was <u>not</u> the case for the original field commutators (2.16b), as we remarked, but only after gauge choices were imposed to leave a reduced set of variables for which positive relations such as (2.16a) hold. Therefore we may indeed trust (5.7); at least modulo all the formal quantum field theoretical problems associated with products of operators. This one-line proof of course contains a lot of physics - for example, the whole delicate process of splitting the strictly vanishing constraints into linear and remainder parts and showing that the integrals of the former really obey the global algebra just where the total quantities vanish. All this depends on boundary conditions, the residual global group and the significance of the local constraints as generators of gauge transformations.

Once established for simple supergravity, the basic statement (5.7) is easily seen to follow also when (supersymmetric) matter is included. This is because there is no formal difference, when (bounded) sources are present, in our earlier discussion. It is still the linearized parts of the constraints which obey (5.7), although their values of course now depend on the matter present. (That energy is positive for globally supersymmetric systems has naturally been known as long as the algebra itself; for simple systems, e.g. free fields it is manifest there. But when complicated couplings are introduced, one only knows that $P^0 \geq 0$ by virtue of (5.7), where the generators are volume integrals and not fluxes in the global case; it is not in general possible to prove $E > 0$ explicitly). Finally, one can also take (5.7) over to the most general examples of extended O(N), $N \leq 8$ supergravity, with or without matter couplings, simply because the only possible change in the global algebra there is known to involve at most the presence of central charges (Z_{ij}, Z'_{ij}), the equivalent of (5.6) being

$$\{\bar{Q}_i^\alpha, Q_j^\beta\} = \delta_{ij}\gamma_\mu^{\alpha\beta}P^\mu + Z_{ij}\mathbb{1}^{\alpha\beta} + Z'_{ij}\gamma_5^{\alpha\beta} \tag{5.8}$$

Here the i indices run over the internal space; but since $\mathrm{tr}\,\gamma^0 = \mathrm{tr}\,\gamma^0\gamma_5 = 0$, (5.7) still holds, i.e. $P^0 = 1/2N \sum_{\alpha,i} Q_i^\alpha \cdot Q_i^\alpha$.

Although our result applies so far only to the full quantum supergravity it actually does imply that classical Einstein gravity also has positive energy, although it says nothing about the intermediate step, quantum gravity. The argument goes in two parts: the first consists in considering only expectation values of the energy operator between states from which external spin $3/_2$ particles are absent. The second removes all loop contributions by taking the classical limit $\hbar \to 0$. It is in this step that we also lose quantum gravity, for we cannot remove closed loops involving spin $3/_2$ without also dropping pure graviton loops. So the proof starts from the fact that $\langle|P^0|\rangle \geq 0$ for any state. In particular, choose those states in which there are no external (on-shell) spin $3/_2$ particles. Next take the tree limit, $\hbar \to 0$, of these matrix elements. What is left is just the pure tree classical gravity Hamiltonian, and it is therefore also non-negative (and all the formal operator problems of Q_α^2 become irrelevant in this limit!). Furthermore, the proof also applies to classical gravity coupled to arbitrary spin 1 or 0 bose fields, since we may start with an extended supergravity model involving the latter, as well as possible supermatter sources. Now again set all external fermion lines to zero and take $\hbar \to 0$; QED. It suffices then, in showing that classical gravity has positive energy, that it could be (when quantized) part of a quantum supergravity system!

In addition to establishing that $P^0 \geq 0$, one may also show that there are no tachyonic ($P^\mu P_\mu = -m^2 > 0$) solutions. For if there were, we could find an asymptotic Lorentz frame in which $P^0 = 0$; but then Q^α would vanish by (5.7), and so therefore would all the P^μ. (This fact should also be demonstrable directly by "squaring" (5.6).) Likewise, there are probably no null ($P^2 = 0$) solutions with finite energy, since plane waves are not normalizable in infinite space. This would imply that, as physically required, all bounded solutions of supergravity (and by the previous arguments, of classical gravity) have future timelike four-momentum. This, together with the fact that the propagation anomalies of higher spin systems coupled to gravity have been shown [12] to be avoided in supergravity, speaks for the physical consistency of this new gauge theory.

The only systems for which the energy theorems fail to apply are those with cosmological constant. This is simply because asymptotically flat solutions are excluded there; at best there is a residual global DeSitter rather than Poincaré group. But P^μ and $P^\mu P_\mu$ are not constants in DeSitter space. So there is no energy to prove theorems about, either for super or ordinary gravity. Note also that for spaces of Euclidean signature neither Majorana spinors nor a preferred time direction or Hamiltonian are available. Perhaps one could make models in terms of an enlarged set of complex spinors, with statements about positive action replacing positive energy.

VI. CONCLUSIONS

We have seen that supergravity is a perfectly normal (double) gauge theory from the canonical point of view and that it has the properties necessary for a consistent physical interpretation. Fortunately for these lectures, we have not had to deal with the difficult questions we do not yet understand. These include a) symmetry breaking, since all particles in nature do not seem to have zero mass, b) particle systematics, i.e., can all "fundamental" particles be accommodated within the restricted multiplicities implicit in supergravity and c)renormalizability, or finiteness of at least some models. It is reassuring however, that the basic theory from which one begins is, despite (or rather because of) its peculiar bose/fermi symmetry, a very elegant start towards these problems.

Acknowledgements

I thank Dr. D. W. Sciama for hospitality at the Astrophysics Department, Oxford, while these notes were being prepared. This work was supported in part by a grant from the National Science Foundation.

REFERENCES

1) L. D. Faddeev & V. N. Popov, Sov. Phys. Usp. 16, 777 (1974); E. S. Fradkin & M. A. Vasiliev, Phys. Lett. 72B, 70 (1977).
2) D. Z. Freedman, P. van Nieuwenhuizen & S. Ferrara, Phys. Rev. D13, 3214 (1976); S. Deser & B. Zumino, Phys. Lett. 62B, 335 (1976).
3) S. Deser, J. H. Kay & K. S. Stelle, Phys. Rev. D16, 2448 (1977).
4) M. Pilati, Nucl. Phys. B132, 138 (1978).
5) S. Deser & C. Teitelboim, Phys. Rev. Lett. 39, 249 (1977).
6) M. T. Grisaru, Phys. Lett. 73B, 207 (1978).
7) C. Teitelboim, Phys. Rev. Lett. 38, 1106 (1977); Phys. Lett. 69B, 240 (1977); R. Tabensky & C. Teitelboim, Phys. Lett. 69B, 453 (1977).
8) R. Arnowitt, S. Deser & C. W. Misner in Gravitation: ed. L. Witten (Wiley, N.Y. 1962).
9) S. Deser & C. J. Isham, Phys. Rev. D14, 2505 (1976); J. E. Nelson & C. Teitelboim, Phys. Lett. 69B, 433 (1977).
10) J. Schwinger, Particles, Sources and Fields (Addison-Wesley, Reading, Mass. 1970).
11) D. Z. Freedman & A. Das, Nucl. Phys. B120, 221 (1977); P. K. Townsend, Phys. Rev. D15, 2802 (1977).
12) S. Deser & B. Zumino, Phys. Rev. Lett. 38, 1433 (1977).
13) K. Kuchar, J. Math. Phys. 18, 1589 (1977).
14) A. Das & D. Z. Freedman, Nucl. Phys. B114, 271 (1977); D. Senjanovic, Phys. Rev. D16, 307 (1977).
15) H. Van Dam & M. Veltman, Nucl. Phys. B22, 297 (1970); D. Boulware & S. Deser, Phys. Rev. D6, 3368 (1972).
16) D. Boulware, S. Deser & J. H. Kay, Physica (May 1979).
17) L. Brink, S. Deser, B. Zumino, P. Di Vecchia and P. Howe, Phys. Lett. 64B, 435 (1976).
18) C. Aragone & S. Deser, Brandeis preprint (1979).

EXTENDED SUPERSYMMETRY AND EXTENDED SUPERGRAVITY THEORIES

J. Scherk

Laboratoire de Physique Théorique de l'Ecole Normale

Supérieure - 24, rue Lhomond 75231 PARIS CEDEX 05

A INTRODUCTION

The failure of quantized gravity in interaction with quantized matter fields (whether scalar, spinor of vector fields) to provide renormalizable results even at the one loop level [1] has led to a situation analoguous to the puzzle of the V-A theory of weak interactions before the advent of gauge theories. The V-A theory of weak interactions was known to be correct at the tree approximation, but having a coupling constant whose dimension was the inverse of a mass squared ($G_F = 1.01\ 10^{-5} GeV^{-2}$) it was giving divergent and non renormalizable answers even at the one loop level. In a similar fashion, classical general relativity is known to describe gravity accurately, but quantum gravity leads to divergent results, except in the one loop case and for pure gravity [1]. The analogy with the V-A theory of weak interactions is even stronger if one remarks that the coupling constant of gravity also has the dimension of the inverse of a mass ($K \sim (G_N)^{1/2} \sim .8\ 10^{-19}\ GeV^{-1}$). One can think that the trouble with quantized gravity is due to the perturbative approach rather than with the theory itself. The other possibility is to look for alternative theories of gravity which would be free of divergences.

The mildest modifications of gravity which might work are the so-called extended supergravity theories [2 - 10], in which the graviton is a member of a finite supermultiplet of fields. This approach allows one to achieve Einstein's dream of unification of gravity with gauge fields, and also with spinors and scalar fields. It also introduces a set of new, and hitherto unobserved particles of spin 3/2 (gravitinos) which are essential for the cancellations which may insure renormalizability. Extended supergravity theories are indeed an improvement over the usual theory of quantum gravity interacting with matter fields since convergent results are obtained for one and two loop diagrams[11] although the presence of possible counterterms[12] at the three- and higher loop levels makes one doubt whether these models are convergent at all orders or not.

If extended supergravity theories fail to provide us with a fully renormalizable version of quantum gravity, a more drastic change would be to modify Einstein's theory at short distances, thereby introducing into the theory a parameter playing the role of a fundamental length. Only one example of such a model is known which does not violate some fundamental principles like : absence of ghosts (particles of negative metric) or tachyons (particles of imaginary masses), namely the spinning string model[13, 14] where the graviton is a member of an infinite multiplet of particles of increasing masses and spins. Actually this model is not in contradiction with the extended supersymmetry models since it connects smoothly with them in the limit where the fundamental length is put to zero[15, 16].

A useful technique to derive extended supersymmetric theories (i.e. theories having several spinorial charges) is to notice that they can be obtained by dimensional reduction of simple super-symmetric theories (i.e. having only one spinorial charge) in higher space-time dimensions, the extra dimensions one adds being of the Bose type[16 - 17]. This technique, and its applications to supersymmetry and supergravity will be discussed in the first section. In the next section we will discuss extended supergravity theories in 4 dimensions, namely their classification, the problem of field assignment versus particle content, their construction, their algebraic properties and in particular the curious antigravity-like effect observed in the O(2) theory coupled to a massive O(2) scalar multiplet [18 - 19], the general form of the fermionic equations of motion, the existence of a rigid U(N) invariance in an O(N) supergravity theory[18,6], and the gauging of the O(N) symmetry. Finally in the last section, we shall briefly discuss the spinning string model of gravity.

B EXTENDED SUPERSYMMETRY AND DIMENSIONAL REDUCTION

a) Generalities

Let us consider a 4 + N dimensional Minkowskian space-time, with one time-like direction and 3 + N space-like directions, the metric being $\eta_{\mu\nu}$ = diag (+ --...-). Further, let us consider a Poincaré invariant theory in that space-time, for instance a scalar $\lambda\,\phi^4$ theory :

$$S = \int d^{4+N}x \left\{ \frac{1}{2}\, \partial_\alpha \phi\, \partial_\alpha \phi - \frac{1}{2}\mu_0^2\, \phi^2 - \frac{1}{4!}\lambda_0\, \phi^4 \right\} \qquad \text{B.1}$$

To make sense of such a theory, at least classically, we can choose the N extra space-like dimensions to be circles of length L_1, ... L_N [20] . This is equivalent to assume that the field ϕ obeys the periodic boundary conditions :

$$\phi[x_{3+i} + L_i] = \phi[x_{3+i}] \qquad \text{B.2}$$

We can then expand ϕ in Fourier series :

$$\phi[x_\mu, x_{3+i}] = \frac{1}{(L_1 \cdots L_N)^{1/2}} \sum_{\{n_i\}} \phi_{\{n_i\}}(x_\mu) \exp 2\pi i \sum_1^N \frac{x_{3+i}\, n_i}{L_i} \qquad \text{B.3}$$

where the coefficients $\phi_{\{n_i\}}(x_\mu)$ are fields depending only upon the first four space-time coordinates, satisfying the reality condition $\phi^*_{\{-n_i\}} = \phi_{\{n_i\}}$. We can now integrate upon the extra x_{3+i} coordinates over one period and obtain the following result :

$$S = \int d^4x \left\{ \frac{1}{2}\, \partial_\mu \phi^*_{\{n_i\}}\, \partial^\mu \phi_{\{n_i\}} - \frac{1}{2} m^2_{\{n_i\}}\, \phi^*_{\{n_i\}}\, \phi_{\{n_i\}} \right.$$

$$\left. - \frac{\lambda}{4!}\, \phi_{\{n_i^1\}}\, \phi_{\{n_i^2\}}\, \phi_{\{n_i^3\}}\, \phi_{\{n_i^4\}} \prod_{i=1}^N \delta(n_i^1 + n_i^2 + n_i^3 + n_i^4) \right. \qquad \text{B.4}$$

where $\lambda = \dfrac{\lambda_0}{L_1 \ldots L_N}$ and

$$m^2_{\{n_i\}} = \mu_0^2 + 4\pi^2 \sum_1^N \frac{n_i^2}{L_i^2} \qquad \text{B.5}$$

So we see that in this theory where space-time has been partly compactified, there is an infinite sequence of particles of increasing masses. There are N abelian conserved charges, integrally quantized and the mass of the particles as a function of these charges is given by B.5 . This formula, in the case of N = 2 and $\mu_0 = 0$ is the formula obtained [21 - 24] for the particles and solitons of a supersymmetric gauge theory and suggests that the electric and magnetic charges can be considered as quantized fifth and sixth momenta, as was originally suggested for the electric charge by Kaluza and Klein[25] .

Dimensional reduction can be achieved smoothly by letting $L_i \to 0$ with λ fixed in which case only the $\phi_{\{0\}} = \phi$ field keeps a finite mass and we obtain the usual $\lambda \phi^4$ theory in 4 dimensions:

$$S \to S_R = \int d^4x \left\{ \frac{1}{2} \partial_\mu \phi \, \partial^\mu \phi - \frac{1}{2} \mu_0^2 \phi^2 - \frac{\lambda}{4!} \phi^4 \right\} \qquad \text{B.6}$$

In this case the reduced theory keeps no trace of its reduction from a 4 + N dimensional theory. In fact, one could have obtained it directly by assuming that $\phi[x_\mu, x_{3+i}]$ is independent of the x^{3+i} coordinates, i.e. setting

$$\partial_{3+i} \phi = 0$$

A less trivial example is obtained if we start from the Maxwell action in 4 + N dimensions and repeat the procedure :

$$S = -\frac{1}{4} \int d^{4+N}x \; F_{\alpha\beta} F^{\alpha\beta} \qquad \text{B.7}$$

where

$$F_{\alpha\beta} = \partial_\alpha A_\beta - \partial_\beta A_\alpha \qquad \text{B.8}$$

Skipping the intermediate stage of compactification, we get

directly the reduced theory by assuming A_α to be independent upon x^{3+i} and dropping the x^{3+i} integration. However, now splits into one vector A_μ and N scalars : $A_\alpha = (A_\mu, \phi_1, \ldots, \phi_N)$ and the reduced action is

$$S_R = -\frac{1}{4}\int d^4x\, F_{\mu\nu}F^{\mu\nu} + \frac{1}{2}\int d^4x \sum_1^N \partial_\mu \phi_i \partial^\mu \phi_i \qquad \text{B.9}$$

Here the resulting theory recalls the signature of the 4 + N dimensional metric since if some of the extra dimensions had been time like, the corresponding ϕ_i would have the wrong sign for the kinetic term. Also the Poincaré invariance in 4 + N dimensions has been broken down to the Poincaré invariance in 4 dimensions times O(N), A_μ being an O(N) singlet, the ϕ_i being in the vector representation of O(N). The O(N) invariance is really a property of the reduced theory, rather than of the compactified theory which has $U(1)^N$ invariance, and arises because in the $L_i \to 0$ limit, only the fields independent of x^{3+i} survive thus ensuring invariance under rotations in the N extra directions.

Note that both vectors and scalars are massless, however there is no transformation which rotates them into each other, and therefore it would be wrong to consider that B.9 is a unified model of vectors and scalars. Besides, this would contradict the the Coleman-Mandula [26] theorem.

Really non trivial examples of dimensional reduction are obtained if the original action is invariant under a simple supersymmetry algebra [27] in the 4 + N dimensional space-time.

$$\{Q_{\hat\alpha}, \bar{Q}_{\hat\beta}\} = 2(\Gamma^{\hat\nu})_{\hat\alpha\hat\beta} P_{\hat\nu} \qquad \text{B.10}$$

The Lorentz index $\hat\nu$ runs over all space-time indices, while the $\hat\alpha$, $\hat\beta$ are the indices of the Dirac matrices obeying the Clifford algebra in 4 + N dimensions :

$$\{\Gamma^{\hat\mu}, \Gamma^{\hat\nu}\} = 2\eta^{\hat\mu\hat\nu} \qquad \text{B.11}$$

If D = 4 + N the representation of this algebra has matrices of dimensions $2^{[D/2]}$ and here the dimension of the original space-time enters crucially. (The bracket designates the integral part of D/2).

In most theories, dimensional reduction ensures that P_{3+L} vanishes in the limit $L_i \to 0$ although it would be interesting to find examples where some of them do not vanish and yield central charges. We shall assume for simplicity that they do vanish. Also one can always construct a representation of the Clifford algebra in the form of a tensor product of 4 X 4 Dirac matrices (γ^μ, γ^5 or 1) by $2^{[N/2]} \times 2^{[N/2]}$ "internal symmetry" matrices such that for $\mu = 0,1,2,3$

$$\Gamma^\mu = \gamma^\mu \otimes 1 \tag{B.12}$$

The Dirac index α splits into a product of an ordinary Dirac index α and an internal symmetry index i so that after dimensional reduction we obtain the extended supersymmetry algebra :

$$\{ Q_{\alpha i} , \bar{Q}_{\beta j} \} = 2(\gamma^\nu)_{\alpha\beta} \delta_{ij} P_\nu \tag{B.13}$$

The resulting reduced theory will be invariant under an O(N) group and under M spinorial charges, where M is roughly $2^{[N/2]}$ up to factors of 1/2 due to the kind of conditions imposed upon the Q_α in the original space-time. The number M is conventionally defined with respect to Majorana spinors Q_α^i (i = 1,...M) in 4 dimensions so that to obtain the rank of the extended supersymmetry we have to consider carefully the problem of defining Majorana and Weyl spinors in 4 + N dimensions, which is the subject of the next section.

b) Majorana and Weyl Spinors in Arbitrary Space-Time Dimensions

For even space-time dimensions D one can introduce a matrix Γ^{D+1} which anticommutes with all the previous Dirac matrices Γ^0, Γ^1, ..., Γ^{D-1} and is given by $\Gamma^{D+1} = \eta \Gamma^0 \Gamma^1 \ldots \Gamma^{D-1}$ where η is chosen such that $(\Gamma^{D+1})^2 = 1$. Then $\eta^2 = (-1)^{(D-2)/2}$ (More generally if we have s space components and t time components $\eta^2 = (-1)^{(s-t)/2}$). Left and right handed Weyl spinors are defined by the conditions : $\Gamma^{D+1} \Psi_{L,R} = \pm \Psi_{L,R}$, which is compatible with Dirac's equation only in the case of massless spinors.

A Dirac spinor represents $2^{D/2}$ degrees of freedom while a Weyl spinor represents $1/2\ 2^{D/2}$ degrees of freedom.

For odd space-time dimensions, no Weyl spinor exists. If D = d + 1 where d is even, a representation of the Clifford algebra in d + 1 dimensions is obtained by adding to the previous matrices the space-like matrix $i\Gamma^{d+1}$.

Majorana spinors, on the other hand exist only for some special space-time dimensions. One can show the following theorem [15, 28]:

Th : For and only for D = 2,3,4 mod 8 there exists a Majorana representation of the Γ matrices, i.e. a representation where all Γ matrices are pure imaginary. For those dimensions, massive or massless Majorana spinors can be defined and in the Majorana representation they are pure real.

The proof goes as follows : Since Γ^μ and $-\Gamma^{\mu *}$ (complex conjugate) satisfy the same Clifford algebra and the representation of the Γ matrices is irreducible, there exists a matrix B which one may call the complex conjugation matrix, such that

$$\Gamma^{\mu *} = - B \Gamma^\mu B^{-1}$$

B has involutive properties which ensure that B and B^{*-1} are proportional to each other $BB^* = \epsilon I$ and one can scale B such that $|\epsilon| = 1$. Further, one can show that ϵ is real and can take only two values +1 or -1. The value of ϵ is independent upon the representation of the Γ matrices chosen.

If Ψ is a spinor satisfying the Dirac equation

$$((i \not\partial - e \not A) - m) \Psi = 0 \qquad \text{B.14}$$

$B^{-1}\Psi^*$ satisfies the same Dirac equation with e changed into -e and thus represents the antiparticle of the particle described by the spinor Ψ. Majorana spinors are defined as being their own antiparticles which implies e = 0 and $\Psi = B^{-1}\Psi^*$. This condition which relates the real and imaginary parts of cannot always be imposed as it implies

$$\Psi^* = B \Psi \qquad \text{B.15}$$

and by complex conjugation $\Psi^* = (B^{-1})^* \Psi$ so that

$$\Psi = B^* B \Psi = \epsilon \Psi \qquad \text{B.16}$$

which is possible only for $\epsilon = +1$

For $\epsilon = -1$ as is the case for instance in Euclidean 4 dimensional space-time one needs a different definition of Majorana spinors [29].

To compute $\epsilon(D)$ one has to bring into focus the signature of the metric which here is + --...-

The Γ matrices have hermiticity or antihermiticity properties :

$$\Gamma^{0\dagger} = \Gamma^{0} \qquad ; \quad \Gamma^{i\dagger} = - \Gamma^{i} \tag{B.17}$$

and therefore the matrix of hermitian conjugation is Γ^0 itself :

$$\Gamma^{\mu\dagger} = \Gamma^{0}\,\Gamma^{\mu}\,\Gamma^{0} \tag{B.18}$$

We can now define the charge conjugation matrix C which relates the matrices Γ^μ and $-\Gamma^{\mu T}$, which have the same algebra :

Let
$$B^T = C\,\Gamma^0 . \tag{B.19}$$

then $-\Gamma^{\mu T} = C\,\Gamma^\mu C^{-1}$ where ^{T} designates the transposition operation.

Computing $(\Gamma^\mu)^T$ in two different ways, i.e. as $(\Gamma^{\mu *})^\dagger$ or as $(\Gamma^{\mu\dagger})^*$ one obtains two different expressions namely in the first way $-(B^+)^{-1}\ \Gamma^0\,\Gamma^\mu\ (B^{+-1}\,\Gamma^0)^{-1}$ and in the second way $-(B\,\Gamma^0)\,\Gamma^\mu\,(B\,\Gamma^0)^{-1}$.

This implies that B and $(B^+)^{-1}$ are proportional to each other, and because of the normalization of B that B is a unitary matrix : $BB^+ = B^+B = 1$. Since we know that $BB^* = \epsilon\, I$ this implies that

$$B = \epsilon\, B^T \tag{B.20}$$

$$C^T = -\epsilon\, C \tag{B.21}$$

so that the matrices B and C are either symmetric or antisymmetric. To fix ϵ one counts in two different ways the number of $2^{D/2} \times 2^{D/2}$ independent antisymmetric matrices (assume here D to be even). Obviously their number is $1/2\ 2^{D/2}(2^{D/2} - 1)$.

We can also count them by introducing a complete basis for these matrices by forming all antisymmetrized products of Γ matrices :

$$\Gamma^{(0)} = 1 \quad , \quad \Gamma^{(1)\mu} = \{\Gamma^\mu\} \quad , \quad \Gamma^{(2)\mu\nu} = \frac{1}{2}[\Gamma^\mu, \Gamma^\nu]$$
$$\ldots \quad \Gamma^{(D)\mu_1\ldots\mu_D}$$

There are $\binom{D}{n}$ independent matrices of the $\Gamma^{(n)}$ type. It is easy now to show that :

$$C\,\Gamma^{(n)}C^{-1} = (-1)^n (-1)^{\frac{n(n-1)}{2}}\ \Gamma^{(n)T} \tag{B.22}$$

which implies that $C\,\Gamma^{(n)}$ is either symmetric or antisymmetric :

$$(C\,\Gamma^{(n)})^T = \epsilon\,(-1)^{\frac{(n-1)(n-2)}{2}}\,C\,\Gamma^{(n)}$$

B.23

The number of antisymmetric matrices can be counted by introducting an operator which counts 1 for each antisymmetric matrix and 0 for a symmetric matrix :

$$N = \sum_{n=0}^{D} \frac{1}{2}\left[1 - \epsilon(-1)^{\frac{(n-1)(n-2)}{2}}\right]\binom{D}{n} = \frac{1}{2}\,2^{D/2}\,(2^{D/2}-1)$$

B.24

The summation can be done noticing that

$$(-1)^{\frac{(n-1)(n-2)}{2}} = -\frac{1}{2}\left[(1+i)\,i^n + (1-i)(-i)^n\right]$$

B.25

which reduces it to a geometric series.

Finally this yields $\epsilon = -\sqrt{2}\cos\frac{\pi}{4}(D+1)$

So $\epsilon = +1$ for D = 2,4 mod 8

$\epsilon = -1$ for D = 0,6 mod 8. For D = 2,4 mod 8 one can also show that a pure imaginary representation of the Γ matrices exists in which case B = 1, C = Γ^0 and a Majorana spinor is pure real.

In D = 0,6 mod 8 no pure imaginary representation of the matrices exists and Majorana spinors cannot be defined (at least with this definition of Majorana spinors ; one can also consider more general cases where the spinors bear an internal symmetry index and define Majorana spinors taking this internal symmetry into account) [30].

For odd D, setting D = d + 1 where d is even, one can enquire whether there exists a pure imaginary representation of the Γ matrices. In d + 1 dimensions, the Γ matrices have the same dimension as in d dimensions and are given by :

$$\Gamma^0, \Gamma^1, \ldots, \Gamma^{d-1} \quad \text{and} \quad \Gamma^{D-1} = i\eta\,\Gamma^0 \ldots \Gamma^{d-1}$$

where the i has been introduced so as to make the last matrix space-like. For d = 2,4 mod 8 the first d matrices can be chosen pure imaginary. $\Gamma^0 \ldots \Gamma^{d-1}$ is real and if η is real we have a Majorana representation of the Γ matrices in d + 1 dimensions, and hence Majorana spinors exist. This implies that $\eta^2 = (-1)^{d/2-1} = +1$ hence d/2 must be odd so that D = 3 mod 8 works while D = 5 mod 8 does not.

Finally one can enquire whether Majorana-Weyl massless spinors can exist. D must be even and $\Psi = \Psi^*$ must be compatible with $\Gamma^{D+1}\Psi = \Psi$ i.e. $\Gamma^{D+1} = \eta\,\Gamma^0 \ldots \Gamma^{D-1}$ must be a real matrix. This implies again $\mathcal{C}^2 = +1$, i.e. D/2 odd and D = 2,4 mod 8. Hence Majorana (real) Weyl spinors can be defined for D = 2 mod 8 only.

This theorem can be extended to the case where instead of one time dimension, one has t time directions and s spatial directions. For s + t even one finds that a pure imaginary representation of the Γ matrices and Majorana spinors exists if s - t = 0,2 mod 8.

Also in the case of massless spinors, a more general definition of Majorana spinors can be given since the Dirac equation reduces to $i\,\alpha^\mu \partial_\mu \Psi = 0$ where $\alpha^0 = 1$, $\alpha^i = \Gamma^0 \Gamma^i$, $\{\alpha^i, \alpha^j\} = 2\,\delta^{ij}$. If one can find a representation of the α matrices which is pure real, Majorana spinors can be defined to be pure real spinors [31]. It turns out that for D = 6,8 mod 8 one can also, in this case, find a real representation of the α^i, and define Majorana spinors. In general massless Majorana spinors can be shown to be equivalent to Weyl spinors, the only interesting case being when a spinor can be both Majorana and Weyl, which occurs for D = 2 mod 8.

We can now determine the number M of supersymmetry charges which is obtained by dimensional reduction from D dimensions to 4 dimensions from a simple supersymmetry algebra. The charge $Q_{\hat{a}}$ decomposes into M Majorana spinor charges Q_α^i, each of these spinors representing two real degrees of freedom. $Q_{\hat{a}}$ itself represents $2^{D/2}$ r degrees of freedom where r is a reduction coefficient, taking into account the nature of the spinorial charge : r = 1 for Dirac charges ; r = 1/2 for Majorana or Weyl charges ; r = 1/4 for Majorana-Weyl charges.

It follows that the number of Majorana spinorial charges of the extended supersymmetry algebra in 4 dimensions is given by $M = 1/2\ 2^{[D/2]}\ r$.

We plot the various values of M which emerge from D = 4 up to D = 12.

Type \ D	4	5	6	7	8	9	10	11	12
Dirac	2	2	4	4	8	8	16	16	32
Majorana	1	-	-	-	-	-	8	8	16
Weyl	1	-	2	-	4	-	8	-	16
Maj-Weyl	-	-	-	-	-	-	4	-	-
G	-	-	O(2)	O(3)	O(4)	O(5)	O(6)	O(7)	O(8)

On the last line we have plotted the obvious symmetry group which results from dimensional reduction. The real group of invariance of the reduced theory may however be bigger, as for instance in the case of massless theories the superconformal group may be a symmetry after reduction.

Thus M increases rapidly with D. As we shall see in section c) representations of extended supersymmetry algebra exist with

$J_{MAX} = \frac{1}{2}$ for $M \leq 2$
$J_{MAX} = 1$ for $M \leq 4$
$J_{MAX} = 2$ for $M \leq 8$

So multiplets with $J_{MAX} = 1/2$ and an explicit mass term can be obtained only from D = 4,5. Multiplets with $J_{MAX} = 1$ (extended supersymmetric Yang-Mills theories) can exist up to D = 10 for Majorana-Weyl spinors [32]. Supergravity theories, which have $J_{MAX} = 2$ can exist up to D = 11 [32], for Majorana spinors. Beyond D = 11 it seems impossible to construct non free supersymmetric theories as for D = 12 we need already to introduce spin 4 particles.

This algebraic argument only suggests that these theories may exist. To see whether they actually exist, one has to construct them explicitly. We give below two interesting examples : the supersymmetric Yang-Mills theory in D = 10, which upon dimensional reduction gives a representation of M = 4 supersymmetry with O(6) invariance, and the supergravity theory in D = 11 dimensions which gives a supergravity theory with M = 8 spinorial charges.

c) Supersymmetric Yang-Mills Theory for D = 10 [16]

Let us consider the Yang-Mills theory of an arbitrary compact gauge group G, coupled to Majorana-Weyl spinors in the adjoint representation of the group G, for D = 10 dimensions.

The generators X_i of the gauge group G obey the algebra $[X_k, X_l] = i f_{klm} X_m$ where the structure constants f_{klm} are real and completely antisymmetric. Defining the vector and spinor matrices :

$$A_\mu = A_\mu^i X^i$$
$$\lambda = \lambda^i X^i \qquad \text{B.26}$$

We have

$$G_{\mu\nu} = \partial_\mu A_\nu - \partial_\nu A_\mu + ig[A_\mu, A_\nu]$$
$$D_\mu \lambda = \partial_\mu \lambda + ig[A_\mu, \lambda]$$

and the action reads

$$S = \int d^{10}x \, \mathrm{Tr}\left(-\frac{1}{4} G_{\mu\nu} G^{\mu\nu} + \frac{1}{2} i \bar{\lambda} \Gamma^\mu D_\mu \lambda\right) \qquad \text{B.27}$$

To see that such an action has a chance to be supersymmetric invariant, we count the number of degrees of freedom of bosons and fermions. Since they are both in the adjoint representation of the group, the group index is irrelevant. Each vector boson on shell is purely transverse because of gauge invariance and masslessness, and so describes D - 2 = 8 physical degrees of freedom. Each Majorana-Weyl spinor describes also $2^{D/2}$ 1/4 = 8 degrees of freedom, as it should be.

One can indeed verify that the action is invariant under the supersymmetry transformation laws :

$$\delta A_\mu = i \bar{\epsilon} \Gamma_\mu \lambda \qquad \text{B.28}$$

$$\delta \lambda = G_{\mu\nu} \sigma^{\mu\nu} \epsilon \qquad \text{B.29}$$

where $\sigma^{\mu\nu} = \frac{1}{4}[\Gamma^\mu, \Gamma^\nu]$

The proof of the invariance of the action is rather simple. Two kinds of terms arise from the variation of the fields. First there are terms of the type Tr($\bar{\epsilon}$ λ D G) which arise from the variation of G^2 and of λ in the kinetic term of the spinor field. These terms cancel by integration by part, and use of the identity :

$$\Gamma^\mu \sigma^{\alpha\beta} = \frac{1}{2}(\eta^{\mu\alpha}\Gamma^\beta - \eta^{\mu\beta}\Gamma^\alpha) - \frac{1}{7!}\epsilon^{\mu\nu\beta\lambda_1\ldots\lambda_7}\Gamma^{11}\Gamma^{\lambda_1}\ldots\Gamma^{\lambda_7} \qquad \text{B.30}$$

and of the Bianchi identity :

$$\epsilon^{\mu\lambda_1\ldots\lambda_7\alpha\beta} D_\mu G_{\alpha\beta} = 0 \qquad \text{B.31}$$

together with the Majorana property $\bar{\epsilon}\Gamma^\beta\lambda = -\bar{\lambda}\Gamma^\beta\epsilon$.
The second kind of terms are cubic in the Fermi fields and arise from the variation of A_μ in the minimal coupling of A_μ to the spinors. This term turns out to be proportional to X = $f_{ijk}\,\bar{\epsilon}\Gamma_\mu\lambda_i \times \bar{\lambda}_j\Gamma^\mu\lambda_k$. Its vanishing can be shown by use of a Fierz transformation :

$$X = \frac{1}{2^{D/2}} f_{ijk}\,\bar{\epsilon}\Gamma_A\lambda_i\,\bar{\lambda}_j\Gamma^\mu\Gamma^A\Gamma_\mu\lambda_k \qquad \text{B.32}$$

where the antisymmetry property of the f_{ijk} is used. The Γ^A are all the antisymmetrized products of Γ matrices normalized such that $\Gamma_A^2 = \pm 1$, $\Gamma_A\Gamma^A = +1$. Since the λ_i are Weyl spinors of the same handedness (left or right), only $\Gamma_A = \Gamma_\alpha, \Gamma_{\mu\nu\rho}, \Gamma_{\mu\nu\rho\sigma\lambda}, \Gamma_{\mu\nu\rho}\Gamma^{11}, \Gamma_\mu\Gamma^{11}$ contribute and the last two double up the contribution of the first two. A general result for even D is that $\Gamma^\mu\Gamma_{\alpha_1\ldots\alpha_{D/2}}\Gamma_\mu = 0$ so that the $\Gamma^{(5)}$ contribution vanishes. Finally the $\Gamma^{(3)}$ contribution vanishes too because :

$$\bar{\lambda}_j\Gamma^{(3)}_{\mu\nu\rho}\lambda_i = -\bar{\lambda}_i\Gamma^{(3)}_{\rho\nu\mu}\lambda_j = +\bar{\lambda}_i\Gamma^{(3)}_{\mu\nu\rho}\lambda_j$$

This shows that $\bar{\lambda}_j \Gamma^{(3)}_{\mu\nu\rho} \lambda_i$ is symmetric in i, j and its contraction with f_{ijk} gives zero. The final result of the Fierz transformation is X = -1/2 X = 0.

The same action is invariant under supersymmetry transformation for D = 6 and Weyl spinors and for D = 4 with Majorana or Weyl spinors [17].

It is now easy to reduce the theory from 10 to 4 dimensions and obtain the largest extended supersymmetric theory with $J_{MAX} \leq 1$, having 4 spinorial charges.

We introduce 6 real, antisymmetric 4 x 4 matrices obeying the SU(2) x SU(2) algebra :

$$\begin{aligned} \{\alpha^i, \alpha^j\} &= \{\beta^i, \beta^j\} = -2\,\delta^{ij} \\ [\alpha^i, \beta^j] &= 0 \\ [\alpha^i, \alpha^j] &= -2\,\epsilon^{ijk}\alpha^k \\ [\beta^i, \beta^j] &= -2\,\epsilon^{ijk}\beta^k \end{aligned} \qquad \text{B.33}$$

A Majorana representation of the Clifford algebra in D = 10 is given by

$$\begin{aligned} \Gamma^\mu &= \gamma^\mu \otimes \begin{pmatrix} I_4 & 0 \\ 0 & -I_4 \end{pmatrix} \\ \Gamma^{3+j} &= i\otimes\beta^3 \begin{pmatrix} 0 & \alpha^j \\ \alpha^j & 0 \end{pmatrix} \\ \Gamma^{6+i} &= \gamma^5 \otimes \begin{pmatrix} \beta^i & 0 \\ 0 & \beta_i \end{pmatrix} \end{aligned} \qquad \text{B.34}$$

where $\beta^i = \beta_i$ except for i = 3 where $\beta_3 = -\beta^3$

$$\Gamma^{11} = \Gamma^0 \ldots \Gamma^9 = 1 \otimes \begin{pmatrix} 0 & -\beta_3 \\ \beta_3 & 0 \end{pmatrix}$$

In this representation a Majorana-Weyl spinor in 10 dimensions is of the form $\Psi = \begin{pmatrix} \psi_K \\ \beta_3\psi_K \end{pmatrix}$ where K = 1,2,3,4 and ψ_K are four Majorana (real) spinors.

Dropping all dependence upon X^{3+i}, and relabelling $A_{\hat\mu} = (A_\mu, A_i, B_i)$ i = 1,2,3 the extra components of $A_{\hat\mu}$ appear as three scalars and three pseudoscalars. The reduced action in D = 4 reads

$$S_R = \int d^4x \, Tr \{ -\frac{1}{4} G_{\mu\nu} G^{\mu\nu} + \frac{1}{2} (D_\mu A^i)^2 + \frac{1}{2} (D_\mu B^i)^2 + \frac{1}{2} i \bar{\lambda}_k \gamma^\mu D_\mu \lambda_k + \frac{1}{2} g \bar{\lambda}_k [(\alpha^j_{kl} A_j + i \gamma^5 \beta^j_{kl} B_j), \lambda_l] + \frac{1}{4} g^2 ([A_i, A_j]^2 + [B_i, B_j]^2 + 2[A_i, B_j]^2)) \}$$

B.35

This theory has non minimal couplings of the Yukawa and ϕ^4 type, proportional to g and g^2 respectively. The potential has valleys so that one can break spontaneously the O(6) symmetry by giving non zero constant vacuum expectation values to the scalar fields. Further it has remarkable properties as it turns out that the Gell-Mann-Low function vanishes for 1 and 2 loops [33], so that this model could be exactly scale invariant, which is a fascinating conjecture.

In addition to being invariant under four separate supersymmetry transformations, the model is also invariant under global ("rigid") SO(6) $\sim$ SU(4) transformations :

$$\begin{aligned} \delta A_\mu &= 0 \\ \delta \lambda_k &= -\frac{1}{4} [\epsilon^{ijm} \beta^m \Lambda_{ij} + \epsilon^{ijm} \alpha^m \Lambda'_{ij} + i \gamma^5 \alpha^i \beta^j \tilde{\Lambda}_{ij}]_{kl} \lambda_l \\ \delta A_i &= \Lambda'_{ij} A_j - \tilde{\Lambda}_{il} B_l \\ \delta B_i &= \Lambda_{ij} B_j + \tilde{\Lambda}_{li} A_l \end{aligned}$$

B.36

where the Λ_{ij} , Λ'_{ij} matrices are antisymmetric. The subgroup of SU(4) which commutes with parity is simply O(4).

By truncating this theory, one obtains also easily the M = 2 and M = 1 cases which are asymptotically free.

The M = 2 model, which can also be obtained from the D = 6 case [17] is of special interest because it provides a supersymmetric extension of the Georgi-Glashow model and admits classical solutions that are the supersymmetric generalizations of the magnetic monopole and dyon. The quantum corrections to the monopole mass seem to vanish at all orders [23,24] .

d) Simple Supergravity Theory for D = 11

As we have seen from algebraic arguments D = 11 is the limiting space-time dimension for which we can possibly construct a supergravity ($J_{MAX} = 2$) theory. To construct it explicitly [34] we should first find the field content. Obviously the theory must contain the graviton, described by an "elfbein" $V^a_\mu(x)$, and a Majorana spin-vector $\psi_\mu(x)$. However, another field must be introduced as we can see directly from counting the degrees of freedom described by the fields :

V^a_μ, because of coordinate invariance and local Lorentz invariance, describes on-shell a transverse, symmetric, traceless tensor in D - 2 dimensions. So it represents 1/2 (D - 2)(D - 1) - 1 = 2 degrees of freedom for D = 4 but 44 degrees of freedom for D = 11.

ψ_μ, in flat space satisfies the generalized Rarita-Schwinger equation $\Gamma^{\mu\nu\rho}\partial_\nu\psi_\rho = 0$, invariant under $\psi_\rho \to \psi_\rho + \partial_\rho \epsilon$. On the mass shell ($p^2 = 0$) the equation can be shown to reduce to $p.\psi = 0$, $\not p\,\psi_\mu = 0$ after the gauge choice $\gamma\cdot\psi = 0$ [16].

This shows that as far as the μ index is concerned ψ_μ behaves like a massless vector A_μ and describes D - 2 degrees of freedom. The Dirac index implies $2^{[D/2]}$ r degrees of freedom. In addition we have to take into account that the gauge condition $\gamma\cdot\psi = 0$ eliminates the spin 1/2 degree of freedom. So the number of degrees of freedom described by a massless Rarita-Schwinger field is given on shell by :

$$(D-3)\;2^{[D/2]}\,r \qquad \text{B.37}$$

For D = 4 and for Majorana or Weyl spinors, this number is 2 so that a representation of supersymmetry may (and does) exist with just ψ_μ, V^a_μ as basic fields : this is the well-known "ordinary" simple supergravity theory.

For D = 11 this number is 128 so that 84 bosonic degrees of freedom need to be introduced. Since we are dealing with massless particles, the group O(D - 2) = O(9) classifies [32] the transverse components of the fields which represent the physical degrees of freedom. There is indeed an irreducible representation of O(9) which is 84 dimensional obtained by considering a tensor A_{ijk} i,j,k = 1,...9 completely antisymmetric in the i, j, k, indices. Indeed, this represents $\binom{9}{3} = 84$ degrees of freedom. The existence of this tensor as a basic field can also be deduced from arguments drawn from the dual spinor model [34].

To have a covariant description, we introduce a completely antisymmetric massless gauge potential $A_{\mu\nu\rho}$, analogous to the vector potential A_μ of electromagnetism and require an analogous abelian gauge invariance of the action under

$$\delta A_{\mu\nu\rho} = \partial_\mu \zeta_{\nu\rho} + \partial_\nu \zeta_{\rho\mu} + \partial_\rho \zeta_{\mu\nu} \qquad \text{B.38}$$

where $\zeta_{\mu\nu}(x) = - \zeta_{\nu\mu}(x)$ is the gauge parameter. In the kinetic term $A_{\mu\nu\rho}$ will appear only through the 4-form field strength : $F_{\mu\nu\rho\sigma} = 4\, \partial_{[\mu} A_{\nu\rho\sigma]}$ where $[\ ]$ indicates the antisymmetrized sum over all permutations of indices divided by 4 !

To be certain that V_μ^a, ψ_μ, $A_{\mu\nu\rho}$ is the correct field content of D = 11 simple supergravity, and that for instance 84 scalar fields would not work we can reduce the field content to D = 4. We know that we must obtain a supergravity theory with 8 spinorial Majorana charges, and the particle content of the M = 8 theory is known from arguments which will be derived in the next section. Comparing the field contents gives the following table, where we indicate the number of spin 2, 3/2,...0 fields obtained from reducing to D = 4 V_μ^a, ψ_μ, $A_{\mu\nu\rho}$ and the total and comparing it to the M = 8, D = 4 extended supergravity field content.

J	V_μ^a	$A_{\mu\nu\rho}$	ψ_μ	Total	M = 8
2	1	0	0	1	1
$\frac{3}{2}$	0	0	8	8	8
1	7	21	0	28	28
$\frac{1}{2}$	0	0	56	56	56
0	28	35+7	0	70	35+35

As we see, the particle content matches. An obvious problem however in comparing the two theories is that the M = 8 theory which has been constructed perturbatively up to order K^2 has a built-in O(8) invariance, while the theory obtained by reducing from D = 11 to D = 4 has only a manifest O(7) invariance. Both have 8 spinor charges so that it is unlikely that they are two different theories, but complicated field redefinitions have to be done to relate the two formulations [49].

The theory is constructed by the usual Noether procedure which stops at order K^2 in the action and K in the transformation laws. In second order formalism where the connection $\omega_{\mu ab}$ is given explicitly in terms of V_μ^a, and ψ_μ, one obtains :

$$\mathcal{L} = -\frac{V}{4K^2} R(\omega) - \frac{i}{2} V \bar{\Psi}_\mu \Gamma^{\mu\nu\rho} D_\nu \left(\frac{\omega+\hat{\omega}}{2}\right) \Psi_\rho$$

$$- \frac{V}{48} F_{\mu\nu\rho\sigma} F^{\mu\nu\rho\sigma} + \frac{2\kappa}{(144)^2} \epsilon^{\alpha_1 \dots \alpha_8 \mu\nu\rho} F_{\alpha_1 \dots \alpha_4} F_{\beta_1 \dots \beta_4} A_{\mu\nu\rho}$$

$$+ \frac{\kappa V}{192} \left(\bar{\Psi}_\mu \Gamma^{\mu\nu\alpha\beta\gamma\delta} \Psi_\nu + 12 \bar{\Psi}^\alpha \Gamma^{\gamma\delta} \Psi^\beta \right) \left(F_{\alpha\beta\gamma\delta} + \hat{F}_{\alpha\beta\gamma\delta} \right) \qquad \text{B.39}$$

The transformation laws of the fields are :

$$\delta V_\mu^a = -i\kappa \bar{\epsilon} \Gamma^a \Psi_\mu \qquad \text{as in } D = 4$$

$$\delta A_{\mu\nu\rho} = \frac{3}{2} \bar{\epsilon} \Gamma_{[\mu\nu} \Psi_{\rho]}$$

$$\delta \Psi_\mu = \frac{1}{\kappa} \hat{D}_\mu \epsilon \qquad \text{B.40}$$

where

$$\hat{D}_\mu = \partial_\mu + \frac{1}{4} \hat{\omega}_{\nu ab} \Gamma^{ab} \Psi_\mu + \frac{i\kappa}{144} \left(\Gamma^{\alpha\beta\gamma\delta}{}_\mu - 8 \Gamma^{\beta\gamma\delta} \delta_\mu^\alpha \right) \hat{F}_{\alpha\beta\gamma\delta}$$

$\hat{\omega}_{\nu ab}$ is the supercovariant (i.e. it transforms without $\partial\epsilon$ terms) extension of $\omega^\circ_{\nu ab}(V)$ and has the same expression as in D = 4 :

$$\hat{\omega}_{\nu ab} = \omega^\circ_{\nu ab} + \frac{i\kappa^2}{2} \left(\bar{\Psi}_\mu \Gamma_b \Psi_a - \bar{\Psi}_\mu \Gamma_a \Psi_b + \bar{\Psi}_b \Gamma_\mu \Psi_a \right) \qquad \text{B.41}$$

It differs from the value one would obtain for ω by varying in the first order formalism ω independently (replacing $(\omega + \hat{\omega})/2$ by ω in the kinetic term of the Ψ field) and which is given by :

$$\omega_{\mu ab} = \hat{\omega}_{\mu ab} - \frac{i\kappa^2}{4} \bar{\Psi}_\alpha \Gamma_{\mu ab}{}^{\alpha\beta} \Psi_\beta \qquad \text{B.42}$$

and is not a supercovariant quantity. Hence the first order formalism is of little use in this model, except to group terms in a certain way so that all K^2 quartic terms are hidden either in ω, $\hat{\omega}$ or $\hat{F}_{\mu\nu\rho\sigma}$, which is the supercovariant extension of $F_{\mu\nu\rho\sigma}$ and is given by :

$$\hat{F}_{\mu\nu\rho\sigma} = F_{\mu\nu\rho\sigma} - 3 \bar{\Psi}_{[\mu} \Gamma_{\nu\rho} \Psi_{\sigma]} \qquad \text{B.43}$$

In this theory the Ψ equation of motion is very simple and reads

$$\Gamma^{\mu\nu\rho} \hat{D}_\nu \Psi_\rho = 0 \qquad \text{B.44}$$

The geometrical interpretation of $A_{\mu\nu\rho}$ as a gauge field in this theory is still unclear. If V^a_μ gauges the translations P^a and Ψ_μ^α the supersymmetry transformations Q^α, one does not know what transformations (if any) $A_{\mu\nu\rho}$ gauges.

The algebra closes on shell as usual, giving coordinate transformations, gauge transformations on $A_{\mu\nu\rho}$ and field dependent supersymmetry transformations. An interesting problem would be to find a set of auxiliary fields allowing the closure of the algebra without use of the equations of motion.

The dimensional reduction is rather complicated and the relation to the O(8) theory needs field redefinitions.[49] One piece of the reduction which is rather easy is the reduction of Einstein's action from 4 + N dimensions to 4 dimensions. One assumes the metric g_{AB} to be independent of the x^{3+i} extra coordinates and parametrizes it in the following way :

$$g_{AB}(x) = \begin{pmatrix} g_{\mu\nu} + A^i_\mu A^j_\nu g_{ij} & A^j_\mu g_{ij} \\ A^i_\mu g_{ij} & g_{ij} \end{pmatrix} \qquad \text{B.45}$$

Under coordinate transformations in the 3+i directions, but x^{3+i} independent, A^i_μ transform as N abelian vector fields :

$$\delta A^i_\mu = \partial_\mu \zeta^{4+i}(x)$$

and the result of the reduction is :

$$V_{(4+N)} = V_{(4)} \left(\det g_{ij}\right)^{1/2} \qquad \text{B.46}$$

$$R_{4+N} = R_4 - \frac{1}{4} g_{ij} F^i_{\mu\nu} F^{j\mu\nu}$$
$$+ \frac{1}{4} g^{\rho\sigma} \left(g^{ik} g^{jl} - g^{il} g^{jk} \right) \partial_\rho g_{ik} \partial_\sigma g_{jl} \qquad \text{B.47}$$

where g^{ij} is the inverse of the matrix g_{ij} and 4 dimensional indices are raised and lowered by the metric $g_{\mu\nu}$. This piece of the action describes ordinary gravity in interaction with N abelian vector fields and N(N+1) /2 scalars. It has obvious O(N) invariance, and in the case $g_{ij} = \delta_{ij}$ and N = 1, reduces to the usual Kaluza-Klein [25] formulation of Einstein gravity coupled to electromagnetism.

Now we turn to the subject of extended supersymmetry and supergravity theories in D = 4 as they have been directly constructed without considering them as subcases of 4+N simple supersymmetric and supergravity theories.

C EXTENDED SUPERSYMMETRY AND SUPERGRAVITY IN FOUR DIMENSIONS

a) The Representation Content of Extended Supersymmetry

Let us work out the representation content of the simple supersymmetry algebra :

$$\{ Q_\alpha , \bar{Q}_\beta \} = 2(\gamma^\nu)_{\alpha\beta} P_\nu \qquad \text{C.1}$$

$$[Q_\alpha , M_{\mu\nu}] = i(\sigma^{\mu\nu})_{\alpha\beta} Q_\beta \qquad \text{C.2}$$

in D = 4 dimensions, the Q_α being Majorana charges, and for the case of zero mass representations. We can use the following representation of the γ matrices

$\gamma^i = \begin{pmatrix} 0 & \sigma^i \\ -\sigma^i & 0 \end{pmatrix}$ $\gamma^0 = \begin{pmatrix} 0 & 1 \\ 1 & 0 \end{pmatrix}$ which is constructed such that

$$\gamma^5 = \begin{pmatrix} 1 & 0 \\ 0 & -1 \end{pmatrix}$$

Any spinor can then be decomposed into left and right handed parts (dotted and undotted indices) :

$$\Psi = \begin{pmatrix} \Psi_\alpha \\ \Psi_{\dot\alpha} \end{pmatrix} \qquad \alpha , \dot\alpha = 1,2$$

and the algebra becomes

$$\{ Q_\alpha , \bar{Q}_\beta \} = \{ Q_{\dot\alpha} , \bar{Q}_{\dot\beta} \} = 0 \qquad \text{C.3}$$

$$\{ Q_\alpha , \bar{Q}_{\dot\beta} \} = 2(\sigma^\nu)_{\alpha\dot\beta} P_\nu \qquad \text{C.4}$$

where $\sigma^\nu = (1 , \sigma^i)$

In this representation $C = i\gamma^2\gamma^0$ and a Majorana spinor satisfies $\Psi = C\gamma^0 \Psi^* = i\gamma^2\Psi^*$

or $$\begin{pmatrix} \Psi_\alpha \\ \Psi_{\dot\alpha} \end{pmatrix} = \left(\begin{array}{cc|cc} & 0 & 0 & 1 \\ & & -1 & 0 \\ \hline 0 & -1 & & 0 \\ 1 & 0 & & \end{array}\right) \begin{pmatrix} \Psi_\alpha^* \\ \Psi_{\dot\alpha}^* \end{pmatrix} \qquad \text{C.5}$$

This equation relates dotted and undotted indices by

$$\Psi_\alpha = \epsilon_{\alpha\dot\beta} \Psi_{\dot\beta}^* \qquad \text{C.6}$$

So that the algebra can be written as :

$$\{ Q_\alpha , Q_\beta \} = 0 \qquad \text{C.7}$$

$$\{ Q_\alpha , Q_\beta^* \} = 2(\sigma^\nu)_{\alpha\beta} P_\nu \qquad \text{C.8}$$

Thus the Q_α appear as two annihilation operators and the Q^*_α as two creation operators.

In the zero mass case we can choose $P_\mu = (1,0,0,1)\,|P|$ and since

$$\sigma^0 + \sigma^3 = 2\begin{pmatrix}1 & 0\\ 0 & 0\end{pmatrix} \qquad \text{C.9}$$

we obtain

$$\{Q_i\,,\,Q_j\} = 0 \qquad \text{C.10}$$

$$\{Q_2, Q_2^*\} = \{Q_1\,,\,Q_2^*\} = \{Q_2, Q_1^*\} = 0 \qquad \text{C.11}$$

$$\{Q_1\,,\,Q_1^*\} = 4\,|P| \qquad \text{C.12}$$

Therefore Q_2 creates zero norm states and should be ignored in the counting of physical states which have positive norm and we have to deal with just one creation operator Q_1. The helicity operator is M_{12}. Using the algebra of the Q_α with $M_{\mu\nu}$ and the explicit representation of the γ matrices it is easy to see that :

$$[Q_1\,,\,M_{12}] = \frac{1}{2}\,Q_1 \qquad \text{C.13}$$

Therefore Q_1 lowers the helicity λ of a state defined by $M_{12}\,|\lambda\rangle = \lambda\,|\lambda\rangle$ by 1/2 unit. So in an irreducible representation of massless supersymmetry we have the two states $|\lambda_{MAX}\rangle$, $Q_1|\lambda_{MAX}\rangle$ of helicities λ_{MAX}, $\lambda_{MAX} - 1/2$. To include CPT so as to have a field theoretical representation one doubles up the representation to have $\pm\lambda_{MAX}$, $\pm(\lambda_{MAX} - 1/2)$.

So we obtain the well-known multiplets of N = 1 supersymmetry :

$\lambda\,,\,A\,,\,B$	scalar multiplet
$A_\mu\,,\,\lambda$	vector multiplet
$\psi_\mu\,,\,A_\mu$	spin 3/2 multiplet
$V^a_\mu\,,\,\psi_\mu$	the multiplet of supergravity

For extended supersymmetry, we have the following general algebra. Let T_K be the generators of an internal symmetry group G : Instead of one Majorana charge Q_α, we introduce N of them which transform as spinors under the group G. The extended supersymmetry algebra reads :

$$[\, Q_\alpha^i \,,\, T_K \,] = i(S^{ij})_K \, Q_\alpha^j \tag{C.14}$$

$$\{\, Q_\alpha^i \,,\, Q_\beta^j \,\} = i\, \epsilon_{\alpha\beta}\, Z^{ij} \tag{C.15}$$

$$\{\, Q_\alpha^i \,,\, Q_\beta^{*j} \,\} = 2\, \delta^{ij} (\sigma^\nu)_{\alpha\beta}\, P_\nu \tag{C.16}$$

The Z^{ij} are central charges, i.e. they commute with all generators of the algebra. For massive representations the group G is not fixed, however in the massless case, Haag, Lopuszanski and Sohnius [35] showed that G had to be U(N) except for N = 4 when it could be either U(4) or SU(4). In the massless case the central charges Z^{ij} which have the dimension of a mass vanish and it is easy to work out the representation content of extended supersymmetry of rank N. We have N generators Q_1^i (i = 1,...N) which act as creation operators of states with positive norm. Each of them decreases the helicity by 1/2, so the helicity content of a massless irreducible multiplet is given by $|\lambda|, |\lambda| - 1/2, \ldots |\lambda| - N/2$.

So we see that matter multiplets with possible explicit mass terms, containing only spin 0 and 1/2 exist up to N = 2. Gauge multiplets with $|\lambda|_{MAX} = 1$ exist up to N = 4. Supergravity theories $|\lambda|_{MAX} = 2$ exist up to N = 8

The representation content of an irreducible multiplet is obtained by letting the Q_1^i act on the state of highest helicity $|\lambda_{MAX}\rangle$. So we get the states $Q_1^{\alpha_1} \ldots Q_1^{\alpha_K} |\lambda_{MAX}\rangle$ of helicity $\lambda_{MAX} - K/2$, and there are $\binom{N}{K}$ of them. The representation is self conjugate if $\lambda_{MIN} = \lambda_{MAX} - N/2 = -\lambda_{MAX}$ i.e. $\lambda_{MAX} = N/4$. If it is not self conjugate, one must add to these states those obtained by adding the CPT conjugate states $Q_1^{*\alpha_1} \ldots Q_1^{*\alpha_K} |\lambda_{MIN}\rangle$. For instance, for the supersymmetric Yang-Mills theories we have :

λ	N = 1	N = 2	N = 3	N = 4
1	1	1	1	1
$\frac{1}{2}$	1	2	3 + 1	4
0		1 + 1	3 + 3	6
$-\frac{1}{2}$	1	2	1 + 3	4
-1	1	1	1	1

So we have essentially 3 possible types of gauge theories since the N = 3 and 4 cases have the same number of fields. All these cases are explicitly known and are obtained from dimensional reduction of the 10 and 6 dimensional theories [16-17].

Similarly, for supergravity theories ($\lambda_{MAX} = 2$) we have :

λ	N = 1	2	3	4	5	6	7	8
2	1	1	1	1	1	1	1	1
$\frac{3}{2}$	1	2	3	4	5	6	7+1	8
1		1	3	6	10	15+1	21+7	28
$\frac{1}{2}$			1	4	10+1	20+6	35+21	56
0				1 + 1	5+5	15+15	35+35	70
$-\frac{1}{2}$			1	4	1+10	6+20	21+35	56
-1		1	3	6	10	1+15	7+21	28
$-\frac{3}{2}$	1	2	3	4	5	6	1+7	8
-2	1	1	1	1	1	1	1	1

So there are essentially 7 different supergravity theories as the field contents of the M = 7 and 8 theories are identical. All these theories can be constructed so that they have a manifest O(M) invariance.

The field assignment to this particle content is fixed unambiguously up to M = 3

M = 1 $\qquad V_\mu^a \quad , \quad \Psi_\mu$

M = 2 $\qquad V_\mu^a \quad , \quad \Psi_\mu^i \quad , \quad A_\mu \qquad$ i = 1,2.

M = 3 $\qquad V_\mu^a \quad , \quad \Psi_\mu^i \, , \, A_\mu^i \, , \, \chi \qquad$ i = 1,2,3.

For M = 4, since O(4) = SU(2) x SU(2) an ambiguity arises. At least two theories exist.

The O(4) theory (5-7) with field content V_μ^a, Ψ_μ^i, $A_\mu^{ij} = -A_\mu^{ji}$ (i,j = 1,...4), A (scalar), B (pseudoscalar). The SU(4) theory [8] : V_μ^a, Ψ_μ^i (i = 1,...4), A_μ^n (n = 1,...3) (vectors), B_μ^n (n = 1,...3) (axial vectors), Φ (scalar), B (pseudoscalar).

The existence of this second theory can be inferred from the following chain of reductions :

Starting from D = 11 simple supergravity (V^a_μ , Ψ_μ , $A_{\mu\nu\rho}$) one reduces it to D = 10 and obtains extended M = 2 supergravity V^a_μ , A_μ , ϕ , $A_{\mu\nu\rho}$, $A_{\mu\nu}$, Ψ^i_μ , χ^i (i = 1,2, Ψ_μ, χ Majorana-Weyl spinors). This can further be reduced to simple supergravity (M = 1) in D = 10 with field content V^a_μ , $A_{\mu\nu}$, ϕ , Ψ_μ, χ Reducing that theory to 4 as was done for the D = 10 Yang-Mills theory would give extended M = 4 supergravity in 4 dimensions with the above field content, coupled to 6 M = 4 vector multiplets of the type discussed above. The SU(4) $\sim$ O(6) invariance should be manifest from the reduction used.

Indeed both theories have been constructed fully and are in fact equivalent [8] , as long as the O(4) group is not gauged, in which case they are unequivalent [36] .

For M = 8 the simplest assignment compatible with O(8) invariance is [9] :

$$V^a_\mu \ , \ \Psi^i_\mu \ , \ A^{[ij]}_\mu , \ \chi^{[ijk]} \ , \ A^{[ijkl]} \ , \ B^{[ijkl]}$$

where i,... = 1,...8 and the bracket indicates antisymmetrization with respect to all indices. A and B are self-dual and anti-self-dual fields in the internal symmetry indices and represent 35 degrees of freedom each. So far this theory has been constructed only up to order K^2, while one may hope to obtain it at all orders in K by dimensional reduction of the D = 11 supergravity theory. 49

To construct these theories, in the absence of a tensor calculus [37] for M $\geqslant$ 2 and of superspace techniques [38] (which however have recently been able to recover the M = 1,...3 cases) and of group theoretical techniques (which have recovered the M = 1,2 cases) [39] the method used has been to construct the action and transformation laws by an order by order expansion in K (the square root of Newton's constant)

$$\mathcal{L} = \mathcal{L}_0 + K\,\mathcal{L}_1 + K^2\,\mathcal{L}_2 + \cdots$$

where $\mathcal{L}_0$ is the kinetic term of the field content properly covariantized with respect to local Lorentz transformations and coordinate transformations.

For instance for O(2) supergravity :

$$\mathcal{L}^0 = -\frac{1}{4K^2} V R(\omega^0) - \frac{1}{2}\epsilon^{\lambda\rho\mu\nu}\,\overline{\Psi}^i_\lambda \gamma_5 \gamma_\mu D_\nu(\omega^0)\Psi^i_\rho - \frac{1}{4} V g^{\mu\rho} g^{\nu\sigma} F_{\mu\nu} F_{\rho\sigma} \qquad \text{C.17}$$

and the problem consists in finding $\mathcal{L}_1, \mathcal{L}_2$ ... and the expression of the transformation laws which read at lowest order in K

$$\delta V_{a\mu} = -iK \bar{\epsilon}^i \gamma_a \Psi_\mu \quad ; \quad \delta \Psi^i_\mu = \frac{1}{K} D_\mu \epsilon^i + \ldots \qquad \text{C.18}$$

Fortunately, the type of terms which can appear in the action is limited by consideration of invariance and of dimensionality[40]. The vector fields for instance appear only through F and F^2 since the theory contains at most two derivatives. The Fermi fields appear only in even powers and at most, quartic terms need to be introduced. Remarkably enough, this result is valid for any space-time dimensions. In D dimensions K has in units of mass the dimension - 1/2 (D - 2). The gravitino fields have dimension 1/2 (D - 1). If we consider the scattering of n gravitinos, n - 1 gravitons are exchanged between them at the tree level, so that the coupling constant K occurs with the power 2(n - 1) in such graphs. Because of gauge invariance, the S matrix at the tree level must be invariant under the substitution of the polarization tensor of a gravitino ϵ_μ by its momentum times an arbitrary spinor $\epsilon_\mu \to \epsilon_\mu + \alpha P_\mu$. If it is not, we need to add to the action a contact term, of the form $\lambda_{2n}\, \partial^p (\bar{\Psi}\Psi)^n$. The dimension of λ_{2n} is given by $[\lambda_{2n}] = D - n(D - 1) - p$, but λ_{2n} must also be proportional to $K^{2(n-1)}$. So we get the equation

$$-2(n-1)\frac{D-2}{2} = D - (D-1)n - p \qquad \text{C.19}$$

or

$$n = 2 - p \qquad \text{C.20}$$

In this equation, the space-time dimension drops out and has only one non trivial solution, namely p = 0 and n = 2. So while quartic terms can (and do) appear in the action, Ψ^6 terms are forbidden. A similar analysis can be applied to spin one half fields χ^i which have the same dimensionality as the Ψ fields and also only quartic terms are allowed. This implies that the extended supergravity theories up to N = 3 can be constructed in a finite number of steps. On the other hand, scalar fields which appear in N $\geqslant$ 4 extended supergravity model introduce a non trivial difficulty since K A, where A is a scalar field, is dimensionless and therefore non polynomial functions of the scalar fields can appear in the action and transformation laws, leading to an infinite series in K. To circumvent this difficulty, it is always possible to introduce everywhere in the action and transformation laws arbitrary functions of the scalar fields, and write down the system of differential equations which express the invariance of the action under supersymmetry transformations. In practice, this is rather difficult, as it involves a large number of arbitrary functions, and writing down and solving the system of equations is a rather cumbersome and formidable problem. Fortunately, some short cutting devices can be found which help in reducing the number of functions to be determined.

For instance, one help in the construction is the reduction to previously known subcases. Obviously, the O(N + 1) theory must contain as a subcase the O(N) theory. For instance, the O(3) theory contains as a consistent reduction (i.e. the variation of the fields put to zero does vanish) the simple supergravity theory (2, 3/2) coupled to the Maxwell (1, 1/2) multiplet [41], and this was useful for the construction of the O(3) theory [4]. Similarly, the O(4) and SU(4) theories contain, as a subcase, simple supergravity coupled to the scalar multiplet (χ , A, B). In simple supergravity, this coupling involves an arbitrary function [42] G(KA, KB), while in O(4) and SU(4) supergravity this function becomes fixed. For instance, in the O(4) theory [7] one finds that the kinetic term of the A and B fields is indeed non-polynomial and reads :

$$\mathcal{L}^{\circ}_{sc} = \frac{1}{2} V \frac{(\partial_\mu A)^2 + (\partial_\mu B)^2}{(1 - K^2 (A^2 + B^2))^2} \qquad \text{C.21}$$

This has the annoying feature that the range of the A and B fields is limited to $K^2 (A^2 + B^2) \leq 1$.

Other useful tools to construct these theories and which we shall discuss in the next sections are the supersymmetry algebra of the transformation laws, the simplicity of the fermionic equations of motion and the global ("rigid") U(N) invariance of these theories.

b. Extended Supersymmetry Algebra

Let us take as a reference point the M = 2 extended supergravity theory [3], which for definiteness reads :

$$\mathcal{L} = -\frac{1}{4K^2} V R(\hat{\omega}) - \frac{1}{2} \epsilon^{\lambda\rho\mu\nu} \bar{\Psi}^i_\lambda \gamma_5 \gamma_\mu D_\nu(\hat{\omega}) \Psi^i_\rho$$
$$- \frac{1}{4} V F_{\mu\nu} F^{\mu\nu} - \frac{V}{2} K \epsilon^{ij} \bar{\Psi}^i_\mu (F^{\mu\nu} - i \gamma^5 \tilde{F}^{\mu\nu}) \Psi^j_\nu$$
$$- \frac{V}{4} K^2 \epsilon^{ij} \epsilon^{k\ell} \bar{\Psi}^i_\mu \Psi^j_\nu [\bar{\Psi}^{\mu k} \Psi^{\nu \ell} - i \widetilde{\bar{\Psi}^{\mu k} \gamma^5 \Psi^{\nu \ell}}] \qquad \text{C.22}$$

where

$$D_\nu(\hat{\omega}) \Psi^i_\rho = (\partial_\nu + \frac{1}{2} \omega_{\nu ab} \sigma^{ab}) \Psi^i_\rho \qquad \text{C.23}$$

and
$$\tilde{F}^{\mu\nu} = \frac{1}{2V} \epsilon^{\mu\nu\rho\sigma} F_{\rho\sigma} \qquad \text{C.24}$$

and some quartic terms have been eliminated by use of $\hat{\omega}$ which is the supercovariant version of ω and can also be obtained by first order formalism :

$$\hat{\omega}_{\mu ab} = \omega^{o}_{\mu ab}(V, \partial V) + \frac{i\kappa^2}{2}(\bar{\Psi}^{i}_{\mu} \gamma_b \Psi^{i}_{a} - \bar{\Psi}^{i}_{\mu} \gamma_a \Psi^{i}_{b} + \bar{\Psi}^{i}_{b} \gamma_\mu \Psi^{i}_{a}) \qquad \text{C.25}$$

The transformation laws of the fields under supersymmetry read :

$$\delta V^{a}_{\mu} = -i\kappa\, \bar{\epsilon}^{i} \gamma^{a} \Psi^{i}_{\mu} \qquad \text{C.26}$$

(which is a universal form, valid for all extended supergravity theories).

$$\delta \Psi^{i}_{\rho} = \frac{1}{\kappa} D_\rho(\hat{\omega})\,\epsilon^{i} - \frac{i}{2} \epsilon^{ij} \sigma^{\mu\nu} \hat{F}_{\mu\nu} \gamma_\rho \epsilon^{j} \qquad \text{C.27}$$

$$\delta A_\mu = -\epsilon^{ij} \bar{\epsilon}^{i} \Psi^{j}_{\mu} \qquad \text{C.28}$$

where a simplification in the transformation law of Ψ has been brought by using $\hat{\omega}$ and the supercovariant version of the Maxwell tensor

$$\hat{F}_{\mu\nu} = F_{\mu\nu} + \kappa\, \epsilon^{ij} \bar{\Psi}^{i}_{\mu} \Psi^{j}_{\nu} \qquad \text{C.29}$$

The supersymmetry algebra can be studied by computing the commutator of two transformations of supersymmetry $[\delta_1, \delta_2]$ of parameters ϵ_1, ϵ_2. One finds that the algebra closes exactly on Bose fields, but only up to equations of motion on Fermi fields [43,18] This is a general feature of extended supergravity theories and a set of auxiliary fields closing the algebra has still to be found.

The transformations which arise from the commutation are of two types :
a) field independent transformations, which do not vanish when the fields are set to zero
b) field dependent transformations, which do vanish for vanishing fields. The last class contains local Lorentz transformations, supersymmetry transformations and gauge transformations.
The field independent transformations are very simple and exhibit the structure of the algebra. They consist of :

- a coordinate transformation of parameter $\xi^\mu(x) = i(\bar{\epsilon}^i_2 \gamma^\mu \epsilon^i_1)$ of universal form, for all N and this is the expression that the anticommutator of $\{Q^i_\alpha, \bar{Q}^j_{\dot\beta}\}$ contains $(\gamma^\mu)_{\alpha\beta} \delta^{ij} P_\mu$, P_μ being the generator of translations.
- a gauge transformation on $A_\mu(x)$ of parameter proportional to $1/K \; \partial_\mu (\bar{\epsilon}^i_2 \epsilon^j_1) \epsilon^{ij}$. More generally in O(N) $(N \geq 2)$ theories one finds that $\delta A^{ij}_\mu \sim 1/K \; \partial_\mu \bar{\epsilon}^{[i}_2 \epsilon^{j]}_1$. Algebraically, this means that the A^{ij}_μ are $N(N-1)/2$ gauge fields associated with $N(N-1)/2$ central charge operators Z^{ij}.

In the case of pure O(N) supergravity, these Z^{ij} charges vanish, just as in a pure gauge theory the charges vanish. An interesting case is when we couple an O(N) supergravity theory to a "matter" multiplet having non zero central charges. There are only two known examples of multiplets having non zero central charges and both occur in M = 2 supersymmetry:
- The vector gauge multiplet (A^a, λ^a_i (i = 1,2), S^a, P^a) which has topological non vanishing central charges for monopole-like solutions [24].
- The scalar multiplet (χ^i, A^i, B^i) i = 1,2, [44] which has a non-vanishing central charge in the massive case [18] (central charges have the dimension of a mass). Indeed, one finds in flat space that the transformation laws of supersymmetry on this multiplet imply that

$$[\delta_1, \delta_2]\phi^i = i\, \delta^{ab}\, \bar{\epsilon}^a_2 \gamma^\mu \epsilon^b_1 \partial_\mu \phi^i + m\, \epsilon^{ab}\, \bar{\epsilon}^a_2 \epsilon^b_1\, \epsilon^{ij} \phi^j \qquad \text{C.30}$$

where $\phi^i = \chi^i$, A^i, or B^i. The first term is the usual translation, while the second rotates the multiplets ϕ^1 and ϕ^2 into each other. Hence, here $Z^{ij} = \epsilon^{ij} Z$ where Z is the electric charge operators whose eigenvectors are the fields $\phi^1 \pm i\phi^2 = \phi_\pm$. Oddly enough this electric-like charge is proportional to the mass.

When we go to curved space and couple this multiplet to M = 2 supergravity, ϵ becomes x -dependent and $[\delta_1, \delta_2]$ on the action of the matter system produces a non-zero variation proportional to the divergence of the electric Noether current, i.e. it is proportional to $m\, \epsilon^{ab}\, \bar{\epsilon}^a_2 \epsilon^b_1\, \partial_\mu J^\mu_N(x)$ where

$$J^\mu_N(x) = \epsilon^{ij} [z^i \overleftrightarrow{\partial_\mu} \bar{z}^j - i \bar{\chi}^i \gamma^\mu \chi^i] \qquad \text{C.31}$$

$$z^i = A^i + i\gamma^5 B^i \qquad \text{C.32}$$

To compensate this term, since $[\delta_1, \delta_2]$ on $A_\mu(x)$ produces a gauge transformation of parameter $1/K\ \partial_\mu \epsilon^{ij}(\bar{\epsilon}_2^i\ \epsilon_1^j)$ one must add to the action a Noether coupling proportional to $Km\ A_\mu\ J^\mu_N$ [18] where the proportionality constant is fixed by the algebra only. The full coupling of O(2) supergravity to this multiplet has actually been fully worked out, at all orders in K by Zachos [19].

A consequence of this Noether coupling is that if one considers the one particle exchange graphs between particles of the matter multiplet, not only can they exchange a graviton which in the static limit gives rise to a Newtonian attractive force proportional to k^2m^2/r^2, but also the massless vector particle which like the photon gives a Coulomb force e_1e_2/r^2. Here, $e_i \sim \pm Km$ and one finds that between particles of the same charges the vector exchange exactly cancels the tensor exchange, while between particles of opposite charges it exactly doubles it. This is the first model where such an antigravity-like effect arises purely out of algebraic considerations, although more realistic models should be considered including the super Higgs effect before one can predict whether antigravity belongs still to the realm of science fiction writers or could be experimentally tested.

Cancellation of forces between exchanges of particles of different spin are not uncommon in field theory. For instance, in particular models, the force between magnetic monopoles of the same charges is known to vanish at long distance [45] while there is an attractive Newtonian potential between monopoles of opposite charges. In the same model, the electric repulsion between charged vector fields of the same electric charges is exactly cancelled by the exchange of a massless scalar, while it is doubled in the case of opposite charges [22]. So, it is not a complete surprise that such an effect arises in O(2) supergravity where a massless vector particle is in the same multiplet as the graviton.

c. Fermionic Equations of Motion

The fermionic equations of motion in simple and extended supergravity have a simple form when written in terms of supercovariant objects. As they contain only one derivative, their variation cannot contain a $\partial\epsilon$ term since its coefficient, having no derivatives, could not be proportional to an equation of motion. So a fermionic equation can always be written in terms of purely supercovariant objects. For instance, in the Maxwell-Einstein (2, 3/2) coupled to (1, 1/2), or in the O(3) extended supergravity theory, the spin 1/2 equation of motion reads $\gamma^\mu \hat{D}_\mu \chi = 0$ where $\hat{D}_\mu$ is the supercovariant derivative acting on χ which is known when $\delta\chi$ is known.

Similarly, for the spin 3/2 gauge fields Ψ_μ^i one can define a supercovariant gauge field strength [7] which transforms without $\partial\epsilon$ once $\delta\Psi$ is known.

For M = 1 (simple) supergravity, one has $\delta\Psi_\mu = \frac{1}{\kappa} D_\mu(\hat\omega)\epsilon$ and $\Psi_{\nu\rho} = D_\nu(\hat\omega)\Psi_\rho - D_\rho(\hat\omega)\Psi_\nu$ is supercovariant. Similarly, for M = 2 $\delta\Psi_\rho = \frac{1}{\kappa} D_\rho^{ij}\epsilon^j$ where

$$D_\rho^{ij} = \delta^{ij} D_\rho(\hat\omega) - \frac{i\kappa}{2}\epsilon^{ij}\sigma^{\mu\nu}\gamma_\rho \hat F_{\mu\nu} \qquad \text{C.33}$$

and the supercovariant field strength $\Psi_{\nu\rho}^i$ is given by

$$\Psi_{\nu\rho}^i = D_\nu^{ij}\Psi_\rho^j - D_\rho^{ij}\Psi_\nu^j \qquad \text{C.34}$$

In terms of these gauge field strengths, the spin 3/2 equations of motion simply read

$$\epsilon^{\lambda\mu\nu\rho}\gamma_5\gamma_\mu\Psi_{\nu\rho}^i = 0 \qquad \text{for M = 1,2} \qquad \text{C.35}$$

Supercovariant field strengths can be constructed also for M = 3,4. One of the uses of the fact that fermionic equations can always be written in a supercovariant form is to construct the invariant action of these theories. Once one knows the first Noether couplings which arise in a fermionic equation of motion and the transformation laws of the fields, one supercovariantizes it and one can deduce the action from which it is derived. This procedure was of great use to construct the O(4) and SU(4) models.

d. U(N) Invariance in Extended Supergravity Theories

According to the Haag-Lopuzanski-Sohnius theorem [35] the global invariance group of an extended supersymmetric theory with M spinor charges and massless particles should be U(M) except for M = 4 where it can be either U(4) or SU(4). An example of the last case is the M = 4 supersymmetric Yang-Mills theory, which is invariant under SU(4) (O(6)), but not under U(4).

Extended supergravity theories, on the other hand are constructed so as to have a manifest global O(M) invariance, so that there should be a way to discover that this global O(M) invariance can be extended to U(M) [18, 6]

Indeed the simple M = 1 supergravity theory is invariant under global U(1) chiral transformations :

$$\delta_U V_\mu^a = 0 \qquad \delta_U \Psi_\mu = i\Lambda\gamma_5\Psi_\mu \qquad \text{C.36}$$

where Λ is an infinitesimal constant

M = 2 supergravity has a built-in O(2) invariance which can easily be extended to SU(2) by chiral transformations :

$$\delta V_\mu^a = 0 \quad ; \quad \delta \Psi_\mu^i = (C^{ij} + i\gamma^5 \Lambda^{ij})\,\Psi_\mu^j$$

$$\delta A_\mu = 0 \qquad \text{C.37}$$

where $C^{ij} = -C^{ji} \;;\; \Lambda^{ij} = \Lambda^{ji} \;;\; Tr\Lambda = 0$

To enlarge SU(2) to U(2), we need an additional U(1) transformation.

$$\delta_{U_1} V_\mu^a = 0 \qquad \delta_{U_1} \Psi_\mu^i = -\frac{1}{2}\, i\,\gamma_5\,\Lambda\,\Psi_\mu^i \qquad \text{C.38}$$

works for the kinetic term of the Ψ field but the Noether coupling is not invariant as can be seen in eq. C.22 unless one performs on $F_{\mu\nu}$ a simultaneous duality transformation :

$$\delta_{U_1} F_{\mu\nu} = \Lambda\,\tilde{F}_{\mu\nu} + \cdots \qquad \text{C.39}$$

Duality transformations must however be used with care. For instance, in the free Maxwell system, if we set $\delta F_{\mu\nu} = \Lambda \tilde{F}_{\mu\nu}$ and require that a corresponding transformation exists for the vector potential, it implies that $\Lambda \tilde{F}_{\mu\nu} = \partial_\mu \delta A_\nu - \partial_\nu \delta A_\mu$ is a curl and thus that $\partial_\mu(\Lambda \tilde{F}\quad) = -\Lambda \partial_\mu F^{\mu\nu} = 0$, i.e. that the equation of motion of A_μ is satisfied.

So duality transformations cannot be used on the action directly, but require that the vector equations of motion are satisfied.

In O(2) extended supergravity theory, since A_μ does not appear explicitly, the vector equation of motion reads :

$$D_\mu G^{\mu\nu} = 0 \qquad \text{C.40}$$

where $G^{\mu\nu} = F^{\mu\nu}$ + bilinears in Ψ and

$$D_\mu \tilde{F}^{\mu\nu} = 0 \qquad \text{C.41}$$

which is the curl condition.
If we set

$$\delta_{U_1} F^{\mu\nu} = \Lambda\,\tilde{G}^{\mu\nu} \qquad \text{C.42}$$

this is obviously consistent with the curl condition and it is a small piece of algebra to show that the variation of the first equation of motion is proportional to the curl condition.

All other equations of motion are indeed U(2) invariant and this can be generalized to all known O(M) extended supergravity theories, which exhibit a U(M) global invariance, realized by a mixture of chiral and dual transformations. This invariance has been used extensively to construct the O(4) theory and also to generate more terms in the perturbative approach used to construct the O(8) theory.

Under duality transformation, the supercovariant $\hat{F}_{\mu\nu}$ transforms very simply :

$$\delta_{U_1} \hat{F}^{\mu\nu} = \Lambda \tilde{\hat{F}}^{\mu\nu} \qquad \text{C.43}$$

The action of O(2) supergravity can be further contracted if one writes it in term of $G_{\mu\nu}$ (which defines the vector equation of motion) and $F_{\mu\nu}$, in such a way that all quartic terms are absorbed either in $\hat{\omega}$ or in $\hat{F}$:

$$\mathcal{L} = -\frac{1}{4K^2} V R(\hat{\omega}) - \frac{1}{2}\epsilon^{\lambda\rho\mu\nu} \bar{\Psi}_\lambda^i \gamma_5 \gamma_\mu D_\nu(\hat{\omega})\Psi_\rho^i$$
$$-\frac{1}{4} V F^{\mu\nu} G_{\mu\nu} - \frac{1}{4}\kappa V \epsilon^{ij} \bar{\Psi}_\mu^i (\hat{F}^{\mu\nu} - i\gamma_5 \tilde{\hat{F}}^{\mu\nu})\Psi_\nu^j \qquad \text{C.44}$$

In general, the transformations which extend O(N) to U(N) involve duality transformations and can be carried out only when the equations of motion of the vector fields are satisfied. One exception is the O(2) theory where the invariance is extended to SU(2) without use of duality transformations. Another exception is the N = 4, so called SU(4) theory [8] with field content V_μ^a, Ψ_μ^i (i = 1,...4) $A_\mu^{(n)}$, $B_\mu^{(n)}$ (n = 1,2,3), Φ, B which has an SU(4) invariance realized without duality transformations. The first few terms of its action are given by [8]

$$\mathcal{L} = -\frac{1}{4K^2} V R(\hat{\omega}) - \frac{1}{2}\epsilon^{\lambda\mu\nu\rho} \bar{\Psi}_\lambda^i \gamma_5 \gamma_\mu D_\nu(\hat{\omega})\Psi_\rho^i$$
$$-\frac{1}{4} V \left(A_{\mu\nu}^{(n)} A^{\mu\nu(n)} + B_{\mu\nu}^{(n)} B^{\mu\nu(n)} \right) \exp -2\kappa\phi$$
$$-\frac{1}{2} V \kappa B \left(A_{\mu\nu}^{(n)} \tilde{A}^{\mu\nu(n)} + B_{\mu\nu}^{(n)} \tilde{B}^{\mu\nu(n)} \right)$$
$$+\frac{1}{2} i V \bar{\chi}^i \gamma^\mu \hat{D}_\mu \chi^i$$
$$+\frac{1}{2} V g^{\mu\nu} [\partial_\mu \Phi \partial_\nu \Phi + \exp 4\kappa\Phi\, \partial_\mu B \partial_\nu B] \qquad \text{C.45}$$
$$+ \cdots$$

where

$$A_{\mu\nu}^{(n)} = \partial_\mu A_\nu^{(n)} - \partial_\nu A_\mu^{(n)}, \quad B_{\mu\nu}^{(n)} = \partial_\mu B_\nu^{(n)} - \partial_\nu B_\mu^{(n)} \qquad \text{C.46}$$

At first glance the O(4) and SU(4) theories are very different. The O(4) theory has a manifest O(4) invariance, is non polynomial in A, B, and the range of the scalar fields is restricted by $K^2(A^2 + B^2)$ 1. The SU(4) theory has a manifest SU(4) invariance, is non polynomial in ϕ and polynomial in B, and the range of and B is infinite. In fact, one can show that these theories are classically equivalent in the sense that by a combination of chiral transformations, of scalar field redefinitions, and duality transformations, the equations of motions of these two theories can be transformed into each other. It is however not completely clear that they are quantum mechanically equivalent, due to possible anomalies, and to the fact that duality transformations are non local transformations on the vector potentials of these theories.

Similar (and more complicated) transformations are met also when one tries to relate the O(8) theory with the dimensional reduction of the D = 11 theory which has 8 spinor charges too, but only a manifest O(7) invariance.[49]

A surprise, ill understood still now, is that for N = 4 the U(4) invariance can be extended to SU(4) x SU(1,1). In the SU(4) theory there is a hidden U(1) invariance which mixes ϕ and B into each other in a complicated way, but there are also two simple transformations which together with U(1) generate a SU(1,1) group. These two transformations are

- a translation $B(x) \rightarrow B(x) + \beta$ β: constant C.47

- a scale transformation $B(x) \rightarrow \alpha^2 B(x)$ α: constant

$$A_\mu^{(n)}, B_\mu^{(n)} \rightarrow \alpha^{-1} A_\mu^{(n)}, \alpha^{-1} B_\mu^{(n)}$$

$$\phi \rightarrow \phi - \frac{1}{K} \ln \alpha \qquad \text{C.48}$$

In the O(4) theory these transformations are easily seen on the kinetic term of the scalar fields which reads

$$\mathcal{L} = \frac{V}{2K^2} \frac{\partial_\mu Z \, \partial_\mu \bar{Z}}{(1 - Z\bar{Z})^2} \qquad \text{C.49}$$

where $Z = K(A + iB)$ and $\mathcal{L}$ is invariant under

$$Z \rightarrow Z' = \frac{aZ + \bar{c}}{cZ + \bar{a}}$$

where $|a^2| - |c^2| = 1$ C.50

It is still unclear how this invariance generalizes for higher extended supergravity theories, in particular in the case of O(8).

e. Gauged O(N) Theories

So far, the extended supergravity theories we have discussed had only one coupling constant K and the N(N - 1)/2 vector fields A_μ^{ij} were abelian. Since there is just the right number of them to gauge the O(N) group, one can try to replace $F_{\mu\nu}^{ij}$ by the Yang-Mills curvature, thus introducing a new coupling constant g, of dimension zero [10,36].

For instance, in the N = 2 theory one can gauge O(2) by replacing $D_\nu(\hat{\omega})\,\psi_\rho^i$ by $D_\nu(\hat{\omega})\,\psi_\rho^i - g\,\epsilon^{ij} A_\nu\,\psi_\rho^j$. Here A_ν enters explicitly so that the U(N) global invariance is broken down to O(N) which however is made local. This indeed works, provided that one adds to the Lagrangian the following terms :

$$\mathcal{L}' = -\frac{g}{K} V \bar{\Psi}_\mu^i \sigma^{\mu\nu} \Psi_\nu^i + \frac{3}{2}\frac{g^2}{K^4} V \qquad \text{C.51}$$

which consist of a mass-like term for the spin 3/2 and a cosmological constant. The sign of the cosmological constant is fixed and corresponds to an O(3,2) de Sitter universe. A similar procedure works for O(3) and provides a beautiful unification of the Einstein and Yang-Mills theories.However, the problem with this is that the size of the universe is very small unless $g^2 < 10^{-20}$! One can hope to cancel this cosmological constant by adding matter fields as indeed happens in the super Higgs effect in the N = 1 theory [42], but the problem here is that extended supergravity theories coupled to extended super matter multiplets are presumably not renormalizable even for one-loop.

Further problems arise with the M = 4 theories. The O(4) and SU(4) theories become inequivalent when the SU(2) x SU(2) group is gauged and give rise to field dependent cosmological constants [36]

$$\lambda_{O(4)} = \frac{g^2}{2K^4}\left(1 + \frac{2}{1 - z\bar{z}}\right) \qquad \text{C.52}$$

$$\lambda_{SU(4)} = \frac{g_A^2 + g_B^2}{8K^4}\exp 2K\phi \qquad \text{C.53}$$

Both potentials $(-\lambda)$ are unbounded below and exhibit no signs of spontaneous symmetry breaking. The second one does not even have a local extremum. Similar problems are expected in the O(8) theory if it can be gauged. The O(8) theory itself does not seem to be a fully realistic theory as it is too small to contain,as a subgroup of the gauge group, $SU(3)_c$ x SU(2) x U(1). It only contains [46]as a subgroup, $SU(3)_c$ x U(1) x U(1), and thus makes the prediction that if the 8 gluons, the photon and the Z^0 are elementary, the W^+ and W^- should arise as soliton excitations in the model. Similarly, the theory contains 5 quark flavors, a neutral octet of heavy leptons, the electron and its neutrino, while the muon, τ and their neutrinos are missing.

D. POSSIBLE MODIFICATIONS AT SHORT RANGE OF GRAVITY

Extended supergravity theories provide a beautiful unification between Einstein's theory of gravity and the field theories considered as relevant to particle physics, i.e. the gauge theories, the Dirac theory of the electron and the Klein-Gordon theory of scalar fields which are needed in gauge theories for the Higgs mechanism. All these fields appear in representation of the extended supersymmetry algebra as soon as $M \geq 4$. Yet some unphysical features appear with the gauging of O(M) and renormalizability is not guaranteed beyond two loops. If these theories are indeed not renormalizable, is there still a hope to achieve a finite theory of quantum gravity in interaction with matter fields ?

May be gravity (and its supersymmetric versions) should be modified at short distances just as the V - A theory was modified at short distances as seen as a particular approximation of the Weinberg-Salam model. One would then expect that $G_N \sim K^2$ would be a phenomenological coupling constant rather than a fundamental one and should be expressible as $G_N \sim g^2/\Lambda^2$ where g is a fundamental dimensionless coupling constant and Λ a new energy scale, or inverse of a distance, at which Einstein's theory of gravity, which was based on long-range observations, stops to be valid. In the Weinberg-Salam model the natural cut-off Λ is the mass of the W boson.

In a similar vein, one may expect that the graviton exchange should be accompanied by the exchange of heavy, short range particles, which would have hitherto escaped detection. (The graviton itself has not yet been detected !). Examples of such theories of gravity are the models of the type $R + R^2$ where the propagator of the graviton falls like $1/p^4$,and are indeed renormalizable [47] . However, in these models, the massive particles which accompany the graviton are either tachyons ($m^2 \leq 0$) or ghosts (negative metrics).

There is so far only one example of model where the graviton is a massless member of a set of particles having desirable properties (i.e. being neither ghosts nor tachyons), and this is the dual spinor model (or spinning string model) [13,14] . In this model, instead of considering spinning point particles moving in a Minkowskian flat space-time, one considers one dimensional objects (strings) moving in flat space-time, each point of the string having in addition a spin degree of freedom of its own. The length of the string is not fixed, and is a dynamical variable, however, its order of magnitude is given by a parameter having the dimension of a length $(\alpha')^{1/2}$, α' being also the inverse of a $(\text{mass})^2$. The classical action of the string is in fact supersymmetric in the two dimensional space swept by the string throughout its evolution in space-time. Classically, the more a string is spinning, the heavier it becomes, and one can prove that $J \leq \alpha' M^2$, where M is the mass of the string and J its angular momentum.

When one quantizes canonically the string system, its vibration modes give rise to an infinite sequence or tower of particles lying on straight Regge trajectories. J becomes integrally or half-integrally quantized and so does $\alpha' M^2$. As there are two types of strings, open and closed, two quantized sectors of the theory emerge. To the open strings, one can attach quantum numbers of an arbitrary internal symmetry group U(N) while the closed strings have to be singlets under U(N). All states of the quantized open strings are in the adjoint representation of the U(N) group.

The condition of compatibility between Lorentz invariance and canonical quantization is very strong and implies that the model can exist only in a D = 10 Minkowskian space-time. However, the 6 extra dimensions can be compactified [20] as in the field theories described in section B)a) so that if the six lengths $L_1 \ldots\ldots L_6$ are small enough, no incompatibility with everyday experience arises.

Free quantized strings can be made to interact by introduction of a coupling constant g which describes the breaking or joining of strings.

The quantized spectrum of open and closed strings starts at M = 0. If α' is small enough, the excited states having $(\text{masses})^2$ of order $1/\alpha'$ can be unobservable. One can show that at each mass level the numbers of bosonic and fermionic states are equal, which is a necessary requirement for supersymmetry. Further, none of these states are ghosts or tachyons due to the existence in the model of an infinite graded Lie algebra of gauge operators which eliminates the ghosts.

An interesting limit is to study the classical interaction among massless particles of the model in the limit where all energies are small compared to the cut-off energy $1/(\alpha')^{1/2}$. The classical interaction is described by the tree diagrams of the theory which are known explicitly. Taking $1/(\alpha')^{1/2}$ to be very big, one can, from the tree diagrams, reconstruct a phenomenological action from which they can be derived in that limit.

The zero mass states of the quantized open string sector are a vector particle and a Majorana-Weyl fermion in the adjoint representation of U(N). In the limit α' negligible, they are found to interact just as in the D = 10 supersymmetric Yang-Mills theory which was described in section B)c) [15,16].

The zero mass particles of the closed string sector have spin 2, 3/2 and lower and interact just as fields in a D = 10 supergravity theory do [15,16]. In particular, the interaction between the spin 2 particles is just described by Einstein's action [48]. It is quite remarkable that although defined in flat space-time this model naturally brings curvature in, as it contains, within its spectrum, gravitons which interact just as in Einstein's theory at low energy.

In this model, the graviton coupling constant and the gauge coupling constant are related as the only parameters in the theory are g, α' and the 6 parameters $L_1 \ldots\ldots L_6$. In the simple case where L_i are all of the same order of magnitude as $(\alpha')^{1/2}$, one finds [48] that $G_N \sim K^2 \sim g^4 \alpha'$ so that $1/\alpha' = \Lambda^2$ appears as a cut off analogous to the W mass in the Weinberg-Salam model. Moreover, the g^4, rather than g^2, power expresses the remarkable fact that in this theory, the whole graviton sector (closed strings) is obtained as a bound state of the Yang-Mills sector (open strings) already at the one loop level.

Typically, $(\alpha')^{1/2}$ would be of the order of Planck's length which would indeed make the higher excited states unobservable.

Finally, the short distance behaviour introduced by the cut-off parameter $\Lambda^2 = 1/\alpha'$ tends to smooth out the interaction at high energy between particles. For instance, in Einstein's theory of gravity, the graviton-graviton scattering at fixed angle in the center of mass grows like E^2 where E is the energy in the centre of mass and this violates unitarity at high enough energy. In the dual spinor model, this amplitude decreases exponentially like $\exp - \alpha' E^2 f(\theta)$ and we see that α' damps the interaction at high energy. Similarly, for loop diagrams, one finds that no ultraviolet divergences arise, but rather that there are infrared divergences. Loop diagrams containing only closed strings are finite because of topological reasons while loop-diagrams containing open strings need renormalization of α'.

The advantage of this model over extended supergravity theories is that it still allows a coupling between gauge fields of an arbitrary U(N) group with gravity and is not in contradiction with them since in the low energy limit it reproduces extended supergravity theories coupled to the M = 4 supersymmetric Yang-Mills theory. If extended supergravity theories are not fully renormalizable, may be this could be a starting point for a finite theory of quantum gravity.

REFERENCES

[1] G. 't Hooft and M. Veltman, Ann. Inst. H. Poincaré 20 (1974) 69
S. Deser, H.S. Tsao and P. Van Nieuwenhuizen, Phys. Rev. D10 (1974) 3337
For a general review of results obtained in quantum gravity, see P. Van Nieuwenhuizen in Proceedings of the Marcel Grossmann Meeting (1976) published by North Holland Co.

[2] D.Z. Freedman, P. Van Nieuwenhuizen and S. Ferrara, Phys. Rev. D13 (1976) 3214
S. Deser and B. Zumino, Phys. Lett. 62B (1976) 335

[3] S. Ferrara and P. Van Nieuwenhuizen, Phys. Rev. Letters 37 (1976) 1669

[4] D.Z. Freedman, Phys. Lett. 38 (1977) 105
S. Ferrara, J. Scherk and B. Zumino, Phys. Lett. 66B (1977) 35

[5] A. Das, Phys. Rev. D15 (1977) 2805

[6] E. Cremmer, J. Scherk and S. Ferrara, Phys. Lett 68B (1977) 234

[7] E. Cremmer and J. Scherk, Nucl. Phys. B127 (1977) 259

[8] E. Cremmer, J. Scherk and S. Ferrara, Phys. Lett. 74B (1978) 61

[9] B. de Wit and D.Z. Freedman, Nucl. Phys. B130 (1977) 105

[10] D.Z. Freedman and A. Das, Nucl. Phys. B120 (1977) 221

[11] M.T. Grisaru, P. Van Nieuwenhuizen and J.A.M. Vermaseren, Phys. Rev. Lett. 37 (1976) 1662
S. Deser, J. Kay and K. Stelle, Phys. Rev. Lett. 38 (1977) 527
P. Van Nieuwenhuizen and J.A.M. Vermaseren, Phys. Rev. D16 (1977) 298
For a review see P. Van Nieuwenhuizen, CERN preprint TH 2473 (1978) invited talk at the Orbis Scientiae Coral Gables, 1978

[12] S. Deser and J.H. Kay, **Phys. Lett. 76B (1978) 400**

[13] P. Ramond, Phys. Rev. D3 (1971) 2415
A. Neveu and J.H. Schwarz, Nucl. Phys. B31 (1971) 86, Phys. Rev. D4 (1971) 1109
A. Neveu, J.H. Schwarz and C.B. Thorn, Phys. Lett. 35B (1971) 521

[14] "Dual Theory" edited by M. Jacob, North Holland Pub. Co. (1974)
J. Scherk, Rev. Mod. Phys. 47 (1975) 123

[15] F. Gliozzi, J. Scherk and D. Olive, Phys. Lett. 65B (1976) 282

[16] F. Gliozzi, J. Scherk and D. Olive, Nucl. Phys. B122 (1977) 253

[17] L. Brink, J.H. Schwarz and J. Scherk, Nucl. Phys. B121 (1977) 77

[18] S. Ferrara, J. Scherk and B. Zumino, Nucl. Phys. B121 (1977) 293

[19] C.K. Zachos, Phys. Lett. 76B (1978) 329 and unpublished work

[20] E. Cremmer and J. Scherk, Nucl. Phys. B103 (1976) 399

[21] E.B. Bogolmony, Sov. J. Nucl. Phys. 24 (1976) 449
S. Coleman, S. Parke, A. Neveu and C.M. Sommerfield, Phys. Rev. D15 (1977) 544

[22] C. Montonen and D. Olive, Phys. Lett. 72B (1977) 117

[23] A. D'Adda, R. Horsley and P. Di Vecchia, Phys. Lett. 76B (1978) 298

[24] E. Witten and D. Olive, HUTP 78 IA013 preprint (1978)

[25] T. Kaluza, Sitzungber. Prenss. Akad. Wiss. Berlin Math-Phys KA (1921) 1966
O. Klein, Z. Phys. 37 (1926) 895
W. Thirring, Acta Physica Austriaca, Supp. IX (1972) 266 and references therein
[26] S. Coleman and J. Mandula, Phys. Rev. 159 (1967) 1251
[27] For reviews of supersymmetry algebras, see B. Zumino, Proc. 17th Intern. Conf. on High Energy Physics, London 1974 (Science Research Council, Didcot, 1974) p I-254
J. Wess, Lecture Notes in Physics (Springer, Berlin 1974) vol. 37, p 352
L. Corwin, Y. Ne'eman and S. Sternberg, Rev. Mod. Phys. 47 (1975) 573
A. Salam and J. Strathdee, Phys. Rev. D11 (1975) 521
L. O'Raifertaigh, Comm. Dublin Inst. for Advanced Studies, Series A, Theor. Phys. Nr 22 (1975)
B. Zumino, Proc. Conf. on Gauge Theories and Modern Field Theory, Boston 1975 (MIT press, Cambridge, Mass 1976) p 255
A. Salam and J. Strathdee, ICTP preprint, IC176122 (1976)
S. Ferrara, Rivista Nuovo Cimento 6 (1976) 105
P. Fayet and S. Ferrara, Phys. Rep. 32 (1977) 251
S. Ferrara, CERN preprint TH 2514 (1978
[28] J. Tits, Tabellen zu den Einfachen Lie Gruppen und chre Darstellungen, Lecture Notes in Mathematics 40 (Springer 1967)
[29] See S.W. Hawking, Cargèse Lecture Notes (1978)
[30] M.F. Sohnius, Nucl. Phys. B138 (1978) 109
[31] D. Olive, private communication (unpublished)
[32] W. Nahm, Nucl. Phys. B135 (1978) 149
[33] D.R.T. Jones, Phys. Letters 72B (1977) 199
E.C. Poggio and H.N. Pendleton, Phys. Lett. 72B (1977) 200
[34] E. Cremmer, B. Julia and J. Scherk, Phys. Lett. 76B (1978) 409
[35] R. Haag, J.T. Lopuszanski and M. Sohnius, Nucl. Phys. B88 (1975) 257
[36] D.Z. Freedman and J.H. Schwarz, Nucl. Phys. B137 (1978) 333
[37] S. Ferrara and P. Van Nieuwenhuizen, Phys. Lett. 74B (1978) 333
K.S. Stella and P.C. West, Phys. Lett. 74B (1978) 330
S. Ferrara and P. Van Nieuwenhuizen, CERN preprint TH 2484 (1978) to be published in Phys. Lett. B, Ecole Normale Sup. Report LPTENS 78/14 (1978)
[38] A. Salam and J. Strathdee, Nucl. Phys. B76 (1974) 477
S. Ferrara, J. Wess and B. Zumino, Phys. Lett. B51 (1974) 239
J. Wess and B. Zumino, Phys. Lett. 66B (1977) 361
R. Grimm, J. Wess and B. Zumino, Phys. Lett. 74B (1978) 51
L. Brink, M. Gell-Mann, P. Ramond and J.H. Schwarz, Phys. Lett. 74B (1978) 336 and Caltech report 68-649 (1978)
[39] S.W. Mac Dowell and F. Mansouri, Phys. Rev. Lett. 38 (1977) 739
P.K. Townsend and P. Van Nieuwenhuizen, Phys. Lett. B67 (1977)439
Y Ne'mann and T. Regge, Phys. Lett. B76 (1978) 54
[40] S. Ferrara, F. Gliozzi, J. Scherk and P. Van Nieuwenhuizen,

Nucl. Phys. B117 (1976) 333
[41] S. Ferrara, J. Scherk and P. Van Nieuwenhuizen, Phys.Rev. Lett. 37 (1976) 1035
[42] E. Cremmer, B. Julia, J. Scherk, P. Van Nieuwenhuizen, S. Ferrara and L. Girardello, Ecole Normale Supérieure preprint 78/17 (1978)
[43] D.Z. Freedman and P. Van Nieuwenhuizen, Phys. Rev. D, Vol. 14 (1976) 912
[44] P. Fayet, Nucl. Phys. B113 (1976) 135
[45] N.S. Manton, Nucl. Phys. B126 (1977) 525
[46] M. Gell-Mann, invited paper at the Washington Meeting of Amer. Phys. Soc. April 1977, to be published
[47] K.S. Stelle, Phys. Rev. D, Vol. 16, 4 (1977) 953
[48] J. Scherk and J.H. Schwarz, Nucl. Phys. B81 (1974) 118, Phys. Lett. 57B (1975) 463
T. Yoneya, Nuovo Cimento Lett. 8 (1973) 951 ; Prog. of Theor. Phys. Vol. 51 (1974) 1907
J.H. Schwarz, CALT preprint 68-637 (1978) talk presented at Orbis Scientiae 1978, and to be published
[49] The O(8) theory has been recently obtained by dimensional reduction for the D = 11 supergravity theory by E. Cremmer and B. Julia LPTENS 78/23 preprint (1978)

LECTURES IN SUPERGRAVITY THEORY

P. van Nieuwenhuizen

Institute for Theoretical Physics
State University of New York at Stony Brook
Long Island, New York, N.Y. 11974

1. Introduction

Supergravity[1-3] is general relativity with an extra symmetry: Fermi-Bose symmetry, also called supersymmetry. In field theories in curved space, this symmetry must be local. Conversely, local Fermi-Bose symmetry can only be realized by field theories in curved spacetime. This explains the name supergravity.

These lectures are devoted to the recently established tensor calculus for O(1) supergravity[4-6] (for nomenclature see below). The need for such a calculus was stressed from the beginning of supergravity two years ago, but it took a long time before it was found because, as it now turns out, first another well-known problem had to be solved: the minimal set of auxiliary fields which are needed to close the gauge algebra.[7,8]

The importance of a tensor calculus can best be illustrated by realizing how familiar we are in ordinary relativity with manipulations with four-vectors. Up till the advent of the tensor calculus, we were still working in supergravity with individual components, so to speak. By means of the tensor calculus one can now immediately write down invariant actions, and tedious Noether calculations are no longer necessary. We will give several examples. Also, the problem of renormalizability or, rather, the question of which invariants of general relativity can be extended to locally supersymmetric invariants, has been solved at the general n-loop level in O(1) supergravity. For the O(n) models, neither the auxiliary fields nor a tensor calculus is known at this moment. This constitutes one of the most important problems in supergravity.

It is perhaps good to compare the tensor calculus and the superspace approach. Both are based on superfields of global supersymmetry, but whereas the tensor calculus uses only the fields which are used in ordinary supergravity (and thus works with component fields in a given superspace gauge), the superspace approach ends up with an enormous number of redundant fields and must eliminate them afterwards by choosing a suitable gauge. The complexities of superspace have prevented, for example, the reformulation in superspace of N>1 supergravity theories, and, as far as I know, no new theories have been found up till now in that approach, but it is interesting in itself, and might lead to new theories.

The tensor calculus was originally developed by Ferrara and the author for scalar multiplets.[4,5] Analogous results for vector multiplets were obtained by Stelle and West.[6] We will present their combined work in a simplified form. In the lectures the quantization and geometrical and group theoretical aspects of Einstein and conformal supergravity were also discussed. These subjects will not be included in the present notes, but will be treated in more detail in forthcoming publication.[9] Readers who want a simple introduction to supergravity might start with a recent Scientific American article,[10] followed by any of the many conference proceedings.[11]

NOMENCLATURE

Supergravity theories containing the Einstein action R are called Einstein or also Poincare supergravities. Extensions of the conformal action $R_{\mu\nu}^2 - 1/3R^2$ are called conformal or also Weyl supergravities. As in general relativity, one can further classify them by distinguishing theories with and without matter couplings; the latter are called pure supergravities. The pure Einstein supergravities are also called pure O(n) theories. There are only eight of them viable, $1 \leq n \leq 8$, since for $n>8$ one has several gravitons, and particles with spin $J>2$. A given O(n) theory contains (for $n\leq 8$)one graviton, n gravitinos (= real massless spin 3/2 field) $\frac{1}{2}n(n-1)$ real spin 1 fields, etc. The name is due to the local O(n) symmetry, which rotates,for example, the n gravitinos into each other. A given O(n) theory has in addition many global symmetries (at least a U(n) group of which the O(n) group is a subgroup). The field theories for n= 1,2,3,4, 8 and are known explicitly.[12]

The simplest Einstein supergravity theory is the pure O(1) theory, containing only two particles (graviton and gravitino), but which, as we shall see, should be formulated in terms of five gauge fields: the spin 2 vierbein field e_μ^a , the spin 3/2 Rarita-Schwinger field ψ_μ and three auxiliary fields S, P and A_a.

The Weyl supergravities have a local U(n) symmetry. Of the pure U(n) theories, only the U(1) theory is explicitly known.[13]

CONVENTIONS

We will use the positive Pauli metric: $\delta_{\mu\nu}=(+,+,+,+)$, $\not\partial=\gamma_\mu\partial_\mu$ $\{\gamma_a,\gamma_b\}=2\delta_{ab}$ for a = 1,4; $\gamma_5=\gamma_1\gamma_2\gamma_3\gamma_4$ is hermitian and $\gamma_5^2 = 1$. A Majorana spinor satisfies $\lambda = C\bar{\lambda}^T$ where $C^{-1}=C^T=-C$ and $C\gamma_a C^{-1} = -\gamma_a^T$. The Fierz rearrangement relations for four-component spinors are

$$(\bar{\psi} M \chi)(\bar{\varphi} N \lambda) = -\tfrac{1}{4}(\bar{\psi} O_j \lambda)(\bar{\varphi} N O_j M \chi)\,\sigma(j)$$

with

$$\sigma(j) = +1 \quad \text{for} \quad O_j = 1, \gamma_5, \gamma_a$$

$$\sigma(j) = -1 \quad \text{for} \quad O_j = \gamma_a\gamma_5$$

$$\sigma(j) = -2 \quad \text{for} \quad \sigma_{ab} \equiv \tfrac{1}{4}[\gamma_a, \gamma_b]$$

Part 1. PRELIMINARIES

In the first sections, the action and transformation laws of pure O(1) Einstein supergravity without auxiliary fields are given. The commutator algebra is shown not to close off-shell, and the minimal set of auxiliary fields, needed to close the gauge algebra, is given. The result is that the gauge fields of pure Einstein supergravity are the graviton e_μ^a, the gravitino ψ_μ, an axial vector A_a, a scalar S and a pseudoscalar P.

Review of pure O(1) Einstein supergravity

The gauge actions of pure Einstein supergravity without auxiliary fields is the sum of the Einstein action with torsion and the Rarita-Schwinger action for spin 3/2 fields ψ_μ (Majorana 4-component spinors with a vector index).

$$\mathcal{L} = -\tfrac{1}{2} e\kappa^{-2} R - \tfrac{1}{2}\epsilon^{\mu\nu\rho\sigma}\bar{\psi}_\mu \gamma_5 \gamma_\nu D_\rho \psi_\sigma$$

$$D_\rho \psi_\sigma = \left(\partial_\rho + \tfrac{1}{2}\omega_{\rho ab}\sigma^{ab}\right)\psi_\sigma$$

$$R = e_a{}^\nu e_b{}^\mu R_{\mu\nu}{}^{ab}(\omega)$$

$$R_{\mu\nu}{}^{ab}(\omega) = \partial_\mu \omega_\nu{}^{ab} + \omega_\mu{}^{ac}\omega_{\nu cb} - \mu \leftrightarrow \nu$$

It is invariant under the following transformation rules of local supersymmetry [1-3]

$$\delta e^a{}_\mu = \kappa \bar{\epsilon} \gamma^a \psi_\mu \qquad \delta\psi_\mu = 2\kappa^{-1} D_\mu \epsilon$$

The spin connection field $\omega_\mu{}^{ab}$ is not an independent field but is expressed in terms of the vierbein field $e^a{}_\mu$ and ψ_μ by solving its field equation $\delta I/\delta\omega_\mu{}^{ab} = 0$, which happens to be algebraic in Einstein (but not in Weyl) supergravity. The result contains torsion

$$\omega_\mu{}^{ab} = \omega_\mu{}^{ab}(e) + \frac{\kappa^2}{4}\left(\bar{\psi}_\mu \gamma^a \psi^b - \bar{\psi}_\mu \gamma^b \psi^a + \bar{\psi}^a \gamma_\mu \psi^b\right)$$

The non-torsion part of the spin connection, $\omega_\mu{}^{ab}(e)$, is given by solving the vierbein-postulate

$$\partial_\mu e^a{}_\nu + \omega_\mu{}^{ab}(e)\, e_{b\nu} - \Gamma^\alpha_{\mu\nu}(g)\, e_{a\alpha} = 0$$

An explicit proof of invariance of this action is trivial if one uses that one need not vary ω in the action, since $\delta I/\delta\omega = 0$[9] (this obvious but useful trick is called 1.5 order formalism).

Since supergravity contains fermions, one needs the vierbein field e^a_μ rather than its square, the metric $g_{\mu\nu}$, to describe gravity. Its determinant e is not needed in the spin 3/2 part since $\varepsilon^{\mu\nu\rho\sigma}$ is already a density. The derivative D_ρ acts only on Lorentz indices (such as the index a in $e^a{}_\mu$) but not world indices (such as the σ in ψ_σ). Since the action is a sum of the spin2 and spin 3/2 curvatures, only the curl $D_\rho\psi_\sigma - D_\sigma\psi_\sigma$ appears in the latter. Clearly, this curl is a good world tensor. One may differentiate and integrate partially with the derivative D_ρ.

<u>Exercise:</u> derive the spin 3/2 field equation

$$e R^\mu = \epsilon^{\mu\nu\rho\sigma} \gamma_5 \gamma_\nu D_\rho \psi_\sigma = 0$$

Hint: use 1.5 order formalism and the vierbein postulate

$$\left[D_\mu e^a{}_\nu - \tfrac{1}{4}\kappa^2 \bar\psi_\mu \gamma^a \psi_\nu\right] \epsilon^{\mu\nu\rho\sigma} = 0$$

Because supergravity contains torsion, the usual identities become modified. For example, the Ricci tensor is no longer symmetric.

Exercise: Prove that supergravity is consistent, i.e., prove that on-shell not only $R^\mu = 0$ but also $D_\mu R^\mu = 0$. Hint: use

$$\gamma_\nu = e^a{}_\nu \gamma_a \quad \text{and} \quad D_\mu \gamma_\nu = \gamma^a D_\mu e^a{}_\nu$$

and derive the torsion identity

$$\epsilon^{\mu\nu\rho\sigma} R_{\mu\nu\rho\tau} = -(\bar\psi_\lambda \gamma_\tau D_\mu \psi_\nu)\, \epsilon^{\lambda\sigma\mu\nu}$$

The important point is that the commutator $[D_\mu, D_\nu]\psi_\sigma$ yields the Einstein tensor which satisfies the spin 2 equation, and not, for example, the Riemann tensor.

Non-closure of the gauge algebra.[3]

The action of the preceding section is invariant under general coordinate transformations G, local Lorentz rotations L and local supersymmetry transformations S. If one evaluates the commutator of any two symmetries one finds on the right hand side always a sum of local symmetries with field dependent coefficients. For example

$$[\delta_S(\epsilon_1), \delta_S(\epsilon_2)]e^a{}_\mu = \delta_G(\xi^\alpha)\, e^a{}_\mu + \delta_L(\xi^\alpha \omega_{\alpha ab})\, e^a{}_\mu + \delta_S(-\tfrac{1}{2}\xi^\alpha \psi_\alpha)\, e^a{}_\mu \quad \text{with } \xi^\alpha = 2\bar\epsilon_2 \gamma^\alpha \epsilon_1$$

where the parameters are indicated between curly brackets. However, for ψ_μ there is one commutator which yields extra terms, proportional to R^{μ}

$$[\delta_S(\epsilon_1), \delta_S(\epsilon_2)]\psi_\mu = (\text{as on } e^a{}_\mu) + (\bar\epsilon_1 \gamma^\alpha \epsilon_2)\left(\tfrac{1}{4}\gamma_\alpha g_{\mu\rho} + \tfrac{e}{2}\epsilon_{\mu\alpha\rho\tau}\gamma_5\gamma^\tau\right)R^\rho + (\bar\epsilon_1 \sigma^{\rho\sigma}\epsilon_2)\left(\tfrac{1}{2}\sigma_{\rho\sigma} g_{\mu\tau} + g_{\mu\rho} g_{\sigma\tau} + \tfrac{e}{2}\epsilon_{\rho\sigma\tau\mu}\gamma_5\right)R^\tau$$

Thus, the gauge algebra only closes on-shell where $R^{\mu} = 0$ and the symmetries of the action are not good symmetries as far as the algebra is concerned. Since closure of the algebra is needed for the tensor analysis (and to prove Slavnov-Taylor identities in the quantum theory) this is an undesirable state of affairs.

It is well known in field theory, that one can destroy off-shell closure of a global algebra by eliminating auxiliary fields.

<u>Exercise</u>: Show the invariance of the action, and the on- and off-shell closure of the algebra, of the following globally supersymmetric system

$$I = -\frac{1}{2}\int d^4x \left[(\partial_\mu A)^2 + (\partial_\mu B)^2 + \bar{X}\not\partial X - F^2 - G^2 \right]$$

$$\delta A = \bar{\epsilon} X, \quad \delta B = -i\bar{\epsilon}\gamma_5 X, \quad \delta F = \bar{\epsilon}\not\partial X, \quad \delta G = i\bar{\epsilon}\gamma_5\not\partial X$$

$$\delta X = \not\partial (A - iB\gamma_5)\epsilon + (F + i\gamma_5 G)\epsilon$$

Show that elimination of F and G by inserting their field equations into action and algebra destroys off-shell closure.

The appearance in the gauge algebra of terms proportional to a field equation suggests that addition of auxiliary fields to the transformation rules might restore off-shell closure. The problem is now to find these fields. A hint is given by super fields.[14] A general vector superfield $V_a(x,\Theta)$ can be expanded in the anti-commuting Majorana parameter Θ as follows

$$V_a(x,\theta) = C_a + \bar{\theta}X_a + \frac{1}{2}\bar{\theta}\theta H_a + \frac{1}{2}\bar{\theta}i\gamma_5\theta K_a + \frac{1}{2}\bar{\theta}i\gamma_c\gamma_5\theta B_{ac} + \bar{\theta}(\psi_a + \frac{1}{2}\not\partial X_a)\bar{\theta}\theta + \frac{1}{4}(\bar{\theta}\theta)^2 (A_a + \frac{1}{2}\Box C_a)$$

At the global linear level there are two kinds of symmetries which the future action is supposed to possess: global linearized supersymmetry transformations and local linear gauge transformations. (For the spin 3/2 field they are respectively $\delta\psi_\mu = \Omega_\mu{}^{ab}\sigma_{ab}\varepsilon$ with $\Omega_\mu{}^{ab}$ the linearized part of $\omega_\mu{}^{ab}$ and $\delta\psi_\mu = 2\kappa^{-1}\partial_\mu\varepsilon$. At the nonlinear local level both combine to yield the rule $\delta\psi_\mu = 2\kappa^{-1}D_\mu\varepsilon$). Expressed on $V_a(x,\Theta)$, these symmetries have the form

$$\delta_S V_a = (\bar{\epsilon} G) V_a \qquad G_\alpha = \partial/\partial\bar{\theta}^\alpha - (\gamma\theta)_\alpha$$

$$\delta_G V_a = \bar{D} i \gamma_a \gamma_5 \Lambda \qquad D_\alpha = \partial/\partial\bar{\theta}^\alpha + (\gamma\theta)_\alpha$$

where Λ_α is a general spinor superfield which must satisfy a constraint ($\bar{D}(1+\gamma_5)D\,\bar{D}(1-\gamma_5)\,\Lambda = 0$, otherwise δ_G contains also local conformal symmetries). One can now choose Λ such that in $V_a+\delta_G V_a$ all components are zero except essentially

$$e_{ab} \equiv B_{ab} + B_{ba} \,,\ \psi_a \,,\ A_a \,,\ S \equiv \partial_a H_a \,,\ P \equiv \partial_a K_a$$

The original global supersymmetry transformations satisfy the global algebrawith $[\delta_S\ \delta_S] \sim \not{\partial}$. If one now defines a new global supersymmetry transformation $\delta_{S'} = \delta_S+\delta_G$ such that δ_G rotates away the fields which were initially gauged away but were produced in the meantime by δ_S, then also this $\delta_{S'}$ satisfies the global algebra with $[\delta_{S'},\delta_{S'}] \sim \not{\partial}$ (because $\{D_\alpha, G_\beta\} = 0$). Therefore, at the global linear level, one has a closed algebra in terms of the vierbein, the gravitino and three more fields: a scalar S, a pseudoscalar P and an axial vector A_a.

It is tempting to conjecture that S, P and A_a are the auxiliary fields which are needed in the local case to close the algebra. Under $\delta_{S'}$ the global rules read[14]

$$\delta_{S'}\psi_a = \left(\tfrac{1}{\kappa}\omega_a{}^{bc}\sigma_{bc} + iA_a\gamma_5\right)\epsilon + \tfrac{1}{3}\gamma_a\left(S - i\gamma_5 P - i\not{A}\gamma_5\right)\epsilon$$

$$\delta_{S'}e_{ab} = \tfrac{\kappa}{2}\left(\bar{\epsilon}\gamma_a\psi_b + \bar{\epsilon}\gamma_b\psi_a\right)$$

$$\delta_{S'}S = \tfrac{1}{2}\bar{\epsilon}\gamma\cdot R$$

$$\delta_{S'}P = -\tfrac{i}{2}\bar{\epsilon}\gamma_5\gamma\cdot R$$

$$\delta_{S'}A_a = \tfrac{3i}{2}\bar{\epsilon}\gamma_5\left(R_a - \tfrac{1}{3}\gamma_a\gamma\cdot R\right)$$

with on the right hand side only linearized expressions and constant ε.

The minimal set of auxiliary fields

From the example in the previous section involving F and G, and the hint from superfields, we begin by postulating the action

$$I = \int d^4x \left[\mathcal{L}^2 + \mathcal{L}^{3/2} - \frac{e}{3}\left(S^2 + P^2 - A_a^2 \right) \right]$$

Hence, in addition to the previous action we only have squares of S, P and A_a. (The normalization is arbitrary.) This guarantees that the extra fields are non-physical. For the transformation laws we conjecture

$$\delta e^a{}_\mu = \bar{\epsilon}\gamma^a\psi_\mu\,, \qquad \eta = -\tfrac{1}{3}\left(S - i\gamma_5 P - i\not{A}\gamma_5\right)$$

$$\delta\psi_\mu = 2\kappa^{-1}\left(D_\mu + \tfrac{i\kappa}{2}A_\mu\gamma_5\right)\epsilon - \gamma_\mu\eta\epsilon$$

$$\delta S = \tfrac{1}{2}\bar{\epsilon}\gamma\cdot R^{cov}$$

$$\delta P = \tfrac{-i}{2}\bar{\epsilon}\gamma_5\gamma\cdot R^{cov}$$

$$\delta A_a = \tfrac{3i}{2}\bar{\epsilon}\gamma_5\left(R_a^{cov} - \tfrac{1}{3}\gamma_a\gamma\cdot R^{cov}\right)$$

For this simple result, it is essential that the index of A_a is flat. Inserting the field equations $S = P = A_a = 0$, one finds back the previous results, while the linearized level agrees with the last section. The symbol cov denotes that in the curl $D_\rho\psi_\sigma - D_\sigma\psi_\rho$ which is present in R^μ, one has to take the covariant derivative.

Exercise: if $\delta A = \bar{\varepsilon}B$, prove that $D_\mu A = \partial_\mu A - \frac{\kappa}{2}\bar{\epsilon}B$ is covariant, i.e., that $\delta(D_\mu A)$ does not contain derivatives of ε, if δB is covariant.

Due to the curl structure, one need not covariantize the terms. In more detail

$$R^{\mu,cov} = \epsilon^{\mu\nu\rho\sigma}\gamma_5\gamma_\nu\left[D_\rho\psi_\sigma - \tfrac{i\kappa}{2}A_\sigma\gamma_5\psi_\rho + \tfrac{\kappa}{2}\gamma_\sigma\eta\psi_\rho\right]$$

Explicit evaluation now reveals that the commutator algebra closes uniformly, i.e. for $\phi = e^a{}_\mu, \psi_\mu$, S, P or A_a one has the same commutators, without field equations.

$$[\delta_S(\epsilon_1), \delta_S(\epsilon_2)]\phi = [\delta_G(\xi^\alpha) + \delta_S(-\tfrac{1}{2}\xi^\alpha\psi_\alpha) + \delta_L(\xi^\alpha\omega_{\alpha ab} - \tfrac{i\kappa}{3}\epsilon_{abcd}\xi^c A^d + \tfrac{4\kappa}{3}\bar{\epsilon}_2\sigma^{ab}(S^{-}i\gamma_5 P)\epsilon_1)]\phi$$

(It is very easy to verify this result for the vierbein field). Although this solution is due to Ferrara and the author and Stelle and West, Breitenlohner[15] was the first to find a closed gauge algebra. He required, however, that there were no tensor terms $\bar{\epsilon}_2\sigma_{ab}\epsilon_1$ in the above commutator, and consequently needed many more auxiliary fields.

That an axial vector field was necessary as an auxiliary field was suspected long ago.[16] (See also the exercise below). *A posteriori* one can also easily see that a scalar and pseudoscalar auxiliary field are necessary, too. Consider the coupling of the scalar multiplet (the example in the previous section). Starting with $\mathcal{L} = eF$ and $\delta F = \bar{\epsilon}\not{D}^P X$ with supercovariant derivative D^P_μ, the Noether method shows that one must add a term $\frac{\kappa}{2}\bar{\psi}\gamma X$ to the action in order to obtain invariance at the order κ^0. At the order κ level, one then is left with a term of the form $\kappa\bar{\epsilon}\sigma_{\mu\nu}D_\mu(A-iB\gamma_5)\psi_\nu$. Partially integrating, the term with $D_\mu\epsilon$ is easily cancelled by adding a term of the form $\bar{\psi}_\mu\sigma^{\mu\nu}(A-iB\gamma_5)\psi_\nu$ to the action. However, the term $\kappa\bar{\epsilon}(A-iB\gamma_5)\sigma_{\mu\nu}D^\mu\psi^\nu$ is proportional to the spin 3/2 field equation $(\sigma_{\mu\nu}D_\mu\psi_\nu = \frac{1}{4}\gamma\cdot R)$ One could now add a term of the form $\kappa\gamma_\mu(A-iB\gamma_5)\epsilon$ to $\delta\psi_\mu$ but this would introduce matter fields in the laws for gauge fields, which is to be avoided. (The converse, gauge fields in the laws for matter fields, is frequently the case; for example, in general relativity). One can get around this problem by introducing two auxiliary fields S and P. Adding to the action a term AS+BP one can cancel the last order-κ variation, and S and P will enter into the transformation law of ψ_μ in order that $\delta(\bar{\psi}\cdot\gamma X)$ cancel $(\delta A)S + (\delta B)P$

<u>Exercise:</u> Try to repeat this analysis for the spin $(1,\frac{1}{2})$

Maxwell system with fields B_μ, λ, D. This demonstrates the existence of the third auxiliary field, the axial vector A_a.

It is most remarkable that simple covariantization of the flat-space, globel linear results leads to the complete solution at the local nonlinear level.[17] This might be helpful in finding the auxiliary fields for the extended supergravities. One should view the complete set of $S,P,A_a,e^a{}_\mu$ and ψ_μ as the gauge fields of supergravity, and not consider S,P,A_a as less important. In Einstein supergravity they are indeed nonpropagating, but in conformal supergravity A_a is propagating (and gauges chiral rotations) while in higher derivative theories also S and P can be propagating. Instead of all these different dynamical models, it is rather the local algebra which characterizes supergravity.

Exercise: Show that if $S=P=A_a = 0$ then for a field $\phi_{b\ldots}$ with only local Lorentz indices one simply has

$$[\delta(\epsilon_1), \delta(\epsilon_2)]\,\phi_{b\ldots} = 2(\bar{\epsilon}_2 \gamma^\alpha \epsilon_1)\,(D^P_\alpha \phi_{b\ldots})$$

where D^P_α is the completely covariant derivative, covariant with respect to the super-Poincare group.

PART II. TENSOR CALCULUS

In this part the scalar and vector multiplets of supergravity are defined, multiplication rules are given and it is shown how one can extract an invariant action from any multiplet. As applications, we first rederive the known models of O(1) supergravity and show how to construct without any work, arbitrarily many new actions. Then we give the solution for the general n-loop counter terms of O(1) supergravity, thus reproducing and extending the results on renormalization.

Scalar multiplets[4,5]

A scalar multiplet is any collection of five objects

$$\Sigma = (A, B, X, F', G')$$

which transform under local supersymmetry into each other as follows

$$\delta A = \bar{\epsilon}\chi\,, \qquad \delta B = -i\bar{\epsilon}\gamma_5\chi$$
$$\delta\chi = \not{D}^P(A - iB\gamma_5)\epsilon + (F' + i\gamma_5 G')\epsilon$$
$$\delta F' = \bar{\epsilon}(\not{D}^P - \tfrac{i\kappa}{2}\not{A}\gamma_5)\chi + \kappa\bar{\epsilon}\eta\chi$$
$$\delta G' = i\bar{\epsilon}\gamma_5(\not{D}^P - \tfrac{i\kappa}{2}\not{A}\gamma_5)\chi - i\kappa\bar{\epsilon}\eta\gamma_5\chi$$

We recall that $\eta_P = -\frac{1}{3}(S - i\gamma_5 P - i\not{A}\gamma_5)$.
The derivatives D^P_μ are covariant with respect to the full local symmetry group, i.e., the super-Poincaré group. From the definition of covariant derivatives, one derives

$$D^P_\mu A = \partial_\mu A - \tfrac{\kappa}{2}\bar{\psi}_\mu\chi$$
$$D^P_\mu B = \partial_\mu B + \tfrac{i\kappa}{2}\bar{\psi}_\mu\gamma_5\chi$$
$$D^P_\mu\chi = D_\mu\chi - \tfrac{\kappa}{2}\left[\not{D}^P(A - iB\gamma_5)\right]\psi_\mu - \tfrac{\kappa}{2}(F' + i\gamma_5 G')\psi_\mu$$

For a given scalar multiplet Σ, the following is an invariant action

$$I_S(\Sigma) = \int d^4x\, e\left[F' + \tfrac{\kappa}{2}\bar{\psi}\cdot\gamma\,\chi + \tfrac{\kappa^2}{2}\bar{\psi}_\mu\sigma^{\mu\nu}(A - iB\gamma_5)\psi_\nu + \kappa(SA + PB)\right]$$

From the transformation rules of the gauge fields and of Σ, one may check that the integrand changes under variation by a total derivative.

Two scalar multiplets can be multiplied to give yet another scalar multiplet S

$$S(\Sigma\otimes\sigma) = S\left[(A,B,\chi,F',G')\otimes(a,b,\chi,f',g')\right] =$$
$$\left[Aa - Bb,\ Ab + aB,\ (A + iB\gamma_5)\chi + (a + ib\gamma_5)\chi,\right.$$
$$\left. Af' + aF' - Bg' - bG' - \bar{\chi}\chi,\ Ag + aG + Bf + bF + i\bar{\chi}\gamma_5\chi\right]$$

This multiplication rule is the same as in global supersymmetry.

These results were first derived using conformal supergravity notions. For conformal supergravity, the tensor calculus is obtained by simply replacing ordinary derivatives ∂_μ by superconformal derivatives D^c_μ in the tensor calculus of global supersymmetry. In that case, δF contains explicitly the Weyl weight $\lambda(\Sigma_c)$ of the conformal multiplet Σ_c to which F belongs, and in the product $\Sigma_c \otimes \sigma_c$ the sum of the weights of Σ_c and σ_c appears. If one wants to extend this tensor calculus to Einstein supergravity, one must find first a formulation in which $\lambda(\Sigma_c)$ is not explicitly present, such that Weyl multiplets with different λ's have the same Einstein transformation laws. But that is not all; also the product of two Weyl multiplets, as defined above, should transform according to again these same rules. If that is possible, then one can also define these rules to be the rules for Einstein multiplets for which no λ can be defined at all. This was shown to be possible by considering for conformal multiplets as last components

$$F' = F - \frac{\lambda}{3}(AS + PB), \quad G' = G - \frac{\lambda}{3}(BS - PA)$$

in which case one obtains the λ-independent results given above. These final results hold thus both for Einstein and for conformal multiplets. For historical reasons (and to prevent confusion) we will keep writing F', G' (rather than omitting the prime) for the components of Σ, although there are no F, G components for a general Einstein multiplet. Thus, a general Einstein multiplet is given by $\Sigma = (A, B, X, F', G')$ and transforms as indicated above.

A more direct though more laborious derivation is to start from $\mathcal{L} = eF'$, assuming the rules for the gauge fields, and to find the extra terms in $\delta A, \ldots, \delta G'$ and $\mathcal{L}$ needed to obtain complete invariance. This Noether method is not unique. For example, one need not introduce S and P but can choose F' appropriately at each order in κ. This would not be a useful solution, since the Jacobi identities guarantee that whatever complicated expression one has for $A,\ldots G'$ in terms of basic fields, these components must always realize the same uniform local algebra as the elementary constituants. Thus $\delta A,\ldots,\delta G'$ may not contain $\delta\varepsilon$-terms, and one needs at least everywhere covariant derivatives. Under this assumption, the Noether procedure is probably unique.

In proving that $S(\Sigma \otimes \sigma)$ is again a scalar multiplet, one uses that covariant derivatives satisfy the Leibnitz rule. Also this is not sufficient. For example, one could have defined scalar multiplet by taking two conformal multiplets $\Sigma_c=(A,B,X,F,G)$ and $\sigma_c=(a,b,\chi,f,g)$ with, say $\lambda=1$, and then defined the general real multiplet by the transformation rules of Σ_c (and σ_c). What would go wrong is that in $\Sigma_c \otimes \sigma_c$ different rules are needed (obtained from the general conformal rules by replacing $\lambda(\Sigma_c \otimes \sigma_c)$ by 2). It is only the rules we have given above which are consistent with the simple multiplication rule above.

One last remark concerns multiplets with external indices,[17] for example a spinor multiplet $\Sigma_\alpha = (A_\alpha,\ldots\ldots)$. The commutator algebra has to take into account the extra spinor index α, so that there are extra terms in $\delta X_{\beta\alpha}$. However, the determination of $X_{\beta\alpha}$ from δA_α, and of F_α and G_α from $\delta X_{\beta\alpha}$ is unchanged since the extra terms do not appear in the ε and $\gamma_5\varepsilon$ terms in $\delta X_{\beta\alpha}$.

The kinetic scalar multiplet[5]

In global supersymmetry one can obtain from a global scalar multiplet $\Sigma = (A,B,X,F,G)$ another scalar multiplet involving derivatives of Σ, denoted henceforth ty $T(\Sigma)$ and given by $T(\Sigma) = (F, -G, \not\partial X, \Box A, -\Box B)$. It is clear that F' and $-G'$ are not the first two components of the local version of $T(\Sigma)$, since in

$$\delta F' = \bar{\epsilon}\left(\not{D}^P - \tfrac{i\kappa}{2}\not{A}\gamma_5\right)X + \bar{\epsilon}\eta X$$

$$\delta G' = i\bar{\epsilon}\gamma_5\left(\not{D}^P - \tfrac{i\kappa}{2}\not{A}\gamma_5\right)X - i\bar{\epsilon}\gamma_5\eta X$$

the η -terms have opposite signs, so that $\delta F'$ and $\delta G'$ do not yield the same third component for $T(\Sigma)$. However, a consistent choice is

$$\tilde{A} = F' + \tfrac{\kappa}{3}(SA + PB)\ , \quad \tilde{B} = -G' - \tfrac{\kappa}{3}(SB - PA)$$

By varying $\tilde{A}$ and $\tilde{B}$, one finds the same third component

$$\tilde{X} = \left(\not{D}^P - \tfrac{i\kappa}{6}\not{A}\gamma_5\right)X + \tfrac{\kappa}{6}(A - i\gamma_5 B)\,\gamma\cdot R^{cov}$$

Varying $\tilde{X}$ one finds the last two components (while one might also check that the remaining terms have the correct form of the variation of a third component. They do.) Thus one finds

$$\tilde{F}' = \Box^c A - \tfrac{2}{3}(F'S - G'P) - \tfrac{2}{9}A\,(S^2 + P^2)$$

$$\tilde{G}' = -\Box^c B + \tfrac{2}{3}(G'S + F'P) + \tfrac{2}{9}B\,(S^2 + P^2)$$

where $\Box^c$ is the completely covariant superconformal derivative, evaluated as if $\lambda(\Sigma) = 1$.[5]

Clearly, $T(\Sigma)$ is not uniquely determined. For example, since the sum of scalar multiplets is a scalar multiplet, one could have added to $\tilde{A}$ the first component of any other scalar multiplet and still have determined a complete scalar multiplet. Our choice has the virtue that for $\kappa = 0$ one reobtains the global multiplet $T(\Sigma)$.

For a scalar multiplet without external indices, it follows from the uniform closure of the gauge algebra that once $\tilde{A}$ and $\tilde{B}$ determine the same $\tilde{X}$, they determine a scalar multiplet, and one need not check that $\delta\tilde{X}$ is consistent with the variation of a third component.* However, if Σ has external indices (see next section), then this is no longer so.

---

*I thank Dr. Breitenlohner for a discussion on this point.

Examples of scalar multiplets

The simplest scalar multiplet is the unit multiplet E_S, defined by $S(\Sigma \otimes E_S) = \Sigma$. (We will denote by S, Σ etc., scalar multiplets, and by $\otimes$ any product operation. No confusion with the auxiliary field S seems likely.) The reader may verify that

$$E_S = (1, 0, 0, 0, 0)$$

is indeed a scalar multiplet. The associated action is $(S + \frac{1}{2}\kappa\bar{\psi}_\mu\sigma^{\mu\nu}\psi_\nu)\kappa^{-2}$ and describes the de-Sitter extension of O(1) Einstein-supergravity.[18]

A second multiplet is obtained by differentiation. Defining $U = 3\kappa^{-1}T(E_S)$ one finds[4]

$$\mathcal{U} = \left(S, P, \tfrac{1}{2}\gamma\cdot R^{cov}, \tfrac{1}{2}R^{cov} - \tfrac{2\kappa^2}{3}(S^2+P^2+\tfrac{1}{2}A_a^2), -D_aA^a\right)$$

This simple form of U is found in ref. (17). This multiplet contains the scalar curvature. Its associated action is the gauge action of pure Einstein supergravity

$$I_S(\mathcal{U}) = -\int d^4x \left[\mathcal{L}^2 + \mathcal{L}^{3/2} - \tfrac{e}{3}(S^2 + P^2 - A_a^2)\right]$$

It is remarkable that the tensor analysis not only reproduces matter couplings, but also gauge actions.

From any local scalar multiplet Σ, one can obtain ita parity reversed companion[6]

$$\text{par}\,\Sigma = (B, -A, -i\gamma_5 X, G', -F')$$

which is also a local scalar multiplet as one easily verifies. The associated action now starts with eG'.

The most general scalar multiplet with second (first) derivatives on bosons (fermions) is given by

$$S = \sum_{m,n=0}^{\infty} a_{mn}\Sigma^m \otimes T(\Sigma^n) + \sum_{n=0}^{\infty} b_n \Sigma^n$$

All nonimproved and improved couplings of the spin (o,½) system are contained in this general result,(see the contribution of Dr. Ferrara in these proceedings).

A multiplet with an external spinor index can be constructed from the fields A_μ, λ, D of the spin (1,½) system.[4] These fields transform as

$$\delta A_\mu = -\bar{\epsilon}\gamma_\mu\lambda \ , \quad \delta\lambda = \left(\sigma^{\mu\nu} F^P_{\mu\nu} + i\gamma_5 D\right)\epsilon$$
$$\delta D = i\bar{\epsilon}\gamma_5\left(\not{D}^P + \tfrac{i}{2}\not{A}\gamma_5\right)\lambda$$

The index P denotes, as always, supercovariant derivatives. In terms of them, one may define

$$\Sigma_\alpha = \Big[\ \bar{\lambda}_\alpha\ ,\ -i\left(\bar{\lambda}\gamma_5\right)_\alpha ,\ \left(-\sigma^{\mu\nu}F^P_{\mu\nu} + i\gamma_5 D\right)_{\beta\alpha}\ ,$$
$$\left\{+\bar{\lambda}\left(-\overleftarrow{\not{D}}^P + \tfrac{i\kappa}{2}\not{A}\gamma_5\right) - \tfrac{1}{2}\bar{\lambda}\left(S - i\gamma_5 P\right)\right\}_\alpha ,$$
$$\left\{i\bar{\lambda}\left(\overleftarrow{\not{D}}^P\gamma_5 - \tfrac{i\kappa}{2}\not{A}\gamma_5\right) + \tfrac{i}{2}\bar{\lambda}\gamma_5\left(S - i\gamma_5 P\right)\right\}_\alpha\ \Big]$$

Note, however, that although one still defines $X_{\beta\alpha}$ by $\delta\bar{\lambda}_\alpha = \bar{\epsilon}_\beta X_{\beta\alpha}$, $\delta X_{\beta\alpha}$ contains extra terms in order that $\delta_L\bar{\lambda}_\alpha$ have the required form. The scalar and pseudoscalar terms define again uniquely F' and G' by $\delta X_{\beta\alpha} = \cdots + (F_\alpha\epsilon_\beta + i(\gamma_5\epsilon)_\beta G_\alpha)$. Squaring this multiplet and contracting the indices by the metric in spinor space (the charge conjugation matrix $C^{\alpha\beta}$), one obtains a scalar multiplet without external indices.

$$\Sigma = \Sigma_\alpha \otimes \Sigma_\beta\, C^{\alpha\beta}$$

A multiplet with three external indices, two Lorentz and one spinor index,[5] can be constructed from the curvatures of the superconformal group.[13] It reads

$$W_{ab\alpha} = \Big[\ R_{ab}(Q)_\alpha\ ,\ -i\left(R_{ab}(Q)\gamma_5\right)_\alpha ,$$
$$\left(\sigma_{cd} W_{abcd} + i\gamma_5 R_{cd}(A)\, T_{cdab}\right)_{\beta\alpha} ,$$
$$\left(\hat{R}_{cd}(S)\, T_{cdab}\right)_\alpha ,\ \left(i\hat{R}_{cd}(S)\gamma_5 T_{cdab}\right)_\alpha\ \Big]$$

where

$$T_{cdab} = \frac{1}{3}\left(\delta_{ac}\delta_{bd} - \frac{1}{2}\epsilon_{abcd}\gamma_5 + 2\sigma_{ac}\delta_{bd}\right)_{\underset{\vee}{ab},\,\underset{\vee}{cd}}$$

projects onto self-dual tensor which are traceless in spinor space and W_{abcd} is the Weyl tensor, constructed from the covariantized Riemann tensor

$$\hat{R}_{\mu\nu ab} - \frac{\kappa}{2} R_{ab}(Q)(\gamma_\mu \psi_\nu - \gamma_\nu \psi_\mu)$$

($\hat{R}_{\mu\nu ab}$ is the Lorentz-curvature minus the terms with the conformal gauge field.) $\hat{R}(S)$ is the covariant curvature of conformal supersymmetry transformations.

Vector multiplets[6]

A local vector multiplet contains the seven components of a general scalar superfield $V(x,\theta)$. Thus, with

$$V(x,\theta) = C + \bar{\theta}Z + \frac{1}{2}\bar{\theta}\theta H + \frac{1}{2}\bar{\theta}i\gamma_5\theta K$$
$$+ \frac{1}{2}\bar{\theta}\, i\gamma_a\gamma_5\theta\, B_a + \bar{\theta}\theta\,\bar{\theta}(\Lambda + \frac{1}{2}\not{D}^P Z) +$$
$$\frac{1}{4}(\bar{\theta}\theta)^2 (D + \frac{1}{2} D_a^P D_a^P C)$$

a vector multiplet is defined as

$$V = (C, Z, H, K, B_a, \Lambda, D)$$

The transformation rules are given by

$$\delta C = i\bar{\epsilon}\gamma_5 Z$$
$$\delta Z = (i\gamma_5 H - K - \not{B} + \not{D}^P C\, i\gamma_5)\epsilon$$
$$\delta H = i\bar{\epsilon}\gamma_5(\not{D}^P - \frac{i}{2}\not{A}\gamma_5)Z + i\bar{\epsilon}\gamma_5\Lambda - i\bar{\epsilon}\eta\gamma_5 Z$$
$$\delta K = -\bar{\epsilon}(\not{D}^P - \frac{i}{2}\not{A}\gamma_5)Z - \bar{\epsilon}\Lambda - \bar{\epsilon}\eta Z$$
$$\delta B_a = -\bar{\epsilon}(D_a^P - \frac{i}{2}A_a\gamma_5)Z - \bar{\epsilon}\gamma_a\Lambda + \frac{1}{2}\bar{\epsilon}\eta\gamma_a Z$$
$$\delta\Lambda = (\sigma^{ab}\hat{F}_{ab} + i\gamma_5 D)\epsilon$$
$$\delta D = i\bar{\epsilon}\gamma_5(\not{D}^P + \frac{i}{2}\not{A}\gamma_5)\Lambda$$

These rules are the covariantizations of the global supersymmetry transformations, given by $\delta_S V(x,\theta) = (\bar{\epsilon}G)V(x,\theta)$ except for the η terms and except that $\hat{F}_{ab}$ is given by

$$\hat{F}_{ab} = \Big(D_a B_b + \tfrac{\kappa}{2} \bar{\psi}_a \big(D^P_b - \tfrac{i}{2} A_b \gamma_5 \big) Z + \tfrac{\kappa}{2} \bar{\psi}_a \gamma_b \Lambda$$
$$- \tfrac{\kappa}{2} \bar{Z} e_b{}^{\mu} \big(D_a + \tfrac{i}{2} A_a \gamma_5 \big) \psi_\mu \Big) - (a \leftrightarrow b)$$

The last term is not obtained from δB_a by the $\varepsilon \to -\frac{1}{2}\psi_\mu$ precription. Nevertheless, $\hat{F}_{ab}$ is covariant.* The variation $\delta\psi_b = \frac{2}{\kappa}(D_b + \frac{i}{2}A_b\gamma_5)\varepsilon$ leads to a curl, which is covariant by itself, while the $(D_a\bar{\varepsilon})\eta\gamma_a Z$ term is covariantized by $\delta\psi_b = -\gamma_b\eta\varepsilon$. For curls there are thus two inequivalent covariant derivatives.

The η-term in the transformation rules and the unusual covariant derivative in $\hat{F}_{ab}$ are obtained by requiring, just as in global supersymmetry, that the following object, constructed from the components of a scalar multiplet Σ, is a vector multiplet

$$V(\text{par}\,\Sigma) = \big(-A,\ i\gamma_5 X,\ F',\ G',\ D^P_a B,\ 0,\ 0 \big)$$

It is an instructive exercise to check this.
From the transformation rules of Σ one finds the η terms in the transformation rules for $V(\Sigma)$, and thus for general V as well. The vanishing of Λ and D in $V(\Sigma)$ must be consistent, i.e., also $\delta\Lambda$ and δD should vanish. Thus if $V = V(\Sigma)$ one should require that $\hat{F}_{ab}$ vanishes. This is the case only if one uses the expression for $\hat{F}_{ab}$ given above, but not if one would have used the $\varepsilon \to -\frac{1}{2}\psi_\mu$ prescription.

The appearance of supercovariant derivatives D^P_μ is necessary, for example in order that the com mutator algebra does not contain $\partial\varepsilon$ terms. The chiral connections enter the stage because in $\delta\not{D}^P C$ there is a term $\delta\psi_\mu = 2\kappa^{-1}(D_\mu + \frac{i}{2} A_\mu \gamma_5)\varepsilon$.

Multiplets are multiplied as in global supersymmetry. The precise rule is most easily obtained by working out the product of two superfields $V(x,\theta)$ and $\mathcal{V}(x,\theta)$ and identifying the coefficients of a given product of θ's as the components of the product. One finds

---

*In ref.(8) it is written as $D_a\psi_b$.

$$V(V\otimes v) = V[(C,Z,H,K,B_a,\Lambda,D)\otimes(c,z,h,k,b_a,\lambda,d)]$$
$$=[Cc,\ Cz+cZ,\ Ch+cH-\tfrac{1}{2}\bar{Z}z,\ Ck+cK+\tfrac{1}{2}\bar{z}i\gamma_5 Z,$$
$$Cb_a+cB_a-\tfrac{1}{2}\bar{Z}i\gamma_a\gamma_5 Z,\ C\lambda+c\Lambda+\tfrac{1}{2}\{-(\not{D}^P C)z$$
$$-(\not{D}^P c)Z+Zh+zH-i\gamma_5(Zk+zK)-i\gamma_a\gamma_5(Zb_a+zB_a)\},$$
$$Cd+cD-(\bar{Z}\lambda+\bar{z}\Lambda)+Hh+Kk-B_a b_a-(D_a^P C)(D_a^P c)$$
$$-\tfrac{1}{2}\bar{Z}(\not{D}^P-\tfrac{i}{2}\not{A}\gamma_5)z-\tfrac{1}{2}\bar{z}(\not{D}^P-\tfrac{i}{2}\not{A}\gamma_5)Z\,]$$

A new feature is the appearance of derivatives in this product rule. They are due to our choice of writing the components of V as $\lambda+\frac{1}{2}\not{D}^P C$ and $D+\frac{1}{2}D_a^P D_a^P C$, which was done to simplify the formulae for $\delta H,\ldots,\delta D$.

An invariant action for any vector multiplet V is given by[6]

$$I_V(V)=\int d^4x\, e\,[\,D-\tfrac{\kappa}{2}\bar{\psi}\cdot\gamma\ i\gamma_5\chi-\tfrac{2\kappa}{3}(HS-KP-B_aA_a)$$
$$-\tfrac{\kappa}{3}\bar{Z}i\gamma_5\gamma\cdot R+\tfrac{i\kappa^2}{4}\epsilon^{mnrs}(\bar{\psi}_m\gamma_n\psi_r)(B_s-\tfrac{\kappa}{2}\bar{\psi}_s Z)$$
$$-\tfrac{2}{3}\kappa^2 C\,\mathcal{L}(\text{gauge action})\,]$$

One may obtain this result from the Noether procedure. It is instructive to check the result for the special cases that only D or Λ are nonzero, or the case V=V(S) (see next sections).

The combination $B'_\sigma = B_\sigma - \frac{\kappa}{2}\bar{\psi}_\sigma Z$ varies under supersymmetry into $-\bar{\epsilon}\gamma_\sigma\lambda-\partial_\sigma(\bar{\epsilon}Z)$. One can define a local gauge transformation by $\delta_G V=V(\Sigma)$ with arbitrary Σ. In that case $\delta_G B'_\sigma = \partial_\sigma A$ so that if one happens to have a gauge-invariant action, one can define a new supersymmetry transformation $\delta_{S'}=\delta_S+\delta_G$ such that

$$C=Z=H=K=0\,,\quad \delta_{S'}(C,Z,H,K)=0$$
$$\delta_{S'}B'_\sigma=-\bar{\epsilon}\gamma_\sigma\Lambda\,,\quad \delta_{S'}D=i\bar{\epsilon}\gamma_5(\not{D}^P+\tfrac{i}{2}\not{A}\gamma_5)\Lambda$$
$$\delta_{S'}\Lambda=\delta_S\Lambda=2\sigma^{\mu\nu}\epsilon(\partial_\mu B'_\nu+\tfrac{\kappa}{2}\bar{\psi}_\mu\gamma_\nu\Lambda)+i\gamma_5 D\epsilon$$

These are the usual rules for the supersymmetric Maxwell system which is indeed gauge invariant, and it is B'_σ which is to be identified with the Maxwell field and not B_σ.[6]

Examples of vector multiplets

Also for vector multiplets a unit element E_v exists, satisfying $V(E_v \otimes v) = v$. It is given by

$$E_v = (1,0,0,0,0,0,0)$$

and again one may verify that this is a multiplet. The associated action is proportional to the complete gauge action of O(1) supergravity, depending on e_μ^a, ψ_μ, S, P, A_a

$$I_V(E_v) = -\frac{2}{3} I(\text{gauge action})$$

Another vector multiplet starts with the auxiliary field A_a (with flat index!). After variation one finds the other components. At the linear level[14]

$$W_a = \Big[A_a\,,\ \chi_a = \tfrac{3}{2}\big(R_a - \tfrac{1}{3}\gamma_a\gamma\cdot R\big)\,,\ \partial_a P,\ \partial_a S,$$

$$\tfrac{3}{2}\big(R_{ab} - \tfrac{1}{6}\, g_{ab} R\big) + \tfrac{i}{2}\epsilon_{abcd}F^{cd}\,,\ \partial_a \gamma\cdot\chi - \not\partial\chi_a\,,\ -\partial_b F_{ba}\Big]$$

with $F_{ab} = \partial_a A_b - \partial_b A_a$. This multiplet has an external Lorentz vector index a, which leads to extra terms in the variation rules. For example[20]

$$\delta\chi_a = (\text{as without index a}) - \frac{2}{3}\sigma_{ab}(i\gamma_5 S + P) A_b\epsilon$$

At this moment the complete nonlinear multiplet and transformation rules are unknown.

All other vector multiplets except W_a can also be written as scalar multiplets, as we will discuss in the next section. The multiplet which starts with A_a is called the Einstein multiplet. The extra terms in vector multiplets interfere with the determination of the respective components. For example, in δX_a the components H_a, K_a and B_{ac} reside in (pseudo)scalar and vector structures, while the extra terms have a tensor structure and a pseudoscalar structure. For scalar multiplets there is no such interference.

Relations between scalar and vector multiplets

Every scalar multiplet $\Sigma = (A,B,X,F',G')$ can be written, as we have seen, as a vector multiplet $V(\Sigma) = (B,-X,-G',F',D_a^P A,o,o)$. The converse is also true: from any vector multiplet $V=(X,Z,H,K,B_a,\Lambda,D)$ one can construct a scalar multiplet according to[6]

$$S(V) = \Big[H - \tfrac{2}{3}CS,\ -K - \tfrac{2}{3}CP,\ i\gamma_5\big(\not{D}P + \tfrac{i}{6}\not{A}\gamma_5 - \tfrac{1}{3}S - \tfrac{i}{3}\gamma_5 P\big)Z + i\gamma_5\Lambda - \tfrac{1}{3}C\gamma\cdot R^{cov},\ -D + \cdots,\ -D_a B^a + \cdots \Big]$$

Note that from $V(\Sigma)$ one can retrieve Σ, but that from $S(V)$ one cannot retrieve V; (for example, only the gauge part of B^a appears in $S(V)$). As an application one may verify that

$$S(E_v) = -\tfrac{2}{3}\,\mathcal{U}$$

If one first rewrites a scalar as a vector multiplet, and then applies the above formula to obtain again a scalar multiplet, one finds the derivative multiplet $T(\Sigma)$[21]

$$S\big(V(\mathrm{par}\,\Sigma)\big) = T(\Sigma) + \tfrac{1}{3}S(\mathcal{U}\otimes\Sigma)$$

Multiplication of two scalar multiplets S_1 and S_2 can yield various different multiplets, because one can apply the operations $\mathrm{Par}(S_1)$ and/or $V(S_1)$ and then use the multiplication rule for scalar or vector multiplets. For example

$$V\big[V(S_1)\otimes V(\mathrm{par}\,S_2)\big) + (1\leftrightarrow 2) = V\big(\mathrm{par}\,\Sigma(S_1\otimes \mathrm{par}\,S_2)\big)$$

from which it follows that $S_1 \otimes \mathrm{par}(S_2)$ is symmetric in 1 and 2 (as one easily verifies directly). In this way one can reduce products of multiplets into multiplets, just as in ordinary group theory.

The action for a vector multiplet V can be obtained directly by applying the action formula for vector multiplets, or via the scalar projectin S(V). In both cases onefinds the same result.[6]

$$I_S\,(S(V)) = -I_V\,(V)$$

For the converse, using the vector projection of a scalar multiplet, one finds*

$$I_V[V(\mathrm{par}\,\Sigma)] = -\tfrac{2}{3}\, I_S[S(\mathcal{U}\otimes\Sigma)]$$

Since Par Par $\Sigma = -\Sigma$, one also finds a result for $I_V(V(\Sigma))$. For the special case $\Sigma = E_S$, one finds back a familiar result. We recall

$$V(\mathrm{par}\,\Sigma) = [-A, i\gamma_5 X, F', G', D_a^{\;p} B, 0, 0]$$

In Einstein gravity, there is no pseudoscalar action linear in curvatures. The same is true in supergravity. One might have thought that the relation $I_S(S(V)) = -I_V(V)$ has a parity reversed companion, but

$$I_S[\mathrm{par}\; S(V)] = 0$$

Indeed, Par(S(V)) starts with a term $-\partial\cdot B$ and the density is a total derivative.[22]

<u>Renormalizability</u>

The two main properties a quantum field theory should have, are unitarity (conservation of probability) and renormalizability. In terms of the fields e_μ^a, ψ_μ, A_a S and P supergravity satisfies the usual Becchi-Rouet-Stora invariance, from which one easily proves unitarity for any n-particle cut separately. (The only non-trivial thing is unity of the determinant, but this plays no role in quantum field theory when using dimensional regularization.) Without the

---

*Hint: partially integrate the term in I_V with the ε-symbol.

auxiliary field A_a, S and P the BRS-rules are modified, but also then unitarity for any n-particle cut can be proven directly.[25]

The ultraviolet divergences in Einstein supergravity cannot be renormalized away by rescaling of the physical parameters, just as in the case in general relativity. All one can hope for is that on-shell the divergences cancell "miraculously". The renormalizability question of supergravity was first studied by considering a model (O(2) supergravity) and showing (by explicit calculations and by theoretical proofs) that various processes were one-loop finite.[27] At the two-loop level it was shown that also these various processes have no local divergences,[28] where as it has been argued (but perhaps an additional rigorous proof could be given) that there are no nonlocal divergences.[29] In fact, extensions of these results to any number of in - and outgoing particles have been published,[30] using Noether-type analysis (some details have not been published up to now, which still might be useful, even though the tensor calculus now yields simpler proofs). At the three loop level, however, it was shown that the purely bosonic Bel-Robinson invariant (which is nonzero on-shell) could be extended to a supersymmetric invariant as far as the order ψ^2 terms. An argument that this can actually be done to all order (ψ^{12}) was provided by superspace methods,[14] and as we shall see, the tensor calculus proves this by providing the explicit complete nonlinear invariant[5] (even off-shell as well as on-shell). In fact, the tensor calculus proves that there are arbitrarily many dangerous counter terms, the higher in loop-number the more.[5]

Thus the status of renormalizability is the following. At any loop-level there are supersymmetric counter terms which do not vanish on-shell. This holds probably for all O(n) models[31] (although this has only been proven for the O(1) model, since only here a tensor calculus exists so far). Thus the important question is whether the coefficients of all these counter terms vanish. In principle this could be decided, at the three-loop level to begin with, by an explicit calculation. This is however hard, much harder than the explicit calculations performed to date. Another possibility is that the extra global symmetries which keep being found in the O(n) models, rule out these possible dangerous counter terms.

As an application of the tensor calculus we will now go through the proofs that O(1) supergravity is one- and two-loop finite, but could contain counter terms for 3 and higher loops.

The basic ingredients to construct counter-terms are the Weyl tensor, the Ricci tensor and the scalar curvature. They reside, as we have seen in the multiplets

$$R \text{ in } W, \quad R_{\mu\nu} \text{ in } W_a, \quad R_{\mu\nu}{}^{ab} \text{ in } W_{ab\alpha}$$

On-shell $W=W_\alpha=0$; thus we consider invariants constructed from $W_{ab\alpha}$ only.

A general n-loop invariant has the generic form $\kappa^{2(n-1)} R^{n+1}_{\mu\nu\alpha\beta}$ + terms involving ψ_μ. (of course one can also trade one Riemann tensor $R_{\mu\nu\alpha\beta}$ for two derivatives, since they have the same dimension.)

One-loop. The bosonic part of a one-loop counter term can only be proportional to $R_{\mu\nu\alpha\beta}^{\;2}$, but in the action this vanishes, due to the Gausz-Bonnet theorem

$$R_{\mu\nu\alpha\beta}^{\;2} - 4 R_{\mu\nu}^{\;2} + R^2 = \text{total divergence}$$

In fact, it has been shown be direct computation that the U(1) action (which starts with the square of the Weyl tensor) vanishes on-shell.[13] This action is thus $I_S(W_{ab\alpha} W_{ab\beta} C^{\alpha\beta})$ according to the tensor calculus.

Two-loops. The bosonic part of any two-loop invariant has been shown to be always proportional to[32]

$$\int d^4x \sqrt{-g} \left[R_{\mu\nu\alpha\beta} R^{\alpha\beta\rho\sigma} R_{\rho\sigma}{}^{\mu\nu} \right]$$

Thus, since on-shell the Riemann tensor coicides with the Weyl tensor, one must make an invariant from three multiplets $W_{ab\alpha}$. This is impossible, since for any odd numbers of $W_{ab\alpha}$, the spinorial indices can never be completely contracted away.

Three-loops. Also here a nonvanishing invariant exists.[3c]

$$\int d^4x \sqrt{-g} \left[R_{\mu\nu\alpha\beta} R^{\alpha\beta\rho\sigma} R_{\rho\sigma\kappa\lambda} R^{\kappa\lambda\mu\nu} \right]$$

Again, on-shell all other invariants are proportional to this one. Now, however, it is easy to find a supersymmetric extension. Without any explicit algebra one can write it down at once, to all orders and even off-shell

$$I_S \left[(W_{ab\alpha} W_{ab\beta} C^{\alpha\beta}) \otimes (T \{ W_{cd\gamma} W_{cd\zeta} C^{\gamma\zeta} \}) \right]$$

N-loops. It is now clear that the tensor calculus provides arbitrarily many supersymmetric invariants which do not vanish on-shell; for example (omitting indices)

$$I_S \left[(WW) \otimes TWW \ldots\ldots \otimes TWW \right]$$

In $W \otimes W$ the product of two Weyl-tensors resides in the F'-component, the T-operation maps this product into the first component, and the first component of a product containing the product of first components, it follows that one ends up with a scalar multiplet in whose F'-components one has arbitrarily many Weyl-tensors. The action formula then yields a supersymmetric action containing the F'-term.

General remarks

A few unrelated remarks follow. For more details the reader might consult the literature.

The gauge action of the simplest conformal supergravity theory,[13] the U(1) model, contains the fields e_μ^a and ψ_μ with higher derivatives, and a propagating axial vector field A_a. One might view this action as arising from the corresponding Einstein O(1) action, after putting an extra Dalembertian in the kinetic terms and dropping the S and P fields. The origin of S and P can be seen by coupling a scalar multiplet Σ in a conformally invariant way to the U(1) action[24] (the scalar multiplet for this coupling is $\Sigma T(\Sigma)$).[5] The total system has in addition to the symmetries of Einstein supergravity, the following extra local invariances: dilatational invariance, chiral invariance (=the U(1) symmetry), conformal supersymmetry S, while under conformal boosts all fields are inert. These extra symmetries can be removed by choosing a gauge in which $\Sigma=(1,0,0,F',G')$. The resulting action is the Einstein O(1) supergravity action, and F' and G' are just the auxiliary fields S and P. This shows clearly the different role S and P play as compared to A_a.

Conformal field theories for spin 2 and 3/2 are obtained by sandwiching the spin 2 and spin 3/2 projection operators between the corresponding gauge fields. Thus such theories have on- and off-shell only spin 2 and 3/2 fields present. Since these projection operators have $\Box^{-2}$ singularities, one must multiply the kinetic terms with $\Box^2$ to obtain a local theory, so that conformal spin 2 and 3/2 theories have ghosts. One can take square roots of the spin 2 and 3/2 projection operators which have only $\Box^{-1}$ singularities. This leads uniquely to the Fierz-Pauli and Rarita-Schwinger actions (with mass and, in the limit, without mass).[33] On-shell these theories still have only spin 2 and 3/2 fields, but off-shell there are also lower spins (gauge components). The Bose-Fermi symmetry holds off-shell for the sum of all (physical and auxiliary) field, where one should not forget that gauge invariances remove field components from an action off-shell.[34] For example, the O(1) action we have discussed, contains 12 boson fields $6(e_\nu^a)+4(A_a)+1(S)+1(P)$ and also 12 fermion fields (12 components of ψ_μ).

There is a deep relation between the generators of translations P_a, which are part of the group which is gauged, and general coordinate transformations. By construction the theory has the latter invariance (because one starts in general with curvatures if one gauges a group), and thus the P-invariance cannot be valid at the same time (for example, the commutator of two local supersymmetries yields a general coordinate transformation and not a local translation). The manner in which each particular theory achieves this, is by requiring constraints. For Einstein (super) gravity the constraint turns out to be the torsion equation $\delta I/\delta\omega_{\mu ab} = 0$, but in Weyl supergravity there are four constraints: self duality and irreducibility of the Q-curvature, vanishing of the P-curvature (which is in Einstein (super)gravity equal to the torsion equation according to Palatini formalism, but not in Weyl supergravity), and finally vanishing of the difference of the Einstein and chiral curvatures.[17]

One can also express this result as follows. From the identity

$$\xi^{\lambda}{}_{,\mu} h^{A}{}_{\lambda} + \xi^{\lambda} h^{A}{}_{\mu,\lambda} = (\xi\cdot h^{A})_{,\mu} + f^{A}{}_{BC} h^{B}{}_{\mu}\, \xi\cdot h^{C} + \xi^{\lambda} R^{A}{}_{\mu\lambda}(h)$$

it follows that a general coordinate transformation with a parameter ξ^{λ} on a gauge field h^{A}_{μ} is equal to a sum of all gauge transformations of the group to be gauged (each with parameter $\varepsilon^{A}=\xi\cdot h^{A}$), plus a term containing the curvature of the gauge field

$$\delta_{\text{gen. coord.}}(\xi)\, h^{A}_{\mu} = D_{\mu}(\xi\cdot h^{A}) + \xi^{\lambda} R^{A}_{\mu\lambda}$$

This yields the precise relation between general coordinate transformation, P-gauge transformation, and (covariant) translations in the base-manifold.

Another revealing relation is between Einstein supersymmetry transformations δ^{P}_{Q}, and the two supersymmetry transformations δ^{C}_{Q} and δ^{C}_{S} of conformal supersymmetry (the square roots of the translation generator P_a and the conformal boost generator K_a)[7]

$$\delta^{P}_{Q}(\epsilon) = \delta^{C}_{Q}(2\epsilon) + \delta^{C}_{S}(\eta\epsilon + \not{b}\epsilon)$$

and b_{μ} is the dilation where we recall $\eta = -\frac{1}{3}(S - i\gamma_5 P - i\not{A}\gamma_5)$. It follows that any conformal supergravity theory is also an Einstein supergravity theory (just as in ordinary relativity). The great thing is that in conformal U(1) (and perhaps also in U(n)) theories all fields are gauge fields on which Q- and S-transformations are defined. Thus, one can find immediately δ^{P}_{Q} for e^{a}_{μ}, ψ_{μ} and A_a. Indeed, with $\delta^{C}_{Q}(\epsilon)\psi_{\mu} = (D_{\mu} - \frac{i}{2}A_{\mu}\gamma_5)\epsilon$ and $\delta^{C}_{S}\psi_{\mu} = -\gamma_{\mu}\epsilon$ one has

$$\delta^{P}_{Q}(\epsilon)\psi_{\mu} = 2\kappa^{-1}(D_{\mu} - \frac{i}{2}A_{\mu}\gamma_5\epsilon) - \gamma_{\mu}\eta\epsilon$$

and the dilation terms in $\omega_{\mu}{}^{ab}$ cancel those in $\delta_{S}{}^{C}$.

Outlook

Although supergravity is a fascinating theory itself, ultimately it must correspond to reality if it is to remain part of physics. To this end one might use the tensor calculus to tackle the problem of gauge invariances in systems with a super-Higgs[35] effect. The tensor calculus for O(n) and U(n) theories for $n>1$ is still unknown but clearly this is one of the most important problems.

For the O(1) theories, the tensor calculus reproduces all models: the gauge actions of both O(1) and U(1) supergravity (!), and the matter couplings to O(1) and U (1) supergravity. This brings supergravity to the same level as general relativity: the structure is due to multiplets (in ordinary relativity: four-vectors) which can be multiplied and from which an action can be constructed (in general relativity: by multiplying a scalar by $(-g)^{\frac{1}{2}}$).

With this tensor calculus one can now tackle many problems in supergravity, which have been solved in general relativity. For example, topological invariants, classification schemes, etc. One might also try to find exact solutions (the only known solutions are either supersymmetry rotations of solutions of general relativity or plane waves). Undoubtedly, the geometers will start to rephrase the obtained results in their language. All this work is interesting in itself, but what we really need is an indication that our world at the (sub?) quark level has a local Fermi-Bose symmetry.

Acknowledgement

The author is grateful for many useful discussions with: F. Berends, P. Breitenlohner, E. Cremmer, B. deWit, S. Ferrara, M. Fischler, D. Freedman, M. Grisaru, J. van Holten, B. Julia, M. Kaku M. Rocek, J. Scherk, K. Stelle, P. Townsend, P. West and B. Zumino. He also acknowledges the hospitality of CERN, Ecole Normale Superieure at Paris and het Instituut Lorentz te Leiden.

References

[1] D.Z.Freedman, P.vanNieuwenhuizen and S. Ferrara, Phys. Rev. D13, 3214 (1976).
[2] S. Deser and B. Zumino, Phys. Lett. 62B, 335 (1976).
[3] D.Z. Freedman and P. van Nieuwenhuizen, Phys. Rev. D14, 912 (1976).
[4] S. Ferrara and P. van Nieuwenhuizen, Phys. Lett. 76B, 404 (1978).
[5] idem, Phys. Lett. 78B 573 (1978).
[6] K. Stelle and P.C.West, Phys. Lett. 77B, 376 (1978).
[7] S. Ferrara and P. van Nieuwenhuizen, Phys. Lett. 74B, 333 (1978).
[8] K. Stelle and P. C. West, Phys. Lett. 74B, 331 (1978).
[9] P. van Nieuwenhuizen, Phys.Report, to be published.
[10] D.Z.Freedman and P. van Nieuwenhuizen, February issue 1978.
[11] A list can be found in S. Ferrara, CERN preprint Th-2514.
[12] O(2) model: S.Ferrara and P.van Nieuwenhuizen, Phys. Rev. Letters 37, 1669 (1976).
O(3) model: D.Z.Freedman, Phys. Rev. Lett. 38, 105 (1977) and S. Ferrara, J. Scherk and B. Zumino, Phys. Lett. 66B, 35 (1977).
O(4) model: E. Cremmer and J. Scherk, Nucl. Phys. B127, 259 (1978) E. Cremmer, S. Ferrara and J. Scherk, Phys. Lett. B74, 61 (1978).
O(8) model: E. Cremmer and B. Julia, Ecole Normale preprint, September 1978.
[13] M. Kaku, P. K. Townsend and P. van Nieuwenhuizen, Phys. Lett. 69B, 304 (1977), Phys. Rev. Lett. 39, 1109 (1977), Phys. Rev. D 17, 3179 (1978).
[14] S. Ferrara and B. Zumino, Nucl. Phys. B134, 301 (1978)
For superspace see the lectures by Dr. B. Zumino or A. Salam and J. Strathdee, Phys. Rev. D11, 1521 (1975).
[15] P. Breitenlohner, Phys. Lett. 67B, 49 (1977) and Nucl. Phys. B124, 500 (1977).
B. de Wit and M. T. Grisaru, Phys. Lett. 74B, 57 (1978).
S. Ferrara, J. Scherk and P. van Nieuwenhuizen, Nucl. Phys. B117, 333 (1977).
[17] P. K. Townsend and P. van Nieuwenhuizen, to be published.
[18] P. K. Townsend, Phys. Rev. D15, 2802 (1977).
[19] P. Breitenlohner, S. Ferrara, D.Z.Freedman, F. Gliozzi, J. Scherk, P. van Nieuwenhuizen, Phys. Rev. D15, 1013 (1977).
[20] If C_a is not given by A_a, more extra terms are needed (see ref. (17).)
[21] B. Zumino, private communication.
[22] M. Roček, private communication.
[23] S. Ferrara, J. Scherk and P. van Nieuwenhuizen, Phys. Rev. Lett. 37 1035 (1976)
[24] M. Kaku and P.K.Townsend, Phys. Lett. 76B, 54 (1978).
[25] G. Sterman, P.K.Townsend and P. van Nieuwenhuizen, Phys. Rev. D17, 1501 (1978).
[26] BRS-invariance in supergravity has been discussed in ref (25) and P.K.Townsend and P. van Nieuwenhuizen, Nucl. Phys. B120, 301 (1977)
R. Kallosh, Zh. ETF Pis'ma 26, 575 (1978)
K. Stelle and P.C.West, Nucl. B140, 285 (1978).
B. de Wit and J. van Holten, Physics Letters, to be published.
[27] M.T.Grisaru, P.van Nieuwenhuizen and J.Z.M.Vermaseren, Phys. Rev. Lett. 37, 1662 (1976).

[28] M.T.Grisaru, Phys. Lett. 66B, 75 (1977).
[29] E. Tomboulis, Phys. Lett. 67B, 417 (1977).
[30] S. Deser, J. Kay and K. Stelle, Phys. Rev. Lett. 38, 527 (1977).
[31] One- and two-loop of O(n) theories have been verified by explicit calculations and by proofs using either helicity arguments[27] or Noether-type constructions.[30] At the three-loop level, the existence of a Bel-Robinson type invariant in O(2) and O(3) theory has been shown up to order ψ^2 (S. Deser, J. Kay, Phys. Lett. 76B (1978) 400
[32] C.C.Wu and P.van Nieuwenhuizen, J.Math. Phys. 18, 182 (1977)
[33] V.I. Ogievetski and E.Sokatchev, J. Phys. A10, 2021 (1977).
[34] M. Sohnius, Karlsruhe thesis (1976) on global supersymmetry.
[35] E. Cremmer, S. Ferrara, L. Girardello, B. Julia, J. Scherk, P. van Nieuwenhuizen, Phys. Lett. B and CERN preprint.

IRREDUCIBLE REPRESENTATIONS OF SUPERSYMMETRY

Daniel Z. Freedman

Institute for Theoretical Physics
State University of New York
Stony Brook, New York 11794

ABSTRACT

The irreducible representations of Poincaré supersymmetry are derived for both massless and massive particles. Some recently discovered modifications of the representations when central charges appear in the supersymmetry algebra are also discussed.

INTRODUCTION

For almost any application of supersymmetry and supergravity it is useful to know what kind of particles are described by field theories with this symmetry. These particles transform in representations of the supersymmetry algebra. In the first part of this lecture we will derive and discuss the irreducible representations of ordinary Poincaré supersymmetry. This material is quite well known and appeared first in work of Salam and Strathdee[1] (massive representations) and of Gell-Mann and Ne'eman[2] (massless representations). Later we will discuss some recently discovered interesting modifications of the representations when there are central charges in the algebra.

The basic Poincaré supersymmetry algebra consists of N-Majorana spinor charges Q_α^i (where $\alpha = 1,2,3,4$ and $i = 1,\ldots,N$) and the Poincare generators P^μ and $M^{\mu\nu}$. The graded commutation relations are

$$\{Q_\alpha^i\ ,\ \bar{Q}_\beta^j\} = \delta^{ij}(\gamma^\mu)_{\alpha\beta}\, P_\mu \tag{1}$$

$$[M^{\mu\nu}, Q_\alpha{}^i] = -i(\sigma^{\mu\nu})_{\alpha\beta} Q_\beta{}^i \quad (2)$$

$$[P^\mu, Q_\alpha{}^i] = 0 \quad (3)$$

with the conventions: $g^{\mu\nu} = (+---)$,

$$\{\gamma^\mu, \gamma^\nu\} = 2g^{\mu\nu}, \sigma^{\mu\nu} = \frac{1}{4}[\gamma^\mu, \gamma^\nu], \gamma_5 = i\gamma^0\gamma^1\gamma^2\gamma^3$$

$$C = -C^T, C\gamma^\mu C^{-1} = -\gamma^{\mu T}.$$

The symmetric form of the spinor anti-commutator suggests that the spinor charges transform under an internal symmetry, so we enlarge the algebra by considering Hermitean internal symmetry generators T^a (Lorentz scalars) and T_5^a (pseudoscalars) under which

$$[T^a, Q_\alpha^i] = -(t^a)^{ij} Q_\alpha{}^j \quad (4)$$

$$[T_5{}^b, Q_\alpha^i] = -\gamma_5(s^b)^{ij} Q_\alpha^j \quad (5)$$

where $(t^a)^{ij}$ and $(s^a)^{ij}$ are pure imaginary and pure real, respectively, as required by the Majorana condition, and both are Hermitean. Hermiticity insures that the bilinear $\bar{Q}^i\gamma^\mu \gamma_5 Q^i$ is invariant under (4-5) as is the free kinetic Lagrangian of a set of gravitino fields ψ_ρ^i which are the gauge fields of the supersymmetry charges.

The $(t^a)^{ij}$ constitute an N dimensional real representation of the subgroup of the scalar charges T^a. Since there are at most $\frac{1}{2}N(N-1)$ independent matrices $(t^a)^{ij}$, this subgroup can be no bigger than the group SO(N). It might, however, be any smaller group which has an N dimensional real representation. Similarly, there are at most $\frac{1}{2}N(N+1)$ matrices $(s^b)^{ij}$, and in the maximal situation, the full symmetry group with generators T^a and $T^b{}_5$ is U(N).

In most supersymmetric field theories the maximal SO(N) symmetry is manifest in the Lagrangian. SO(N) is the natural symmetry because it is the internal symmetry subalgebra of the simple (deSitter) Osp(N,4) graded Lie algebra from which Poincaré supersymmetry is obtained as a Wigner-Inonu group contraction. In supergravity theories the γ_5 transformations which enlarge the symmetry from SO(N) to U(N) are realized as combined chiral rotations of fermion fields and dual transformations of vector field strengths.[3] Non-maximal internal symmetries can be realized in field theories when central charges appear, as we will discuss below. Without central charges, massless representations with non-maximal internal symmetry can be constructed by using non-trivial Clifford vacua (see below), but it is not clear whether this situation occurs in field theories. In the N=4 supersymmetric Yang-Mills theory,[4] there is a global SU(4) internal symmetry which cannot be extended to U(4).

We will now derive the irreducible representations of Poincaré supersymmetry using Wigner's method of induced representations or, in other words, a simple generalization of the "little group" method used to discuss representations of the Poincaré group. The representations for massless particles are simpler than those for particles with mass, so we discuss them first. In both cases we will use for simplicity only a scalar internal symmetry group. Thus we search for irreducible representations of the graded Lie algebra specified by (1-4) plus Poincaré and internal symmetry commutators.

Massless Irreducible Representations

For the massless case it is best to introduce a Weyl representation (γ_5 diagonal) of the Dirac algebra

$$\gamma^0 = \begin{pmatrix} 0 & 1 \\ 1 & 0 \end{pmatrix} \quad \gamma^i = \begin{pmatrix} 0 & \sigma_i \\ \sigma_i & 0 \end{pmatrix} \quad \gamma_5 = \begin{pmatrix} -1 & 0 \\ 0 & 1 \end{pmatrix} \quad C = i\gamma^2\gamma^0 \qquad (6)$$

In this representation a Majorana spinor takes the form

$$Q_\alpha = \begin{pmatrix} Q_1 \\ Q_2 \\ Q_2^* \\ -Q_1^* \end{pmatrix} \qquad (7)$$

and upper two components transform irreducibly under Lorentz transformations as a two-component left-handed Weyl spinor. We therefore refer now only to the two-component Q_α^i and their adjoints Q_α^{i*}.

The anti-commutators (1) can then be rewritten as

$$\{Q_\alpha^{\ i}, Q_\beta^{\ j*}\} = \delta^{ij}(\sigma^\mu)_{\alpha\beta} P_\mu$$

$$\{Q_\alpha^i, Q_\beta^j\} = 0$$

$$\{Q_\alpha^{i*}, Q_\beta^{j*}\} = 0 \qquad (8)$$

which involve the 2 x 2 Weyl matrices ($\sigma^\mu = \mathbb{1}, \underset{\sim}{\sigma}$). From (2) one finds the angular momentum commutators

$$[\, \underset{\sim}{J},\ Q_\alpha{}^i\,] = -\tfrac{1}{2}\,\underset{\sim}{\sigma}_{\alpha\beta}\, Q_\beta{}^i \tag{9}$$

and from (4) the internal symmetry relations

$$[T^a,\ Q_\alpha{}^i\,] = -\,(t^a)^{ij}\, Q_\alpha{}^j . \tag{10}$$

We then pick a basis of one-particle helicity states $|\bar{P},\lambda>$ with momentum $\bar{P}^\mu = (\omega,\ 0,0,\omega)$ in the z-direction. On this basis we search for a representation of the "little algebra", i.e., the subalgebra of Q, Q^+, $M^{\mu\nu}$, T^a which leaves this basis invariant. The little algebra consists of the operators Q, Q^*, and T^a, which commute quite generally with P^μ, and the operators J_3 $M^{01} - J_2$, $M^{02} + J_1$, which constitute the well-known E_2 subgroup of the Poincaré group and leave the special momentum $\bar{P}^\mu$ invariant.

On the basis $|\bar{P},\lambda>$ the anti-commutators (8) reduce to

$$\{Q_2{}^i,\ Q_2{}^{j*}\} = 2\omega\delta^{ij} \tag{11}$$

and all other anti-commutators vanish. In particular $\{ Q_1,\ Q_1{}^*\} = 0$ because of a cancellation between the 0 and 3 components of $\sigma^\mu \bar{P}_\mu$, and positivity requires that the operators $Q_1{}^i$ and $Q_1{}^{i*}$ be represented trivially, i.e. by the zero matrix. The algebra therefore reduces to the anti-commutation rules of N independent fermion annihilators and creators $Q_2{}^i$ and $Q_2{}^{i*}$, i.e. a Clifford algebra of N complexors elements. Further (9) tells us that $Q_2{}^i$ raises and $Q_2{}^{i*}$ lowers helicity by ½ unit. It is now duck soup to find the irreducible representations!

We choose the maximum helicity state $|\bar{P},\ \bar{\lambda}>$ of the basis of the representation to have the properties

$$Q_2{}^i\,|\bar{P},\bar{\lambda}> = Q_1{}^i\,|\bar{P},\bar{\lambda}> = Q_1{}^{i*}\,|\bar{P},\bar{\lambda}> = 0 \tag{12}$$

and (temporarily) to be a singlet of the internal symmetry. By applying the creator $Q_2{}^{i*}$ one finds the sequence of states:

$$|\bar{P},\bar{\lambda}>$$
$$|\bar{P},\bar{\lambda}-\tfrac{1}{2}, i> \;=\; Q_2^{i*}\;|\bar{P},\bar{\lambda}>$$
$$|\bar{P},\bar{\lambda}-1,[ij]> = Q_2^{j*}Q_2^{i*}|\bar{P},\bar{\lambda}>$$
$$\text{etc.} \tag{13}$$

States of helicity $\bar{\lambda} - \frac{1}{2}m$ have multiplicity $N!/m!(N-m)!$, and the sequence stops when the singlet state of helicity $\bar{\lambda} - \frac{1}{2}N$ is reached. Anti-symmetry in [ijk...] is automatic. The total dimension of the representation is 2^N as is required by the representation theory of Clifford algebras.

As far as internal symmetry is concerned, the states of helicity $\bar{\lambda} - \frac{1}{2}m$ transform as m-fold antisymmetric products of the representation $(t^a)^{ij}$, i.e., as m^{th} rank totally anti-symmetric tensors in the case SO(N). One also has the possibility of choosing multiple Clifford vacua $|\bar{P}, \bar{\lambda}, t>$ which transform in an irreducible representation of the internal symmetry. In this case the set of states obtained by applying products of Q_2^{i*} in the manner of (13) are a reducible representation of the Clifford elements, but an irreducible representation of the full little algebra. The internal symmetry content of the states is obtained by Clebsch-Gordon decomposition of the group representation of the products of the Q_2^{i*} with the representation of the vacua.

Thus an irreducible representation is a "tower" of helicity states, which contains particles of maximum helicity $\bar{\lambda}$ and minimum helicity $\bar{\lambda} - \frac{1}{2}N$. A local field theory always contains particles in CPT conjugate pairs of helicities $\pm\lambda$. Therefore a supersymmetric field theory generally describes a reducible representation of the algebra in which the CPT conjugate states of reversed helicity are added to the states of the sequence (13). These states are obtained by starting with the CPT conjugate Clifford vacuum $|\bar{P}, -\bar{\lambda}>$ and applying products of the helicity raising operator Q_2^{i}, which is the CPT conjugate of Q_2^{i*}. The CPT conjugate states are already present in the initial sequence (13), if $N=4\bar{\lambda}$, so such representations are CPT self-conjugate. Therefore field theories for $N = 4\bar{\lambda}$ and $N = 4\bar{\lambda} - 1$ always have the same particle content.

There are now many examples of supersymmetric field theories which correspond to the massless representations we have discussed. Still not all representations can be realized even in free field theory where one simply requires a free kinetic Lagrangian invariant under transformation rules which realize the anti-commutator (1). For example, there are no known field theories for representations containing particles of spin $\geqslant 5/2$ due to higher spin difficulties. There is also a curious low spin counterexample which is the self-conjugate N=2 multiplet consisting of helicity states $|\frac{1}{2}\rangle$, $|0\rangle$, $|0\rangle$ and $|-\frac{1}{2}\rangle$. The field content is equivalent to that of the N=1 chiral multiplet, so that one supersymmetry transformation Q^1_α exists. However, given the standard Wess-Zumino model with fields $\chi(x)$, $A(x)$ and $B(x)$, it is not possible to find a second supersymmetry[5] which leaves the Lagrangian invariant and anti commutes with Q^1_α. The reason seems to be related to superconformal invariance. A free theory of spin 1/2 and spin 0 fields is conformal invariant, so that if it satisfied N=2 Poincaré supersymmetry it would also have N=2 conformal supersymmetry and would have internal symmetry group U(2). However, one cannot realize U(2) symmetry on a set of two real spinless fields A(x) and B(x). One can realize N=2 supersymmetry in a field theory containing a doubled chiral multiplet $\chi^i(x)$, $A^i(x)$, $B^i(x)$ with $i = 1,2$ as has been done by Salam and Strathdee[6] and by Fayet.[7]

Finally we note that the minimum spin massless representation of the Poincaré supersymmetry algebra with N spinor charges contains spins from $s = 0$ up to $s_{max} = \frac{1}{4}(N+1)$ for N odd and $s_{max} = \frac{1}{4}N$ for N even. Following a recent review,[8] we include a table which lists the multiplicities of the various spins for all massless irreducible representations with $s_{max} \leqslant 2$. Because of the correlation between s_{max} and N, the case of N = 8 supersymmetry now appears to be a natural barrier in field theory.

Massive Irreducible Representations

The natural Lorentz frame of a massive particle is the rest frame with $\bar{P}^\mu = (M,0,0,0)$, and the little algebra of (1-4) which leaves rest states invariant consists of the Clifford elements $Q_\alpha{}^i$ and $Q_\alpha{}^{i*}$ for $\alpha = 1,2$ together with the three angular momentum operators J_i and the internal symmetry generators T^a. Note that we use the Weyl representation (6) here also so that (7-10) are still valid. However, in the rest frame, one finds from (1) or (8) that

$$\{ Q_1{}^i , Q_1{}^{j*} \} = M\delta^{ij}$$

$$\{ Q_2{}^i , Q_2{}^{j*} \} = M\delta^{ij} . \tag{14}$$

while all other anti-commutators vanish. Thus we now have 2N non-trivial Clifford elements, so that irreducible representations are larger than in the massless case.

The operators

$$\underset{\sim}{\mathcal{J}} = \underset{\sim}{J} - \frac{1}{2M} Q^{i*} \underset{\sim}{\sigma} Q^{i} \tag{15}$$

have the commutation relations of an angular momentum and commute with the Clifford elements, as is easy to verify. Therefore $\mathcal{J}^2$ is a Casimir operator of the little group. Fully analogous properties hold for the generalized internal symmetry operator

$$\mathcal{T}^a = T^a - \frac{1}{M} Q^* t^a Q \quad . \tag{16}$$

We can therefore view the little algebra as a direct product of the bosonic operators $\mathcal{J}_i$ and $\mathcal{T}^a$ with the set of Clifford elements Q_α^i, Q_β^{j*}. For simplicity of notation we now assume that the internal symmetry is SO(3) although the results below clearly generalize to the case of any other internal symmetry.

We choose a set of (2J+1) × (2T+1) Clifford vacua $|\mathcal{J}, \mathcal{J}_3, \mathcal{T}, \mathcal{T}^3\rangle$ which are eigenvectors of the $\mathcal{J}_a \mathcal{T}^a$ system and satisfy

$$Q_\alpha^a |\mathcal{J}, \mathcal{J}^3, \mathcal{T}, \mathcal{T}^3\rangle = 0. \tag{17}$$

Because of (17) these vacua are also eigenstates of ordinary angular momenta and isospin with the indicated quantum numbers. General states of the representation are given by applying products of the creators Q_1^{i*} and Q_2^{*j} (which, respectively raise and lower J_3 by $\frac{1}{2}$ unit):

$$Q_1^{i\,*} \cdots Q_1^{i_m*} \; Q_2^{j_1*} \cdots Q_2^{j_n*} |\mathcal{J}, \mathcal{J}^3, \mathcal{T}, \mathcal{T}^3\rangle \tag{18}$$

where $m < N$ and $n < N$. These states span an irreducible representation of dimension $2^{2N}(2\mathcal{J}+1)(2\mathcal{T}+1)$.

The spin content of the representation is determined from the Clebsch-Gordon addition of the angular momentum carried by the creators with that of the vacua. In general the maximum spin is $s_{max} = \mathcal{J} + \frac{1}{2}N$. Similar remarks apply to the isospin content. Clearly the lowest spin massive supersymmetric field theories correspond to $\mathcal{J} = 0$ for the vacuum. Even then one runs into higher spin difficulties for $N \geqslant 3$.

Central Charges

The concept of central charges in supersymmetry was first discussed by Haag, Lopuszanski and Sohnius.[9] They showed that there is a consistent modification of the Poincaré supersymmetry

Internal Symmetry	Spin	s=2	s =3/2	s=1	s=1/2	$s^P = 0^{\pm}$
N = 1	s_{max} =2	1	1			
	s_{max} = 3/2		1	1		
	s_{max} = 1			1	1	
	s_{max} = 1/2				1	1+1
SO(2)	s_{max} = 2	1	2	1		
	s_{max} = 3/2		1	2	1	
	s_{max} = 1			1	2	1+1
	s_{max} = 1/2				1	2
SO(3)	s_{max} = 2	1	3	3	1	
	s_{max} = 3/2		1	3	3	1+1
	s_{max} = 1			1	3+1	3+3
SO(4)	s_{max} = 2	1	4	6	4	1+1
	s_{max} = 3/2		1	4	6+1	4+4
	s_{max} = 1			1	4	3+3
SO(5)	s_{max} = 2	1	5	10	10+1	5+5
	s_{max} = 3/2		1	5+1	10+5	10+10
SO(6)	s_{max} = 2	1	6	15+1	20+6	15+15
	s_{max} = 3/2		1	6	15	20
SO(7)	s_{max} = 2	1	7+1	21+7	35+21	35+35
SO(8)	s_{max} = 2	1	8	28	56	35+35

algebra in which (1) is replaced by

$$\{ Q_\alpha^{\ i}, \bar{Q}_\beta^{\ j} \} = \delta^{ij} \gamma^\mu_{\ \alpha\beta} P_\mu + i\delta_{\alpha\beta} U^{ij} + \gamma_{5\alpha\beta} V^{ij} \quad . \tag{19}$$

The U^{ij} and V^{ij} are scalar and pseudoscalar central charges which are anti-symmetric as required by the Majorana condition and Hermitean in the conventions used here. By definition central charges commute with each other and with all other elements of the algebra including internal symmetry generators.

This last requirement should be viewed as a restriction on the internal symmetry permitted. To see this one simply calculates the commutator of scalar internal symmetry generators T^a with both sides of (19) using (4). One then finds

$$[T^a, U^{ij}] = -(t^{aik} U^{kj} - U^{ik} t^{akj}) \tag{20}$$

and a similar relation for V^{ij}. For given central charges U^{ij}, the vanishing of $[T^a, U^{ij}] = 0$, which is required for proper Jacobi identities, means that we can only consider internal symmetry generators for which the right side of (20) vanishes. In general this means that the maximal scalar internal symmetry group is smaller than SO(N). There is no restriction for N=2 because central charges must be proportional to the invariant alternating symbol ε^{ij}. Pseudoscalar internal symmetries are restricted for all N.

Since the central charges have dimension 1, they occur only in field theories where there is a dimensional parameter. The two known cases of interest are

(a) the paramter is a mass in the Lagrangian,
(b) the energy scale is introduced via spontaneous breakdown of internal symmetry.

An example of case (a) is the field theory of the N = 2 multiplet of maximum spin ½ which has been known for some time.[6,7] Examples of case (b) are the N = 2 and N = 4 supersymmetric Yang-Mills theories (based on massless representations of $s_{max} = 1$) in which a very rich structure involving central charges is found after spontaneous breakdown. This structure has been pointed out in recent work of Witten and Olive[10] and Fayet.[11]

We will discuss here some modifications of the irreducible representations due to central charges in the N = 2 supersymmetry algebra, which is the most thoroughly studied case. Other aspects of central charges, including an intriguing connection[10] with topological charges carried by solitons, are discussed in the recent literature[10-13] and we suggest that readers consult it.

In the case of N = 2 supersymmetry, (19) can be rewritten as

$$\{Q_\alpha{}^i, \bar{Q}_\beta{}^i\} = \delta^{ij}\gamma^\mu{}_{\alpha\beta}P_\mu + i\varepsilon^{ij}(\delta_{\alpha\beta}U - i\gamma_{5\alpha\beta}V) . \quad (21)$$

Witten and Olive have obtained a lower bound on particle masses in representations of this algebra in terms of the central charges. To obtain it we first note that since U and V are central elements, they must become numerical multiples of the identity matrix in any representation. Thus for a massive representation in the rest frame, the right side of (21) can be viewed as a purely numerical 8×8 matrix, with $P_\mu = (M, 0, 0, 0)$ and

$$U - i\gamma_5 V = r e^{-i\gamma_5\theta} .$$

We make a chiral transformation $Q^i \to e^{i\frac{1}{2}\gamma_5\theta} Q^i$, and note that the new charges satisfy

$$\{Q_\alpha^i, Q_\beta^{j*}\} = \delta^{ij}\delta_{\alpha\beta}M + i\varepsilon^{ij}\gamma^0{}_{\alpha\beta}r \quad (22)$$

The bound follows simply from the positivity of the left side. All eigenvalues must be non-negative. Since the eigenvalues of $i\varepsilon^{ij}\gamma^0$ are ±1 this implies

$$M \geqslant r = \sqrt{U^2 + V^2} . \quad (23)$$

Further insight can be gained by explicit diagonalization of the spinor anti-commutator. Going back to (21), we see that the ij structure is diagonalized by taking complex spinor charges

$$Q_\alpha = \frac{1}{\sqrt{2}}(Q_\alpha{}^1 + i Q_\alpha{}^2)$$

$$\bar{Q}_\alpha = \frac{1}{\sqrt{2}}(\bar{Q}_\alpha^1 - i \bar{Q}_\alpha^2) \quad (24)$$

which satisfy, after the chiral rotation above,

$$\{Q_\alpha, \bar{Q}_\beta\} = \not{P} + r$$

$$\{Q_\alpha, Q_\beta\} = 0 . \quad (25)$$

At this point the standard representation of the γ^μ with

$$\gamma^0 = \begin{pmatrix} 1 & 0 \\ 1 & -1 \end{pmatrix} \qquad \gamma^i = \begin{pmatrix} 0 & \sigma_i \\ -\sigma_i & 0 \end{pmatrix} \tag{26}$$

is more useful than the Weyl representation, so we adopt it. We then find in the rest frame

$$\{ Q_\alpha , Q^*_\beta \} = \begin{pmatrix} M+r & 0 & 0 & 0 \\ 0 & M+r & 0 & 0 \\ 0 & 0 & M-r & 0 \\ 0 & 0 & 0 & M-r \end{pmatrix} \tag{27}$$

We now see that if the bound (23) is not saturated there are four non-trivial Clifford elements in the algebra. If the bound is saturated then two eigenvalues vanish leaving two non-trivial Clifford elements. Therefore one expects that the irreducible representations are smaller (4 states in the Clifford factor rather than 16) if (22) is saturated. This is a curious shortening of massive irreducible representations in the presence of central charges, which has the practical implication that the explosion of spins (i.e. $s_{max} \geq \frac{1}{2}$ N) of the previous section is limited. In particular one can realize massive representation of the N=2 algebra with $s_{max} = \frac{1}{2}$ and of the N = 4 algebra with $s_{max} = 1$, when there are central charges.

The complete structure of the representations[14] can be found using the techniques of previous sections. In particular the relation

$$[\underset{\sim}{J} , Q_\alpha] = - \frac{1}{2} \underset{\sim}{\sigma}_{\alpha\beta} Q_\beta \tag{28}$$

holds separately for the upper and lower pair of components of Q, so that Q_1 and Q_3 are raising operators for J_3 while Q_2^* and Q_4^* are lowering operators. The operators

$$\underset{\sim}{\mathcal{J}} = \underset{\sim}{J} - \frac{1}{2(M+r)} (Q^* \underset{\sim}{\sigma} Q)_{12} - \frac{1}{2(M-r)} (Q^* \sigma\, Q)_{34} \tag{29}$$

$$\tilde{\mathcal{T}} = T - \frac{1}{M+r} (Q^* Q)_{12} - \frac{1}{-r} (Q^* Q)_{34}$$

are generalized angular momentum and internal symmetry operators which commute with Clifford elements. Thus $\mathcal{J}^2$ and $\tilde{\mathcal{T}}$

are Casimir operators of the little algebra. Clifford vacua $|\mathcal{J}\,\mathcal{J}_3, \mathcal{T}\rangle$ may be defined as eigenstates of $\mathcal{J}^2, \mathcal{J}_3, \mathcal{T}$ which satisfy

$$Q_\alpha |\mathcal{J}, \mathcal{J}_3\ \mathcal{T}\rangle = 0 \tag{30}$$

for $\alpha = 1, 2, 3, 4$. For $M \neq r$ a basis of an irreducible representation of dimension $16(2\mathcal{J}+1)$ is obtained by applying monomials in the creation operators Q_α^*. This gives a tower of particles with $s_{max} = \mathcal{J} + 1$, which is no different from the case without central charges.

When $M = r$ the spinor charges Q_3, Q_4, Q_3^* and Q_4^* are represented trivially by null matrices. The operators in (29) now contain only upper spinor components but they still give Casimir operators. One then has the $4(2\mathcal{J}+1)$ dimensional basis

$$\begin{aligned} &|\mathcal{J}, \mathcal{J}_3, \mathcal{T}\rangle \\ &Q_1^* |\mathcal{J}, \mathcal{J}_3\ \mathcal{T}\rangle \\ &Q_2^* |\mathcal{J}, \mathcal{J}_3\ \mathcal{T}\rangle \\ &Q_1^* Q_2^* |\mathcal{J}, \mathcal{J}_3, \mathcal{T}\rangle \end{aligned} \tag{31}$$

which consists of four particles of conventional spin and SO(2) quantum number

$(\mathcal{J}+\tfrac{1}{2}, \mathcal{T}+1)$, $(\mathcal{J}, \mathcal{T})$, $(\mathcal{J}, \mathcal{T}+2)$, $(\mathcal{J}-\tfrac{1}{2}, \mathcal{T}+1)$.

References and Footnotes

[1]A. Salam and J. Strathdee, Nucl.Phys. B80, 499 (1974).

[2]M. Gell-Mann and Y. Ne'eman, unpublished.

[3]S. Ferrara, J. Scherk and B. Zumino, Nucl. Phys. B121, 393 (1977).

[4]F. Gliozzi, J. Scherk and D. Olive, Nucl. Phys. B122, 253 (1977); L. Brink, J. H. Schwarz and J. Scherk, Nucl. Phys. B121, 77 (1977).

[5]D. Z. Freedman and C. Zachos, unpublished.

[6]A. Salam and J. Strathdee, Nucl. Phys. B97, 293 (1975).

[7]P. Fayet, Nucl. Phys. B113, 135 (1976).

[8]S. Ferrara, CERN preprint TH 2514 (1978).

[9]R. Haag, J. Lopuszanski and M. Sohnius, Nucl.Phys. B88, 257 (1975).

[10]E. Witten and D. Olive, Phys. Lett. 78B, 97 (1978).

[11]P. Fayet, Cal.Tech.Preprint CALT-68-668 (1978).

[12]M. Sohnius, Nucl. Phys. B138, 109 (1978).

[13]A. D'Adda, R. Horsley and P. di Vecchia, Phys. Lett. 76B,

[14]The impetus to consider this structure in detail came from conversations with D. Olive and B. Zumino.

MASSIVE GRAVITINOS

S. Ferrara

CERN, Geneva, Switzerland

The recently established tensor calculus for local supersymmetry[1], with a minimal set of auxiliary fields[2] (complete set of gauge potentials) enables us to study the spontaneous symmetry breaking in local supersymmetry in a model-independent way.

The conditions for mass generation of the gravitino (super-Higgs effect) and absence of an induced cosmological constant can be studied for general scalar-matter interactions $(0^+, 0^-, \frac{1}{2}$ matter system) and explicit examples exhibited[3].

This investigation, as a by-product, also answers the question on the degree of arbitrariness of scalar-matter coupling to supergravity[4].

We recall that the complete set of gauge potentials in supergravity is given by

$$(e_{a\mu}, \psi_\mu, A_\mu, S, P) \tag{1}$$

$e_{a\mu}$ and ψ_μ are the familiar vierbein and Rarita-Schwinger fields describing the graviton and the gravitino particles while A_μ

(world axial vector), S (scalar) and P(pseudoscalar) are six auxiliary fields needed to make the local supersymmetry algebra closed.

It is worth while mentioning that the fields A_μ, S and P are non-propagating only in Einstein supergravity. In higher derivative theories, like in the supersymmetric extension of Yang and Weyl theories, they all become propagating.

The introduction of auxiliary fields allows the formulation of a tensor calculus similar to the one existing already in global supersymmetry. An obvious, but invaluable consequence is that one can now obtain actions for complicated systems by simply adding invariant subactions, without the tiresome order-by-order Noether procedure previously needed. Heartwarming calculations, involving complicated Fierz identities, Bianchi identities, integrations by parts, high precision γ matrices algebra, possible only to supergravity practitioners, are finally circumvented.

Let us give some basics of tensor analysis. We will confine our discussion to local scalar (chiral) multiplets only, although general rules for local vector multiplets exist as well[5].

A scalar (chiral) multiplet of global supersymmetry is a set of real fields $\Sigma = (A, B, \chi, F, G)$ which transforms as

$$\begin{aligned} \delta z &= 2\bar{\epsilon}_L \chi_L \\ \delta \chi_L &= \not{\partial} z\, \epsilon_R + H \epsilon_L \\ \delta H &= 2\bar{\epsilon}_R \not{\partial} \chi_L \end{aligned} \tag{2}$$

in which we have used complex (chiral) expressions:

$$z = A + iB \;, \quad \chi_L = \tfrac{1}{2}(1+\gamma_5)\chi \;, \quad H = F + iG$$

The corresponding local scalar multiplet transforms as

$$\begin{aligned} \delta z &= 2\bar{\epsilon}_L \chi_L \\ \delta \chi_L &= \not{\mathcal{D}} z\, \epsilon_R + H\, \epsilon_L \\ \delta H &= 2\bar{\epsilon}_R \not{\mathcal{D}} \chi_L - \frac{2}{3}\bar{\epsilon}_L \chi_L u - \frac{i}{3}\bar{\epsilon}_R \not{A} \chi_L \end{aligned} \tag{3}$$

where $u = S - iP$ and $\mathcal{D}_\mu$ is a supercovariant derivative:

$$\begin{aligned} \mathcal{D}_\mu z &= \partial_\mu z - \bar{\psi}_{\mu L} \chi_L \\ \mathcal{D}_\mu \chi_L &= D_\mu \chi_L - \frac{1}{2} \not{\mathcal{D}} z\, \psi_{\mu R} - \frac{1}{2} H\, \psi_{\mu L} \end{aligned} \tag{4}$$

(we put $k = 1$ from now on). D_μ is the gravitational covariant derivative with spin ½ connection and spin 3/2 torsion

$$D_\mu = \partial_\mu + \frac{1}{2}\hat{\omega}_{\mu ab}(e,\psi)\,\sigma^{ab} \tag{5}$$

and

$$\hat{\omega}_{\mu ab}(e,\psi) = \omega_{\mu ab}(e) + \frac{1}{4}\left(\bar{\psi}_\mu \gamma_a \psi_b - \bar{\psi}_\mu \gamma_b \psi_a - \bar{\psi}_a \gamma_\mu \psi_b\right)$$

It is well known that the last component $H = F + iG$ of a scalar multiplet is a density in global supersymmetry.

In local supersymmetry a local density is given by

$$eH + e\bar{\psi}_R \cdot \gamma \chi_L + e(\bar{\psi}_{R\mu} \sigma^{\mu\nu} \psi_{R\nu} + u) z \tag{6}$$

the real and imaginary part of Eq. (6) is a scalar (pseudoscalar) density, respectively.

Beyond Eqs. (3) and (6) one can extend all operations of global supersymmetry to local supersymmetry[6)]. Of particular relevance is the local kinetic multiplet $T(\Sigma)$ which associates to any multiplet Σ a new multiplet containing derivatives, like the supercovariant extensions of the Dirac and Klein-Gordon expressions $\not{\partial}\chi$, $\Box A$.

In addition one can construct scalar (chiral) multiplets out of the gauge potential multiplet (1) like, for instance, the scalar curvature multiplet $\mathcal{U}$.

We return to the question of spontaneous symmetry breaking.

Let us recall how spontaneous breaking takes place in global supersymmetry[7)]. A necessary condition is that one of the auxiliary fields of a global multiplet acquires a non-vanishing expectation value. For scalar multiplets this means that

$$\langle 0|H|0\rangle \neq 0 \tag{7}$$

If this happens, then there is a (spin ½) fermion in the theory for which (see Eq. (2))

$$\delta_\epsilon \chi = \frac{1}{a}\epsilon + \text{q-number terms} \tag{8}$$

which in turn implies

$$\langle 0|\{Q_\alpha, \chi_\beta\}|0\rangle \neq 0 \quad \text{i.e.,} \quad Q_\alpha|0\rangle \neq 0 \tag{9}$$

χ is called a Goldstone fermion (Goldstino) and is a spin ½ massless particle of the spectrum of a spontaneously broken supersymmetric theory.

In specific dynamical models $H = H(z)$ (because H is an auxiliary field), $z = A + iB$ being a given set of fundamental (propagating) scalar fields of the theory. Therefore

$$\langle H(z) \rangle = H(\langle z \rangle) \tag{10}$$

which may or may not imply that $\langle z \rangle \neq 0$. We remark, "en passant", that $\langle z \rangle \neq 0$ can be compatible with no breaking of supersymmetry if $H(\langle z \rangle) = 0$. In particular, this happens in models with spontaneous breakdown of an internal symmetry but no breaking of supersymmetry. Furthermore, it seems that today's particle table does not contain Goldstone fermions, so that we must call upon the Higgs mechanism to absorb them.

Super Higgs effect: when supersymmetry is promoted to a local symmetry (supergravity) the Higgs-Kibble mechanism may occur, namely the Goldstino can be eaten by the massless gravitino, the gauge fermion of local supersymmetry; after such a banquet it becomes massive[8)].

In Ref. 9), using a non-linear realization of supersymmetry (the Volkov-Akulov Lagrangian) it has been shown, to lowest order in the gravitational coupling constant k, that the spin 3/2 gravitino can become massive with vanishing cosmological term (the last requirement being a natural assumption from present experimental evidence). Moreover, the gravitino mass is related to the symmetry breaking parameter as follows:

$$\delta\chi = \frac{1}{a}\epsilon + \dots \qquad , \; m^2_\psi = \frac{1}{6a^2} \tag{11}$$

χ being the Goldstino field.

In Ref. 3) the occurrence of spontaneous symmetry breaking and super-Higgs effects has been proven in a class of interactions of the scalar-matter multiplet $\Sigma = (z, \chi_L, H)$ with the gauge multiplet of supergravity, cf. Eq. (1).

The most general Lagrangian describing this coupling is obtained by taking the real part of Eq. (6) applied to the following scalar (chiral) multiplet

$$a_{nm}\,\Sigma^n\,T(\Sigma^m) + b_n\,\Sigma^n \tag{12}$$

(summation over repeated indices is understood). Parity conservation requires the coefficients b_n to be real and a_{nm} real symmetric.

If we define the functions

$$\begin{aligned} \Phi(z,z^*) &= a_{nm}\,z^n\,z^{*m} \\ g(z) &= b_n\,z^n \end{aligned} \tag{13}$$

the starting terms in the Lagrangian obtained by (12) are

$$\begin{aligned} \mathcal{L} = {} & \frac{e}{6}\Phi R - e\,\Phi_{,zz^*}|\partial_\rho z|^2 - e\,\Phi_{,zz^*}\bar{\chi}\,\not{\partial}\chi + \\ & + \frac{1}{6}\Phi\,\varepsilon^{\mu\nu\rho\sigma}\bar{\psi}_\mu\gamma_5\gamma_\nu D_\rho\psi_\sigma + \dots\dots \end{aligned} \tag{14}$$

the complete bosonic part is

$$\mathcal{L}(\mathrm{BOS}) = \mathcal{L}(\mathrm{KIN}) + \mathcal{L}(\mathrm{POT}) \tag{15}$$

where

$$\begin{aligned}\mathcal{L}(\mathrm{KIN}) = {} & \frac{e}{6}\Phi R - e\Phi_{,zz^*}|\partial_\mu z|^2 - \frac{e}{9}\Phi A_\mu^2 + \\ & + \frac{e}{3} i A^\mu (\Phi_{,z}\partial_\mu z - \Phi_{,z^*}\partial_\mu z^*)\end{aligned} \tag{16}$$

$$\begin{aligned}\mathcal{L}(\mathrm{POT}) = {} & \frac{e}{9}\Phi|u|^2 + \frac{e}{3}\Phi_{,z^*}H^* u + \\ & + \frac{e}{3}\Phi_{,z} H u^* + e\Phi_{,zz^*}|H|^2 + \\ & + e\,\mathrm{Re}(g' H + g u)\end{aligned} \tag{17}$$

The over-all Lagrangian $\mathcal{L}(\mathrm{TOT})$ is given by

$$\mathcal{L}(\mathrm{TOT}) = \mathcal{L}(\mathrm{BOS}) + \mathcal{L}(\mathrm{FERM}) \tag{18}$$

We remark that the auxiliary fields u, H, A_μ are also present in $\mathcal{L}(\mathrm{FERM})$ but they couple only linearly to fermions. One can further eliminate u, H and A_μ from $\mathcal{L}(\mathrm{TOT})$ by using their equations of motion.

Under certain regularity conditions, namely,

$$\Phi < 0 \quad , \quad (\log -\Phi)_{,zz^*} < 0$$

it is possible to perform a Weyl rescaling on the vierbein field $e_{a\mu}$ and a field redefinition on the fermion fields ψ_μ, χ such that the final Lagrangian takes a very simple and compact form (for details, see Ref. 3)).

If one defines the Weyl parameter λ, $\lambda = -\frac{1}{2} \log -\Phi/3$ and the function $\mathcal{G} = \log (-\Phi/3)^3/|g|^2/4$, the redefined fields are

$$\begin{aligned} e^N_{a\mu} &= e_{a\mu} \exp -\lambda \\ z^N &= z \\ \chi^N_L &= \left(\frac{g}{g^*}\right)^{1/4} \chi_L \sqrt{-2\mathcal{G}_{,zz^*}} \exp \frac{\lambda}{2} \\ \psi^N_{\mu L} &= \left(\frac{g^*}{g}\right)^{1/4} \psi_{\mu L} \exp -\frac{\lambda}{2} - \frac{2}{\sqrt{-2\mathcal{G}_{,zz^*}}} \gamma^N_\mu \chi^N_R \lambda_{,z^*} \end{aligned} \qquad (19)$$

In terms of these new variables $\mathcal{L}$ (TOT) becomes

$$\mathcal{L}(\mathrm{TOT}) = \mathcal{L}(\mathrm{SG}) + \mathcal{L}(\mathrm{MATTER}) \qquad (20)$$

where $\mathcal{L}$ (SG) is the pure supergravity Lagrangian and $\mathcal{L}$ (MATTER) is given by

$$\mathcal{L}(\text{MATTER}) = \mathcal{L}_B + \mathcal{L}_F \tag{21}$$

We observe at this point that the final Lagrangian (see below) depends only on the real function $\mathcal{G}$ of two real variables (A,B). There is a simple superspace[10)] argument why this is the case. If one performs a super-Weyl rescaling on the supervielbein, the Weyl (chiral) superparameter can be chosen such that the potential $g(\Sigma)$ is scaled to one (supersymmetric cosmological term). Then the first term in Eq. (12) becomes a function of $\mathcal{G}(\Sigma,\Sigma^*)$ because the superdeterminant rescales as $E \to E|g(\Sigma)|^{-2/3}$.

The bosonic part of the Lagrangian (Eq. (21)) is

$$\mathcal{L}_B = e\,\mathcal{G}_{,zz^*}|\partial_\mu z|^2 + e\exp(-\mathcal{G})\left(3 + \frac{|\mathcal{G}_{,z}|^2}{\mathcal{G}_{,zz^*}}\right) \tag{22}$$

and for the fermionic part of the Lagrangian we have ($\hat{z} = A + i\gamma_5 B$):

$$\begin{aligned}\mathcal{L}_F = &-\frac{e}{2}\bar{\chi}\not{D}\chi + e\exp(-\mathcal{G}/2)\Big[\bar{\psi}_\mu\sigma^{\mu\nu}\psi_\nu \\ &-(-2\mathcal{G}_{,zz^*})^{-1/2}\bar{\psi}\cdot\gamma\hat{\mathcal{G}}_{,z}\chi + (2\mathcal{G}_{,zz^*})^{-1}\bar{\chi}\Big(\frac{\hat{\mathcal{G}}_{,zzz^*}\hat{\mathcal{G}}_{,z}}{\mathcal{G}_{,zz^*}} \\ &+\hat{\mathcal{G}}^2_{,z} - \hat{\mathcal{G}}_{,zz}\Big)\chi\Big] + \text{(four-fermion terms)} \\ &+ \text{(two-fermion terms containing } \partial_\mu z, \partial_\mu z^*).\end{aligned} \tag{23}$$

The terms that we have not written explicitly are not relevant for the super Higgs effect. The potential term in Eq. (22) is

$$-V(z, z^*) = \exp(-\mathcal{G})\left(3 + \frac{|\mathcal{G}_{,z}|^2}{\mathcal{G}_{,zz^*}}\right) \tag{24}$$

If we want to have an absolute minimum with vanishing cosmological term, V must be non-negative which is best expressed by the condition

$$\left[\exp\left(\mathcal{G}/3\right)\right]_{,zz^*} \geqslant 0 \tag{25}$$

The conditions for spontaneous symmetry breaking and absence of cosmological term are

$$\exp(-\mathcal{G}/2)\,\frac{\mathcal{G}_{,z^*}}{\sqrt{-\mathcal{G}_{,zz^*}/2}} \neq 0\ ,\quad V_{,z} = V = 0 \quad \text{at} \quad z = z_0 \tag{26}$$

the first condition implies that $\langle H \rangle \neq 0$ at the minimum, the second one that the minimum occurs at zero potential.

Conditions (26) imply

$$|\mathcal{G}_{,z}|^2 = -3\,\mathcal{G}_{,zz^*} \quad \text{at} \quad z = z_0 \tag{27}$$

which is a consistent requirement because we must have $\mathcal{G}_{,zz} < 0$.

By inspection of formula (22) we see that this condition also implies that the scalar particles are not ghosts.

Let us consider the fermionic sector (Eq. (23)). At the minimum, after chiral redefinition of χ, the fermionic mass matrix becomes

$$\mathcal{L}_F(\text{MASS}) = \exp\left(-\frac{G_0}{2}\right)\left(\bar{\psi}_\mu \sigma^{\mu\nu}\psi_\nu - \sqrt{\frac{3}{2}}\,\bar{\psi}\cdot\gamma\chi - \bar{\chi}\chi\right) \tag{28}$$

Finally the substitution

$$\psi_\mu \to \psi_\mu - \frac{1}{\sqrt{6}}\gamma_\mu\chi - \sqrt{\frac{2}{3}}\exp\left(\frac{G_0}{2}\right)\partial_\mu\chi$$

eliminates χ altogether and leaves only the mass and kinetic term for ψ_μ with mass

$$m_\psi = \exp(-G_0/2) \quad , \quad G_0 = G\big|_{z=z_0} \tag{29}$$

In the class of models for which

$$\log -\frac{\Phi}{3} = -\frac{1}{6}\,|f(z)|^2 \quad , \quad g \text{ arbitrary} \tag{30}$$

(canonical (pseudo-) scalar kinetic terms), one obtains the following mass formula

$$m_A^2 + m_B^2 = 4\,m_\psi^2 \tag{31}$$

Condition (30) is fulfilled, for example, by having

$$\Phi = -3\exp\left(-\tfrac{1}{6}zz^*\right) \quad \text{i.e.,} \quad G_{,zz^*} = -\tfrac{1}{2} \tag{32}$$

then from Eqs. (26) and (27) it follows that

$$|G_{,zz}| = |G_{,zz^*}| = \tfrac{1}{2} \;,$$

hence

$$m_A^2 + m_B^2 = 4\,V_{,zz^*}\big|_{z=z_0}$$

which is nothing but Eq. (31).

Let us conclude by giving an explicit example (given in Ref. 11)) which satisfies all previous requirements. We take Φ as given by Eq. (32) and $g(z) = \lambda(z + b)$ with arbitrary $\lambda \neq 0$ and $b = 2\sqrt{2} - \sqrt{6}$. Then the minimum occurs at $z = \sqrt{6} - \sqrt{2}$. The scalar and pseudoscalar masses are easily computed to be $m_A^2 = 2\sqrt{3}\, m_\psi^2$, $m_B = 2(2 - \sqrt{3})\, m_\psi^2$. The gravitino mass m_ψ depends on the arbitrary scale λ.

Let us state some conclusions and outlooks. We have shown that there are existing models of spontaneously broken supersymmetry exhibiting the Higgs effect for spin 3/2 particles. The crucial point is that this phenomenon can occur without induced cosmological term, a physical constraint which was not obvious to fulfil from the beginning.

The existence of models with vanishing cosmological constant permits a conventional quantization of these theories without the problems connected with quantization in a De Sitter background metric. One could hope that these models share better quantum properties than other models.

Finally, this investigation opens the door to the study of more realistic models with spontaneously broken local supersymmetry encompassing all fundamental interactions.

REFERENCES

1) S. Ferrara and P. van Nieuwenhuizen, Phys. Letters 76B (1978) 404; LPTENS 78/14 (to appear in Phys. Letters B).

2) S. Ferrara and P. van Nieuwenhuizen, Phys. Letters 74B (1978) 333;
K. Stelle and P.C. West, Phys. Letters 74B (1978) 330.

3) E. Cremmer, B. Julia, J. Scherk, P. van Nieuwenhuizen, S. Ferrara and L. Girardello, LPTENS 78/17; CERN preprint TH.2554 (1978).

4) E. Cremmer and J. Scherk, Phys. Letters 69B (1977) 97;
A. Das, M. Fischler and M. Rocek, Phys. Letters 69B (1977) 186; Phys. Rev. D16 (1977) 3427.

5) K. Stelle and P.C. West, Reports ICTP/77-78/15; ICTP/77-78/24 (1978).

6) See the contribution of P. van Nieuwenhuizen to these Proceedings.

7) For a review, see: P. Fayet and S. Ferrara, Phys. Rep. 32C (1977) 250.

8) D.V. Volkov and V.A. Soroka, JETP Letters 18 (1973) 312.
B. Zumino, Proc. 17th Scottish Universities Summer School, Edinburgh, 1976 (SUSSP Publications, Edinburgh, 1977) p. 549; Deep pathways in high-energy physics, Proc. Orbis Scientiae, Coral Gables, 1977 (eds. A. Perlmutter and L.F. Scott), Plenum Press, N.Y., 1977, p. 259.

9) S. Deser and B. Zumino, Phys. Rev. Letters 38 (1977) 1433.

10) J. Wess and B. Zumino, Phys. Letters 74B (1978) 51 and CERN preprint TH.2553 (1978).
B. Zumino, to be published in this volume.
P.S. Howe and R.W. Tucker, CERN preprint TH.2524 (1978), to be published in Phys. Letters B.

11) J. Polonyi, Budapest preprint KFKL-1977-93.

ANOMALIES IN SUPERSYMMETRIC THEORIES

M.T. Grisaru

CERN, Geneva, Switzerland and

Brandeis University, Waltham, MA, USA

It was pointed out some time ago by Ferrara and Zumino[1)] that in a supersymmetric theory the axial current j^5_μ, the improved energy-momentum tensor $\theta_{\mu\nu}$ and the improved supersymmetry current S_μ can be identified with the components of a superfield V_μ, so that an intimate relation exists between them. At the classical level this superfield satisfies certain relations reflecting the conservation and trace relations (modulo mass terms)

$$\partial_\mu S^\mu = \partial_\mu \theta^{\mu\nu} = 0 \tag{1}$$

$$\partial_\mu j^{5\mu} = 0 \tag{2a}$$

$$\theta^\mu_\mu = \partial_\mu (x_\nu \theta^{\mu\nu}) = 0 \tag{2b}$$

$$\gamma^\mu S_\mu = \partial_\mu (\gamma . x\, S^\mu) = 0 \tag{2c}$$

We note that the improved supersymmetry current is obtained by adding terms to the Noether current so as to satisfy Eq. (2c) while maintaining Eq. (1)[1)].

At the quantum level the matrix elements of j_μ^5, $\theta_{\mu\nu}$, S_μ no longer satisfy Eqs. (2). Instead, they become equal to expressions which can be identified with matrix elements of operators constructed out of the basic fields. These are the anomalies (chiral, trace or supercurrent). We summarize in this lecture what is known about these anomalies and what their existence might imply, both in models of rigid (global) supersymmetry and in models containing supergravity fields.

For the scalar or vector multiplet of rigid supersymmetry, Eqs. (2) are associated with invariance of the models under conformal supersymmetry[2)] with parameter $\alpha = \alpha(x) = \alpha_0 + \gamma\cdot x\,\alpha_1$ (α_0, α_1 are constant Majorana parameters). For instance, the massless scalar multiplet model (with weight $n = \frac{1}{2}$ in the notation of Ref. 2)) is invariant under

$$\delta A = i\bar{\alpha}\chi \;, \qquad \delta B = i\bar{\alpha}\gamma_5\chi$$

$$\delta\chi = \partial_\mu(A-\gamma_5 B)\gamma^\mu\alpha + (A-\gamma_5 B)\gamma^\mu\partial_\mu\alpha + (F+\gamma_5 G)\alpha \;, \tag{3}$$

$$\delta F = i\bar{\alpha}\gamma^\mu\partial_\mu\chi \;, \qquad \delta G = i\bar{\alpha}\gamma_5\gamma^\mu\partial_\mu\chi$$

with α given above. Furthermore, the commutator of two such supersymmetry transformations acting on the fields gives

$$[\delta_2,\delta_1]A = \xi^\mu\partial_\mu A + \frac{1}{4}(\partial_\mu\xi^\mu)A + \frac{1}{2}\eta B$$

$$[\delta_2,\delta_1]B = \xi^\mu\partial_\mu B + \frac{1}{4}(\partial_\mu\xi^\mu)B - \frac{1}{2}\eta A \tag{4}$$

$$[\delta_2,\delta_1]\chi = \xi^\mu\partial_\mu\chi + \frac{3}{8}(\partial_\mu\xi^\mu)\chi + \frac{1}{2}\partial_\mu\xi_\nu\sigma^{\mu\nu}\chi - \frac{1}{4}\eta\gamma_5$$

with

$$\xi_\mu = 2i\bar{\alpha}_1\gamma_\mu\alpha_2 \quad , \quad \eta = i\left[\partial_\mu\bar{\alpha}_1\gamma_5\gamma_\mu\alpha_2 - 1 \leftrightarrow 2\right] \tag{5}$$

Thus the model is invariant under a set of transformations containing rigid supersymmetry, conformal supersymmetry, conformal transformations and chiral transformations ($A \to B$, $B \to -A$, $\chi \to -\frac{1}{2}\gamma_5\chi$). The generators $D_\mu = x^\nu\theta_{\mu\nu}$, $I_\mu = \gamma\cdot x\, S_\mu$, j^5_μ go into each other under the action of the algebra, in particular under rigid supersymmetry transformations.

We note for future reference that in the case of the vector multiplet (with $n = 1$) the commutator of two supersymmetry transformations has the following effect on the spinor field[2)]

$$\begin{aligned}[\delta_2, \delta_1]\lambda &= \xi^\mu\partial_\mu\lambda + \tfrac{3}{8}(\partial_\mu\xi^\mu)\lambda \\ &\quad + \tfrac{1}{2}\partial_\mu\xi_\nu\sigma^{\mu\nu}\lambda + \tfrac{3}{4}\eta\gamma_5\lambda\end{aligned} \tag{6}$$

Thus the chiral charge of the spinor is -3 times that of the spinor in the scalar multiplet. (These chiral charge assignments can also be read off from the coupling of these fields to the axial vector auxiliary field A_μ of supergravity[3)]).

The anomalies in these theories are associated with a breaking of the (super-) conformal part of the algebra, with ordinary rigid supersymmetry remaining unbroken. A logical way to study the anomalies is to examine quantum modifications of the Ward identities for the superfield current V_μ associated with the superconformal transformations and identify the anomaly superfield. This would have the advantage of yielding all the anomalies at once, and would show that they are related to each other as the components of a superfield. This procedure has been carried out by Piguet and Schweda, and Clark, Piguet and Sibold[4)] for the case of the self-interacting scalar multiplet, in the framework of BPHZ supersymmetric renormalization. It has not been carried out for any other system (vector multiplet, scalar or vector multiplet in interaction with supergravity fields) where a pedestrian approach which simply calculates the (one-loop) anomalies and compares them has been used. We shall now give the results of such calculations for three examples.

Example 1 : Scalar multiplet in a background vector multiplet[5)]

$$\begin{aligned} \mathcal{L} = & -\tfrac{1}{2}(D_\mu A)^2 - \tfrac{1}{2}(D_\mu B)^2 - \tfrac{i}{2}\bar{\chi}^a \gamma^\mu (D_\mu \chi)^a \\ & -\tfrac{1}{2} g^2 [A^2 B^2 - (A^a B^a)^2] \\ & - g f^{abc} \bar{\lambda}^a (A^b + \gamma_5 B^b) \chi^c \\ & - [\tfrac{1}{4} v_{\mu\nu}^2 + \tfrac{i}{2} \bar{\lambda}^a \gamma^\mu (D_\mu \lambda)^a] \end{aligned} \tag{7}$$

where (λ^a, v^a_μ) form a Yang-Mills supersymmetric multiplet and D_μ is the Yang-Mills covariant derivative. The last two terms may be omitted if one wishes to think of (λ^a, v^a_μ) as being arbitrary background fields transforming into each other as vector multiplet fields.

The quantities of interest are

$$\begin{aligned} S_\mu = & \gamma^\lambda \gamma_\mu [D_\lambda (A + \gamma_5 B)]^a \chi^a \\ & + \tfrac{4}{3} \sigma_{\mu\lambda} \partial^\lambda [(A^a + \gamma_5 B^a) \chi^a] \end{aligned} \tag{8a}$$

$$\begin{aligned} \theta_{\mu\nu} = & D_\mu A D_\nu A + D_\mu B D_\nu B + \tfrac{i}{4} \bar{\chi} (\gamma_\mu D_\nu + \gamma_\nu D_\mu) \chi \\ & - \eta_{\mu\nu} \mathcal{L} - \tfrac{1}{6} (\partial_\mu \partial_\nu - \eta_{\mu\nu} \Box)(A^2 + B^2) \end{aligned} \tag{8b}$$

$$j_\mu^5 = A \overleftrightarrow{D}_\mu B - \frac{i}{4} \bar{\chi} \gamma_5 \gamma_\mu \chi \tag{8c}$$

The form of the chiral current follows from the chiral properties implicit in Eqs. (4).

We look at one-loop matrix elements of the above quantities between the vacuum and states of the vector multiplet (λ, v_μ). The anomaly of j_μ^5 is the usual chiral anomaly[6], as obtained from the diagram of Fig. 1a. Only the spinor part of the current contributes.

The spinor current anomaly can be calculated from the diagrams in Figs. (1b), (1c) (1d)[7], while the trace anomaly is known also[5]. One finds, with the external fields on shell:

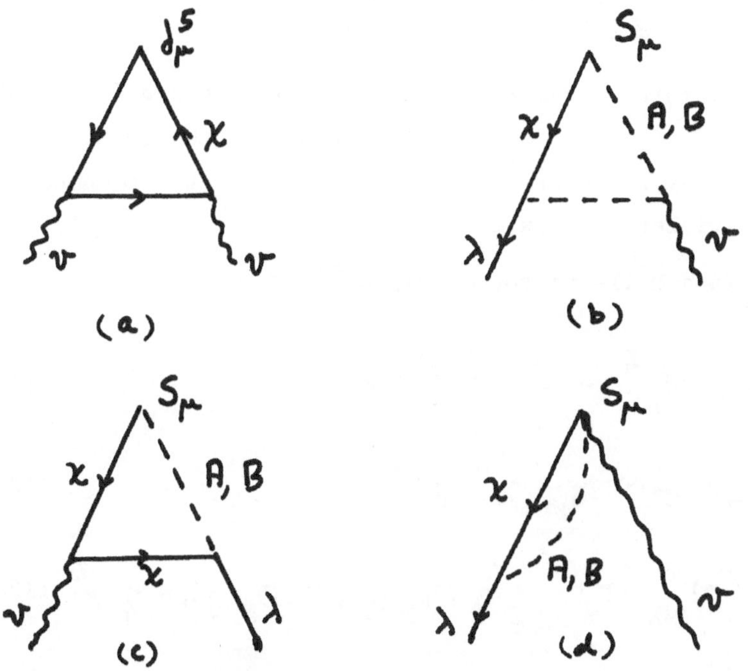

Fig. 1 : Diagrams for one-loop chiral and supercurrent anomalies.

$$a: \qquad c\, 2\bar{\lambda}\lambda$$

$$b: \qquad c\, 2\bar{\lambda}\gamma_5\lambda$$

$$\begin{aligned} x: &\quad \gamma^{\mu} S_{\mu} = c\, 4 v_{\mu\nu}\,\sigma^{\mu\nu}\lambda \\ f: &\quad \theta_{\mu}^{\ \mu} = c\, v_{\mu\nu}\, v^{\mu\nu} \\ g: &\quad \partial_{\mu} j^{5\mu} = -c\, \tilde{v}_{\mu\nu}\, v^{\mu\nu} \end{aligned} \tag{9}$$

$$c = -\frac{g^2}{32\pi^2}\, c_V \tag{10}$$

with c_V the Casimir operator for the Yang-Mills group (for 0(3) $c_V = 2$). We have adjoined to the three anomalies the quantities a,b. Together they transform into each other like the components of a scalar multiplet (a,b,x,f,g).

Example 2 : Yang-Mills vector multiplet[8)]

$$\mathcal{L} = -\frac{1}{4} v_{\mu\nu}^2 - \frac{i}{2}\bar{\lambda}\gamma.D\lambda \tag{11}$$

$$S_{\mu} = -i\,\sigma^{\lambda\nu}\gamma_{\mu}\lambda^a v_{\lambda\nu}^a \tag{12a}$$

$$\theta_{\mu\nu} = v_{\mu\lambda} v_{\nu}^{\ \lambda} - \frac{i}{4}\bar{\lambda}(\gamma_{\mu} D_{\nu} + \gamma_{\nu} D_{\mu})\lambda - \eta_{\mu\nu}\mathcal{L} \tag{12b}$$

$$j_\mu^5 = \frac{3i}{4} \bar{\lambda}^a \gamma_5 \gamma_\mu \lambda^a \tag{12c}$$

The coefficient in j_μ^5 follow from the chiral properties implicit in Eq. (6). The trace and chiral anomalies are standard (the latter being three times the Adler anomaly because of the factor of 3 in j_μ^5), while the supercurrent anomaly has been calculated by a number of authors[9]. One finds the same results as in Eq. (9) with $c \to -3c$.

Three comments are in order: As is well known, in the case of the chiral anomaly a new current can be defined whose divergence is zero:

$$\tilde{j}_\mu^5 = j_\mu^5 + \frac{3g^2 c_V}{32\pi^2} \varepsilon_{\mu\nu\rho\sigma} v^{\nu\rho} v^\sigma \tag{13}$$

However, this current is not gauge invariant. Similarly one can define a new supercurrent

$$\tilde{S}_\mu = S_\mu - \frac{3g^2 c_V}{8\pi^2} v_{\rho\mu} \gamma^\rho \lambda \tag{14}$$

which satisfies $\gamma \cdot \tilde{S} = 0$, but $\partial_\mu \tilde{S}^\mu \neq 0$. In fact in an explicit calculation one will find either $\gamma \cdot s = 0$ or $\partial \cdot S = 0$ depending on what kind of regularization scheme one uses. For instance, dimensional regularization leads to $\partial \cdot S = 0$ but $\gamma \cdot S \neq 0$, while a point-splitting calculation has given $\gamma \cdot S = 0$ but $\partial \cdot S \neq 0$ [10]. In general it seems preferable to maintain the rigid supersymmetry relation $\partial \cdot S = 0$ and give up the $\gamma \cdot S = 0$ condition.

Our second comment relates to higher order effects. The Adler-Bardeen theorem asserts that the anomaly in $\partial_\mu j_\mu^5$ receives no contributions from higher orders. It is known that this is not the case for θ^μ_μ. (The β function is not zero in higher order for this model.) We must conclude therefore that either the multiplet pattern of Eq. (9) does not hold in higher order or that in a calculation that preserved rigid supersymmetry at each step the Adler-Bardeen or β function result (or both) would be modified. This is suggested by our next example.

Our third comment stems from the observation that for a system consisting of three scalar multiplets and one vector multiplet interacting as in the above examples the one-loop anomalies cancel[11]. Furthermore, it has been shown[12] that also the two-loop β function (trace anomaly) is zero for a model, equivalent to the above one at the one-loop level, but with additional interactions which make it O(4) symmetric.

Example 3 : Self-interacting scalar multiplet[13].

$$\mathcal{L} = -\frac{1}{2}(\partial_\mu A)^2 - \frac{1}{2}(\partial_\mu B)^2 - \frac{i}{2}\bar{\chi}\gamma.\partial\chi + \frac{1}{2}F^2 + \frac{1}{2}G^2 + g\left[FA^2 - FB^2 + 2GAB - i\bar{\chi}(A-\gamma_5 B)\chi\right] \tag{15}$$

This system is interesting because naïvely one would expect $\partial_\mu j^5_\mu = 0$ while $\theta^\mu_\mu \neq 0$ at the one-loop level. This is because, according to the usual lore, external vector mesons are needed to get an anomaly, while here we are taking matrix elements to external spin zero or one-half particles. However, the actual result is regularization dependent. Using a supersymmetric regularization procedure (e.g., Pauli-Villars), one finds[14] at the one-loop level:

$$a: \quad d\,[AF - BG]$$

$$b: \quad d\,[AG + BF]$$

$$\chi: \quad \gamma . S = d\,[(A - \gamma_5 B)\gamma.\partial\chi + (F - \gamma_5 G)\chi]$$

$$f: \quad \theta^{\mu}_{\mu} = d\,[A\Box A + B\Box B - i\,\bar{\chi}\gamma.\partial\chi + F^2 + G^2]$$

$$g: \quad \partial_\mu j^{5\mu} = d\,[\partial_\mu (A \overleftrightarrow{\partial_\mu} B + \tfrac{i}{2}\,\bar{\chi}\gamma_5\gamma_\mu\chi\,]$$

$$d = -\frac{g^2}{\pi^2} \tag{16}$$

Unlike the Adler anomaly case, one could here redefine the current j^5_μ by subtracting the quantity $d(A \overleftrightarrow{\partial}_\mu B + i/2\ \bar{\psi}\ \gamma_5\gamma_\mu\psi)$, so that the new current is anomaly free, since no gauge invariance requirements exist. However, it turns out that the resulting current $\tilde{j}^5_\mu$ would not be a finite operator (Green's functions involving $\tilde{j}^5_\mu$ would be infinite even after renormalization), so that this is not a desirable procedure. The above results, to all orders, have also been obtained in Ref. 4) using supersymmetric BPHZ renormalization techniques.

Thus for the self-interacting scalar multiplet the anomaly situation seems quite clear. It is possible, by insisting on rigid supersymmetry, to define quantum currents such that the anomalies transform as members of a scalar multiplet to all orders. For the vector multiplet, where (Yang-Mills) gauge invariance must also be imposed, the higher loop situation remains unclear. As we mentioned above, if the anomalies are to form a multiplet either the β function results or the Adler-Bardeen theorem must be changed in a completely supersymmetric calculation. Otherwise, one should understand why the higher-order calculations do not lead to supersymmetric results.

We turn now to the anomalies of the scalar or vector multiplets in the presence of external supergravity fields. Explicit calculations have been carried out for the supercurrent[15] while the trace and chiral anomalies can be obtained from known results, which are summarized in the following table given by Christensen and Duff[16].

Table

Spin	$180(4\pi)^2\ \theta^\mu_\mu =$	$180(4\pi)^2\ \partial_\mu j^{\mu 5} =$
0	$R_{\mu\nu\alpha\beta}R^{\mu\nu\alpha\beta}$	$0 \times \tilde{R}_{\mu\nu\alpha\beta}R^{\mu\nu\alpha\beta}$
$\frac{1}{2}$	$\frac{7}{4} \times$ " "	$\frac{-15}{2} \times$ " "
1	-13 " "	0 " "
$\frac{3}{2}$	$\frac{-233}{4}$ " "	$\frac{315}{2}$ " "
2	212 " "	0 " "

(here j^5_μ is defined with unit chiral charge). For instance, the trace anomaly for the scalar multiplet is given by twice the spin zero contribution plus the spin ½ contribution leading to a coefficient of 15/4, which matches that of the chiral anomaly. Similarly for the vector multiplet the sum of spinor and vector contributions gives -45/4, to be compared to (-3) × (-15/2) for the chiral anomaly (remembering that the spinor has chiral charge -3). We find, for the scalar multiplet, with external supergravity fields on shell

$$
\begin{aligned}
&a: && c\,\bar{\psi}_{\nu,\alpha}(\psi_{\nu,\alpha}-\psi_{\alpha,\nu}) + \ldots \\
&b: && c\,\bar{\psi}_{\nu,\alpha}\gamma_5(\psi_{\nu,\alpha}-\psi_{\alpha,\nu}) + \ldots \\
&x: && \gamma . S = c\,R^{\alpha\nu\beta\lambda}\sigma_{\beta\lambda}\psi_{\nu,\alpha} + \ldots \\
&f: && \theta_\lambda^{\ \lambda} = c\,R^{\alpha\nu\beta\lambda}R_{\alpha\nu\beta\lambda} + \ldots \\
&g: && \partial_\mu j^{5\mu} = c\,R^{\alpha\nu\beta\lambda}\tilde{R}_{\alpha\nu\beta\lambda} + \ldots
\end{aligned}
\tag{17}
$$

with

$$c = \frac{\varkappa^2}{768\pi^2}$$

For the vector multiplet one obtains the same results times a factor of -3.

Finally we turn to the anomalies of the supergravity multiplet[3)] itself. The trace and chiral anomalies can be read from the above table and give factors of 615/4 (the sum of spin 3/2 and spin 2 trace anomalies) and 945/4, respectively. (We note that the spin 3/2 field carries a chiral charge -3, just like the spinor of the vector multiplet. This can be seen most easily from its coupling to the axial vector auxiliary field. In fact the fermion of the scalar multiplet is exceptional in carrying unit chiral charge, apparently because it belongs to a chiral superfield.) The anomaly of the spinor current has not yet been calculated but the numbers above are hard to reconcile with the supposition that the anomalies should belong to a multiplet. A priori there is no reason why they should; ordinary Poincaré supergravity is not (super-) conformally invariant and the currents do not satisfy conservation equations. A study of the quantum Ward identities is hampered by the off-shell non-renormalizability of the theory. The situation may be different in conformal supergravity[3)] but the anomaly situation has not been investigated there.

Finally, we discuss some implications of the anomalies. First of all, one can look at their topological significance. For the chiral anomalies this is well known, and instanton effects appear to be present in supersymmetric theories[17)]. The trace anomalies have similar well-known significance. In this context it is interesting to look at models with vanishing trace anomalies. We have already mentioned the O(4) supersymmetric Yang-Mills theory which has a vanishing one- and two-loop β function. The same model clearly has vanishing (one-loop) anomalies when coupled to external supergravity fields. Finally, it is easy to check from the table above that in O(3) supergravity (one spin 2, 3 spin 3/2, 3 spin 1 and one spin ½) the trace anomaly is zero[16)18)].

At the present time it is unclear what topological significance, if any, the supercurrent anomaly might have either in ordinary space-time, or as part of the anomaly superfield in superspace.

In the context of conventional field theory, one has to worry about the effect of anomalies on renormalization. In rigid supersymmetry the currents with anomalous properties do not appear in any couplings so the anomalies have no effect there. In supergravity the same is the case (with one exception that we discuss below). While one cannot talk about renormalizability in the conventional sense, an anomaly in the $\partial_\mu S^\mu = 0$ equation would destroy rigid supersymmetry and affect finiteness arguments beyond one-loop. If the anomaly is present only in the $\gamma \cdot S = 0$ equation it presumably has no effect.

The anomalies seem to play a curious role in two models of extended supergravity with SU(4) invariance[19]. One of the models is SO(4) supergravity[20] in which the action is SO(4) invariant and only the equations of motion exhibit SU(4) invariance. In the other one[21], recently constructed, the action itself is SU(4) invariant.

At the classical level the two theories are equivalent in the sense that their equations of motion can be transformed into each other by a combination of point field and duality transformations. At the quantum level the equivalence is not clear. In fact, in the SU(4) theory decay of the pseudoscalar meson of the model into gravitons appears possible because it couples to the divergence of chiral currents of the spin ½ and spin 3/2 fields, and also to a particular combination of spin 1 fields. In the SO(4) model the couplings do not permit such decays. Remarkably enough, the strength of the couplings and the numerical values of the chiral anomalies are such that the total decay amplitude is zero in SU(4) theory. Thus, at least for this process "anomalies" are necessary in order to maintain the equivalence between SO(4) and SU(4) supergravity[19].

In SU(4) supergravity there are also couplings of scalars and spin ½ fermions to objects related to θ^μ_μ and $\gamma \cdot S$. The precise role these couplings play is under investigation.

We conclude with the following observation. For the case of the scalar or vector multiplets, the fact that the anomalies themselves form multiplets could have been used to obtain one from the other, if the numbers were not already known. To a certain extent also, rigid supersymmetry "explains" the relative size of the numbers 1, 7/4, -13 and -15/2 in the table. ($2 \times 1 + 7/4$ must match $1/2 \times 15/2$, and $7/4 - 13$ must match $3/2 \times 15/2$.)

Another relation, between the Adler anomaly and the spin ½ axial anomaly in the presence of gravitons is revealed by examining the situation for the coupling of the vector multiplet to off-shell external supergravity, including the auxiliary fields[3]. In this case it is best to think of the external supergravity fields as being fields of conformal supergravity. The couplings are the same but the system is also invariant under chiral transformations of the spin ½ field, together with gauge transformations on the external axial vector field A_μ (and chiral transformations of ψ_μ). The anomaly $\partial_\mu j^{\mu 5}$ must be proportional to the G component of a scalar multiplet and it is known[3] that the gravitational and axial fields A_μ must appear in the combination

$$c\left[\tilde{R}_{\mu\nu\alpha\beta}R^{\mu\nu\alpha\beta} - \frac{1}{6}\tilde{F}_{\mu\nu}(A)F^{\mu\nu}(A)\right]$$

But c and $c/6$ is what one would compute by conventional Feynman diagrams, with two external gravitons, or spin one mesons, respectively (the fact that A_μ couples with $\gamma_5\gamma_\mu$ rather than γ_μ simply reverses the sign compared to the usual Adler anomaly). Taking into account the unusual normalization of the field A_μ, and the strength of its coupling to the spinor field, one obtains a result for the ratio of Adler to gravitational anomaly which agrees with the conventional values.

Thus, if one were to rewrite history, one could argue that a knowledge of the Adler anomaly, together with rigid and local supersymmetry is all that is required in order to compute almost all the one-loop anomalies of matter fields (spin 0, ½, 1).

I wish to acknowledge many discussions with B. Zumino, as a consequence of which my understanding of the anomaly situation was considerably enhanced.

REFERENCES

1) S. Ferrara and B. Zumino, Nuclear Phys. B87 (1975) 207.

2) J. Wess and B. Zumino, Nuclear Phys. B70 (1974) 39.

3) See, for instance, P. van Nieuwenhuizen, this volume.

4) O. Piguet and M. Schweda, Nuclear Phys. B92 (1975) 334; T.E. Clark, O. Piguet and K. Sibold, University of California report UCB-PTH-78/1.

5) S. Ferrara and B. Zumino, Nuclear Phys. B79 (1974) 413.

6) See, for example, S. Adler, Phys. Rev. 111 (1969) 2426.

7) L.F. Abbott, M.T. Grisaru and H.J. Schnitzer, Phys. Letters 71B (1977) 161.

8) A. Salam and J. Strathdee, Nuclear Phys. B16 (1974) 411. S. Ferrara, J. Wess and B. Zumino, Phys. Letters 51B (1974) 239.

9) L.F. Abbott, M.T. Grisaru and H.J. Schnitzer, Phys. Rev. D16 (1977) 2295; T. Curtright, Phys. Letters 71B (1977) 185.

10) H. Inagaki, Phys. Letters 77B (1978) 56.

11) L.F. Abbott, M.T. Grisaru and H.J. Schnitzer, Phys. Letters 71B (1977) 161.

12) E. Poggio and H.N. Pendleton, Phys. Letters 72B (1978) 200; D.R.T. Jones, Phys. Letters 72B (1978) 199.

13) J. Wess and B. Zumino, Phys. Letters 49B (1974) 52.

14) M.T. Grisaru, J. Wess and B. Zumino (unpublished).

15) L.F. Abbott, M.T. Grisaru and H.J. Schnitzer, Phys. Letters 73B (1978) 71.

16) S.M. Christensen and M.J. Duff, Phys. Letters 76B (1978) 571.

17) L.F. Abbott, M.T. Grisaru and H.J. Schnitzer, Phys. Rev. D15 (1977) 3002;
S.W. Hawking and C.N. Pope, Symmetry breaking by instantons in supergravity (Univ. of Cambridge report, 1978).

18) T. Yoneia, CCNY report, HEP 77-13 (1977).

19) M.T. Grisaru, Phys. Letters (to be published).

20) A. Das, Phys. Rev. D15 (1977) 2805;
E. Cremmer and J. Scherk, Nuclear Phys. B121 (1977) 259.

21) E. Cremmer, J. Scherk and S. Ferrara, Phys. Letters 74B (1978) 61.

INDEX